Cell Cycle Control

METHODS IN MOLECULAR BIOLOGY™

John M. Walker, Series Editor

308. **Therapeutic Proteins:** *Methods and Protocols,* edited by *C. Mark Smales and David C. James, 2005*
307. **Phosphodiesterase Mehtods and Protocols,** edited by *Claire Lugnier, 2005*
306. **Receptor Binding Techniques:** *Second Edition,* edited by *Anthony P. Davenport, 2005*
305. **Protein–Ligand Interactions:** *Methods and Protocols,* edited by *G. Ulrich Nienhaus, 2005*
304. **Human Retrovirus Protocols:** *Virology and Molecular Biology,* edited by *Tuofu Zhu, 2005*
303. **NanoBiotechnology Protocols,** edited by *Sandra J. Rosenthal and David W. Wright, 2005*
302. **Handbook of ELISPOT: Methods and Protocols,** edited by *Alexander E. Kalyuzhny, 2005*
301. **Ubiquitin–Proteasome Protocols,** edited by *Cam Patterson and Douglas M. Cyr, 2005*
300. **Protein Nanotechnology:** *Protocols, Instrumentation, and Applications,* edited by *Tuan Vo-Dinh, 2005*
299. **Amyloid Proteins:** *Methods and Protocols,* edited by *Einar M. Sigurdsson, 2005*
298. **Peptide Synthesis and Application,** edited by *John Howl, 2005*
297. **Forensic DNA Typing Protocols,** edited by *Angel Carracedo, 2005*
296. **Cell Cycle Control:** *Mechanisms and Protocols,* edited by *Tim Humphrey and Gavin Brooks, 2005*
295. **Immunochemical Protocols,** *Third Edition,* edited by *Robert Burns, 2005*
294. **Cell Migration:** *Developmental Methods and Protocols,* edited by *Jun-Lin Guan, 2005*
293. **Laser Capture Microdissection:** *Methods and Protocols,* edited by *Graeme I. Murray and Stephanie Curran, 2005*
292. **DNA Viruses:** *Methods and Protocols,* edited by *Paul M. Lieberman, 2005*
291. **Molecular Toxicology Protocols**, edited by *Phouthone Keohavong and Stephen G. Grant, 2005*
290. **Basic Cell Culture Protocols**, *Third Edition,* edited by *Cheryl D. Helgason and Cindy L. Miller, 2005*
289. **Epidermal Cells**, *Methods and Applications,* edited by *Kursad Turksen, 2005*
288. **Oligonucleotide Synthesis**, *Methods and Applications,* edited by *Piet Herdewijn, 2005*
287. **Epigenetics Protocols**, edited by *Trygve O. Tollefsbol, 2004*
286. **Transgenic Plants:** *Methods and Protocols,* edited by *Leandro Peña, 2005*
285. **Cell Cycle Control and Dysregulation Protocols:** *Cyclins, Cyclin-Dependent Kinases, and Other Factors,* edited by *Antonio Giordano and Gaetano Romano, 2004*
284. **Signal Transduction Protocols,** *Second Edition,* edited by *Robert C. Dickson and Michael D. Mendenhall, 2004*
283. **Bioconjugation Protocols,** edited by *Christof M. Niemeyer, 2004*
282. **Apoptosis Methods and Protocols,** edited by *Hugh J. M. Brady, 2004*
281. **Checkpoint Controls and Cancer, Volume 2:** *Activation and Regulation Protocols,* edited by *Axel H. Schönthal, 2004*
280. **Checkpoint Controls and Cancer, Volume 1:** *Reviews and Model Systems,* edited by *Axel H. Schönthal, 2004*
279. **Nitric Oxide Protocols,** *Second Edition,* edited by *Aviv Hassid, 2004*
278. **Protein NMR Techniques,** *Second Edition,* edited by *A. Kristina Downing, 2004*
277. **Trinucleotide Repeat Protocols,** edited by *Yoshinori Kohwi, 2004*
276. **Capillary Electrophoresis of Proteins and Peptides,** edited by *Mark A. Strege and Avinash L. Lagu, 2004*
275. **Chemoinformatics,** edited by *Jürgen Bajorath, 2004*
274. **Photosynthesis Research Protocols,** edited by *Robert Carpentier, 2004*
273. **Platelets and Megakaryocytes, Volume 2:** *Perspectives and Techniques,* edited by *Jonathan M. Gibbins and Martyn P. Mahaut-Smith, 2004*
272. **Platelets and Megakaryocytes, Volume 1:** *Functional Assays,* edited by *Jonathan M. Gibbins and Martyn P. Mahaut-Smith, 2004*
271. **B Cell Protocols,** edited by *Hua Gu and Klaus Rajewsky, 2004*
270. **Parasite Genomics Protocols,** edited by *Sara E. Melville, 2004*
269. **Vaccina Virus and Poxvirology:** *Methods and Protocols,*edited by *Stuart N. Isaacs, 2004*
268. **Public Health Microbiology:** *Methods and Protocols,* edited by *John F. T. Spencer and Alicia L. Ragout de Spencer, 2004*
267. **Recombinant Gene Expression:** *Reviews and Protocols, Second Edition,* edited by *Paulina Balbas and Argelia Johnson, 2004*
266. **Genomics, Proteomics, and Clinical Bacteriology:** *Methods and Reviews,* edited by *Neil Woodford and Alan Johnson, 2004*
265. **RNA Interference, Editing, and Modification:** *Methods and Protocols,* edited by *Jonatha M. Gott, 2004*
264. **Protein Arrays:** *Methods and Protocols,* edited by *Eric Fung, 2004*
263. **Flow Cytometry,** *Second Edition,* edited by *Teresa S. Hawley and Robert G. Hawley, 2004*
262. **Genetic Recombination Protocols,** edited by *Alan S. Waldman, 2004*
261. **Protein–Protein Interactions:** *Methods and Applications,* edited by *Haian Fu, 2004*

METHODS IN MOLECULAR BIOLOGY™

Cell Cycle Control

Mechanisms and Protocols

Edited by

Tim Humphrey

MRC Radiation and Genome Stability Unit,
Harwell, Didcot, United Kingdom

Gavin Brooks

School of Pharmacy, University of Reading
Whiteknights, Reading, United Kingdom

HUMANA PRESS ✳ TOTOWA, NEW JERSEY

Softcover reprint of the hardcover 1st edition 2005
999 Riverview Drive, Suite 208
Totowa, New Jersey 07512

www.humanapress.com

This publication is printed on acid-free paper. ∞
ANSI Z39.48-1984 (American Standards Institute)

Permanence of Paper for Printed Library Materials.

Cover design by Patricia F. Cleary
Cover illustration: Subcellular localization of Aurora A and B during the cell cycle (Chapter 22, Fig. 1; *see* complete figure, caption, and discussion on pp. 372–373).

For additional copies, pricing for bulk purchases, and/or information about other Humana titles, contact Humana at the above address or at any of the following numbers: Tel.: 973-256-1699; Fax: 973-256-8341; E-mail: humana@humanapr.com; or visit our Website: www.humanapress.com

10 9 8 7 6 5 4 3 2 1

eISBN 978-1-59259-857-1

ISSN 1064-3745

Library of Congress Cataloging-in-Publication Data

Cell cycle control: mechanisms and protocols / edited by Tim Humphrey, Gavin Brooks.
p. ; cm. -- (Methods in molecular biology ; 296)
Includes bibliographical references and index.
ISBN 978-1-61737-385-5
1. Cell cycle--Laboratory manuals.
[DNLM: 1. Cell Cycle--physiology. 2. Eukaryotic Cells--physiology. 3. Molecular Biology--methods. QH 605 C3927 2005] I. Humphrey, Tim (Tim Carter) II. Brooks, Gavin. III. Series: Methods in molecular biology (Clifton, N.J.) ; v. 296.
QH605.C445 2005
571.8'4--dc22 2004008991

To Anna, Katie, Lucy and Sophie (GB)

and

To Deborah and Georgina (TH)

Preface

The fundamental question of how cells grow and divide has perplexed biologists since the development of the cell theory in the mid-19th century, when it was recognized by Virchow and others that "all cells come from cells." In recent years, considerable effort has been applied to the identification of the basic molecules and mechanisms that regulate the cell cycle in a number of different organisms. Such studies have led to the elucidation of the central paradigms that underpin eukaryotic cell cycle control, for which Lee Hartwell, Tim Hunt, and Paul Nurse were jointly awarded the Nobel Prize for Medicine and Physiology in 2001 in recognition of their seminal contributions to this field.

The importance of understanding the fundamental mechanisms that modulate cell division has been reiterated by relatively recent discoveries of links between cell cycle control and DNA repair, growth, cellular metabolism, development, and cell death. This new phase of integrated cell cycle research provides further challenges and opportunities to the biological and medical worlds in applying these basic concepts to understanding the etiology of cancer and other proliferative diseases.

As more investigators from different fields undertake such integrative research, it is surprising that there are so few books that provide an inclusive account of eukaryotic cell cycle control mechanisms and/or cell cycle study methods. This book aims to provide such information and uniquely combines overviews of cell cycle control in well-studied organisms together with a comprehensive set of protocols for studying the eukaryotic cell cycle and its key regulatory molecules. As such, it is hoped that this volume will be a useful resource for both new and experienced cell cycle researchers alike.

Given the extraordinary volume of research in the cell cycle area, we trust that the members of the cell cycle community will understand those faults that resulted from space constraints.

Tim Humphrey
Gavin Brooks

Acknowledgments

We thank all our coauthors, without whose help and valuable contributions this volume would not have been possible. In addition, we are grateful to Professor John Walker for commissioning this work and for his advice and help during the preparation stages. We also thank various members of our respective research groups for their contributions and valuable input during the writing stages. Finally, we wish to give special thanks to our families for their constant understanding, encouragement, and support.

Contents

Contributors

PETER D. ADAMS • *Fox Chase Cancer Center, Philadelphia, PA*
KATRINA A. BICKNELL • *School of Pharmacy, University of Reading, Reading, UK*
LYDIA BRIMAGE • *Department of Zoology, University of Oxford, Oxford, UK*
GAVIN BROOKS • *School of Pharmacy, University of Reading, Reading, UK*
SORAB N. DALAL • *Biochemistry and Cell Biology, ACTREC, Navi Mumbai, India*
CLAIRE DITCHFIELD • *School of Biological Sciences, University of Manchester, Manchester, UK*
JOHN H. DOONAN • *John Innes Centre, Norwich, UK*
BRIAN DOVE • *School of Biochemistry and Microbiology, University of Leeds, Leeds, UK*
ROBERT J. DURONIO • *Department of Biology, University of North Carolina, Chapel Hill, NC*
STEVAN R. EMMETT • *School of Biochemistry and Microbiology, University of Leeds, Leeds, UK*
JANE ENDICOTT • *Laboratory of Molecular Biophysics, Department of Biochemistry, University of Oxford, Oxford, UK*
ROBERT P. FISHER • *Molecular Biology Program, Memorial Sloan-Kettering Cancer Center, New York, NY*
SUSAN A. GERBI • *Department of Molecular Biology, Cell Biology and Biochemistry, Brown University, Providence, RI*
JANE V. HARPER • *School of Animal and Microbial Sciences, University of Reading, Reading, UK*
RÜDIGER VON HARSDORF • *Department of Cardiology, Campus Virchow Clinic, Charité Humboldt University, Berlin, Germany*
LUDGER HAUCK • *Max-Delbrück Center for Molecular Medicine, Berlin, Germany*
JULIAN A. HISCOX • *School of Biochemistry and Microbiology, University of Leeds, Leeds, UK*
TIM HUMPHREY • *MRC Radiation and Genome Stability Unit, Harwell, Didcot, UK*
STEPHEN E. KEARSEY • *Department of Zoology, University of Oxford, Oxford, UK*
NICHOLAS KEEN • *Cancer and Infection Research Area, AstraZeneca Pharmaceuticals, Mereside, UK*
JEREMY KUPSCO • *Department of Biology, University of North Carolina, Chapel Hill, NC*
STÉPHANE LAROCHELLE • *Molecular Biology Program, Memorial Sloan-Kettering Cancer Center, New York, NY*
WALTER F. LEISE III • *Department of Molecular Genetics and Cell Biology, University of Chicago, Chicago, IL*
LAURA MAHONEY • *School of Biochemistry and Microbiology, University of Leeds, Leeds, UK*

Mark L. McCleland • *Department of Biochemistry and Molecular Genetics, University of Virginia Medical School, Charlottesville, VA*
Thomas J. McGarry • *Division of Cardiology, Department of Medicine, Northwestern University Medical School, Chicago, IL*
René H. Medema • *Division of Molecular Biology, Netherlands Cancer Institute, Amsterdam, The Netherlands.*
Paul R. Mueller • *Department of Molecular Genetics and Cell Biology, University of Chicago, Chicago, IL*
Mandana Namdar • *Department of Zoology, University of Oxford, Oxford, UK*
Carmela Palermo • *Joint Graduate Program in Cellular and Molecular Pharmacology, UMDNJ–Graduate School for Biomedical Sciences, and Rutgers, The State University of New Jersey, Piscataway, NJ*
Amanda Pearce • *MRC Radiation and Genome Stability Unit, Harwell, Didcot, UK*
Anna Philpott • *Hutchison/MRC Research Centre, Department of Oncology, Addenbrooke's Hospital, University of Cambridge, Cambridge, UK*
Emma Ralph • *Department of Zoology, University of Oxford, Oxford, UK*
P. Todd Stukenberg • *Department of Biochemistry and Molecular Genetics, University of Virginia Medical School, Charlottesville, VA*
Lisa Swanhart • *Department of Biology, University of North Carolina, Chapel Hill, NC*
Stephen S. Taylor • *School of Biological Sciences, University of Manchester, Manchester, UK*
Sander van den Heuvel • *Massachusetts General Hospital Cancer Center and Harvard Medical School, Boston, MA*
Marcel A. T. M. van Vugt • *Division of Molecular Biology, Netherlands Cancer Institute, Amsterdam, The Netherlands*
Melanie Volkening • *Cell Cycle Control and Carcinogenesis, German Cancer Research Center (DKFZ), Heidelberg, Germany*
Nancy C. Walworth • *Department of Pharmacology, UMDNJ–Robert Wood Johnson Medical School, Piscataway, NJ*
Julie Welburn • *Laboratory of Molecular Biophysics, Department of Biochemistry, University of Oxford, Oxford, UK*
Torsten Wurm • *School of Animal and Microbial Sciences, University of Reading, Reading, Berkshire, UK*
Xiaowen Yang • *Department of Zoology, University of Oxford, Oxford, UK*
P. Renee Yew • *Department of Molecular Medicine, Institute of Biotechnology, University of Texas Health Science Center at San Antonio, San Antonio, Texas*

I

Overviews of Cell Cycle Control in Different Organisms

1

Cell Cycle Molecules and Mechanisms of the Budding and Fission Yeasts

Tim Humphrey and Amanda Pearce

Summary

The cell cycles of the budding yeast *Saccharomyces cerevisiae* and the fission yeast, *Schizosaccharomyces pombe* are currently the best understood of all eukaryotes. Studies in these two evolutionarily divergent organisms have identified common control mechanisms, which have provided paradigms for our understanding of the eukaryotic cell cycle. This chapter provides an overview of our current knowledge of the molecules and mechanisms that regulate the mitotic cell cycle in these two yeasts.

Key Words

Cell cycle; *Saccharomyces cerevisiae*; *Schizosaccharomyces pombe*; fission yeast; budding yeast; review.

1. Introduction

The eukaryotic cell cycle can be considered as two distinct events, DNA replication (S-phase) and mitosis (M-phase), separated temporally by gaps known as G_1 and G_2. These events must be regulated to ensure that they occur in the correct order with respect to each other and that they occur only once per cell cycle. Moreover, these discontinuous events must be coordinated with continuous events such as cell growth, in order to maintain normal cell size (reviewed in **ref.** ***1***). Significant advances in understanding such cell cycle controls have arisen from the study of these yeasts. The use of yeast as a model system for studying the cell cycle provides a number of advantages: yeasts are single-celled, rapidly dividing eukaryotes that can exist in the haploid form. Thus yeast are readily amenable to powerful genetic analyses, and molecular tools are available (reviewed in **refs.** ***2*** and ***3***). Although both yeasts are evolutionarily divergent ***(4)***, common mechanisms control their cell cycles that are conserved throughout eukaryotes (reviewed in **refs.** ***5*** and ***6***). Moreover, following the sequencing of both yeast genomes ***(7,8)***, systematic genetic analyses together with reverse

From: *Methods in Molecular Biology, vol. 296, Cell Cycle Control: Mechanisms and Protocols*
Edited by: T. Humphrey and G. Brooks © Humana Press Inc., Totowa, NJ

genetics are beginning to provide global insights into the cell cycle control of these model organisms, and hence all eukaryotes.

2. Yeast Life Cycles

S. cerevisiae proliferates by budding, during which organelles, and ultimately a copy of the genome, are deposited into a daughter bud, which grows out of the mother cell. The bud grows to a minimal size and after receiving a full complement of chromosomes pinches off from the mother cell in a process called cytokinesis. Budding yeast can exist in a haploid (16 chromosomes) or diploid (32 chromosomes) state (reviewed in **ref.** ***9***).

In contrast, *S. pombe* grows by medial fission, whereby newly born daughter cells grow from the tips of their cylindrical rod shape by a process known as new-end takeoff. Once a mature length is reached, the cell ceases growth and produces a septum that bisects the mother cell into two daughter cells. Fission yeasts exist naturally in a haploid form (one set of three chromosomes), limiting the diploid phase to the zygotic nucleus, which enters meiosis immediately (reviewed in **ref.** ***10***).

Conditions of nitrogen starvation have the same consequences for both yeasts and may result in several developmental fates. If the culture contains cells of a single mating type, then the cell cycle will arrest in stationary phase in G_1 and enter G_0. However, if the opposite mating type is also available, pheromone production will result in conjugation to form diploid cells, which will undergo meiosis and form spores. Budding yeasts are distinct from fission yeasts in that they can arrest in G_1 in the absence of nitrogen starvation and may exist as diploids in the mitotic cell cycle (reviewed in **refs.** ***9*** and ***10***).

3. The Mitotic Cell Cycle of Yeasts

3.1. Budding Yeast

In budding yeast, a point exists in mid-G1 after which the cell becomes committed to the mitotic cell cycle. This point is commonly referred to as *Start* ***(11)***. Start plays an important role in coordinating division with growth. Growth is rate-limiting for the cell cycle, and if a critical size requirement is not reached, cells cannot progress through Start. Prior to Start (in early G_1), cells can respond to the environment. If nutrients are plentiful, they can proceed into the next cell cycle; however, if nutrients are limiting, they can make the decision to enter stationary phase or meiosis. In addition, passage through Start may be inhibited by mating factors from other yeasts; hence if two haploid yeast of the opposite mating types detect each other's pheromones, then they will "schmoo" toward one another, mate and form a diploid. Having passed Start, cells are programmed to complete the cell cycle irrespective of the nutrient state or exposure to pheromones.

Entry into mitosis is classically defined by three physiological events in eukaryotes: the formation of the mitotic spindle, breakdown of the nuclear membrane and chromosomal condensation. Both yeasts undergo what is termed a *closed* mitosis, in which the mitotic nuclear membrane, remains intact. In addition, *S. cerevisiae* is distinct from other eukaryotic cells in that the mitotic spindle begins to form during early

S-phase. Thus *S. cerevisiae* does not have a clear landmark event distinguishing the G_2 and M-phase, and thus the G_2/M transition is difficult to define in this organism (reviewed in **ref.** ***12***).

3.2. Fission Yeast

In fission yeast the G_1 and S-phases are relatively short (each accounting for 10% of the time it takes to complete the cell cycle), whereas G_2 is considerably longer (70% of the time is spent in this phase, in which most growth occurs; reviewed in **ref.** ***10***). Again, a critical Start point exists, and passage through this point is dependent on the prior completion of mitosis in the previous cell cycle and on the cell reaching a critical minimal size ***(13)***. Following spore germination or nutrient starvation, when cells are unusually small, a period of growth before Start is required such that a critical size is obtained. However, under nonlimiting conditions, cells have already achieved a minimal size requirement for passage through G_1. Consequently, G_1 is usually cryptic in logarithmically dividing cultures of *S. pombe*, and S-phase directly follows completion of nuclear division, resulting in cells that are already in G_2 at the time of cell separation ***(14)***.

The G_2/M transition is the major control point in the cell cycle of fission yeast and determines the timing of entry into mitosis (as opposed to *S. cerevisiae*, in which Start in G_1 is the major control point). Entry into mitosis is dependent on the cell having previously completed S-phase; on repairing any DNA damage; and on reaching a critical size. Cells coordinate size such that if G_2 is shortened, G_1 will be lengthened and vice versa (reviewed in **ref.** ***10***).

4. Cell Cycle Molecules

4.1. cdc Mutants

Much of what we know about the cell cycle was discovered through the isolation of temperature sensitive (*ts*), cell division cycle (*cdc*) mutants. In 1970 Hartwell et al. ***(15)*** discovered that a number of these *ts* mutants, upon shifting to the restrictive temperature, arrested the cell population with the same morphology, suggesting that the mutant product was required only at a specific point in the cell cycle. Approximately 60 different *cdc* mutants have been isolated in budding yeast, and approx 30 have been isolated in fission yeasts. In addition to *cdc* genes, a large number of new cell cycle genes have been identified on the basis of interactions with preexisting cell cycle genes (reviewed in **refs.** ***10*** and ***12***).

4.2. Cyclin-Dependent Kinases

A highly conserved class of molecules termed the cyclin-dependent kinases (CDKs) plays a central role in coordinating the cell cycles of all eukaryotes. In both fission and budding yeasts, the cell cycle is controlled both at the G_1/S transition and the G_2/M transition by a single highly conserved CDK, encoded by the *CDC28* and *cdc2*$^+$ genes of *S. cerevisiae* and *S. pombe*, respectively. In budding yeast, *ts* mutations in *CDC28* allowed the definition of Start. The *cdc28ts* mutant blocked budding and cell cycle progression at a point in the G_1-phase at which cells could still enter the sexual cycle instead of proceeding with the mitotic cycle. From this work, Start could be defined genetically as the point in the cell cycle at which budding, DNA replication, and spindle pole body (SPB) duplication become insensitive to loss of Cdc28 function ***(11)***.

In fission yeasts, different mutations in *cdc2*$^+$ result in the cells either elongating ***(16)*** or conversely becoming smaller ***(17)***, a phenotype suggesting that Cdc2 might function in the timing of division. *CDC28* and *cdc2*$^+$ share 63% identity, and both are required for passage through Start as well as mitosis. Indeed, these genes are conserved, with the human *CDC2* gene displaying the same properties, demonstrating conservation of essential features of the cell cycle in all eukaryotes ***(6)***.

Active CDKs generally phosphorylate serine or threonine residues that are followed by a proline and a consensus sequence of K/R, S/T, P, X, K/R (reviewed in **ref. *12***). Although many CDK targets have been identified, a comprehensive analysis of CDK targets remains an important goal.

4.3. Cyclins

All CDKs require positive regulatory partners for activity, known as cyclins ***(1)***, which additionally impart CDK substrate specificity. Cyclins were identified as proteins that oscillated in abundance through the cell cycle in rapidly cleaving early embryonic cells ***(18)***. Not all cyclins show this cell cycle-dependent pattern of synthesis and degradation. However, all cyclins share homology over a domain called the *cyclin box*, a region required for binding and activation of CDKs. In *S. cerevisiae*, a number of cyclins have been identified that associate with Cdc28: G1 cyclins (Cln1, Cln2, and Cln3), S-phase cyclins (Clb5 and Clb6), and G_2 cyclins (Clb1–4. Clb1–6) are all B-type cyclins ***(19)***. *S. pombe* cyclins include Puc1 (a G_1 cyclin), three B-type cyclins (Cig1 and Cig2; S-phase cyclins), and Cdc13 (a G_2 cyclin) (reviewed in **ref. *20***). Cyclins bind to Cdc28/Cdc2, forming an active complex, which is associated with histone H1 kinase activity. In order to bind, cyclins recognize a binding motif present on CDKs known as the PSTAIR motif (corresponding to the conserved amino acids within this domain). Cyclins accumulate at specific times during the cell cycle, leading to overlapping activation of different CDK/cyclin complexes, which in turn regulate the cell cycle (reviewed in **refs. *10*** and ***12***).

5. Regulation of the Yeast CDK/Cyclin Complex

The activity of the CDK/cyclin complex is key to cell cycle progression and can be considered the cell cycle "engine" ***(1)***. Thus CDK/cyclin complexes are subject to a high degree of regulation through a number of posttranslational mechanisms including phosphorylation, inhibition by cyclin-kinase inhibitors, destruction of cyclins, and destruction of the inhibitors at the appropriate time in the cell cycle. These mechanisms ensure that the cell cycle progresses in an orderly fashion. In addition, the periodic activity of particular CDK/cyclin complexes is achieved through feedback loops within the cell cycle: In G_1/S, G_1 cyclins activate the Clb cyclins, which then turn off the G_1 cyclins. Similarly, in mitosis, the mitotic cyclins promote spindle formation and turn on the anaphase-promoting complex (APC), or cyclosome, which then degrades the mitotic cyclins needed for the first step. The molecular basis of these regulatory events in yeast is described below in **Subheadings 5.1.–5.3.** (*see* also **Fig. 1**).

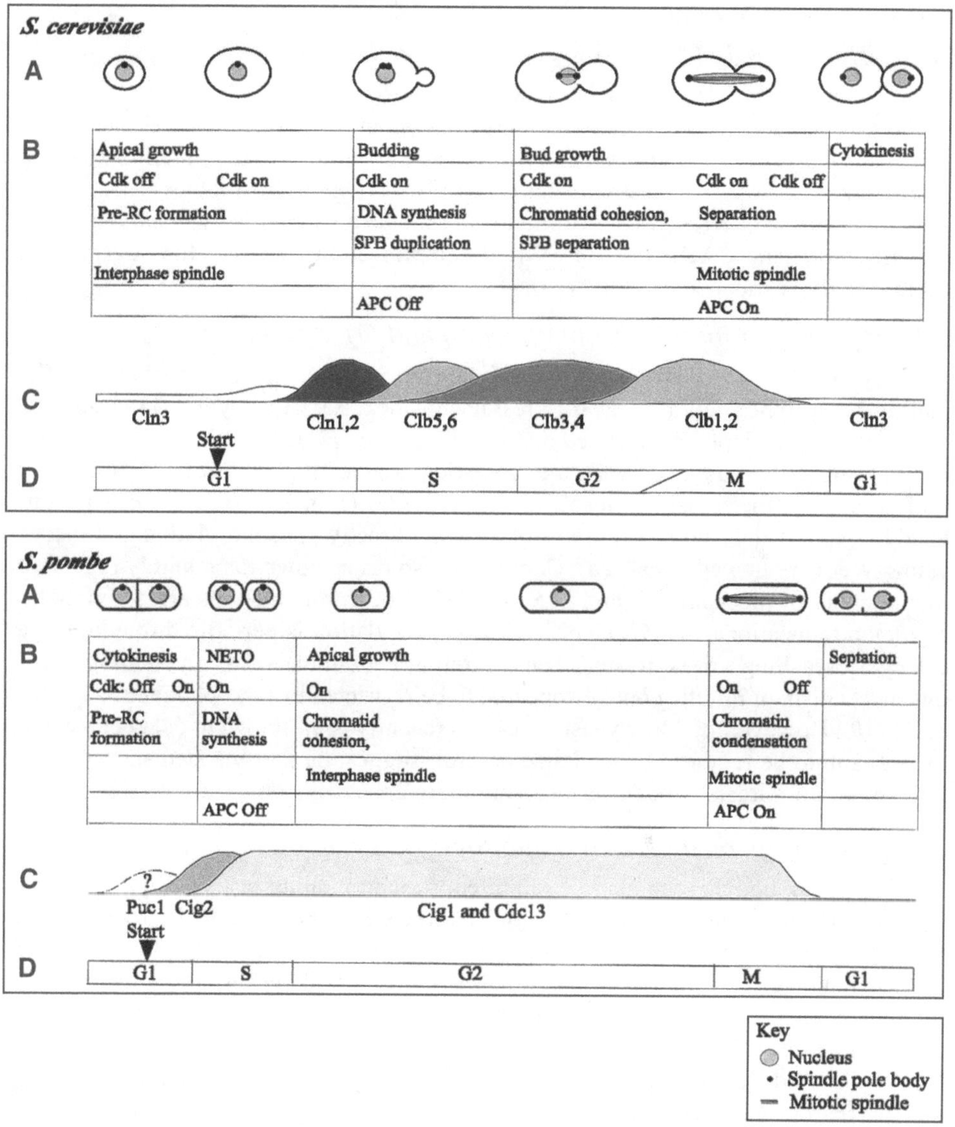

Fig. 1. **(A)** Depiction of cell cycle progression. **(B)** Key cell cycle events. **(C)** Cyclin expression profiles. **(D)** Cell cycle phases of *S. cerevisiae* and *S. pombe*. See text for details and references. APC, anaphase-promoting complex; RC, replication complex; SPB, spindle pole body.

5.1. CDK Phosphorylation

5.1.1. Threonine 161

In fission yeast, Cdc2 is phosphorylated at Thr167 of Cdc2, which corresponds to Thr169 on budding yeast Cdc28 and Thr161 on mammalian Cdc2. In all cases this phosphorylation is essential for activity and results in removal of an inhibitory T-loop from the kinase domain. This phosphorylation is carried out by another CDK, CDK-activating kinase (CAK) (reviewed in **ref. *21***; *see* also Chap. 16). *S. pombe* has two partially redundant CAKs, the Mcs6/Mcs2 complex and Csk1 ***(22)***. In *S. cerevisiae*, CAK activity is encoded by Cak1 ***(23)***.

5.1.2. Cdc2 Tyrosine 15 Phosphorylation and G_2/M Control

Entry into mitosis in fission yeast, and indeed most eukaryotes, is controlled by the inhibitory phosphorylation of the Y15 residue of Cdc2. For Cdc2/cyclin B kinase to be active, it must be dephosphorylated on the Y15 residue ***(24)***. Cdc2/Y15 phosphorylation is principally regulated by the antagonistic tyrosine kinases Wee1 ***(25)*** and Mik1 ***(26)***, as well as the tyrosine phosphatase Cdc25 ***(27)*** (**Fig. 2**). Wee1 is further regulated by Nim1/Cdr1, which promotes mitosis by directly phosphorylating and inactivating Wee1 (reviewed in **ref. *28***). Cdc25 has also been shown to be highly regulated by a number of mechanisms, and in *S. pombe*, Cdc25 protein levels are additionally regulated translationally ***(29)***. Cdc2/Y15 phosphorylation is periodic throughout the cell cycle, reaching a peak in late G2, at the initiation of mitosis ***(24)***. In budding yeast, this mechanism of mitotic control appeared to be restricted to a morphogenesis checkpoint ***(30)***. However, budding yeast Wee1 has recently been shown to delay entry into mitosis and to be required for cell size control, suggesting that mechanisms controlling entry into mitosis in budding yeast are more generally conserved ***(31)***.

5.2. Cyclin-Dependent Kinase Inhibitors

CDK-cyclin activity can also be inhibited through binding of CDK inhibitor proteins. In budding yeast there are potentially three CKIs, Far1p ***(32)***, Sic1p ***(33)***, and Cdc6 ***(34)***. In fission yeast there is one, Rum1 ***(35)***. It is thought that the ability of CKIs to inhibit CDK activity depends on the cyclin. CKIs show periodic accumulation throughout the cell cycle. They are thought to function by restricting access to the active site of the CDK. Far1 specifically inhibits Cdc28/Cln complexes ***(32)***, whereas Sic1 inhibits Cdc28/Clb, G_2 complexes ***(36)***. *FAR1* was isolated in a screen to identify mutants that were defective in pheromone arrest in *S. cerevisiae* ***(37)***. It can only function to inhibit Cdc28/Cln when phosphorylated in response to pheromones in G1 ***(32)***. Sic1 was identified as an in vitro substrate of Cdc28 and associates with Cdc28 in cell extracts ***(33)***. Sic1 coordinates both the G_1/S transition and the M/G_1 transition in budding yeast (reviewed in **ref. *38***). As yeast cells enter G_1, Sic1 is active, inhibiting the Clbs ***(39)***, and thus preventing premature entry into S-phase. As cells proceed into S-phase, destruction of Sic1 is triggered through its phosphorylation by Cdc28/Cln ***(40)***, targeting it for destruction by the Skp1/Cdc53/(cullin) F-box protein complex (SCF) ***(36)***. However, Sic1 phosphorylation is reversed in late mitosis by Cdc14 phos-

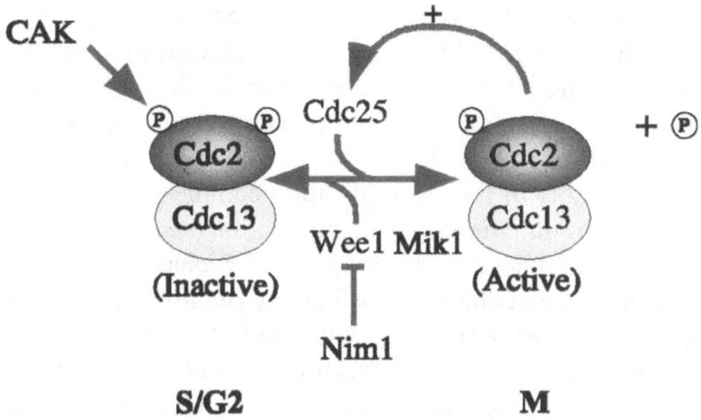

Fig. 2. Regulation of mitotic entry in *S. pombe*. See text for details and references.

phatase, thus promoting Sic1-dependent inhibition of Cdc28/Clb2 and mitotic exit (*see* **Subheading 9.**). Cdc6 also contributes to Cdc28/Clb2 inactivation at the mitotic exit, where it is thought to function in a similar, although less efficient manner to Sic1 ***(34)***. Cdc6 is also involved in DNA replication initiation (*see* **Subheading 7.**).

Fission yeast Rum1 is an inhibitor of Cdc2/Cig2 and Cdc2/Cdc13 and acts like Sic1 ***(41)*** to inhibit Cdc2 kinase activity during G_1. This is important since not all Cdc13 is destroyed at mitosis. Loss of Rum1 can result in cells entering mitosis inappropriately from G_1 ***(35)***. Not only does Rum1 bind Cdc2/Cdc13, it also targets Cdc13 for destruction, probably via the proteolytic machinery ***(42)***.

5.3. Patterns of Cyclin Expression in Yeast

Two *S. cerevisiae* transcription factors, SBF and MBF, control a program of Start-dependent gene activation. SBF (SCB binding factor) recognizes SCB (Swi4/Swi6 cell cycle box) elements and comprises Swi4 and Swi6. MBF (MCB binding factor) recognises MCB (MluI-cell cycle box) elements and is composed of Mbp1 and Swi6. MBF binding is cell cycle-regulated (reviewed in **ref.** ***12***). Targets of MBF and SBF include cyclins, cell wall biosynthesis genes, and genes required for DNA synthesis (reviewed in **ref.** ***43***). *CLN1/2* expression is cell cycle-regulated, peaks in late G_1, and is responsible for Start ***(44)***. Cln3 is less abundant than Cln1 and Cln2, is present throughout the cell cycle, and is regulated through proteolysis via its PEST motifs (corresponding to the conserved amino acids within this domain) ***(45)***. Importantly, Cln3 is also translationally regulated, and links Start to cell growth ***(46)***. Cdc28/Cln3 activates transcription through SBF and MBF (thus driving expression of Cln1 and Cln2, which are required for actin polarization and bud emergence) and subsequently activates Cdc28/Clbs (***47,48***; reviewed in **ref.** ***12***). A global analysis of deletion mutations in *S. cerevisiae* has recently identified a complex network of factors coupling cell growth and Start. These genes, involved in ribosome biogenesis, coordinate cell size with growth by modulating SBF and MBF activity ***(49)***.

Clb5 and Clb6 are required for S-phase. *CLB5/6* activation requires MBF, is positively regulated by Cdc28/Cln3, and occurs in late G_1 (reviewed in **ref. *12***). Cdc28/Clb complexes once formed, are held in an inactive state through Sic1. The activation of Cdc28/Clb complexes and the onset of DNA replication result from Cdc28/Cln-dependent phosphorylation and subsequent destruction of Sic1 (*see* **Subheading 6.1.**). Cdc28/Clbs also block the assembly of the pre-replication complex (pre-RC) after initiation, preventing inappropriate reinitiation of DNA replication (*see* **Subheading 7.**). Mitotic cyclins are subsequently activated, Clb3 and Clb4 in S-phase, which are required for SPB separation, and Clb1 and Clb2 in G_2, which are required for actin depolarization and anaphase (reviewed in **ref. *12***). Cdc28/Clb2 inhibits SBF, thus inhibiting activation of G_1 components in a feedback loop (reviewed in **ref. *19***). Upon entry into mitosis, however, Sic1 levels increase, and *CLB2* trancription levels are reduced, allowing mitotic spindle degradation and exit from mitosis (*see* **Subheading 6.2.2.**).

In fission yeast, an MBF-like activity has also been identified that consists of two distinct complexes: Cdc10-Res1/Sct1, which functions mainly at Start, and Cdc10-Res2/Pct1, which functions in meiosis (reviewed in **refs. *20***). Progression through Start requires Cdc2/Cig2; however, this complex is inhibited by the cyclin kinase inhibitor Rum1 ***(41)*** (*see* **Subheading 5.2.**). To enter S-phase, Rum1 is degraded through a process requiring Cdc2/Cig1 and Cdc2/Puc1 ***(50)***. Cig2 is the main S-phase cyclin, and is both transcriptionally regulated by, and also inhibits MBF, thus forming an autoregulatory feedback-inhibition loop with MBF ***(51)***. Cdc13 is the main B-type cyclin and is required for the onset of M-phase (*see* **Subheading 5.3.**). Prior to S-phase, Cdc2/Cdc13 activity is inhibited through degradation of Cdc13 and through inhibition by Rum1 (*see* **Subheading 5.2.**). Cdc2/Cdc13 additionally functions during replication and G_2, where it binds to replication origins and prevents rereplication ***(52)***. The mitotic cyclins Cdc13 and Cig1 are subsequently degraded in G_1 ***(53)*** (*see* **Subheading 6.**).

6. Proteolysis and Cell Cycle Control

Proteolysis plays a major role in promoting irreversible cell cycle advance. For proteolysis to occur, proteins must first be targeted for destruction by the proteasome. The signal for this is ubiquitylation, which is carried out by specific ubiquitylating enzymes. Ubiquitylation of proteins is imparted through the consecutive action of three classes of enzymes: ubiquitin-activating enzymes (E1), ubiquitin-conjugating enzymes (E2 or UBC), and ubiquitin–protein ligases (E3). Multiubiquitin chains are formed on lysine side chains on the target protein, which bind to a subunit of the 26S proteasome, which is believed to thread the target protein through the central chamber, where it is degraded into peptides (reviewed in **ref. *54***). There are 13 E2s known in *S. cerevisiae* (14 predicted from the *S. pombe* genome), and they provide the first level of specificity in this pathway. There are two important classes of E3 complexes that regulate the cell cycle, the SCF and APC.

6.1. The SCF Complex

The SCF complex catalyzes the phosphorylation-dependent ubiquitylation of a number of cell cycle proteins including G1 cyclins (Cln1 and Cln2), Cdk inhibitors

(Sic1, Far1, and Rum1), and replication proteins (Cdc6 and Cdc18; reviewed in **ref.** ***55***). SCF was first identified in budding yeast, where it was found that mutants in Cdc53, Cdc4, and Cdc34 failed to degrade Sic1p ***(36)***. These proteins form a multiprotein complex, in which Cdc34, an E2 enzyme, is associated with Cdc53, termed a cullin, and Skp1, an F-box binding protein ***(56)***. The SCF complex is constitutively active throughout the cell cycle. Substrate phosphorylation drives capture by specific F-box proteins, which include Cdc4 for phosphorylated Sic1 ***(36)*** and Far1 ***(57)*** and Grr1 for phosphorylated Cln1 and Cln2 ***(58,59)***. In the case of Sic1, following Cln1/2/Cdc28-dependent phosphorylation, phospho-Sic1 is bound by the WD-repeat of the Cdc4 F-box protein and is ubiquitylated by the Cdc34 E2 enzyme ***(60)***. In fission yeast, ubiquitylation of phosphorylated Rum1 and Cdc18 is facilitated by Pop1/2 F-box proteins ***(42)***. F-box proteins recognize substrates through the PEST signal, (rich in Pro, Glu, Ser, and Thr), which can be found in the G1 cyclins Cln2 ***(61)***, Cln3 ***(62)***, and others.

6.2. The APC Complex

The APC is so called for its role in control of the metaphase-to-anaphase transition ***(63)***. The APC is a multimeric complex comprised of at least 12 gene products in *S. cerevisiae*, (reviewed in **ref.** ***38***). and 7 in *S. pombe* ***(64,65)***. The substrates for the APC are targeted by the presence of a destruction box (D-box) motif consisting of nine amino acids ***(66)***.

In yeast, the APC becomes active at anaphase onset in M-phase and persists through G_1 in the next cell cycle ***(67)***. An important mechanism of APC regulation is through association of one of two substrate-specific activators: Cdc20 ***(68)*** and Cdh1/Hct1 ***(69)*** in budding yeast and Slp1 ***(70)*** and Srw1/Ste9 ***(71)*** in fission yeast. These function to direct different substrates to the APC (*see* **Subheadings 6.2.1.** and **6.2.2.**). Cdc20 regulation of the APC is controlled by Cdc28/Clb, which directly phosphorylates Cdc20 and other subunits and appears to stimulate Cdc20–APC activity ***(72)***. Conversely, binding of Emi1 to Cdc20 inhibits APC prior to mitosis ***(73)***. Cdc28-dependent phosphorylation inhibits Cdh1/Hct1, preventing it from binding to the APC before anaphase is complete ***(74,75)***. This phosphorylation is removed by Cdc14, a phosphatase ***(76)***, which is activated by the mitotic exit network (*see* **Subheading 9.**). Cdh1/Hct1-dependent APC activity persists until S-phase and prevents premature expression of Cdc20 ***(77)***. The polo-like kinase Cdc5 appears to be required for Cdh1/Hct1 activation and is itself subject to Cdh1/Hct1-dependent APC destruction ***(78–80)***.

6.2.1. APC and Chromatid Separation

Chromatid separation at the metaphase-to-anaphase transition requires that the cohesin, holding the sister chromatids together, be destroyed. Cohesin consists of four highly conserved subunits, Scc1 (Mdc1) Scc3, Smc1, and Smc3 ***(81,82)***, of which the cleavage of Scc1 (Rad21 in fission yeast) is necessary and sufficient for separation and the onset of anaphase ***(83)***. Cleavage is carried out by a separase (Esp1 in budding yeast ***[84]***; Cut1 in fission yeast ***[85]***). Separase exists in a regulatory complex with a securin (Pds1 ***[84]***; Cut2 ***[85]***) in which securin binds and inhibits separase activity for most of the cell cycle. However, at anaphase onset the APC targets the securin for degradation, allowing the separase to become active ***(86,87)***. APC promotes the anaphase-to-metaphase transition through activation by Cdc20 ***(88)***, which, when coupled to APC, degrades the securin, Pds1/Cut2p, holding the sister chromatids together, thus triggering anaphase ***(86,89)*** (**Fig. 3**).

6.2.2. APC and Mitotic Exit

Inhibition of CDK activity is a prerequisite for mitotic exit and is largely achieved through destruction of mitotic cyclins. Destruction of mitotic cyclins can be driven by both the Cdc20 and the Cdh1/Hct1-dependent APC activities, and Cdc20 is itself regulated by the cell cycle, being destroyed in late mitosis ***(90,91)***. The Cdh1 ortholog in *S. pombe* (Srw1/Ste9) additionally promotes degradation of mitotic cyclins in G1 and is itself later negatively regulated by Cdc2-dependent phosphorylation ***(53,92)***. Cdh1 together with Sic1 are thought to induce the rapid drop in Cdc28 kinase activity required to drive cells out of mitosis and into the next G_1.

7. Regulating DNA Replication

Initiation of DNA replication is regulated such that it occurs precisely once during each cell cycle (**Fig. 3**). Initiation of DNA synthesis involves the assembly of a pre-RC at the origin of replication in G_1 in *S. cerevisiae* ***(93)***, although pre-RC formation may occur earlier, during anaphase in *S. pombe* ***(94)***. This complex is targeted to the origin recognition complex (ORC), which in yeast is associated with DNA throughout the cell cycle ***(95,96)***. During this process, the replication initiation factors Cdc6 (in *S. cerevisiae*) and Cdc18 (in *S. pombe*) bind to ORC ***(97,98)***, where they are required, together with Cdt1, to recruit the minichromosome maintenance complex (MCM) ***(99–102)***. Cdc6 and Cdc18 replication factors are tightly regulated, accumulate in mitosis and G_1, and are targeted for proteolysis at the onset of S-phase ***(103,104)***. The MCM complex is comprised of six highly conserved proteins (Mcm2–Mcm7) ***(105)*** and plays a central role in DNA replication initiation, where it probably acts as a DNA helicase for the growing replication forks (reviewed in **ref.** ***106***). Cdc45 is required for elongation, allowing the MCM complex to leave the origin once it has been converted to a helicase ***(107)***.

Firing of replication origins requires the Dbf4-dependent kinase (DDK), a complex consisting of the Cdc7 kinase (Hsk1 in *S. pombe* ***[108]***) and its regulatory subunit, Dbf4 (Dfp1/Him1 in *S. pombe* ***[108,109]***). DDK activity is cell cycle-regulated and peaks at the G_1/S transition ***(110,111)***. Dbf4 is targeted for degradation by the APC in the M/G_1 phase, and is phosphorylated in a checkpoint-dependent manner ***(112)***. In vitro assays have shown that DDK phosphorylates Mcm2-4, Mcm6, and Cdc45 ***(113,114)***, and phosphorylation of Mcm2 may cause a conformational change resulting in activation of the helicase function of the complex ***(115)***. However, other targets are thought to exist.

CDK activity is also required to trigger replication ***(11)***; in *S. cerevisiae* Cdc28 together with Clb5 and Clb6 are responsible for initiating origin firing ***(116)*** and are required for DDK function ***(114)***. Moreover, S-CDK–dependent phosphorylation of a replication protein, Sld2/Drc1, is required for chromosomal DNA replication ***(117)***.

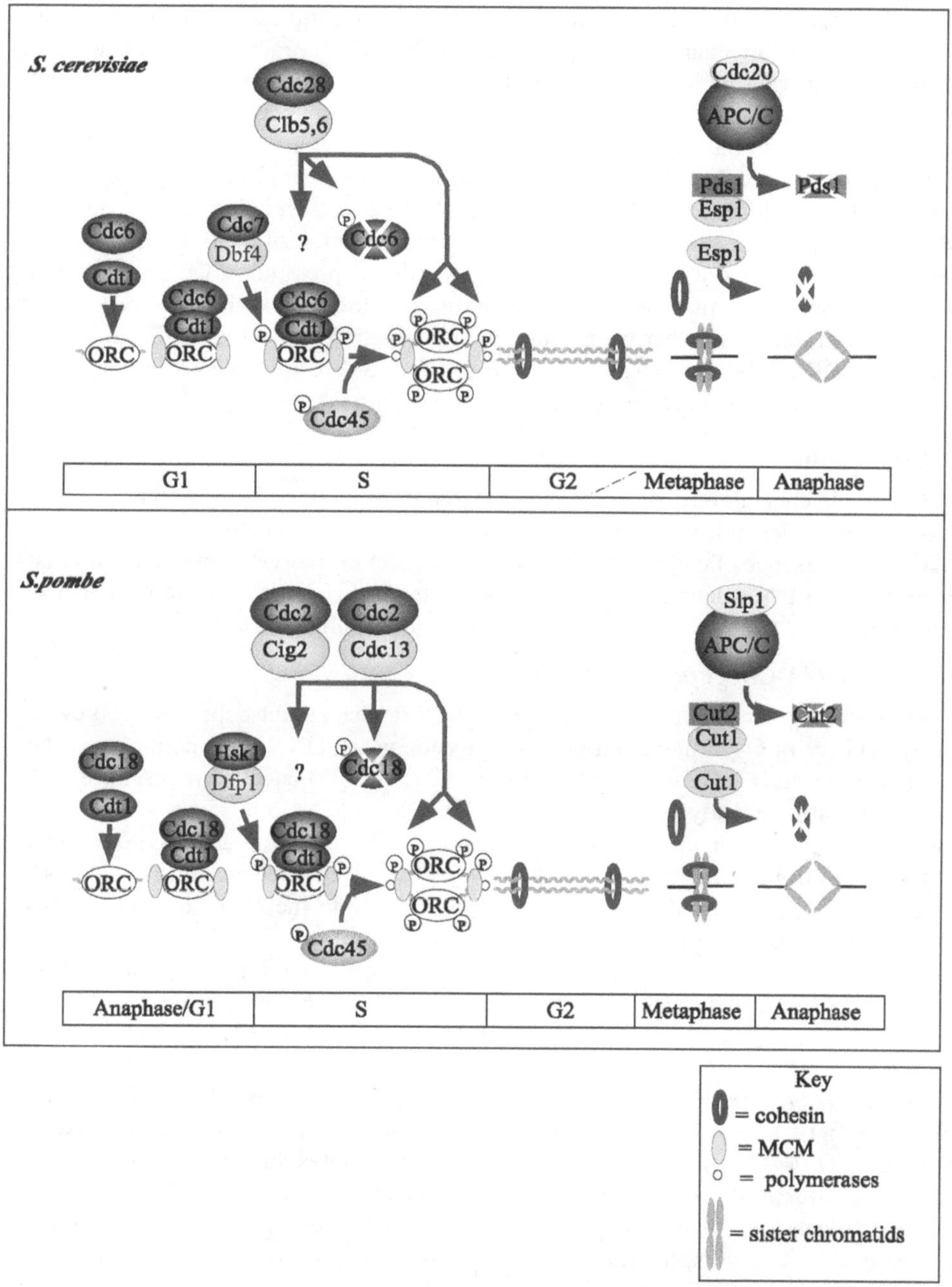

Fig. 3. Key events regulating DNA replication and segregation in *S. cerevisiae* and *S. pombe*. See text for details and references. MCM, minichromosome maintenance complex; ORC, origin recognition complex.

CDK activity additionally functions to block inappropriate replication firing through multiple mechanisms: both Cln and Clb/CDK complexes target Cdc6 for destruction, preventing rereplication ***(118–120)***. Similarly, in *S. pombe*, Cdc13/Cdc2 is responsible for Cdc18 destruction ***(104,121)***. CDK phosphorylation of ORCs additionally blocks reinitiation of DNA replication ***(120,122)***. Furthermore, Cln and Clb/CDK complexes regulate the nuclear localization of a number of budding yeast replication factors, including MCM proteins and Cdt1, which are excluded from the nucleus in G_2 and M-phases ***(102,123 ,124)***. The nuclear localization of the transcription factor Swi5 is also blocked by Cdc28-Clb ***(125)***, so that expression of CDC6 ***(103)*** and subsequent pre-RC formation at origins ***(126)*** occur at the end of mitosis when Cdc28/Clb is inactivated. The latter is mediated both by cyclin degradation and also by the action of CDK inhibitors such as Sic1 and Rum1 (*see* **Subheadings 5.2.** and **6.1.**) in *S. cerevisiae* and *S. pombe*, respectively.

8. Checkpoints

Cell cycle checkpoints are intracellular signal transduction pathways that function to maintain the dependence of later cell cycle events on the completion of earlier events ***(127)***. The presence of cell cycle checkpoints was first formally demonstrated in yeast in response to DNA damage ***(128)***. Here we consider two well-characterized checkpoint pathways, the DNA and spindle-assembly checkpoint pathways.

8.1. The DNA Checkpoint Pathway

DNA damage or a replication block can result in checkpoint-dependent cell cycle delay in G_1, S, or G_2/M in budding yeast. In fission yeast, DNA checkpoints delay the cell cycle in S and G_2 phases (reviewed in **ref.** ***129***). A G_1/M checkpoint response in *S. pombe* has also recently been described ***(130)***. DNA checkpoint responses serve to block cell proliferation until lesions are repaired; thus preventing damaged DNA and other lesions from being inherited by daughter cells. Recent evidence further suggests that the checkpoint machinery may contribute directly to the repair of such lesions (reviewed in **refs.** ***129*** and ***131***).

Accumulating evidence suggests that DNA damage surveillance is performed by three highly conserved checkpoint complexes: a complex comprising Mec1 and Ddc2 in budding yeast ***(132)*** (Rad3 and Rad26 in fission yeast ***[132,133]***); the checkpoint loading complex, comprising Rad24 and replication factor C subunits RFC2–5 in *S. cerevisiae* ***(134)*** (Rad17 and Rfc2-5 in *S. pombe* ***[135]***); and the checkpoint sliding clamp, comprising Rad17, Ddc1, and Mec3 in *S. cerevisiae* (Rad1, Rad9, and Hus1 in *S. pombe*) ***(136–138)***. Both the checkpoint loading complex and the checkpoint sliding clamp structurally resemble the RFC and PCNA components of the replication initiation machinery, respectively (reviewed in **ref.** ***129***). Recent data indicate that the checkpoint loading complex functions to load the sliding clamp complex onto DNA, thus functioning analogously to the replication factor C (RFC) and proliferating cell nuclear antigen (PCNA) complexes ***(139)***. The establishment of replication forks has also been shown to be required for checkpoint activation in response to particular types of DNA damage ***(140)***. Additionally, components of the replication machinery are targeted in response to unreplicated or damaged DNA whereby checkpoints function to block late origin firing and additionally to stabilize stalled replication forks (*141–143*; for review, *see* **ref.** ***129***).

In *S. pombe* the main cell cycle target inhibited in response to damaged or unreplicated DNA is Cdc2/Cdc13 through Cdc2/Y15 phosphorylation ***(144–146)***. This is achieved through Rad3-dependent activation of transduction kinases, Chk1 kinase in response to DNA damage in late S or G_2 ***(147)*** or Cds1 kinase in response to unreplicated DNA or DNA damage during S-phase ***(148)***. These activated transduction kinases subsequently phosphorylate Cdc25 phosphatase ***(149)***, stimulating interaction with 14-3-3 proteins ***(150)***, resulting in either loss of catalytic activity or sequestration into the cytoplasm ***(151,152)***. Cds1 is also required for Wee1 phosphorylation and an increase in Mik1 protein levels following S-phase arrest ***(153)***. Increased levels of Cdc2/Y15 phosphorylation subsequently result in G_2 arrest (*see* **Subheading 5.2.** and **Figs. 2** and **4**).

In *S. cerevisiae*, cell cycle arrest during mitosis is achieved through the concerted effects of two independent pathways, requiring Pds1 and Rad53 ***(154,155)***. Chk1-dependent phosphorylation and stabilization of Pds1 (securin) in response to DNA damage results in inhibition of the metaphase-to-anaphase transition ***(156,157)***. In contrast, Rad53 effects checkpoint control through maintaining activity of Cdc28 kinase, which is achieved through regulation of the Polo-like kinase Cdc5 ***(155)*** (*see* **Fig. 5**).

8.2. The Spindle Assembly Checkpoint Pathway

The spindle assembly checkpoint ensures that during metaphase one chromatid of each pair is attached to microtubules from opposite poles, prior to the onset of anaphase. This checkpoint was first identified in budding yeast, leading to the discovery of highly conserved *MAD* (mitotic arrest deficient) and *BUB* (budding uninhibited by benzamidazol) genes, encoding the spindle-assembly checkpoint machinery ***(158–160)***. The spindle assembly checkpoint machinery can detect a single unattached kinetochore and microtubule defects through either lack of attachment of the microtubules or subsequent tension.

In budding yeast, biochemical analyses indicate that complex formation among Mad1, Bub1, and Bub3 is crucial for spindle checkpoint function ***(161)***. Mad1 additionally binds tightly to Mad2, which may target Mad2 and other checkpoint components to the unattached kinetochores ***(162)***. In response to unattached kinetochores, the spindle assembly checkpoint is thought to arrest cells prior to anaphase through blocking Cdc20/APC activity through interaction of Cdc20 with a complex containing Mad2, Mad3, and Bub3 ***(163)*** (**Fig. 6**). Following attachment of the kinetochore, Mad2 dissociates from Cdc20/APC, thus allowing anaphase to proceed. Similar complexes between Slp1 (Cdc20) and Mad2 have been detected in fission yeast, and disruption of this complex results in failure to arrest in metaphase in response to spindle damage ***(70)***. Differences between the fission yeast and budding yeast spindle checkpoints have been identified, and the Aurora kinase, Ark1, is involved in monitoring unattached kinetochores in fission yeast, ***(164)***, whereas the related kinase, Ipl1, in budding yeast monitors lack of spindle tension ***(165)***.

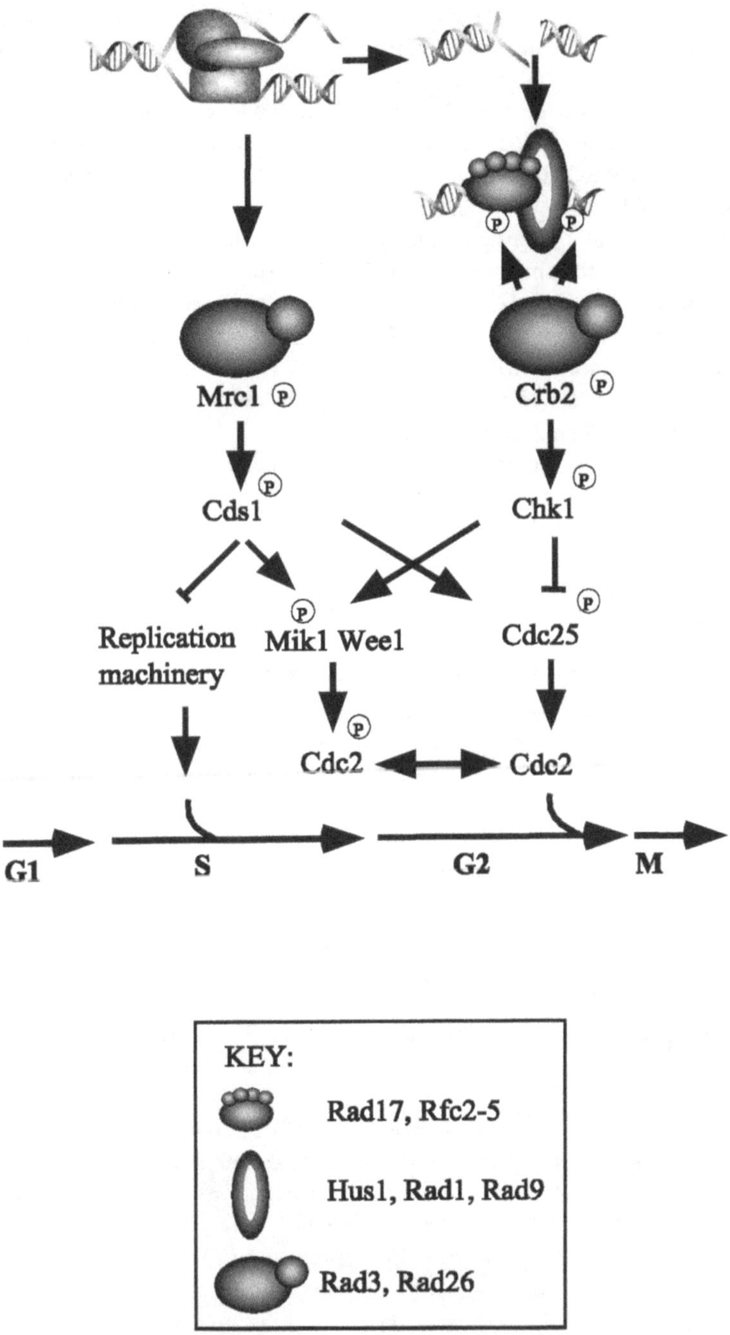

Fig. 4. DNA checkpoints of *S. pombe*. See text for details and references.

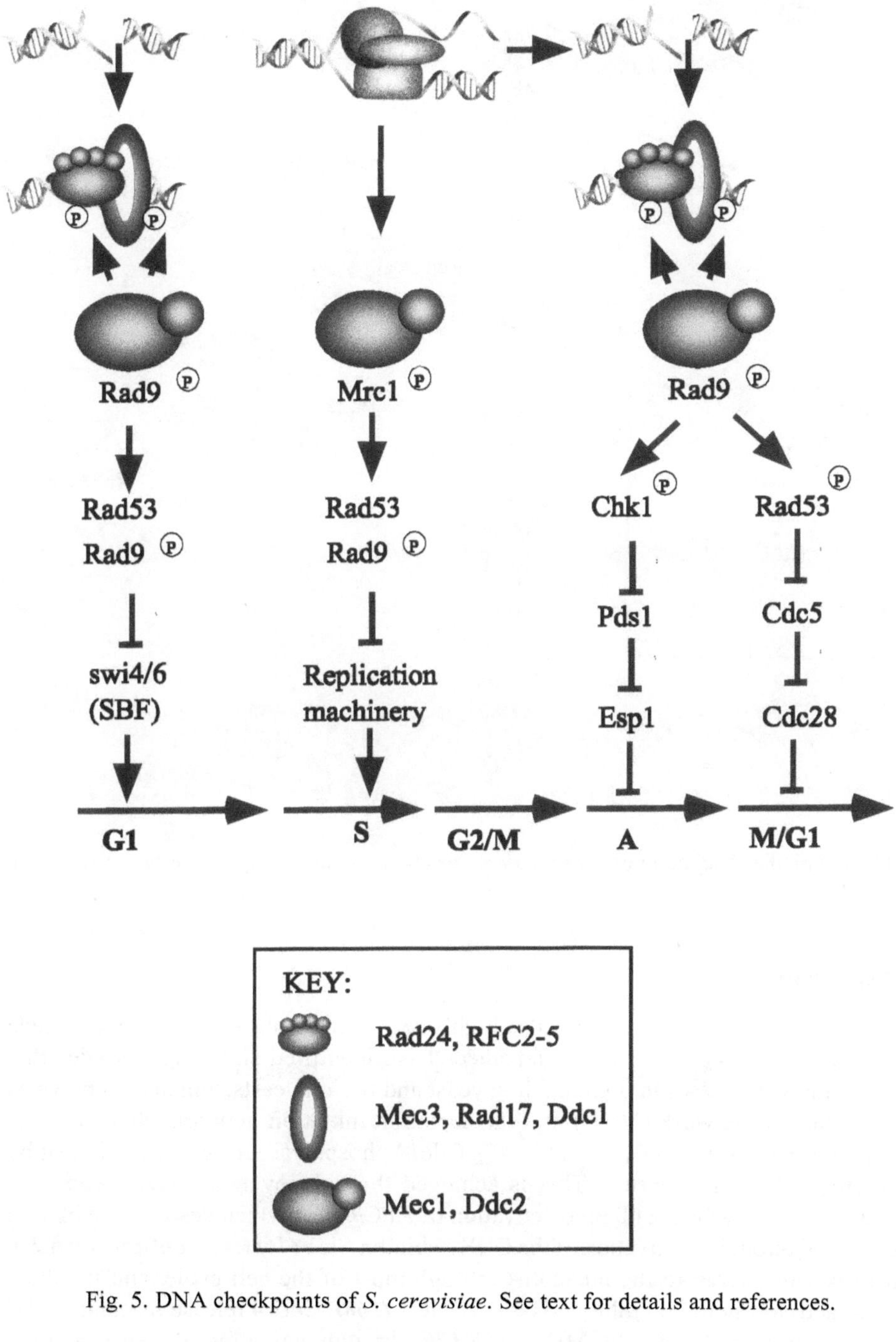

Fig. 5. DNA checkpoints of *S. cerevisiae*. See text for details and references.

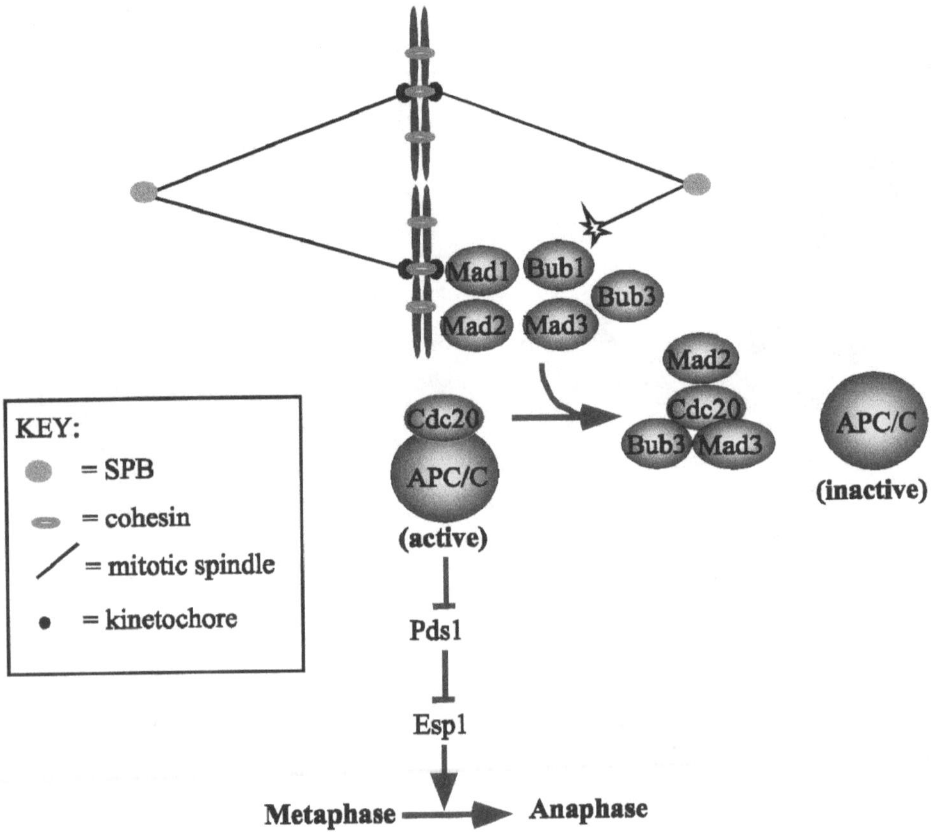

Fig. 6. Spindle checkpoint of *S. cerevisiae*. See text for details and references. SPB, spindle pole body.

9. Exit From Mitosis

Cytokinesis and mitotic exit are also highly regulated to ensure they do not precede chromosomal segregation. Recent advances have identified signaling cascades that regulate these processes in both budding yeast and fission yeasts, which are known as the mitotic exit network (MEN) and the septation initiation network (SIN), respectively (for reviews, *see* **refs. *166*** and ***167***). Cdc14 phosphatase triggers mitotic exit by promoting CDK inactivation. This is achieved through reversing CDK-dependent phosphorylation events, leading to activation of APC/Cdh1, which destroys the mitotic cyclins, and through reactivation of the CDK-inhibitor Sic1 ***(76)*** (*see* **Subheading 6.2.**). Cdc14 is sequestered to the nucleolus through most of the cell cycle, and its phosphatase activity is directly inhibited by Cfi1/Net1 ***(168)***. Cdc14 release from nucleolar sequestration is performed by MEN ***(169,170)*** through activation of Tem1, a small

Ras-like GTPase. Tem1/GTP activation is promoted by Lte1 (guanine nucleotide exchange factor) and inhibited by Bub2/Bfa1, a GTPase-activating complex ***(171,172)***. Activation of MEN appears to be spatially controlled such that mitotic exit is triggered only after the nucleus enters the bud, where Tem1, which is localized to the bud SPB, comes into proximity with its activator Lte1, which is localized to the bud cortex ***(172,173)***. An additional network termed "14 early anaphase release" (FEAR) also regulates Cdc14 release from Cfi/Net1 to the SPB in early anaphase, independently of MEN, which in turn functions to stimulate MEN, thus maintaining Cdc14 release ***(174,175)***.

An additional role for MEN in cytokinesis has also been identified. Mob1, a MEN component, relocalizes from the SPB to the bud neck in late mitosis, where it functions in cytokinesis ***(176)***. Such relocalization requires Cdc14-dependent dephosphorylation of other components of MEN ***(177–179)***.

Fission yeast septum formation is initiated through the activation of the SIN network following entry into mitosis (reviewed in **refs. *166*** and ***167***). An initial trigger for septation appears to be the activation of Spg1, the budding yeast Tem1 homolog ***(180,181)***, which binds and recruits Cdc7 kinase to the SPB ***(182,183)***. Cdc7 then recruits Sid1/Cdc14 to the active SPB, which is thought to facilitate subsequently the translocation of the Sid2/Mob1 kinase complex to the medial ring, where it in turn initiates cell division ***(184,185)***. During interphase, Cdc16/Byr4, a two-component GTPase-activating complex, negatively regulates Spg1 ***(180,181)***. SIN is regulated by both mitotic CDK activity, which must be low for septum formation, and the cytokinesis checkpoint (reviewed in **refs. *166*** and ***167***). A homolog of Cdc14, Clp1/Flp1, is also found in fission yeast, where it appears to regulate mitotic CDKs, through Cdc2/Y15 phosphorylation, by inhibiting Cdc25 and activating Wee1, rather than through cyclin degradation ***(186,187)***. Clp1/Flp1 is localized to the nucleolus during G_1 and S. An active SIN is not required for its release but is required to keep it out of the nucleolus until cytokinesis is complete ***(186,187)***.

The molecular basis of the relationships between mitotic exit and both the spindle and cytokinesis checkpoints are being actively investigated in both yeasts.

10. Conclusions

These fields of study have revealed a striking degree of conservation between the regulatory molecules and mechanisms that control the cell cycles of the evolutionarily divergent budding and fission yeasts. As many areas of yeast cell cycle control have yet to be understood, the application of both classical genetics, together with systematic genomic and proteomic technologies, to these problems is likely to provide important new insights into eukaryotic cell cycle control.

Acknowledgments

We are grateful to Kevin Hardwick, Stephen Kearsey, Karim Labib, David Lydall, Sergio Moreno, Clive Price, and the Humphrey Lab for helpful comments on this chapter. We apologize to the yeast cell cycle community for the oversimplifications and omissions necessary due to space limitations. This chapter is dedicated to the memory of Kristi Forbes Dunfield.

References

1. Murray A. and Hunt, T. (1993) *The Cell Cycle: An Introduction*, 1st ed. W.H. Freeman, New York.
2. Burke, D. D. and Stearns, T. (2000) *Methods in Yeast Genetics: A Cold Spring Harbor Laboratory CouH*harbor, NY.
3. Moreno, S., Klar, A., and Nurse, P. (1991) Molecular genetic analysis of fission yeast *Schizosaccharomyces pombe. Meth. Enzymol.* **194,** 795–823.
4. Heckman, D. S., Geiser, D. M., Eidell, B. R., Stauffer, R. L., Kardos, N. L., and Hedges, S. B. (2001) Molecular evidence for the early colonization of land by fungi and plants. *Science* **293,** 1129–1133.
5. Fantes, P. and Beggs, J. (2000) *The Yeast Nucleus*, Oxford University Press, Oxford.
6. Nurse, P. (1990) Universal control mechanism regulating onset of M-phase. *Nature* **344,** 503–508.
7. The yeast genome directory. (1997) *Nature* **387 (suppl),** 5.
8. Wood, V., Gwilliam, R., Rajandream, M. A., et al. (2002) The genome sequence of *Schizosaccharomyces pombe. Nature* **415,** 871–880.
9. Pringle, J. R. and Hartwell, L. H. (1981) The *Saccharomyces cerevisiae* cell cycle, in *The Molecular Biology of the Yeast* Saccharomyces*: Life Cycle and Inheritance* (J. N. S., eds.), Cold Spring Harbor Laboratory Press, Cold Spring Harbor, NY.
10. MacNeill, S. A. and Nurse, P. (1997) Cell cycle control in fission yeast, in *Yeast III* (Pringle, J. R., Broach, J. and Jones, E. W., eds.), Cold Spring Harbor Labroratory Press, Cold Spring Harbor, NY, pp. 697–763.
11. Hartwell, L. H., Culotti, J., Pringle, J. R., and Reid, B. J. (1974) Genetic control of the cell division cycle in yeast. *Science* **183,** 46–51.
12. Lew, D. J., Weinert, T., and Pringle, J. R. (1997) Cell cycle control in Saccharomyces *cerevisiae*, in *Molecular and Cellular Biology of the Yeast* Saccharomyces (Pringle, J. R., Roach, J. R., and Jones, E. W., eds.), Cold Spring Harbor Laboratory Press, Cold Spring Harbor, NY, pp. 607–695.
13. Nurse, P. (1975) Genetic control of cell size at cell division in yeast. *Nature* **256,** 547–551.
14. Nurse, P. and Thuriaux, P. (1977) Controls over the timing of DNA replication during the cell cycle of fission yeast. *Exp. Cell. Res.* **107,** 365–375.
15. Hartwell, L. H., Culotti, J., and Reid, B. (1970) Genetic control of the cell-division cycle in yeast. I. Detection of mutants. *Proc. Natl. Acad. Sci. USA* **66,** 352–359.
16. Nurse, P., Thuriaux, P., and Nasmyth, K. (1976) Genetic control of the cell division cycle in the fission yeast *Schizosaccharomyces pombe. Mol. Gen. Genet.* **146,** 167–178.
17. Nurse, P. and Thuriaux, P. (1980) Regulatory genes controlling mitosis in the fission yeast *Schizosaccharomyces pombe. Genetics* **96,** 627–637.
18. Evans, T., Rosenthal, E. T., Youngblom, J., Distel, D., and Hunt, T. (1983) Cyclin: a protein specified by maternal mRNA in sea urchin eggs that is destroyed at each cleavage division. *Cell* **33,** 389–396.
19. Futcher, B. (1996) Cyclins and the wiring of the yeast cell cycle. *Yeast* **12,** 1635–1646.
20. Fisher, D. and Nurse, P. (1995) Cyclins of the fission yeast *Schizosaccharomyces pombe. Semin. Cell Biol.* **6,** 73–78.
21. Kaldis, P. (1999) The cdk-activating kinase (CAK): from yeast to mammals. *Cell Mol. Life Sci.* **55,** 284–296.

22. Saiz, J. E. and Fisher, R. P. (2002) A CDK-activating kinase network is required in cell cycle control and transcription in fission yeast. *Curr. Biol.* **12,** 1100–1105.
23. Espinoza, F. H., Farrell, A., Erdjument-Bromage, H., Tempst, P., and Morgan, D. O. (1996) A cyclin-dependent kinase-activating kinase (CAK) in budding yeast unrelated to vertebrate CAK. *Science* **273,** 1714–1717.
24. Gould, K. L. and Nurse, P. (1989) Tyrosine phosphorylation of the fission yeast cdc2+ protein kinase regulates entry into mitosis. *Nature* **342,** 39–45.
25. Russell, P. and Nurse, P. (1987) Negative regulation of mitosis by wee1+, a gene encoding a protein kinase homolog. *Cell* **49,** 559–567.
26. Lundgren, K., Walworth, N., Booher, R., Dembski, M., Kirschner, M., and Beach, D. (1991) mik1 and wee1 cooperate in the inhibitory tyrosine phosphorylation of cdc2. *Cell* **64,** 1111–1122.
27. Russell, P. and Nurse, P. (1986) cdc25+ functions as an inducer in the mitotic control of fission yeast. *Cell* **45,** 145–153.
28. Moser, B. A. and Russell, P. (2000) Cell cycle regulation in *Schizosaccharomyces pombe. Curr. Opin. Microbiol.* **3,** 631–636.
29. Daga, R. R. and Jimenez, J. (1999) Translational control of the cdc25 cell cycle phosphatase: a molecular mechanism coupling mitosis to cell growth. *J. Cell Sci.* **112,** 3137–3146.
30. Sia, R. A., Herald, H. A., and Lew, D. J. (1996) Cdc28 tyrosine phosphorylation and the morphogenesis checkpoint in budding yeast. *Mol. Biol. Cell.* **7,** 1657–1666.
31. Harvey, S. L. and Kellogg, D. R. (2003) Conservation of mechanisms controlling entry into mitosis. Budding yeast wee1 delays entry into mitosis and is required for cell size control. *Curr. Biol.* **13,** 264–275.
32. Peter, M. and Herskowitz, I. (1994) Direct inhibition of the yeast cyclin-dependent kinase Cdc28-Cln by Far1. *Science* **265,** 1228–1231.
33. Mendenhall, M. D., Jones, C. A., and Reed, S. I. (1987) Dual regulation of the yeast CDC28-p40 protein kinase complex: cell cycle, pheromone, and nutrient limitation effects. *Cell* **50,** 927–935.
34. Calzada, A., Sacristan, M., Sanchez, E., and Bueno, A. (2001) Cdc6 cooperates with Sic1 and Hct1 to inactivate mitotic cyclin-dependent kinases. *Nature* **412,** 355–358.
35. Moreno, S. and Nurse, P. (1994) Regulation of progression through the G1 phase of the cell cycle by the rum1+ gene. *Nature* **367,** 236–242.
36. Schwob, E., Bohm, T., Mendenhall, M. D., and Nasmyth, K. (1994) The B-type cyclin kinase inhibitor p40SIC1 controls the G1 to S transition in *S. cerevisiae. Cell* **79,** 233–244.
37. Chang, F. and Herskowitz, I. (1990) Identification of a gene necessary for cell cycle arrest by a negative growth factor of yeast: FAR1 is an inhibitor of a G1 cyclin, CLN2. *Cell* **63,** 999–1011.
38. Zachariae, W. and Nasmyth, K. (1999) Whose end is destruction: cell division and the anaphase-promoting complex. *Genes Dev.* **13,** 2039–2058.
39. Mendenhall, M. D. (1993) An inhibitor of p34CDC28 protein kinase activity from *Saccharomyces cerevisiae. Science* **259,** 216–219.
40. Schneider, B. L., Yang, Q. H., and Futcher, A. B. (1996) Linkage of replication to start by the Cdk inhibitor Sic1. *Science* **272,** 560–562.
41. Martin-Castellanos, C., Labib, K., and Moreno, S. (1996) B-type cyclins regulate G1 progression in fission yeast in opposition to the p25rum1 cdk inhibitor. *EMBO J.* **15,** 839–849.
42. Martin-Castellanos, C., and Moreno, S. (1996) Regulation of G1 progression in fission yeast by the rum1+ gene product. *Prog. Cell Cycle Res.* **2,** 29–35.

43. Breeden, L. (1996) Start-specific transcription in yeast. *Curr. Top. Microbiol. Immunol.* **208,** 95–127.
44. Wittenberg, C., Sugimoto, K., and Reed, S. I. (1990) G1-specific cyclins of *S. cerevisiae*: cell cycle periodicity, regulation by mating pheromone, and association with the p34CDC28 protein kinase. *Cell* **62,** 225–237.
45. Tyers, M., Tokiwa, G., Nash, R., and Futcher, B. (1992) The Cln3-Cdc28 kinase complex of *S. cerevisiae* is regulated by proteolysis and phosphorylation. *EMBO J.* **11,** 1773–1784.
46. Polymenis, M. and Schmidt, E. V. (1997) Coupling of cell division to cell growth by translational control of the G1 cyclin CLN3 in yeast. *Genes Dev.* **11,** 2522–2531.
47. Dirick, L., Bohm, T., and Nasmyth, K. (1995) Roles and regulation of Cln-Cdc28 kinases at the start of the cell cycle of *Saccharomyces cerevisiae. EMBO J.* **14,** 4803–4813.
48. Stuart, D. and Wittenberg, C. (1995) CLN3, not positive feedback, determines the timing of CLN2 transcription in cycling cells. *Genes Dev.* **9,** 2780–2794.
49. Jorgensen, P., Nishikawa, J. L., Breitkreutz, B. J., and Tyers, M. (2002) Systematic identification of pathways that couple cell growth and division in yeast. *Science* **297,** 395–400.
50. Martin-Castellanos, C., Blanco, M. A., de Prada, J. M., and Moreno, S. (2000) The puc1 cyclin regulates the G1 phase of the fission yeast cell cycle in response to cell size. *Mol. Biol. Cell* **11,** 543–554.
51. Ayte, J., Schweitzer, C., Zarzov, P., Nurse, P., and DeCaprio, J. A. (2001) Feedback regulation of the MBF transcription factor by cyclin Cig2. *Nat. Cell Biol.* **3,** 1043–1050.
52. Wuarin, J., Buck, V., Nurse, P., and Millar, J. B. (2002) Stable association of mitotic cyclin B/Cdc2 to replication origins prevents endoreduplication. *Cell* **111,** 419–431.
53. Blanco, M. A., Sanchez-Diaz, A., de Prada, J. M., and Moreno, S. (2000) APC(ste9/srw1) promotes degradation of mitotic cyclins in G(1) and is inhibited by cdc2 phosphorylation. *EMBO J.* **19,** 3945–3955.
54. Hershko, A. and Ciechanover, A. (1998) The ubiquitin system. *Annu. Rev. Biochem.* **67,** 425–479.
55. Tyers, M. and Jorgensen, P. (2000) Proteolysis and the cell cycle: with this RING I do thee destroy. *Curr. Opin. Genet. Dev.* **10,** 54–64.
56. Willems, A. R., Lanker, S., Patton, E. E., et al. (1996) Cdc53 targets phosphorylated G1 cyclins for degradation by the ubiquitin proteolytic pathway. *Cell* **86,** 453–463.
57. Henchoz, S., Chi, Y., Catarin, B., Herskowitz, I., Deshaies, R. J., and Peter, M. (1997) Phosphorylation- and ubiquitin-dependent degradation of the cyclin-dependent kinase inhibitor Far1p in budding yeast. *Genes Dev.* **11,** 3046–3060.
58. Deshaies, R. J., Chau, V., and Kirschner, M. (1995) Ubiquitination of the G1 cyclin Cln2p by a Cdc34p-dependent pathway. *EMBO J.* **14,** 303–312.
59. Barral, Y., Jentsch, S., and Mann, C. (1995) G1 cyclin turnover and nutrient uptake are controlled by a common pathway in yeast. *Genes Dev.* **9,** 399–409.
60. Skowyra, D., Craig, K. L., Tyers, M., Elledge, S. J., and Harper, J. W. (1997) F-box proteins are receptors that recruit phosphorylated substrates to the SCF ubiquitin-ligase complex. *Cell* **91,** 209–219.
61. Lanker, S., Valdivieso, M. H., and Wittenberg, C. (1996) Rapid degradation of the G1 cyclin Cln2 induced by CDK-dependent phosphorylation. *Science* **271,** 1597–1601.
62. Yaglom, J., Linskens, M. H., Sadis, S., Rubin, D. M., Futcher, B., and Finley, D. (1995) p34Cdc28-mediated control of Cln3 cyclin degradation. *Mol. Cell. Biol.* **15,** 731–741.
63. King, R. W., Peters, J. M., Tugendreich, S., Rolfe, M., Hieter, P., and Kirschner, M. W. (1995) A 20S complex containing CDC27 and CDC16 catalyzes the mitosis-specific conjugation of ubiquitin to cyclin B. *Cell* **81,** 279–288.

64. Berry, L. D., Feoktistova, A., Wright, M. D., and Gould, K. L. (1999) The *Schizosaccharomyces pombe* dim1(+) gene interacts with the anaphase-promoting complex or cyclosome (APC/C) component lid1(+) and is required for APC/C function. *Mol. Cell. Biol.* **19,** 2535–2546.
65. Yamashita, Y. M., Nakaseko, Y., Kumada, K., Nakagawa, T., and Yanagida, M. (1999) Fission yeast APC/cyclosome subunits, Cut20/Apc4 and Cut23/Apc8, in regulating metaphase-anaphase progression and cellular stress responses. *Genes Cells* **4,** 445–463.
66. Glotzer, M., Murray, A. W., and Kirschner, M. W. (1991) Cyclin is degraded by the ubiquitin pathway. *Nature* **349,** 132–138.
67. Amon, A., Irniger, S. & Nasmyth, K. (1994) Closing the cell cycle circle in yeast: G2 cyclin proteolysis initiated at mitosis persists until the activation of G1 cyclins in the next cycle. *Cell* **77,** 1037–50.
68. Visintin, R., Prinz, S., and Amon, A. (1997) CDC20 and CDH1: a family of substrate-specific activators of APC-dependent proteolysis. *Science* **278,** 460–463.
69. Schwab, M., Lutum, A. S., and Seufert, W. (1997) Yeast Hct1 is a regulator of Clb2 cyclin proteolysis. *Cell* **90,** 683–693.
70. Kim, S. H., Lin, D. P., Matsumoto, S., Kitazono, A., and Matsumoto, T. (1998) Fission yeast Slp1: an effector of the Mad2-dependent spindle checkpoint. *Science* **279,** 1045–1047.
71. Kitamura, K., Maekawa, H., and Shimoda, C. (1998) Fission yeast Ste9, a homolog of Hct1/Cdh1 and Fizzy-related, is a novel negative regulator of cell cycle progression during G1-phase. *Mol. Biol. Cell* **9,** 1065–1080.
72. Rudner, A. D. and Murray, A. W. (2000) Phosphorylation by Cdc28 activates the Cdc20-dependent activity of the anaphase-promoting complex. *J. Cell Biol.* **149,** 1377–1390.
73. Reimann, J. D., Freed, E., Hsu, J. Y., Kramer, E. R., Peters, J. M., and Jackson, P. K. (2001) Emi1 is a mitotic regulator that interacts with Cdc20 and inhibits the anaphase promoting complex. *Cell* **105,** 645–655.
74. Zachariae, W., Schwab, M., Nasmyth, K., and Seufert, W. (1998) Control of cyclin ubiquitination by CDK-regulated binding of Hct1 to the anaphase promoting complex. *Science* **282,** 1721–1724.
75. Jaspersen, S. L., Charles, J. F., and Morgan, D. O. (1999) Inhibitory phosphorylation of the APC regulator Hct1 is controlled by the kinase Cdc28 and the phosphatase Cdc14. *Curr. Biol.* **9,** 227–236.
76. Visintin, R., Craig, K., Hwang, E. S., Prinz, S., Tyers, M., and Amon, A. (1998) The phosphatase Cdc14 triggers mitotic exit by reversal of Cdk-dependent phosphorylation. *Mol. Cell* **2,** 709–718.
77. Huang, J. N., Park, I., Ellingson, E., Littlepage, L. E., and Pellman, D. (2001) Activity of the APC(Cdh1) form of the anaphase-promoting complex persists until S phase and prevents the premature expression of Cdc20p. *J. Cell Biol.* **154,** 85–94.
78. Charles, J. F., Jaspersen, S. L., Tinker-Kulberg, R. L., Hwang, L., Szidon, A. & Morgan, D. O. (1998) The Polo-related kinase Cdc5 activates and is destroyed by the mitotic cyclin destruction machinery in *S. cerevisiae*. *Curr. Biol.* **8,** 497–507.
79. Cheng, L., Hunke, L., and Hardy, C. F. (1998) Cell cycle regulation of the *Saccharomyces cerevisiae* polo-like kinase cdc5p. *Mol. Cell. Biol.* **18,** 7360–7370.
80. Shirayama, M., Zachariae, W., Ciosk, R., and Nasmyth, K. (1998) The Polo-like kinase Cdc5p and the WD-repeat protein Cdc20p/fizzy are regulators and substrates of the anaphase promoting complex in *Saccharomyces cerevisiae*. *EMBO J.* **17,** 1336–1349.

81. Guacci, V., Koshland, D., and Strunnikov, A. (1997) A direct link between sister chromatid cohesion and chromosome condensation revealed through the analysis of MCD1 in *S. cerevisiae*. *Cell* **91,** 47–57.
82. Michaelis, C., Ciosk, R., and Nasmyth, K. (1997) Cohesins: chromosomal proteins that prevent premature separation of sister chromatids. *Cell* **91,** 35–45.
83. Uhlmann, F., Lottspeich, F., and Nasmyth, K. (1999) Sister-chromatid separation at anaphase onset is promoted by cleavage of the cohesin subunit Scc1. *Nature* **400,** 37–42.
84. Ciosk, R., Zachariae, W., Michaelis, C., Shevchenko, A., Mann, M., and Nasmyth, K. (1998) An ESP1/PDS1 complex regulates loss of sister chromatid cohesion at the metaphase to anaphase transition in yeast. *Cell* **93,** 1067–1076.
85. Kumada, K., Nakamura, T., Nagao, K., Funabiki, H., Nakagawa, T., and Yanagida, M. (1998) Cut1 is loaded onto the spindle by binding to Cut2 and promotes anaphase spindle movement upon Cut2 proteolysis. *Curr. Biol.* **8,** 633–641.
86. Cohen-Fix, O., Peters, J. M., Kirschner, M. W., and Koshland, D. (1996) Anaphase initiation in *Saccharomyces cerevisiae* is controlled by the APC-dependent degradation of the anaphase inhibitor Pds1p. *Genes Dev.* **10,** 3081–3093.
87. Funabiki, H., Yamano, H., Kumada, K., Nagao, K., Hunt, T., and Yanagida, M. (1996) Cut2 proteolysis required for sister-chromatid seperation in fission yeast. *Nature* **381,** 438–441.
88. Fang, G., Yu, H., and Kirschner, M. W. (1998) Direct binding of CDC20 protein family members activates the anaphase-promoting complex in mitosis and G1. *Mol. Cell* **2,** 163–171.
89. Shirayama, M., Toth, A., Galova, M., and Nasmyth, K. (1999) APC(Cdc20) promotes exit from mitosis by destroying the anaphase inhibitor Pds1 and cyclin Clb5. *Nature* **402,** 203–207.
90. Schwab, M., Neutzner, M., Mocker, D., and Seufert, W. (2001) Yeast Hct1 recognizes the mitotic cyclin Clb2 and other substrates of the ubiquitin ligase APC. *EMBO J.* **20,** 5165–5175.
91. Wasch, R. and Cross, F. R. (2002) APC-dependent proteolysis of the mitotic cyclin Clb2 is essential for mitotic exit. *Nature* **418,** 556–562.
92. Yamaguchi, S., Okayama, H., and Nurse, P. (2000) Fission yeast Fizzy-related protein srw1p is a G(1)-specific promoter of mitotic cyclin B degradation. *EMBO J.* **19,** 3968–3977.
93. Diffley, J. F., Cocker, J. H., Dowell, S. J., and Rowley, A. (1994) Two steps in the assembly of complexes at yeast replication origins in vivo. *Cell* **78,** 303–316.
94. Kearsey, S. E., Montgomery, S., Labib, K., and Lindner, K. (2000) Chromatin binding of the fission yeast replication factor mcm4 occurs during anaphase and requires ORC and cdc18. *EMBO J.* **19,** 1681–1690.
95. Aparicio, O. M., Weinstein, D. M., and Bell, S. P. (1997) Components and dynamics of DNA replication complexes in *S. cerevisiae*: redistribution of MCM proteins and Cdc45p during S phase. *Cell* **91,** 59–69.
96. Lygerou, Z. and Nurse, P. (1999) The fission yeast origin recognition complex is constitutively associated with chromatin and is differentially modified through the cell cycle. *J. Cell Sci.* **112,** 3703–3712.
97. Liang, C., Weinreich, M., and Stillman, B. (1995) ORC and Cdc6p interact and determine the frequency of initiation of DNA replication in the genome. *Cell* **81,** 667–676.

98. Grallert, B. and Nurse, P. (1996) The ORC1 homolog orp1 in fission yeast plays a key role in regulating onset of S phase. *Genes Dev.* **10,** 2644–2654.
99. Donovan, S., Harwood, J., Drury, L. S., and Diffley, J. F. (1997) Cdc6p-dependent loading of Mcm proteins onto pre-replicative chromatin in budding yeast. *Proc. Natl. Acad. Sci. USA* **94,** 5611–5616.
100. Tanaka, T., Knapp, D., and Nasmyth, K. (1997) Loading of an Mcm protein onto DNA replication origins is regulated by Cdc6p and CDKs. *Cell* **90,** 649–660.
101. Ogawa, Y., Takahashi, T., and Masukata, H. (1999) Association of fission yeast Orp1 and Mcm6 proteins with chromosomal replication origins. *Mol. Cell. Biol.* **19,** 7228–7236.
102. Tanaka, S. and Diffley, J. F. (2002) Interdependent nuclear accumulation of budding yeast Cdt1 and Mcm2-7 during G1 phase. *Nat. Cell Biol.* **4,** 198–207.
103. Piatti, S., Lengauer, C., and Nasmyth, K. (1995) Cdc6 is an unstable protein whose de novo synthesis in G1 is important for the onset of S phase and for preventing a 'reductional' anaphase in the budding yeast *Saccharomyces cerevisiae*. *EMBO J.* **14,** 3788–3799.
104. Baum, B., Nishitani, H., Yanow, S., and Nurse, P. (1998) Cdc18 transcription and proteolysis couple S phase to passage through mitosis. *EMBO J.* **17,** 5689–5698.
105. Chong, J. P., Mahbubani, H. M., Khoo, C. Y., and Blow, J. J. (1995) Purification of an MCM-containing complex as a component of the DNA replication licensing system. *Nature* **375,** 418–421.
106. Labib, K. and Diffley, J. F. (2001) Is the MCM2-7 complex the eukaryotic DNA replication fork helicase? *Curr. Opin. Genet. Dev.* **11,** 64–70.
107. Tercero, J. A., Labib, K., and Diffley, J. F. (2000) DNA synthesis at individual replication forks requires the essential initiation factor Cdc45p. *EMBO J.* **19,** 2082–2093.
108. Masai, H., Miyake, T., and Arai, K. (1995) hsk1+, a *Schizosaccharomyces pombe* gene related to *Saccharomyces cerevisiae* CDC7, is required for chromosomal replication. *EMBO J.* **14,** 3094–3104.
109. Kitada, K., Johnston, L. H., Sugino, T., and Sugino, A. (1992) Temperature-sensitive cdc7 mutations of *Saccharomyces cerevisiae* are suppressed by the DBF4 gene, which is required for the G1/S cell cycle transition. *Genetics* **131,** 21–29.
110. Jackson, A. L., Pahl, P. M., Harrison, K., Rosamond, J., and Sclafani, R. A. (1993) Cell cycle regulation of the yeast Cdc7 protein kinase by association with the Dbf4 protein. *Mol. Cell. Biol.* **13,** 2899–2908.
111. Yoon, H. J., Loo, S., and Campbell, J. L. (1993) Regulation of *Saccharomyces cerevisiae* CDC7 function during the cell cycle. *Mol. Biol. Cell* **4,** 195–208.
112. Weinreich, M. and Stillman, B. (1999) Cdc7p-Dbf4p kinase binds to chromatin during S phase and is regulated by both the APC and the RAD53 checkpoint pathway. *EMBO J.* **18,** 5334–5346.
113. Lei, M., Kawasaki, Y., Young, M. R., Kihara, M., Sugino, A., and Tye, B. K. (1997) Mcm2 is a target of regulation by Cdc7-Dbf4 during the initiation of DNA synthesis. *Genes Dev.* **11,** 3365–3374.
114. Nougarede, R., Della Seta, F., Zarzov, P., and Schwob, E. (2000) Hierarchy of S-phase-promoting factors: yeast Dbf4-Cdc7 kinase requires prior S-phase cyclin-dependent kinase activation. *Mol. Cell. Biol.* **20,** 3795–3806.
115. Tye, B. K. (1999) MCM proteins in DNA replication. *Annu. Rev. Biochem.* **68,** 649–686.
116. Epstein, C. B. and Cross, F. R. (1992) CLB5: a novel B cyclin from budding yeast with a role in S phase. *Genes Dev.* **6,** 1695–1706.
117. Masumoto, H., Muramatsu, S., Kamimura, Y., and Araki, H. (2002) S-Cdk-dependent phosphorylation of Sld2 essential for chromosomal DNA replication in budding yeast. *Nature* **415,** 651–655.

118. Calzada, A., Sanchez, M., Sanchez, E., and Bueno, A. (2000) The stability of the Cdc6 protein is regulated by cyclin-dependent kinase/cyclin B complexes in *Saccharomyces cerevisiae*. *J. Biol. Chem.* **275,** 9734–9741.
119. Drury, L. S., Perkins, G., and Diffley, J. F. (2000) The cyclin-dependent kinase Cdc28p regulates distinct modes of Cdc6p proteolysis during the budding yeast cell cycle. *Curr. Biol.* **10,** 231–240.
120. Nguyen, V. Q., Co, C., and Li, J. J. (2001) Cyclin-dependent kinases prevent DNA re-replication through multiple mechanisms. *Nature* **411,** 1068–1073.
121. Jallepalli, P. V., Brown, G. W., Muzi-Falconi, M., Tien, D., and Kelly, T. J. (1997) Regulation of the replication initiator protein p65cdc18 by CDK phosphorylation. *Genes Dev.* **11,** 2767–2779.
122. Vas, A., Mok, W., and Leatherwood, J. (2001) Control of DNA rereplication via Cdc2 phosphorylation sites in the origin recognition complex. *Mol. Cell. Biol.* **21,** 5767–5777.
123. Labib, K., Diffley, J. F., and Kearsey, S. E. (1999) G1-phase and B-type cyclins exclude the DNA-replication factor Mcm4 from the nucleus. *Nat. Cell Biol.* **1,** 415–422.
124. Nguyen, V. Q., Co, C., Irie, K., and Li, J. J. (2000) Clb/Cdc28 kinases promote nuclear export of the replication initiator proteins Mcm2-7. *Curr. Biol.* **10,** 195–205.
125. Moll, T., Tebb, G., Surana, U., Robitsch, H., and Nasmyth, K. (1991) The role of phosphorylation and the CDC28 protein kinase in cell cycle-regulated nuclear import of the *S. cerevisiae* transcription factor SWI5. *Cell* **66,** 743–758.
126. Cocker, J. H., Piatti, S., Santocanale, C., Nasmyth, K., and Diffley, J. F. (1996) An essential role for the Cdc6 protein in forming the pre-replicative complexes of budding yeast. *Nature* **379,** 180–182.
127. Hartwell, L. H. and Weinert, T. A. (1989) Checkpoints: controls that ensure the order of cell cycle events. *Science* **246,** 629–634.
128. Weinert, T. A. & Hartwell, L. H. (1988) The *RAD9* gene controls the cell cycle response to DNA damage in *Saccharomyces cerevisiae*. *Science* **241,** 317–322.
129. Nyberg, K. A., Michelson, R. J., Putnam, C. W., and Weinert, T. A. (2002) Toward maintaining the genome: DNA damage and replication checkpoints. *Annu. Rev. Genet.* **36,** 617–656.
130. Synnes, M., Nilssen, E. A., Boye, E., and Grallert, B. (2002) A novel chk1-dependent G1/M checkpoint in fission yeast. *J. Cell Sci.* **115,** 3609–3618.
131. Carr, A. M. (2002) DNA structure dependent checkpoints as regulators of DNA repair. *DNA Repair (Amst)* **1,** 983–994.
132. Edwards, R. J., Bentley, N. J., and Carr, A. M. (1999) A Rad3-Rad26 complex responds to DNA damage independently of other checkpoint proteins. *Nat. Cell Biol.* **1,** 393–398.
133. Paciotti, V., Clerici, M., Lucchini, G., and Longhese, M. P. (2000) The checkpoint protein Ddc2, functionally related to *S. pombe* Rad26, interacts with Mec1 and is regulated by Mec1-dependent phosphorylation in budding yeast. *Genes Dev.* **14,** 2046–2059.
134. Green, C. M., Erdjument-Bromage, H., Tempst, P., and Lowndes, N. F. (2000) A novel Rad24 checkpoint protein complex closely related to replication factor C. *Curr. Biol.* **10,** 39–42.
135. Griffiths, D. J., Barbet, N. C., McCready, S., Lehmann, A. R., and Carr, A. M. (1995) Fission yeast rad17: a homologue of budding yeast RAD24 that shares regions of sequence similarity with DNA polymerase accessory proteins. *EMBO J.* **14,** 5812–5823.
136. Caspari, T., Dahlen, M., Kanter-Smoler, G., et al. (2000) Characterization of *Schizosaccharomyces pombe* Hus1: a PCNA-related protein that associates with Rad1 and Rad9. *Mol. Cell. Biol.* **20,** 1254–1262.

137. Venclovas, C. and Thelen, M. P. (2000) Structure-based predictions of Rad1, Rad9, Hus1 and Rad17 participation in sliding clamp and clamp-loading complexes. *Nucleic Acids Res.* **28,** 2481–2493.
138. Volkmer, E. & Karnitz, L. M. (1999) Human homologs of *Schizosaccharomyces pombe* rad1, hus1, and rad9 form a DNA damage-responsive protein complex. *J. Biol. Chem.* **274,** 567–570.
139. Majka, J. and Burgers, P. M. (2003) Yeast Rad17/Mec3/Ddc1: a sliding clamp for the DNA damage checkpoint. *Proc. Natl. Acad. Sci. USA* **100,** 2249–2254.
140. Tercero, J. A., Longhese, M. P., and Diffley, J. F. (2003) A central role for DNA replication forks in checkpoint activation and response. *Mol. Cell.* **11,** 1323–1336.
141. Tercero, J. A. and Diffley, J. F. (2001) Regulation of DNA replication fork progression through damaged DNA by the Mec1/Rad53 checkpoint. *Nature* **412,** 553–557.
142. Lopes, M., Cotta-Ramusino, C., Pellicioli, A., et al. (2001) The DNA replication checkpoint response stabilizes stalled replication forks. *Nature* **412,** 557–561.
143. Sogo, J. M., Lopes, M., and Foiani, M. (2002) Fork reversal and ssDNA accumulation at stalled replication forks owing to checkpoint defects. *Science* **297,** 599–602.
144. Forbes, K. C., Humphrey, T., and Enoch, T. (1998) Suppressors of cdc25p overexpression identify two pathways that influence the G2/M checkpoint in fission yeast. *Genetics* **150,** 1361–1375.
145. Rhind, N. and Russell, P. (1998) Tyrosine phosphorylation of cdc2 is required for the replication checkpoint in *Schizosaccharomyces pombe*. *Mol. Cell. Biol.* **18,** 3782–3787.
146. Rhind, N., Furnari, B., and Russell, P. (1997) Cdc2 tyrosine phosphorylation is required for the DNA damage checkpoint in fission yeast. *Genes Dev.* **11,** 504–511.
147. Walworth, N. C. and Bernards, R. (1996) rad-dependent response of the chk1-encoded protein kinase at the DNA damage checkpoint. *Science* **271,** 353–356.
148. Lindsay, H. D., Griffiths, D. J., Edwards, R. J., et al. (1998) S-phase-specific activation of Cds1 kinase defines a subpathway of the checkpoint response in *Schizosaccharomyces pombe*. *Genes Dev.* **12,** 382–395.
149. Peng, C. Y., Graves, P. R., Thoma, R. S., Wu, Z., Shaw, A. S., and Piwnica-Worms, H. (1997) Mitotic and G2 checkpoint control: regulation of 14-3-3 protein binding by phosphorylation of Cdc25C on serine-216. *Science* **277,** 1501–1505.
150. Zeng, Y., Forbes, K. C., Wu, Z., Moreno, S., Piwnica-Worms, H., and Enoch, T. (1998) Replication checkpoint requires phosphorylation of the phosphatase Cdc25 by Cds1 or Chk1. *Nature* **395,** 507–510.
151. Graves, P. R., Lovly, C. M., Uy, G. L., and Piwnica-Worms, H. (2001) Localization of human Cdc25C is regulated both by nuclear export and 14-3-3 protein binding. *Oncogene* **20,** 1839–1851.
152. Lopez-Girona, A., Kanoh, J., and Russell, P. (2001) Nuclear exclusion of Cdc25 is not required for the DNA damage checkpoint in fission yeast. *Curr. Biol.* **11,** 50–54.
153. Boddy, M. N., Furnari, B., Mondesert, O., and Russell, P. (1998) Replication checkpoint enforced by kinases Cds1 and Chk1. *Science* **280,** 909–912.
154. Gardner, R., Putnam, C. W., and Weinert, T. (1999) RAD53, DUN1 and PDS1 define two parallel G2/M checkpoint pathways in budding yeast. *EMBO J.* **18,** 3173–3185.
155. Sanchez, Y., Bachant, J., Wang, H., Hu, F., Liu, D., Tetzlaff, M., and Elledge, S. J. (1999) Control of the DNA damage checkpoint by chk1 and rad53 protein kinases through distinct mechanisms. *Science* **286,** 1166–1171.
156. Cohen-Fix, O. and Koshland, D. (1997) The anaphase inhibitor of *Saccharomyces cerevisiae* Pds1p is a target of the DNA damage checkpoint pathway. *Proc. Natl. Acad. Sci. USA* **94,** 14361–14366.

157. Wang, H., Liu, D., Wang, Y., Qin, J., and Elledge, S. J. (2001) Pds1 phosphorylation in response to DNA damage is essential for its DNA damage checkpoint function. *Genes Dev.* **15,** 1361–1372.
158. Li, R. and Murray, A. W. (1991) Feedback control of mitosis in budding yeast. *Cell* **66,** 519–531.
159. Hoyt, M. A., Totis, L., and Roberts, B. T. (1991) *S. cerevisiae* genes required for cell cycle arrest in response to loss of microtubule function. *Cell* **66,** 507–517.
160. Weiss, E. and Winey, M. (1996) The *Saccharomyces cerevisiae* spindle pole body duplication gene MPS1 is part of a mitotic checkpoint. *J. Cell Biol.* **132,** 111–123.
161. Brady, D. M. and Hardwick, K. G. (2000) Complex formation between Mad1p, Bub1p and Bub3p is crucial for spindle checkpoint function. *Curr. Biol.* **10,** 675–678.
162. Chen, R. H., Brady, D. M., Smith, D., Murray, A. W., and Hardwick, K. G. (1999) The spindle checkpoint of budding yeast depends on a tight complex between the Mad1 and Mad2 proteins. *Mol. Biol. Cell* **10,** 2607–2618.
163. Shah, J. V. and Cleveland, D. W. (2000) Waiting for anaphase: Mad2 and the spindle assembly checkpoint. *Cell* **103,** 997–1000.
164. Petersen, J. and Hagan, I. M. (2003) *S. pombe* aurora kinase/survivin is required for chromosome condensation and the spindle checkpoint attachment response. *Curr. Biol.* **13,** 590–597.
165. Biggins, S. and Murray, A. W. (2001) The budding yeast protein kinase Ipl1/Aurora allows the absence of tension to activate the spindle checkpoint. *Genes Dev.* **15,** 3118–3129.
166. Bardin, A. J. and Amon, A. (2001) Men and sin: what's the difference? *Nat. Rev. Mol. Cell. Biol.* **2,** 815–826.
167. McCollum, D. and Gould, K. L. (2001) Timing is everything: regulation of mitotic exit and cytokinesis by the MEN and SIN. *Trends Cell Biol.* **11,** 89–95.
168. Visintin, R., Hwang, E. S., and Amon, A. (1999) Cfi1 prevents premature exit from mitosis by anchoring Cdc14 phosphatase in the nucleolus. *Nature* **398,** 818–823.
169. Shou, W., Seol, J. H., Shevchenko, A., et al. (1999) Exit from mitosis is triggered by Tem1-dependent release of the protein phosphatase Cdc14 from nucleolar RENT complex. *Cell* **97,** 233–244.
170. Straight, A. F., Shou, W., Dowd, G. J., et al. (1999) Net1, a Sir2-associated nucleolar protein required for rDNA silencing and nucleolar integrity. *Cell* **97,** 245–256.
171. Alexandru, G., Zachariae, W., Schleiffer, A., and Nasmyth, K. (1999) Sister chromatid separation and chromosome re-duplication are regulated by different mechanisms in response to spindle damage. *EMBO J.* **18,** 2707–2721.
172. Pereira, G., Hofken, T., Grindlay, J., Manson, C., and Schiebel, E. (2000) The Bub2p spindle checkpoint links nuclear migration with mitotic exit. *Mol. Cell.* 6, 1-10.
173. Bardin, A. J., Visintin, R., and Amon, A. (2000) A mechanism for coupling exit from mitosis to partitioning of the nucleus. *Cell* **102,** 21–31.
174. Pereira, G., Manson, C., Grindlay, J., and Schiebel, E. (2002) Regulation of the Bfa1p-Bub2p complex at spindle pole bodies by the cell cycle phosphatase Cdc14p. *J. Cell Biol.* **157,** 367–379.
175. Stegmeier, F., Visintin, R., and Amon, A. (2002) Separase, polo kinase, the kinetochore protein Slk19, and Spo12 function in a network that controls Cdc14 localization during early anaphase. *Cell* **108,** 207–220.
176. Luca, F. C., Mody, M., Kurischko, C., Roof, D. M., Giddings, T. H. & Winey, M. (2001) *Saccharomyces cerevisiae* Mob1p is required for cytokinesis and mitotic exit. *Mol. Cell. Biol.* **21,** 6972–6783.

177. Frenz, L. M., Lee, S. E., Fesquet, D., and Johnston, L. H. (2000) The budding yeast Dbf2 protein kinase localises to the centrosome and moves to the bud neck in late mitosis. *J. Cell Sci.* **113,** 3399–3408.
178. Song, S. and Lee, K. S. (2001) A novel function of *Saccharomyces cerevisiae* CDC5 in cytokinesis. *J. Cell Biol.* **152,** 451–469.
179. Yoshida, S. and Toh-e, A. (2001) Regulation of the localization of Dbf2 and mob1 during cell division of *Saccharomyces cerevisiae. Genes Genet. Syst.* **76,** 141–147.
180. Schmidt, S., Sohrmann, M., Hofmann, K., Woollard, A., and Simanis, V. (1997) The Spg1p GTPase is an essential, dosage-dependent inducer of septum formation in *Schizosaccharomyces pombe. Genes Dev.* **11,** 1519–1534.
181. Furge, K. A., Wong, K., Armstrong, J., Balasubramanian, M., and Albright, C. F. (1998) Byr4 and Cdc16 form a two-component GTPase-activating protein for the Spg1 GTPase that controls septation in fission yeast. *Curr. Biol.* **8,** 947–954.
182. Fankhauser, C. and Simanis, V. (1994) The cdc7 protein kinase is a dosage dependent regulator of septum formation in fission yeast. *EMBO J.* **13,** 3011–3019.
183. Sohrmann, M., Schmidt, S., Hagan, I., and Simanis, V. (1998) Asymmetric segregation on spindle poles of the *Schizosaccharomyces pombe* septum-inducing protein kinase Cdc7p. *Genes Dev.* **12,** 84–94.
184. Sparks, C. A., Morphew, M., and McCollum, D. (1999) Sid2p, a spindle pole body kinase that regulates the onset of cytokinesis. *J. Cell Biol.* **146,** 777–790.
185. Guertin, D. A., Chang, L., Irshad, F., Gould, K. L., and McCollum, D. (2000) The role of the sid1p kinase and cdc14p in regulating the onset of cytokinesis in fission yeast. *EMBO J.* **19,** 1803–1815.
186. Cueille, N., Salimova, E., Esteban, V., et al. (2001) Flp1, a fission yeast orthologue of the *S. cerevisiae* CDC14 gene, is not required for cyclin degradation or rum1p stabilisation at the end of mitosis. *J. Cell Sci.* **114,** 2649–2664.
187. Trautmann, S., Wolfe, B. A., Jorgensen, P., Tyers, M., Gould, K. L., and McCollum, D. (2001) Fission yeast Clp1p phosphatase regulates G2/M transition and coordination of cytokinesis with cell cycle progression. *Curr. Biol.* **11,** 931–940.

2

The Plant Cell Cycle

An Overview

John H. Doonan

Summary

This chapter reviews the basic features of plant cell proliferation. Although plant cell division seems to be largely similar to animals and fungi, there are a number of peculiarities that are perhaps related to their lifestyle and development. Initial comparisons of animal, fungal, and plant genomes suggest that the central cell division and growth machinery are largely similar but that some key regulatory molecules found in animals appear to be missing from plants. Also, many of the intracellular signaling molecules that coordinate growth during development and the proteins involved in their perception seem to be different between the groups. In some cases, this reflects true divergence in the underlying mechanism, but high rates of gene sequence divergence could explain other examples. However, what is undisputable is that plant and animal cells are highly adapted to different niches and that this is reflected in quite different behavior.

Key Words

Cell cycle; meristem; proliferation; cyclin; cyclin-dependent protein kinase; cdc2.

1. Plant Cells Differ From Animal Cells

1.1. Plant Cells Are Nonmotile

Perhaps one of the most significant differences between plant and animal cells, at least so far as the cell cycle is concerned, is the cell wall. Most animal cells lack a rigid extracellular coat and can move and change their shape relatively freely. Almost all higher plant cells are completely encased in a comparatively rigid carbohydrate-based wall that essentially eliminates cell migration and restricts shape change. Many of the other differences between plant and animal development, it could be argued, follow on from this basic cellular distinction. During plant development, cells are formed and

From: *Methods in Molecular Biology, vol. 296, Cell Cycle Control: Mechanisms and Protocols*
Edited by: T. Humphrey and G. Brooks © Humana Press Inc., Totowa, NJ

differentiate *in situ*, whereas animal cells can migrate relative to one another. The immobility imposed by the cell wall has profound consequences for how plants develop as multicellular organisms. Cellular migration is not an option, so differential proliferation and growth are the main mechanisms by which the plant body is generated. Cell proliferation occurs mainly in specialized regions called meristems, usually placed at the extremities of the plant body. Continuous proliferative activity within meristems provides cells for growth and also maintains the meristem.

As cells are displaced from the meristem, by proliferation and growth, they begin to differentiate and acquire defined cell fates. Cell fate is defined, at least in part, by signals coming from neighboring cells. The almost complete absence of cell migration, combined with the late positional definition of cell fate, means that plants, as a group, are not susceptible to systematic cancers. Indeed, even pathogen-induced cancers tend to be spatially very limited.

The meristematic mode of growth confers another notable characteristic on plant development, allowing the plant to continue to grow and develop throughout its life. Unless the meristem receives a signal to terminate growth and differentiate, it continues to grow and produce new tissues and organs. Animal development, on the other hand, occurs mainly during early embryogenesis, and later changes in body shape are mainly owing to growth of preexisting body parts.

1.2. Plant Cells Can Continue to Grow in the Postmitotic Phase

Plants seem to have partially compensated for the lack of cell mobility by evolving a remarkable ability to regulate their cell size. As in other eukaryotes, cell growth is somehow coupled to cell cycle progression, but many plant cells can grow extensively even when not actively dividing. There appear to be at least two distinct mechanisms for postmitotic growth, one involving endoreduplication of the genome, which therefore can be considered as coupled to cell cycle events, and the other, seemingly independent of cell cycle events, driven by vacuole expansion and under the control of specific plant growth regulators such as gibberellins.

2. Experimental Systems for Cell Cycle Studies in Plants

2.1. Cell Suspensions

A number of fast-growing cell suspensions have been developed over the years, but only a few are widely used. Perhaps the most popular has been the tobacco BY2 (Bright Yellow) line. This line grows as uniform filaments of cells that have lost the ability to differentiate but that can be synchronized to a very high degree for cell cycle studies (for review, *see* **ref. *1***). After release from aphidicolin-induced arrest in early S-phase, cells synchronously progress through G_2 and into mitosis. Up to 70% of cells can be in mitosis at the peak, making this the system of choice for cell cycle studies. The cell line can be readily transformed using *Agrobacterium* to transfer the DNA construct. A variety of well-characterized constitutive and inducible promoters in plant transformation vectors are available for driving gene expression. Codon-modified green fluorescent protein functions well as a tag for following the location and behavior of proteins, or more traditional methods, such as indirect immunofluorescence, have been applied to BY2. Indeed, the large cell size, relatively low autofluorescence (for a plant cell), and optical clarity make it a useful model for cell biological studies.

A number of *Arabidopsis* cell lines have also been developed for similar studies, but these tend to grow more slowly want to have irregular cell shape and size and higher autofluorescence than BY2. They also respond rather poorly to attempts at cell cycle synchronization, perhaps because aphidicolin is not completely reversible. However, a method has recently been published that produces populations enriched for G_1, S, and G_2 cells ***(2)***. Despite these disadvantages, *Arabidopsis* cell lines are likely to increase in popularity as a tool, if only to exploit the molecular and genetic tools developed by the various genome projects focused on this species.

2.2. Whole Plants

Genetic dissection of the cell cycle in fungi and flies provided many of the major insights into cell cycle regulation. This approach in plants is still in its early days, with cell cycle mutants resulting mainly as a byproduct of other screens. Mutants in the core cell cycle regulators, such as the cyclin-dependent kinases (CDKs), are conspicuous by their absence. Gene redundancy may be a factormany genes, including the central cell cycle regulators, belong to large gene families with the potential for functional overlap, so knocking out a single gene may have little effect. However, systematic mutant screens are already very advanced in *Arabidopsis*, and large public collections are available from which insertional mutants produced by T-DNA insertion can be obtained. Lines containing multiple knockouts can be created by crossing, and these may uncover informative phenotypes. Another factor that reduces the likelihood of spotting phenotypes is the ability of plants to compensate for mild defects in cell division by increasing cell expansion and vice versa ***(3)***. Another possible reason could be gametophytic or early embryo lethality. Specific and comprehensive screens have been aimed at isolating such mutants ***(4–6)***, although the genes affected mostly remain to be identified.

Many of the same tools as used in cell suspensions can be applied to the study of cell cycle progression in plants. In addition, increasingly sophisticated imaging techniques are being developed to follow cell behavior in whole plants ***(7,8)***.

3. The Typical Plant Cell Cycle

3.1. The Cell and Microtubule Cycle

As in other eukaryotes, most plant cells sequentially pass through S-phase, when the genome is replicated, and M-phase or mitosis, when the genome is separated. Rapidly dividing meristematic cells might divide every 8–10 h, but most cells are much slower owing to increased Gap phases. The duration of both G_1- and G_2-phases can be increased, and, indeed, differentiated plant tissues can be a mix of cells arrested in either phase. Entry into S-phase has to be studied indirectly (i.e., by flow cytometry), but mitosis leads to cytological changes that reveal some of the interesting differences between plants and animals.

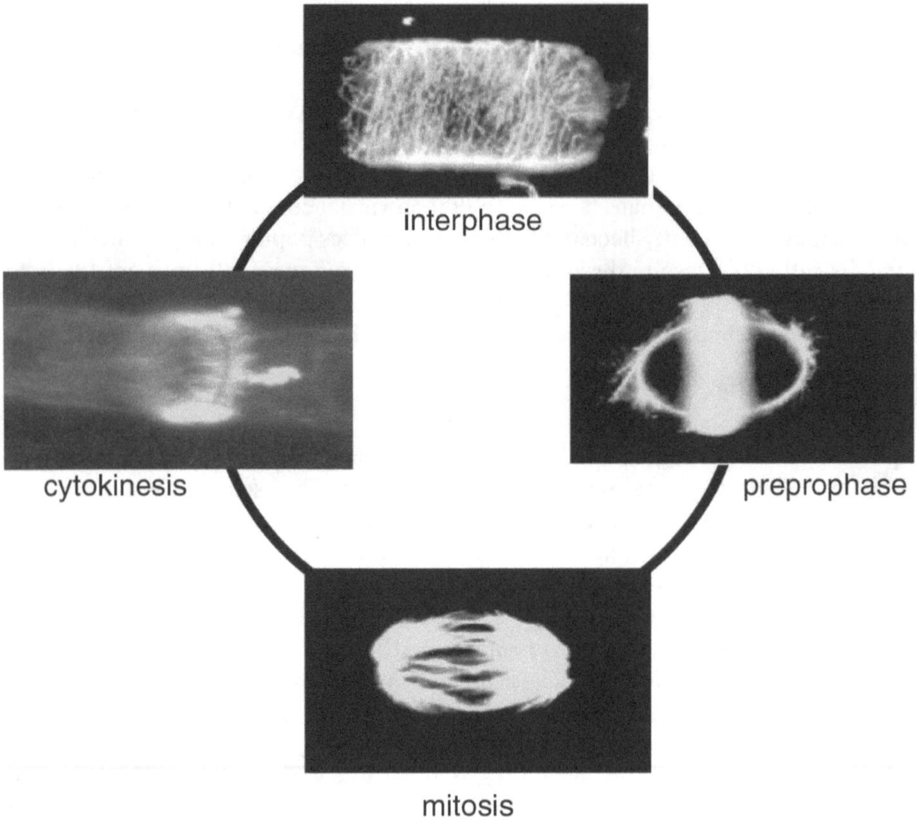

Fig. 1. The microtubule cycle in plant cells as revealed by indirect immunofluorescence. The interphase microtubule array, organized as bundles of microtubules in the cortex, gives way in G_2 to the preprophase band, which marks the site of cell division. The mitotic spindle has broad poles with several foci. The phragmoplast, involved in cell plate deposition during cytokinesis, is composed of a double ring of antiparallel microtubules.

3.1.1. The Microtubule Cycle: Entry Into Mitosis

Microtubules undergo dramatic reorganization during the cell cycle, as illustrated in **Fig. 1**. During interphase, the nucleus normally lies along one edge of the cell, but during the G_2-phase it migrates to the site of nuclear division, typically to the center of the cell if the ensuing division is to give rise to two equal-sized daughters. Coincident with nuclear movement, the cell microtubules begin to rearrange. Many cells have predominantly cortical microtubules during interphase that are organized in short overlapping bundles. As such cells approach mitosis, a much more pronounced band of microtubules, called the preprophase band (PPB), develops in the area of the presumptive division plane. The site of the PPB accurately predicts the division plane, and the correlation has excited much interest over the years. The origin of the PPB is uncertain, but cells that lack organized cortical interphase microtubules (owing to cell type or mutation) rarely have discrete PPBs. One possibility is that cortical microtubules move or collapse toward the presumptive division site and accumulate there, held by an unknown mechanism. Nuclear envelope breakdown is usually coincident with the late PPB and the early stages of spindle assembly.

3.1.2. Assembly of the Mitotic Spindle

Flowering plants completely lack centrioles, and, perhaps as a consequence, spindle assembly and organization appear distinctive. Centrioles play an important part in organizing spindle formation by providing a center for microtubule assembly in animals. In fungi, their place is taken by nuclear or spindle plaques, small multilayered structures that sit on or in the nuclear envelope, which also serve to organize microtubules. In the absence of discrete structured microtubule-organizing centers, spindle initiation appears to occur over the surface of the nucleus; as the spindle forms, the PPB is disassembled. The resulting spindle tends to have broad poles composed of numerous foci. Chromosomes condense and attach to the midzone of the spindle, presumably by mechanisms similar to or analogous to those described in other organisms.

3.1.3. The Phragmoplast: A Novel Plant-Specific Microtubule Array Required for Cell Division

Late in mitosis, another plant specific microtubule array, the phragmoplast, arises in the midzone of the spindle. This is composed of two sets of highly parallel sets of microtubules, each perpendicular to the plane of cell division and on opposite sides. These microtubules form an essential part of the mechanism by which the cross wall is laid down.

3.1.4. Reestablishment of Interphase Cellular Organization

As the nuclear envelope reforms around the nascent daughter nuclei, microtubules arise from the nuclear envelope and appear to spin out toward the cell cortex. At the cortex, these microtubules may be organized into the highly dynamic arrays typical of interphase cells.

4. Molecular Control of the Plant Cell Cycle

This section will briefly review our current understanding of key transitions during the cell cycle. Although the cycle is regulated at numerous stages, extracellular growth signals appear to act at two main points, G_1/S and G_2/M. The nature of these signals and their effects depends on the tissue and developmental stage. Auxin, for example, is the main positive proliferative signal during lateral root formation, but cytokinin is the dominant one in the shoot meristem. Both the signaling pathways and the way they affect the cell cycle are the subject of active research and debate, but some common principles are becoming apparent.

Most animal and fungal cells commit to a round of division at a defined point known as the restriction point or Start in G_1-phase. After cells pass this stage, they are considered unable to respond to signals promoting alternative pathways such as those leading to differentiation. Plants also have a major control point during G_1, reflected by the fact that most differentiated cells arrest in G_1 in response to nutrient limitation or differentiation (reviewed in **ref. 9**). Although yeast and mammal experiments provided the initial insights in this area, studies on plants are interesting from the evolutionary angle, as well as being necessary to understand how the cell cycle responds during the development of a completely different multicellular organism.

4.1. The Cyclin D/E2F/RB Pathway and Entry Into the Cycle

Most differentiated cells, if they still contain a functional nucleus, can be induced by appropriate stimuli to dedifferentiate and reenter the cell cycle. Since one of the most important stimuli is wounding, this may be another adaptation to a sessile lifestyle, allowing repair of various types of damage. The nature of the wound signal is unknown, but exogenous plant growth substances are usually also required, particularly if cell proliferation is to be maintained.

The signal transduction pathway mediating cell cycle reentry is broadly analogous to that of mammalian cells, involving the transcriptional activation of cyclin D genes, inactivation of retinoblastoma (RB), activation of the transcription factor E2F, and production of proteins required for DNA replication (**Fig. 2**). Extracellular signals modulate the activity of an unknown signal cascade involving protein phosphorylation that leads to the synthesis of D-cyclins. Their associated kinase activity results in the phosphorylation and inhibition of an RB-like protein at the G_1/S boundary ***(10)***. The phosphorylated RB protein is thought to release transcription factors such as E2F that then promote the transcription of S-phase genes.

Plants contain an extensive array of cyclin D genes: genome analysis reveals that *Arabidopsis* has at least 10, as opposed to mammals, with 3. All members, in common with mammalian cyclin D proteins, contain a characteristic RB binding motif, LxCxE, near the C-terminus, and this has been shown to bind RB ***(11–13)***. Several plant viral proteins also bind RB, either via a typical LxCxE motif or via some other means, and perhaps modulate this pathway to ensure replication of their DNA. Experimental manipulation of the pathway using viral proteins or overexpression of RB can dramatically alter the potential of plant cells to proliferate ***(14)***.

Structural comparisons suggest that the D-cyclins fall into at least two major groups, the cycD2 (three members) and the cycD3 (three members), but there are also at least four orphans. The limited genetic data available suggest functional redundancy, in that an insertion knockout in the *cycD3;2* gene has no apparent phenotype ***(15)*** and no other mutations have been reported in this family. However, overexpression of various D-cyclins produces a variety of growth phenotypes, supporting the notion that they are limiting for growth. Expression of the *Arabidopsis cycD2;1* gene in tobacco causes faster but normal growth ***(16)*** while over-expression of cycD3;1 in Arabidopsis causes abnormal growth and delayed differentiation in leaves ***(17)***.

Consistent with these observations, the cyclin D genes studied so far are under strict transcriptional and or translational control. In suspension cells, CycD2;1 responds strongly to the availability of a carbon source ***(18)*** whereas *cycD3;1* responds to cytokinin, and expression of this gene can eliminate the requirement for cytokinin

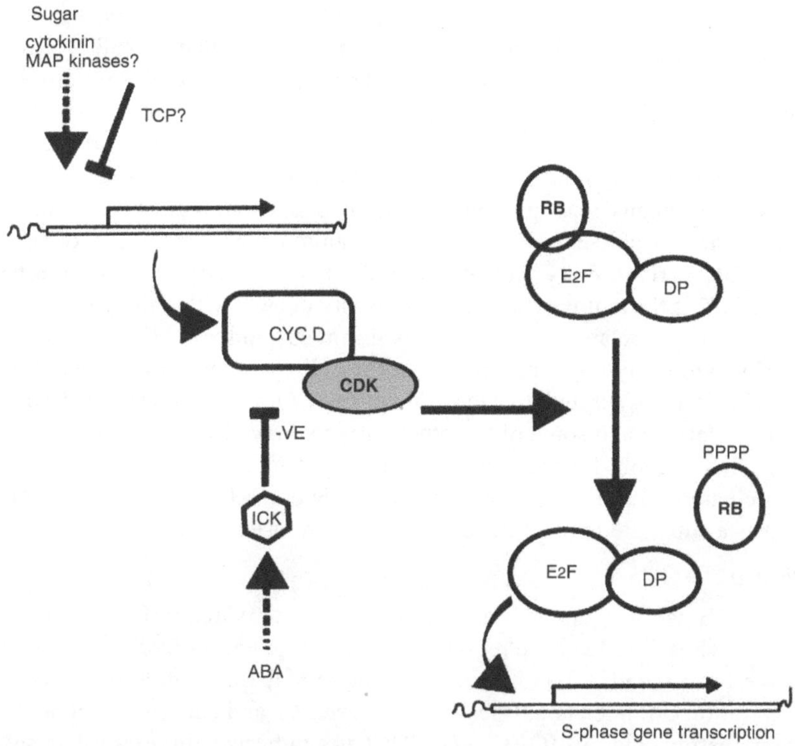

Fig. 2. The E2F/Rb/cyclin D pathway. Extracellular signals feed into the pathway by modulating the synthesis of components of CDK/cycD complexes. Positive signals such as cytokinin or sugar induce cycD transcription, whereas negative signals induce ICK transcription, probably via a signal transduction pathway involving mitogen-activated protein (MAP) kinases. In addition, regulatory transcription factors such as TCP also play a role, either directly or indirectly. The downstream part of the pathway seems largely similar to that in mammalian cells. CYCD, cyclin D; ICK, inhibitor cystine knot; TCP, Teosinte-branched/cycloidea/PCNA regulator.

in leaf explants ***(19)***. Most cyclin D genes are only highly expressed in proliferative regions of the plant, especially the meristem. Some are spatially restricted to certain regions: thus in *Arabidopsis cycD2;2* is expressed mainly in the lateral root ***(20)***, and in *Antirrhinum*, *cycD3A* is expressed only in the lateral organs ***(21)***. Moreover, the spatial domain of expression during floral morphogenesis seems to be regulated by *cycloidea*, a TCP-related transcription factor. The TCP gene family (Teosinte-branched/cycloidea/PCNA regulator) includes a number of developmental regulators, such as *teosinte-branched* from maize and *cincinata* from *Antirrhinum*, that have profound effects on plant morphology by differentially affecting growth within and between organs ***(22–24)***. TCP proteins can act as inhibitors of cell cycle gene expression by binding to *cis*-acting elements in their promoters ***(25)***.

Taken together, these results suggest that the different groups of cycD genes have different functions, probably operating in distinct pathways that tie cell proliferation to developmental or environmental responses. Thus, cycD2 responds to sucrose availability and perhaps tailors plant growth rate to the current conditions, whereas cycD3 acts downstream of pattern determinants, and overexpression has similar effects to that of an oncogene in mammals, leading to tissue disorganization.

As with cyclins, plants contain a complex family of cyclin-dependent protein kinases of which there are six classes, CDKA–CDKF. In animal cells, cyclin D proteins interact with a CDK variant, cdk4, but in plants the evidence suggests that the $p34^{cdc2}$ ortholog, CDKA, is the main partner. CDKA also associates with mitotic cyclins, but it is the major cell cycle-related CDK expressed during G_1. Indeed, CDKA protein levels and transcripts are fairly uniform throughout the cell cycle, and it is believed to play important roles from G_1 through to and within mitosis. Of course, CDK function is only partially characterized and some of the other variants may also play a role in G_1 ***(26)***.

In vivo CDKA/cycD kinase substrates have yet to be conclusively identified. Immunoprecipitated complexes can phosphorylate histone H1 but not Rb ***(27)***, although RB is used as a substrate by complexes assembled in insect cells ***(13)***.

4.2. S-Phase

If the E2F/RB pathway operates as in animals, then activated E2F must switch on a suite of genes whose products are required for DNA replication ***(28)***. A few such candidates have been identified and verified, including ribonucleotide reductase genes ***(29)***; proliferation cell nuclear antigen (PCNA; ***29a,30***), and cdc6, a component of the origin of replication complex (ORC; ***31***); CDC6 is synthesized in response to sucrose, probably by one of the E2F proteins ***(32)***. Combined microarray and bioinformatics surveys of the *Arabidopsis* genome suggest that there is a large number of other E2F targets, as judged by the presence of putative E2F binding sites in the 5' regions of the genes and cell cycle-regulated expression ***(33)***. However, E2F binding sites may depend on context, both genomic and developmental: the E2F sites in the PCNA promoter mediate gene activation in meristematic tissues but repression in differentiated tissues. Whether this means that E2F binds to some sites all the time, and is activation- or repression-regulated through accessory proteins such as RB, or that E2F proteins with different activities compete for the site is not clear. Some plant E2F factors act as activators, and others act as repressors, so both scenarios are possible.

The initiation of DNA replication is controlled by the pre-replication complex (RC), which contains the six proteins of the ORC and the minichromosome maintenance (MCM) proteins. These highly conserved proteins are sequentially recruited onto the OR prior to DNA replication, 'licencing' the origin to commence replication. After replication is initiated, the pre-RC components are inactivated or eliminated from the complex, but the manner in which this occurs varies widely ***(34)***.

4.3. G_2/M

Mitotic entry occurs when cdc2-related protein kinases are activated. Overexpression of a dominant-negative form of CDKA that lacks kinase activity in tobacco plants led to plants that had fewer larger cells, whereas overexpression of cyclin B was found to accelerate root growth in *Arabidopsis* ***(35)***, and local expression of cyclin A3 in tobacco induces local cell proliferation ***(36)***. Unfortunately, no details of the underlying mechanism are available for any of these examples, although one is tempted to assume a G_2-based mechanism.

Cyclin B is tightly regulated at the transcriptional level during G_2 and early M-phase ***(37)***, in which it is probably rate-limiting for entry into M-phase. Transcriptional activation is mediated by small *cis*-elements in the 5' region of the gene that binds myosin binding protein (Myb)-like proteins, resembling c-MYB of animals ***(38)***. Plants appear to contain two classes of c-MYB–like proteins, one that activates and one that represses. The repressor MYB is present throughout the cell cycle, but the activator MYB is transcriptionally activated during G_2 and precedes cyclin B accumulation. Given their expression patterns and their ability to bind to the same site, an antagonistic mechanism has been proposed whereby the activating MYB displaces the inhibitor from the promoter, but this has yet to be proved in vivo. However, expression of the activator MYB gene in cells arrested in S-phase with a DNA synthesis inhibitor will induce cyclin B expression, suggesting that it is a limiting factor for G_2/M-phase progression. Previously, c-MYB had been believed to activate genes only at the G_1/S transition and was implicated in carcinogenesis, but recently a Myb protein has been shown to activate cyclin expression in *Drosophila* ***(39)***. This indicates that the mechanism controlling cyclin B transcription may be conserved between animals and plants. In plants, the activator MYB is only synthesized after S-phase is complete and presumably is under the control of a checkpoint-like signal pathway. The identity of this pathway is currently unknown.

At least some of the proteins involved in the spindle checkpoint are also conserved in plants ***(40)***, including MAD2. In maize, MAD2 is abundant at kinetochores during early mitosis but is barely detectable at kinetochores after the microtubules have attached ***(41)***. The existence of a spindle checkpoint mechanism in plants is further indicated by pharmacological studies. Treatment of synchronized plant cell cultures with microtubule-destabilizing drugs leads to a transient metaphase-like arrest, with highly condensed chromosomes scattered throughout the cell ***(42)***.

Mitotic progression also depends on the anaphase-promoting complex (APC)-mediated proteolysis of key regulatory proteins. The *Arabidopsis* genome contains genes homologous to the components of the APC. The N-terminal domains of both A and B cyclin confer cell cycle stage-specific instability on reporter proteins ***(43)***, suggesting that they contain functional destruction motifs. The proteasome inhibitor MG132, a peptide aldehyde that functions as a substrate analog, inhibits progression past metaphase by inhibiting the APC-dependent proteolysis of cohesion proteins responsible for sister-chromatid separation ***(44)***. Treatment of synchronized BY2 cell cultures with MG132 blocks cells in metaphase with elevated levels of CDK kinase activity and stabilized cyclins ***(43)***. Whether the metaphase arrest observed in plant cells is induced by a similar mechanism as in yeast and in animal cells (involving stabilization of chromatid cohesion proteins) is not known yet, but it seems clear that the mechanisms governing protein turnover during mitosis are largely conserved between plants and animals.

Plant cyclin B1 is degraded at the onset of anaphase and is stabilized during activation of the spindle checkpoint pathway by treatment with microtubule-disrupting drugs, similar to animal B-type cyclins ***(45)***. In contrast, plant cyclin B2 is degraded during mid-prophase, perhaps using a similar degradation mechanism to animal cyclin A. Thus cyclin B1 is not stabilized by activation of the spindle checkpoint, and its N-terminal destruction box contains multiple destruction box elements ***(46)***.

4.4. Mitotic Exit and Cytokinesis

Genetic dissection of early embryo development produced a rich harvest of mutants defective in different steps of cytokinesis and cell plate maturation. Cloning of the corresponding genes identified proteins that function during different events of vesicle trafficking like vesicle formation, transport, and fusion, revealing a highly controlled vesicle trafficking machinery implicated in plant cytokinesis ***(47)***. The phragmoplast consists of short bundles of antiparallel micotubules that are believed to mediate the delivery of Golgi-derived vesicles to the plane of division during the process of cell plate formation. Because the plus ends of phragmoplast microtubules (MTs) overlap at the equator, a plus-end–directed motor such as kinesin is believed to mediate vesicle transport, although no candidate has been identified from among the large family of kinesins described ***(47)***. Two proteins that associate with the cell plate in plant cells are related to the animal large GTPase dynamin, which is involved in endocytosis of synaptic vesicles. Phragmoplastin ***(48)*** and its *Arabidopsis* homolog *ADL1* ***(49)*** both seem to be involved in the formation of fusion tubes during the initial stage of cell plate formation.

The products of two genes, *Knoll E* and *KEULE*, have been found to mediate membrane fusion events concertedly during cytokinesis in the *Arabidopsis* embryo. *KNOLLE* encodes a cytokinesis-specific syntaxin expressed in vesicle-like structures during mitosis and at the phragmoplast ***(50)***. *KEULE* encodes a member of the Sec1 superfamily of proteins that are capable of inducing conformational changes in syntaxins and priming them for interaction with target proteins on vesicle membranes. *KEULE* has been shown to bind *KNOLLE* in vitro, and the synthetic lethality of *knolle/keule* double mutants indicates that the two proteins interact functionally in vivo ***(51,52)***. The precise function of the *KEULE/KNOLLE* complex is not known yet, but it might be involved in integrating cell cycle signals and transducing these to regulate the cytokinetic vesicle fusion machinery.

No homologs of polo-like kinases or aurora kinases, which are all involved in microtubule organization during mitosis and cytokinesis in animals and yeast, have been described in plants.

A mitogen-activated protein kinase (MAPK) kinase kinase (MAPKKK) known as NPK1 is located in the equitorial region of the phragmoplast. Overexpression of a kinase negative mutant form of the MAPKKK disrupts cytokinesis, suggesting a role in phragmoplast expansion toward the cell cortex ***(53)***. NPK1 was also found to interact with a tobacco MAPK kinase NtMEK1, which is known to interact with and activate the tobacco MAPK Ntf6 ***(54)***. Ntf6 is regulated in a cell cycle-specific manner, is activated during anaphase and telophase, and also localizes to the phragmoplast ***(55)***. Thus a MAPK module composed of NPK1, NtMEK1, and Ntf6 is involved in plant cytokinesis, but downstream targets remain to be identified. NPK1 interacts with two kinesin-like proteins, NACK1 and NACK2, that stimulate NPK1 kinase activity and may be required for their localization ***(56)***. Mutations in the *Arabidopsis* homolog of NACK1, *HINKEL*, cause a defect in cytokinesis ***(57)***. The *SCD1* gene is also required for normal cell plate formation in *Arabidopsis*, and it too has homology to animal proteins involved in MAPK signal transduction ***(58)***.

4.5. Molecular Basis of Microtubule Organization in Plants

Microtubule organization is highly dynamic, changing in a characteristic pattern during the plant cell cycle. That microtubules play a crucial role in eukaryotic cell cycle progression is in no doubt~~the~~ role of plant-specific arrays such as the preprophase band and phragmoplast has been less clear until recently. Formal evidence that the phragmoplast plays a role in vesicle trafficking has been provided by mutations in the *PILZ* group of genes that eliminate microtubules and thereby cause a defect in cell plate deposition ***(59)***.

Although microtubule organizational changes have been extensively described, and many of the features seem to be specific to plant cells, the molecular basis of their organization remained obscure until very recently. Microtubule organizing center (MTOC) organization also seems to be radically different in plant cells. Cortical microtubules form parallel groups of overlapping tubules organized around the cell. The orientation of these bundles is highly dynamic, responding very quickly to external stimuli ***(60)***, supporting the idea that they control the orientation of cell growth.

Despite the lack of a recognizable MTOC, plants contain proteins similar to those normally found in the MTOC, but they behave differently. Thus, γ-tubulin, normally discretely localized in the MTOC of fungal and animal cells, is dispersed along interphase microtubules or is associated with other structures such as kinetochores and the nuclear envelope ***(61)***, supporting the idea of a dispersed MTOC activity that assumes various manifestation depending on the stage of the cell cycle ***(62)***. Recently we have shown that EB1, a microtubule plus and minus end-binding protein that plays a crucial role in regulating microtubule dynamics in animals (***[63,64]***, accumulates at the poles of the spindle in surprisingly compact foci ***[64a]***). EB1 also accumulates at the polar (minus) ends of the phragmoplast microtubules and as small dispersed foci at the ends of microtubule bundles in the cell cortex. Microtubules grow from and shrink back to these EB1 foci, providing direct evidence for the dispersed MTOC ideas of Mazia ***(62)***.

The molecular basis of coupling between the cell cycle and microtubule organization in animals seems to involve the phosphorylation of MAPs by cyclin B/CDK1, which then alters microtubule dynamics and perhaps organization. Although there have been several reports of association between CDK related proteins and the cytoskeleton (see third paragraph below), insight into the significance remains elusive. It remains possible that CDK complexes are passive passengers, at least some of the time, and that other regulators mediate changes. A good candidate is PP2A: mutations in TONNEAU2, which encodes for a regulatory subunit of PP2A, lead to the formation of very stunted plants whose cells cannot expand properly and seem to lack PPBs ***(65)***.

Although the plant and animal spindles have a similar overall organization and function, their structure and assembly differ. Plant spindles lack centrosomes, and the multiple spindle poles are usually less focused. Consistent with these observations, the most active MTOC in early mitosis is spread around the nuclear surface ***(66)***. Kinetochores may also serve as MTOCs in that microtubules appear to accumulate around kinetochores early in spindle formation ***(67)***, and at least two plant kinetochore proteins have homology to animal centrosome components. One of these is γ-tubulin ***(61,88)***, which is required for microtubule nucleation and is an important component of centrosomes in animals.

Microtubule motors and other associated proteins (MAPs) are required for spindle assembly and function, but many plant MAPs have been isolated that show little homology to animal MAPs. MAP65 proteins bind to subsets of interphase and mitotic microtubule arrays and bundle microtubules in vitro ***(69–71)***. These proteins belong to a small gene family that shares a remote ancestor with a yeast protein involved in cytokinesis ***(72)*** but have diversified in plants and perhaps acquired new functions. Many of the kinesin related proteins, whose function has been investigated, seem to have a role in phragmoplast formation rather than spindle movements. However, there is a large gene family of kinesin-related proteins in the *Arabidopsis* genome whose functions remain to be elucidated ***(73)***.

CDKA activity is required for normal entry into mitosis and the reorganization of mitotic microtubule arrays. Treatment of root tip cells with the cdk-specific inhibitor roscovitine disrupts mitotic spindle formation and leads to the formation of abnormal or monopolar spindles ***(55)***. Immunolocalization of CDKA and green fluorescent protein (GFP) fusion studies indicate dynamic interaction with the microtubules during mitosis ***(75,76)***. CDKA/GFP is found throughout the cytoplasm and nucleus during interphase but associates specifically with a subset of the PPB microtubules during late G_2. At the metaphase-to-anaphase transition, CDKA moves from the chromatin domain to the anaphase spindle, where it becomes microtubule-associated. The significance of microtubule association is not known, but immunogold localization suggests it is surprisingly intimate ***(73)***.

Furthermore, two mitotic plant cdks, cdkA and cdkB ***(75,76)*** and some plant cyclins, like ZmCycA1;1, ZmCycB1;1, and ZmCycB2;1 are present and are associated with the phragmoplast and the midline of cell divison through anaphase, telophase, and cytokinesis ***(77)***. They might be involved in the regulation of MAPs like MAP65 or molecular motors like TPRK125 ***(78,79)***, which are also located at the midplane of cell division and contain potential cdc2 phosphorylation sites. Several other proteins including protein kinases have been localized to the growing cell plate in plants, but their precise functions remain to be determined. It seems likely that a complex regulatory network oversees the final step of mitosis, which ultimately severs a cell into two.

5. Modified Plant Cell Cycles

Subheading 4. described the typical cell cycle as observed in most meristematic and culture cells, but a number of variant types of cycle are found during normal plant development. These modified cell cycles are associated with specific types of differentiated cells whose genetic dissection has proved, almost incidently, to be a useful source of cell cycle genes.

5.1. Endoreduplication

Cells normally alternate DNA replication with mitosis, thereby keeping the information content of their genome constant across generations. If successive rounds of DNA replication occur in the absence of mitosis, the DNA content of the cell doubles, and this is known as endoreduplication. Endoreduplication is characteristic of many different types of specialized plant cells such as xylem, large epidermal cells known as pavement cells, and, in some species, epidermal hair cells. There is a very strong correlation between DNA content and cell size in such cells ***(80)***, so perhaps endoreduplication provides a means of increasing cell size without the inconvenience of cell division ***(81)***. The correlation extends beyond specific cell types: polyploids tend to have larger cells than diploids in relation to their DNA levels. However, the reverse correlation is not so tight, and the final size of a particular cell is not always correlated with DNA content.

Although endoreduplication may use similar mechanisms for initiating DNA replication, mitosis is suppressed, and the mitosis-associated repression of replication reinitiation is suspended. Multiple rounds of DNA replication within a single cell cycle are usually prevented by the licencing system whereby the DNA RC somehow recognizes that replication has been completed. Therefore, perhaps the key step in the endocycle is reactivation of the DNA replication machinery without going through M. Reactivation is normally prevented by cyclin/CDK complex binding to the RC during G_2 and M, preventing its reactivation. During late M, the cyclin is degraded by the APC, and the RC is available again for reinitiation. In plants, endoreduplication requires either the prevention of cyclin B accumulation or its premature destruction. Activators of APC, such as the CDH1/FIZZY-related *ccs52* gene, fulfil this role ***(82)***. Overexpression of the cyclin B gene in trichomes changes an endocycle to a mitotic cycle ***(83)***, suggesting that suppression of the mitotic kinase is essential for the endocycle.

Arabidopsis mutants with altered levels of endoreduplication have been described by various labs, generally arising from two distinct types of screen that enrich for changes in cell size. Screens for altered trichome size and morphology have identified both positive and negative functions. The *SIAMESE* gene is required to suppress mitotic cell cycles, and its loss leads to ectopic cyclin B expression ***(84)***. Loss of *SPINDLY* ***(85)*** and several other genes ***(86)*** including *KAKTUS*, *RASTAFARI*, and *POLYCHOME* lead to ectopic endocycles in trichomes.

Screens that yield dwarf plants also tend to produce mutants with altered endocycles. A particularly interesting set of mutants in an unusual topoisomerase, *topo*

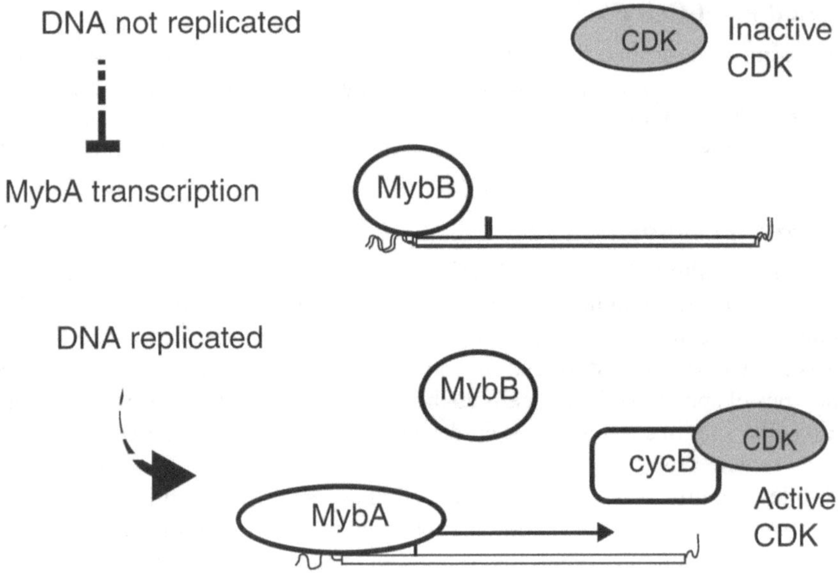

Fig. 3. A model for control of mitotic entry. Unreplicated DNA inhibits the production of activating MybA protein by an unknown mechanism, and inhibitory MybB occupies the promoter, preventing the transcription of M-phase–related genes, including cyclin. When DNA replication is complete, MybA is produced and displaces MybB from the promoter. Cyclin B is transcribed and activates cyclin-dependent kinase A (CDKA), allowing progression into mitosis.

VI, is required to allow cells to achieve very high levels of endoreduplication ***(87,88)***. Topo II normally resolves DNA strands after DNA replication in a process called decatenation. Topo II is required for the downstream separation of sister chromatids during mitosis, and mutations lead to G_2 arrest. This work indicates that the endocycle requires specific components in addition or instead of those required for the normal mitotic cycle.

The mechanism by which cell size and DNA content are linked should provide interesting insights into growth control in plants. Much of the increase in cell and organ size occurs outside the meristem in the postproliferative phase of the cell cycle, either as endoreduplication-associated cell growth or as vacuole-driven cell expansion. In the latter case, the increase in cell size is thought to be driven by primarily vacuole expansion.

5.2. Syncytial Cells

Plants contain a number of different cell types that have multiple nuclei, but perhaps the best studied is the nuclear endosperm. The endosperm arises as a result of double fertilization of a female gamete by two male nuclei. The resultant triploid tissue acts as nurse tissue for its sibling, the diploid embryo, facilitating nutrient transfer in a manner roughly analogous to the mammalian placenta. The endosperm undergoes several rounds of rapid nuclear division and in many species cytokinesis is suppressed, producing a syncytium ***(89)***. This syncytial phase ends with simultaneous partitioning of the multinucleate cytoplasm into individual cells, a process referred to as cellularization ***(90)***. The molecular mechanism that suppresses cytokinesis is not well understood ***(91)***, but genetic dissection of endosperm development has revealed a large number of cell cycle-defective mutants. The genes thus identified include SMC genes ***(92)***, condensins ***(93)***, and tubulin assembly cofactors ***(94)***. Although some of these are specific for the endosperm, for example, the *SPATZLE* gene is specifically required for cellularization, many functions are also required for embryo development ***(90)***. The further study of endosperm development should thus provide useful genetic tools for dissecting the cell cycle generally as well as understanding its developmental modification

6. Conclusions

Plants provide a divergent system in which to study cell cycle regulation. The multicellular state is thought to have arisen independently in animals and plants, so the mechanisms that the two groups have evolved to ensure that cell growth and proliferation are subservient to the needs of the whole organism are likely to differ. One of the main challenges in the future will be to understand the pathways that control cell division during development. Cell cycle regulation also differs radically between different cell types within plants, revealing the extent to which cell division processes can be molded in response to development. The molecular processes underlying the unusual cell cycles should provide new insights into the regular cycle.

References

1. Nagata, T., Nemoto, Y., and Hasezawa, S. (1992) Tobacco BY2 cell line as the HeLa cell in the cell biology of higher plants. *Int. Rev. Cytol.* **132,** 1–30.
2. Menges, M. and Murray, J. A. (2002) Synchronous *Arabidopsis* suspension cultures for analysis of cell-cycle gene activity. *Plant J.* **30,** 203–212.
3. Doonan, J. (2000) Social controls on cell proliferation in plants. *Curr. Opin. Plant Biol.* **3,** 482–487.
4. Meinke, D. W., Meinke, L. K., Showalter, T. C., Schissel, A. M., Mueller, L. A., and Tzafrir, I. (2003) A sequence-based map of *Arabidopsis* genes with mutant phenotypes. *Plant Physiol.* **131,** 409–418.
5. Tzafrir, I., Dickerman, A., Brazhnik, O., et al. (2003) The *Arabidopsis* SeedGenes Project. *Nucleic Acids Res.* **31,** 90–93.
6. McElver, J., Tzafrir, I., Aux, G., et al. (2001) Insertional mutagenesis of genes required for seed development in *Arabidopsis thaliana. Genetics* **159,** 1751–1763.
7. Beemster, G. T. and Baskin, T. I. (1998) Analysis of cell division and elongation underlying the developmental acceleration of root growth in *Arabidopsis thaliana. Plant Physiol.* **116,** 1515–1526.
8. Laufs, P., Grandjean, O., Jonak, C., Kieu, K., and Traas, J. (1998) Cellular parameters of the shoot apical meristem in *Arabidopsis. Plant Cell* **10,** 1375–1390.
9. Fowler, M. R., Eyre, S., Scott, N. W., Slater, A., and Elliott, M. C. (1998) The plant cell cycle in context. *Mol. Biotechnol.* **10,** 123–153.

10. Nakagami, H., Kawamura, K., Sugisaka, K., Sekine, M., and Shinmyo, A. (2002) Phosphorylation of retinoblastoma-related protein by the cyclin D/cyclin-dependent kinase complex is activated at the G1/S-phase transition in tobacco. *Plant Cell* **14,** 1847–1857.
11. Huntley, R., Healy, S., Freeman, D., et al. (1998) The maize retinoblastoma protein homologue ZmRb-1 is regulated during leaf development and displays conserved interactions with G1/S regulators and plant cyclin D (CycD) proteins. *Plant Mol. Biol.* **37,** 155–169.
12. Ach, R. A., Durfee, T., Miller, A. B., et al. (1997) RRB1 and RRB2 encode maize retinoblastoma-related proteins that interact with a plant D-type cyclin and geminivirus replication protein. *Mol. Cell. Biol.* **17,** 5077–5086.
13. Nakagami, H., Sekine, M., Murakami, H., and Shinmyo, A. (1999) Tobacco retinoblastoma-related protein phosphorylated by a distinct cyclin-dependent kinase complex with Cdc2/cyclin D in vitro. *Plant J.* **18,** 243–252.
14. Gordon-Kamm, W., Dilkes, B. P., Lowe, K., et al. (2002) Stimulation of the cell cycle and maize transformation by disruption of the plant retinoblastoma pathway. *Proc. Natl. Acad. Sci. USA* **99,** 11975–11980.
15. Swaminathan, K., Yang, Y., Grotz, N., Campisi, L., and Jack, T. (2000) An enhancer trap line associated with a D-class cyclin gene in *Arabidopsis*. *Plant Physiol.* **124,** 1658–1667.
16. Cockcroft, C. E., den Boer, B. G., Healy, J. M., Murray, J. A. (2000) Cyclin D control of growth rate in plants. *Nature* **405,** 575–579.
17. Dewitte, W., Riou-Khamlichi, C., Scofield, S., et al. (2003) Altered cell cycle distribution, hyperplasia, and inhibited differentiation in *Arabidopsis* caused by the D-type cyclin CYCD3. *Plant Cell* **15,** 79–92.
18. Riou-Khamlichi, C., Menges, M., Healy, J. M., and Murray, J. A. (2000) Sugar control of the plant cell cycle: differential regulation of *Arabidopsis* D-type cyclin gene expression. *Mol. Cell. Biol.* **20,** 4513–4521.
19. Riou-Khamlichi, C., Huntley, R., Jacqmard, A., and Murray, J. A. (1999) Cytokinin activation of *Arabidopsis* cell division through a D-type cyclin. *Science* **283,** 1541–1544.
20. De Veylder, L., de Almeida Engler, J., Burssens, S., et al. (1999) A new D-type cyclin of *Arabidopsis thaliana* expressed during lateral root primordia formation. *Planta* **208,** 453–462.
21. Gaudin, V., Lunness, P. A., Fobert, P. R., et al. (2000) The expression of D-cyclin genes defines distinct developmental zones in snapdragon apical meristems and is locally regulated by the Cycloidea gene. *Plant Physiol.* **122,** 1137–1148.
22. Nath, U., Crawford, B. C., Carpenter, R., and Coen, E. (2003) Genetic control of surface curvature. *Science* **299,** 1404–1407.
23. Takeda, T., Suwa, Y., Suzuki, M., et al. (2003) The OsTB1 gene negatively regulates lateral branching in rice. *Plant J.* **33,** 513–520.
24. Luo, D., Carpenter, R., Copsey, L., Vincent, C., Clark, J., and Coen, E. (1999) Control of organ asymmetry in flowers of *Antirrhinum*. *Cell* **99,** 367–376.
25. Tremousaygue, D., Garnier, L., Bardet, C., Dabos, P., Herve, C., and Lescure, B. (2003) Internal telomeric repeats and TCP domain'protein-binding sites co-operate to regulate gene expression in *Arabidopsis thaliana* cycling cells. *Plant J.* **33,** 957–966.
26. Yamaguchi, M., Kato, H., Yoshida, S., Yamamura, S., Uchimiya, H., and Umeda, M. (2003) Control of in vitro organogenesis by cyclin-dependent kinase activities in plants. *Proc. Natl. Acad. Sci. USA* **100,** 8019–8023.

27. Healy, J. M., Menges, M., Doonan, J. H., and Murray, J. A. (2001) The *Arabidopsis* D-type cyclins CycD2 and CycD3 both interact in vivo with the PSTAIRE cyclin-dependent kinase Cdc2a but are differentially controlled. *J. Biol. Chem.* **276,** 7041–7047.
28. Ramirez-Parra, E., Frundt, C., and Gutierrez, C. (2003) A genome-wide identification of E2F-regulated genes in *Arabidopsis*. *Plant J.* **33,** 801–811.
29. Chaboute, M. E., Clement, B., and Philipps, G. (2002) S phase and meristem-specific expression of the tobacco RNR1b gene is mediated by an E2F element located in the 5' leader sequence. *J. Biol. Chem.* **277,** 17845–17851.
29a. Egelkrout, E. M., Robertson, D., and Hanley-Bowdoin, L. (2001) Proliferating cell nuclear antigen transcription is repressed through an E2F consensus element and activated by geminivirus infection in mature leaves. *Plant Cell* **13,** 1437–1452.
30. Kosugi, S. and Ohashi, Y. (2002) E2F sites that can interact with E2F proteins cloned from rice are required for meristematic tissue-specific expression of rice and tobacco proliferating cell nuclear antigen promoters. *Plant J.* **29,** 45–59.
31. Castellano, M. M., del Pozo, J. C., Ramirez-Parra, E., Brown, S., and Gutierrez, C. (2001) Expression and stability of *Arabidopsis* CDC6 are associated with endoreplication. *Plant Cell* **13,** 2671–2686.
32. de Jager, S. M., Menges, M., Bauer, U. M., and Murra, J. A. (2001) *Arabidopsis* E2F1 binds a sequence present in the promoter of S-phase-regulated gene AtCDC6 and is a member of a multigene family with differential activities. *Plant Mol. Biol.* **47,** 555–568.
33. Menges, M., Hennig, L., Gruissem, W., and Murray, J. A. (2002) Cell cycle-regulated gene expression in *Arabidopsis*. *J. Biol. Chem.* **277,** 41987–42002.
34. Bell, S. P. and Dutta, A. (2002) DNA replication in eukaryotic cells. *Annu. Rev. Biochem.* **71,** 333–374.
35. Doerner, P., Jorgensen, J. E., You, R., Steppuhn, J., and Lamb, C. (1996) Control of root growth and development by cyclin expression. *Nature* **380,** 520–523.
36. Wyrzykowska, J., Pien, S., Shen, W. H., and Fleming, A. J. (2002) Manipulation of leaf shape by modulation of cell division. *Development* **129,** 957–964.
37. Fobert, P. R., Coen, E. S., Murphy, G. J., and Doonan, J. H. (1994) Patterns of cell division revealed by transcriptional regulation of genes during the cell cycle in plants. *EMBO J.* **13,** 616–624.
38. Ito, M., Araki, S., Matsunaga, S., et al. (2001) G2/M-phase-specific transcription during the plant cell cycle is mediated by c-Myb-like transcription factors. *Plant Cell* **13,** 1891–1905.
39. Okada, M., Akimaru, H., Hou, D. X., Takahashi, T., Ishii, S. (2002) Myb controls G(2)/M progression by inducing cyclin B expression in the *Drosophila* eye imaginal disc. *EMBO J.* **21,** 675–684.
40. Starr, D. A., Williams, B. C., Li, Z., Etemad-Moghadam, B., Dawe, R. K., Goldberg, M. L. (1997) Conservation of the centromere/kinetochore protein ZW10. *J. Cell Biol.* **138,** 1289–1301.
41. Yu, H. G., Muszynski, M. G., Kelly, Dawe. R. (1999) The maize homologue of the cell cycle checkpoint protein MAD2 reveals kinetochore substructure and contrasting mitotic and meiotic localization patterns. *J. Cell Biol.* **145,** 425–435.

42. Planchais, S., Glab, N., Inze, D., and Bergounioux, C. (2000) Chemical inhibitors: a tool for plant cell cycle studies. *FEBS Lett.* **476,** 78–83.
43. Genschik, P., Criqui, M. C., Parmentier, Y., Derevier, A., and Fleck, J. (1998) Cell cycle -dependent proteolysis in plants. Identification of the destruction box pathway and metaphase arrest produced by the proteasome inhibitor mg132. *Plant Cell* **10,** 2063–2076.
44. Sherwood, S. W., Kung, A. L., Roitelman, J., Simoni, R. D., Schimke, R. T. (1993) In vivo inhibition of cyclin B degradation and induction of cell-cycle arrest in mammalian cells by the neutral cysteine protease inhibitor N-acetylleucylleucylnorleucinal. *Proc. Natl. Acad. Sci. USA* **90,** 3353–3357.
45. Criqui, M. C., Parmentier, Y., Derevier, A., Shen, W. H., Dong, A., and Genschik, P. (2000) Cell cycle-dependent proteolysis and ectopic overexpression of cyclin B1 in tobacco BY2 cells. *Plant J.* **24,** 763–773.
46. Weingartner, M., Pelayo, H. R., Binarova, P., et al. (2003) A plant cyclin B2 is degraded early in mitosis and its ectopic expression shortens G2-phase and alleviates the DNA-damage checkpoint. *J. Cell Sci.* **116,** 487–498.
47. Nacry, P., Mayer, U., and Jurgens, G. (2000) Genetic dissection of cytokinesis. *Plant Mol. Biol.* **43,** 719–733.
48. Gu, X. and Verma, D.P. (1996) Phragmoplastin, a dynamin-like protein associated with cell plate formation in plants. *EMBO J.* **15,** 695–704.
49. Lauber, M. H., Waizenegger, I., Steinmann, T., et al. (1997) The *Arabidopsis* KNOLLE protein is a cytokinesis-specific syntaxin. *J. Cell Biol.* **139,** 1485–1493.
50. Lukowitz, W., Mayer, U., and Jurgens, G. (1996) Cytokinesis in the *Arabidopsis* embryo involves the syntaxin-related KNOLLE gene product. *Cell* **84,** 61–71.
51. Assaad, F. F., Huet, Y., Mayer, U., and Jurgens, G. (2001) The cytokinesis gene KEULE encodes a Sec1 protein that binds the syntaxin KNOLLE. *J. Cell Biol.* **152,** 531–543.
52. Waizenegger, I., Lukowitz, W., Assaad, F., Schwarz, H., Jurgens, G., Mayer, U. (2000) The *Arabidopsis* KNOLLE and KEULE genes interact to promote vesicle fusion during cytokinesis. *Curr. Biol.* **10,** 1371–1374.
53. Nishihama, R., Ishikawa, M., Araki, S., Soyano, T., Asada, T., Machida, Y. (2001) The NPK1 mitogen-activated protein kinase kinase kinase is a regulator of cell-plate formation in plant cytokinesis. *Genes Dev.* **15,** 352–363.
54. Calderini, O., Glab, N., Bergounioux, C., Heberle-Bors, E., and Wilson, C. (2001) A novel tobacco mitogen-activated protein (MAP) kinase kinase, NtMEK1, activates the cell cycle-regulated p43Ntf6 MAP kinase. *J. Biol. Chem.* **276,** 18139–18145.
55. Calderini, O., Bogre, L., Vicente, O., Binarova, P., Heberle-Bors, E., and Wilson, C. (1998) A cell cycle regulated MAP kinase with a possible role in cytokinesis in tobacco cells. *J. Cell Sci.* **111,** 3091–3100.
56. Nishihama, R., Soyano, T., Ishikawa, M., et al. (2002) Expansion of the cell plate in plant cytokinesis requires a kinesin-like protein/MAPKKK complex. *Cell* **109,** 87–99.
57. Strompen, G., El Kasmi, F., Richter, S., et al. (2002) The *Arabidopsis* HINKEL gene encodes a kinesin-related protein involved in cytokinesis and is expressed in a cell cycle-dependent manner. *Curr. Biol.* **12,** 153–158.
58. Falbel, T. G., Koch, L. M., Nadeau, J. A., Segui-Simarro, J. M., Sack, F. D., and Bednarek, S. Y. (2003) SCD1 is required for cell cytokinesis and polarized cell expansion in *Arabidopsis thaliana. Development* **130,** 4011–4024.
59. Steinborn, K., Maulbetsch, C., Priester, B., et al. (2002) The *Arabidopsis* PILZ group genes encode tubulin-folding cofactor orthologs required for cell division but not cell growth. *Genes Dev.* **16,** 959–971.

60. Yuan, M., Shaw, P. J., Warn, R. M., and Lloyd, C. W. (1994) Dynamic reorientation of cortical microtubules, from transverse to longitudinal, in living plant cells. *Proc. Natl. Acad. Sci. USA* **91,** 6050–6053.
61. Drykova, D., Cenklova, V., Sulimenko, V., Volc, J., Draber, P., and Binarova, P. (2003) Plant gamma-tubulin interacts with αβ-tubulin dimers and forms membrane-associated complexes. *Plant Cell* **15,** 465–480.
62. Mazia, D. (1984) Centrosomes and mitotic poles. *Exp. Cell Res.* **153,** 1–15.
63. Tirnauer, J. S. and Bierer, B. E. (2000) EB1 proteins regulate microtubule dynamics, cell polarity, and chromosome stability. *J. Cell Biol.* **149,** 761–766.
64. Rogers, S. L., Rogers, G. C., Sharp, D. J., and Vale, R. D. (2002) *Drosophila* EB1 is important for proper assembly, dynamics, and positioning of the mitotic spindle. *J. Cell Biol.* **158,** 873–884.
64a. Chan, J. Calder, G. M., Doonan, J. H., and Lloyd, C. W. (2003) EB1 reveals mobile microtubule nucleation sites in *Arabidopsis*. *Nat. Cell Biol.* **5,** 967–971.
65. Camilleri, C., Azimzadeh, J., Pastuglia, M., Bellini, C., Grandjean, O., and Bouchez, D. (2002) The *Arabidopsis* TONNEAU2 gene encodes a putative novel protein phosphatase 2A regulatory subunit essential for the control of the cortical cytoskeleton. *Plant Cell* **14,** 833–845.
66. Lambert, A. M. (1993) Microtubule-organizing centers in higher plants. *Curr. Opin. Cell Biol.* **5,** 116–122.
67. Chan, A. and Cande. W. Z. (1998) Maize meiotic spindles assemble around chromatin and do not require paired chromosomes. *J. Cell Sci.* **111,** 3507–3515.
68. Binarova, P., Cenklova, V., Sulimenko, V., Drykova, D., Volc, J., and Draber, P. (2003) Distribution of gamma-tubulin in cellular compartments of higher plant cells. *Cell Biol. Int.* **27,** 167–169.
69. Chan, J., Jensen, C. G., Jensen, L. C., Bush, M., Lloyd, C. W. (1999) The 65-kDa carrot microtubule-associated protein forms regularly arranged filamentous cross-bridges between microtubules. *Proc. Natl. Acad. Sci. USA* **96,** 14931–14936.
70. Chang-Jie, J. and Sonobe, S. (1993) Identification and preliminary characterization of a 65 kDa higher-plant microtubule-associated protein. *J. Cell Sci.* **105,** 891–901.
71. Smertenko, A., Saleh, N., Igarashi, H., et al. (2000) A new class of microtubule-associated proteins in plants. *Nat. Cell Biol.* **2,** 750–753.
72. Pellman, D., Bagget, M., Tu, Y. H., Fink, G. R., Tu, H. (1995)Two microtubule-associated proteins required for anaphase spindle movement in *Saccharomyces cerevisiae*. *J. Cell Biol.* **130,** 1373–1385.
73. Hemsley, R., McCutcheon, S., Doonan, J., and Lloyd, C. (2001) P34(cdc2) kinase is associated with cortical microtubules from higher plant protoplasts. *FEBS Lett.* **508,** 157–161.
74. Weingartner, M., Binarova, P., Drykova, D., et al. (2001) Dynamic recruitment of Cdc2 to specific microtubule structures during mitosis. *Plant Cell* **13,** 1929–1943.
75. Stals, H., Bauwens, S., Traas, J., Van Montagu, M., Engler, G., and Inze, D. (1997) Plant CDC2 is not only targeted to the pre-prophase band, but also co-localizes with the spindle, phragmoplast, and chromosomes. *FEBS Lett.* **418,** 229–234.
76. Meszaros, T., Miskolczi, P., Ayaydin, F., et al. (2000) Multiple cyclin-dependent kinase complexes and phosphatases control G2/M progression in alfalfa cells. *Plant Mol. Biol.* **43,** 595–605.
77. Mews, M., Sek, F. J., Moore, R., Volkmann, D., Gunning, B. E. S., and John, P. C. L. (1997) Mitotic cyclin distribution during maize cell division: implications for the sequence diversity and function of cyclins in plants. *Protoplasma* **200,** 128–145.

78. Asada, T., Kuriyama, R., and Shibaoka, H. (1997) TKRP125, a kinesin-related protein involved in the centrosome-independent organization of the cytokinetic apparatus in tobacco BY-2 cells. *J. Cell Sci.* **110,** 179–189.
79. Barroso, C., Chan, J., Allan, V., Doonan, J., Hussey, P., and Lloyd, C. (2000) Two kinesin-related proteins associated with the cold-stable cytoskeleton of carrot cells: characterization of a novel kinesin, DcKRP120-2. *Plant J.* **24,** 859–868.
80. Gendreau, E., Orbovic, V., Hofte, H., Traas, J. (1999) Gibberellin and ethylene control endoreduplication levels in the *Arabidopsis thaliana* hypocotyl. *Planta* **209,** 513–516.
81. Traas, J., Hulskamp, M., Gendreau, E., and Hofte, H. (1998) Endoreduplication and development: rule without dividing? *Curr. Opin. Plant Biol.* **1,** 498–503.
82. Cebolla, A., Vinardell, J. M., Kiss, E., et al. (1999) The mitotic inhibitor ccs52 is required for endoreduplication and ploidy-dependent cell enlargement in plants. *EMBO J.* **18,** 4476–4484.
83. Schnittger, A., Schobinger, U., Stierhof, Y. D., and Hulskamp, M. (2002) Ectopic B-type cyclin expression induces mitotic cycles in endoreduplicating *Arabidopsis* trichomes. *Curr. Biol.* **12,** 415–420.
84. Walker, J. D., Oppenheimer, D. G., Concienne, J., and Larkin, J. C. (2000) SIAMESE, a gene controlling the endoreduplication cell cycle in *Arabidopsis thaliana* trichomes. *Development* **127,** 3931–3940.
85. Perazza, D., Vachon, G., and Herzog, M. (1998) Gibberellins promote trichome formation by up-regulating GLABROUS1 in *Arabidopsis. Plant Physiol.* **117,** 375–383.
86. Perazza, D., Herzog, M., Hulskamp, M., Brown, S., Dorne, A. M., Bonneville, J. M. (1999) Trichome cell growth in *Arabidopsis thaliana* can be derepressed by mutations in at least five genes. *Genetics* **152,** 461–476.
87. Sugimoto-Shirasu, K., Stacey, N. J., Corsar, J., Roberts, K., McCann, M. C. (2002) DNA topoisomerase VI is essential for endoreduplication in *Arabidopsis*. *Curr. Biol.* **12,** 1782–1786.
88. Hartung, F., Angelis, K. J., Meister, A., Schubert, I., Melzer, M., and Puchta, H. (2002) An archaebacterial topoisomerase homolog not present in other eukaryotes is indispensable for cell proliferation of plants. *Curr. Biol.* **12,** 1787–1791.
89. Geeta, R. (2003) The origin and maintenance of nuclear endosperms: viewing development through a phylogenetic lens. *Proc. R. Soc. Lond. B Biol. Sci.* **270,** 29–35.
90. Sorensen, M. B., Mayer, U., Lukowitz, W., et al. (2002) Cellularisation in the endosperm of *Arabidopsis thaliana* is coupled to mitosis and shares multiple components with cytokinesis. *Development* **129,** 5567–5576.
91. Brown, R. C. and Lemmon, B. E. (2001) The cytoskeleton and spatial control of cytokinesis in the plant life cycle. *Protoplasma* **215,** 35–49.
92. Siddiqui, N. U., Stronghill, P. E., Dengler, R. E., Hasenkampf, C. A., and Riggs, C. D. (2003) Mutations in *Arabidopsis* condensin genes disrupt embryogenesis, meristem organization and segregation of homologous chromosomes during meiosis. *Development* **130,** 3283–3295.
93. Liu, C. M., McElver, J., Tzafrir, I., et al. (2002) Condensin and cohesin knockouts in *Arabidopsis* exhibit a titan seed phenotype. *Plant J.* **29,** 405–415.
94. Tzafrir, I., McElver, J. A., Liu, C. M., et al. (2002) Diversity of TITAN functions in *Arabidopsis* seed development. *Plant Physiol.* **128,** 38–51.

3

The *C. elegans* Cell Cycle

Overview of Molecules and Mechanisms

Sander van den Heuvel

Summary

The nematode *C. elegans* provides an animal model that allows for genetic dissection of cell cycle regulation in the context of development. Processes such as progression through meiotic prophase, spindle function, chromosome segregation, and cytokinesis are attractive topics to study in the worm, because of their relative visibility and susceptibility to RNA interference. In addition, the reproducible cell lineage greatly helps the identification and characterization of mutants that affect cell cycle entry or exit during larval development. This chapter looks at cell division as an integral part of *C. elegans* development, describes how it can be studied in the worm and summarizes some of the conserved cell cycle components described to date.

Key Words

C. elegans; development; meiosis; G_1 control; chromosome segregation; cytokinesis; cyclin dependent kinases; *lin-35* Rb.

1. Introduction

Studies of cell cycle regulation in the nematode *Caenorhabditis elegans* were initiated little more than a decade ago. Since then, this small round worm gained popularity in the field, especially as a model to study regulation of chromosome segregation and spindle function. This success is explained by several characteristics of *C. elegans* that make it uniquely suited for cell cycle analysis.

C. elegans is nearly as easy to work with as unicellular yeasts: it can be grown on agar plates or in liquid culture, like yeast, and produces an entire generation of progeny in little more than the time needed for a yeast colony to grow. However, unlike yeast, *C. elegans* undergoes a complete program of animal development, allowing for the study of developmental regulation of cell division. Throughout *C. elegans* devel-

From: *Methods in Molecular Biology, vol. 296, Cell Cycle Control: Mechanisms and Protocols*
Edited by: T. Humphrey and G. Brooks © Humana Press Inc., Totowa, NJ

opment, cell divisions can be observed in living animals, as *C. elegans* is transparent. The timing of each and every somatic cell division in this animal is known, since its cell lineage is largely invariant and is described completely. *C. elegans* was originally chosen for its efficient forward genetics, and RNA-mediated interference (RNAi) has provided an efficient new means for targeted gene inactivation. As an important breakthrough, recently developed libraries of cloned gene fragments in bacterial vectors make it possible to screen the entire genome by 'feeding RNAi' in a short period of time. Finally, the community of *C. elegans* researchers provides exceptional support and information and is known for its altruistic and concerted efforts, as demonstrated by the completed sequencing of the *C. elegans* genome before that of any other animal genome.

Several areas of cell cycle control are especially attractive for study in *C. elegans*. These include processes that are particularly visible in the worm, such as progression through meiotic prophase in the germline and centrosome duplication, spindle formation, chromosome segregation, and cytokinesis in the early embryo. Also, other processes that are specific to multicellular organisms and/or involve genes that are not conserved in single cell eukaryotes can be studied in *C. elegans*. Examples of such processes include the Rb/E2F pathway for G_1 control and DNA damage checkpoint mechanisms that trigger apoptosis. Finally, the relative simplicity and detailed knowledge of this animal make it attractive for studies of cell division in the broader context of development. In particular, one can address how cell division is coordinated with cell growth, differentiation, tissue generation, and morphogenesis.

This chapter is intended for the reader with an interest in cell cycle regulation but limited exposure to *C. elegans*. I will briefly review general aspects of *C. elegans* biology, next describe how several cell cycle variations are used throughout *C. elegans* development, and then discuss our current knowledge of the molecules that regulate cell division control in *C. elegans*.

2. *C. elegans* as an Experimental Model Animal

The study of *C. elegans* stems from a search by Sidney Brenner *(1)* for an organism in which development and neuronal function could be subjected to genetic analyses. Important criteria in choosing *C. elegans* included its relatively simple body plan, ease of cultivation, and genetic tractability. The combination of these features suggested that approaches that had proved efficient in the studies of bacteria and bacterial viruses could be applied to *C. elegans*.

C. elegans is a small soil nematode, with adults that are just 1 mm in length and which feeds on microorganisms. In the lab, thousands of animals can be grown on a single Petri dish with *E. coli* bacteria as a food source. Its life cycle is short: in just over 3 d a fertilized egg develops into an adult hermaphrodite or male. Each hermaphrodite produces around 300 progeny by self-fertilization; male–hermaphrodite crosses can generate more than a thousand progeny.

Approximately half of all somatic cells are formed during the first half of embryogenesis *(2)*. Development of the fertilized egg during embryogenesis can be divided into three phases. The initial phase includes establishment of anterior–posterior polarity, formation of a zygote, and generation of the founder cells (blastomeres). The second phase takes place between approx 2 h and 7 h after fertilization and extends from gastrulation to the completion of most cell divisions. The remaining approx 8 h of embryogenesis are characterized by cell differentiation, morphogenesis, and organogenesis. Subsequently, a first-stage larva hatches from the egg and continues growth and development through four larval stages, L1–L4, to the adult stage. Each larval stage culminates in a molt, during which the animal forms a new collagen cuticle and sheds its old cuticle.

Approximately 10% of the cells in the first stage larva maintain the ability to proliferate *(3)*. Divisions of these "somatic blast cells" contribute cells to the hypodermis (skin), nervous system, intestine, musculature, and somatic gonad (**Fig. 1**). In this way, the 558 somatic nuclei present in the early-L1 larva are expanded to a total of 959 in the adult hermaphrodite. Proliferation continues only in the mitotic part of the gonad in adult animals.

As described above, the limited and nearly invariant pattern of somatic divisions combined with the transparency of the animals has allowed a complete description of the *C. elegans* cell lineage *(2,3)*. This lineage provides a unique map of all cell divisions that take place in wild-type animals, which greatly helps in identifying and characterizing animals with defects in cell division. Despite the limited number of cell divisions, most major cell types and tissues are formed, including an extensive neuronal system, muscles, epithelial and intestinal cells, and a germline with gametes. Thus, *C. elegans* provides a model animal that is simple enough to address complex questions, such as how cell division is regulated in the context of animal development.

3. Moving Fast-Forward or Reverse: Classic Genetics and "Genomics"

The most powerful aspect of *C. elegans* research is the animal's genetic tractability. The existence of two different sexes allows for reproduction by either self-fertilization or cross-fertilization, which greatly facilitates genetic manipulation, screens, and strain maintenance *(1)*. For instance, cell cycle mutants usually will not generate viable progeny. However, in a screen, the F_1 progeny of mutagenized animals will be heterozygous for any given mutation. Without the need to set up crosses, such mutant animals will produce homozygous progeny, which can be examined for a cell cycle phenotype. If a mutant of interest is found, recessive mutations can be recovered easily from siblings that are heterozygous, even when the mutation causes death or sterility.

Classic genetic studies start with the identification of mutants that show defects in a biological process of interest. Few such screens have been carried out specifically to find cell cycle mutants, but good examples include a screen for temperature-sensitive embryonic lethal mutants *(4)* and a screen for mutants with defects in G_1 progression *(5)*. In addition, several cell-cycle mutants have been found in general screens for animals with defects in postembryonic development (*see* **Subheadings 5.4.** and **5.6.**).

Reverse genetics starts from a known gene and aims at obtaining a mutant phenotype to reveal its function. Unfortunately, gene knockout by homologous recombination is not efficient in *C. elegans*. The standard procedure is to screen mutagenized populations for a small chromosomal deletion in the selected target by polymerase

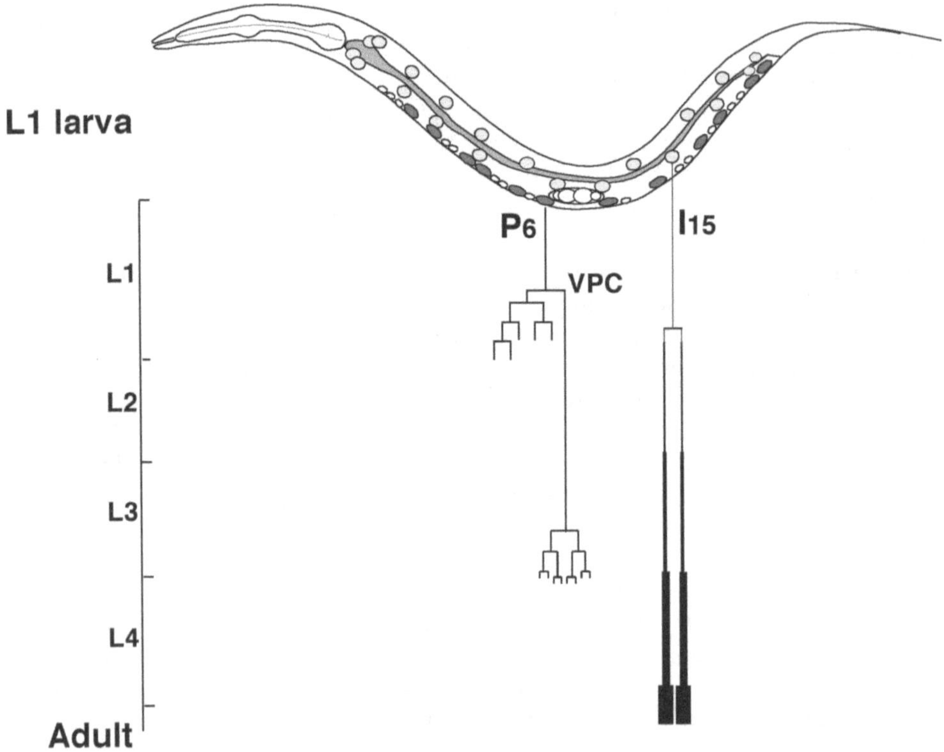

Fig. 1. The timing of cell division is known for all somatic cells throughout *C. elegans* development. Cartoon illustrates a first-stage larva (L1) and some of the precursor cells of the postembryonic lineages (circles). The intestinal nuclei (In) are indicated by circles (light gray) and the precursor cells of the ventral nerve cord (P) as ovals (dark gray). The pattern of divisions is indicated for one of the P cells (P6) and one of the intestinal nuclei (I15). Below the animal, cells are represented by vertical lines and divisions by horizontal lines. The timing of the larval stages L1–L4 is indicated to the left, with the horizontal bars indicating the molts. P6 divides midway through the first larval stage. Subsequently, the anterior daughter cell (left) contributes five neurons to the ventral cord, while the posterior daughter becomes quiescent. This latter cell is a vulval precursor cell (VPC), which can reinitiate cell division and contribute cells to the vulva during the third larval stage. The intestinal cell I15 undergoes a nuclear division followed by a round of DNA replication in the late L1 stage. Additional endoreduplication cycles during each subsequent stage are indicated by thickening of the line and result in nuclei with a 32n DNA content.

chain reaction (PCR) amplification ***(6)***. This method is attractive because apparent genetic null alleles can be selected. However, as the selection is laborious and time consuming, improved screening methods are being developed.

The discovery of RNAi has been a major breakthrough for reverse genetics in *C. elegans (7)*. This approach starts with the introduction of double-stranded (ds) RNA from a gene of interest, either by injection, soaking, or feeding of larvae or adults. This dsRNA is degraded in fragments of approx 21–23 nucleotides that probably provide specificity to an RNAse complex that degrades the message of the corresponding gene. RNAi is particularly efficient in *C. elegans* and has been shown to cause a specific loss-of-function phenotype for many different genes.

Recently, it has become possible to carry out genome-wide RNAi screens *(8)*. Libraries of bacterial plasmids have been developed, with each plasmid expressing dsRNA from a single *C. elegans* gene. These bacteria can be fed to worms grown in 24- or 96-well plates, in order to inhibit the function of many *C. elegans* genes in parallel. The advantage of this approach is enormous. As with classical genetics, it allows the identification of a large number of genes involved in a biological process, yet each gene is identified immediately without the time-consuming process of mapping and cloning.

The major drawback of RNAi is that conclusions about gene function remain uncertain, as RNAi frequently causes incomplete loss of function and sometimes interferes with multiple related genes. To circumvent such issues, the RNAi phenotype may be used as a guidance tool in the identification of mutant alleles. Such alleles could either be previously identified mutations that map to the same genomic region or mutations newly isolated in screens for mutants that resemble a specific RNAi phenotype (*see* **ref.** ***9***). As an additional advantage over reverse genetics, forward genetics allows for isolation of a variety of mutant alleles, including gain-of-function alleles, which can be used in characterizing and ordering gene functions.

4. Developmental Aspects of Cell Division

Specialized cell cycles have evolved to accommodate specific aspects of animal development. Examples are the meiotic cell cycle, with a prolonged arrest in prophase during oocyte formation and maturation; the embryonic cell cycle, which lacks G_1- and G_2-phases; the more typical somatic cell cycle, which can incorporate information from extracellular signals during the G_1-phase; and the endoreduplication cycle, in which S-phase is not followed by M-phase. As an additional level of developmental regulation, cell intrinsic and extracellular signals can affect the position of the mitotic spindle as a means to control the plane of cell division during asymmetric divisions. We will briefly describe how these cycles and mechanisms are used during *C. elegans* development and discuss the contribution of maternal vs zygotic gene function.

4.1. The Germline and Meiotic Cell Cycle

C. elegans is attractive for studies of the meiotic cell cycle. The gonad with germ cells in various stages of meiotic prophase forms a major part of the adult animal (**Fig. 2**). In addition, the germ cell chromosomes can be easily stained and visualized, either in intact animals or within gonads separated from the adults. The hermaphrodite gonad consists of two U-shaped tubular arms, which connect to the uterus through the spermatheca. The distal half of each arm contains a large number of nuclei that are only

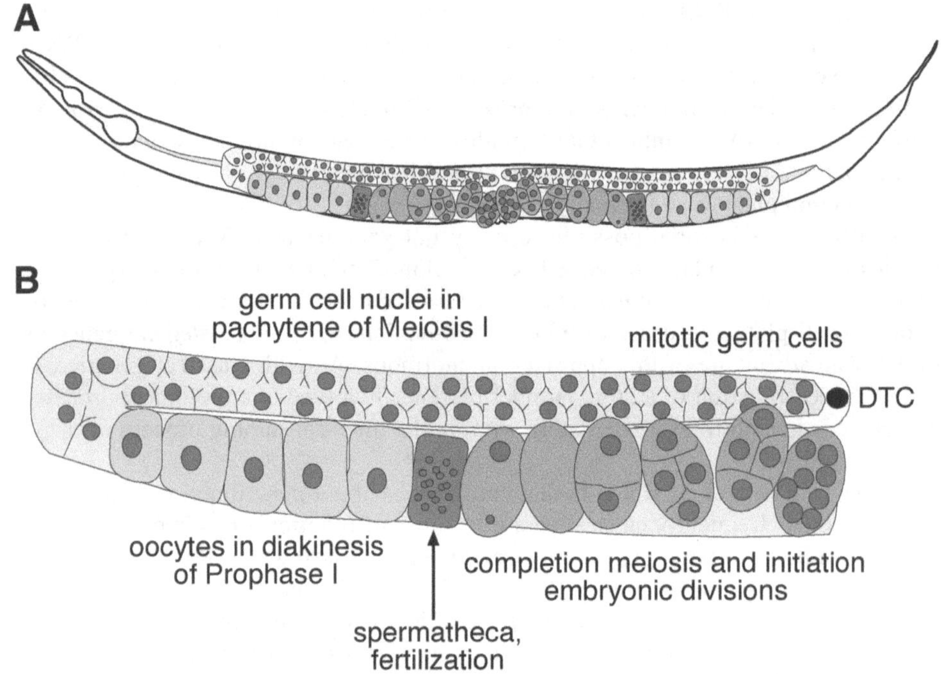

Fig. 2. The *C. elegans* gonad displays a nice temporal arrangement of nuclei in subsequent cell cycle stages. Starting from the distal end, these include: mitotic germ precursor nuclei, nuclei in meiotic prophase, maturing oocytes, fertilized oocytes that complete meiosis, and embryonic cells in mitotic division. **(A)** The two U-shaped gonad arms within the adult hermaphrodite **(B)** Enlargement of one arm. The distal end contains the distal tip cell (DTC) and nuclei in mitosis, and the proximal end is the uterus with embryos.

partly enclosed by a plasma membrane and connect to a common cytoplasm. These nuclei with their surrounding cytoplasm and membranes are commonly referred to as *germ cells*, although they form part of a syncytium ***(10,11)***.

The distal end of each gonad arm harbors a stem cell population that continues mitotic divisions and produces precursor germ cells. These cells enter an extended meiotic prophase when they become removed from the distal tip cell (DTC) and progress through the different stages of prophase while slowly migrating to the proximal end of the germline (**Fig. 2B**). During this process, homologous chromosomes condense in part, pair, and form crossovers. Differentiation initiates in the turn region of the gonad arm, where the germ nuclei cellularize and subsequently continue to grow and mature to oocytes with a typical diakinesis arrangement of the chromosomes.

Further progression through M-phase is induced by the major sperm protein (MSP) in the proximity of the spermatheca. Fertilization is followed by formation of an eggshell and completion of meiosis I and II of the oocyte-derived nucleus, with expulsion of a polar body after each division. Subsequent embryonic divisions follow and continue normally inside the mother until the egg is laid. Thus, the hermaphodite gonad forms a production line with a beautiful temporal arrangement spanning from mitotic germ precursor nuclei to early stage embryos.

The *C. elegans* germline provides ample opportunity to identify genes and mechanisms involved in meiotic processes, such as homolog pairing, crossing over, and meiotic maturation, as well as the decision to enter a meiotic vs mitotic cell cycle.

For example, studies have shown that proliferation in the germline is controlled by a Notch-related signaling pathway that uses the LAG-2 ligand in the DTC and GLP-1 Notch in the distal germ cells (reviewed in **ref. *10***; LAG, *lin-12* and *glp-1* phenotype; GLP, germline proliferation abnormal). Loss-of-function *glp-1* mutants lack germline proliferation, whereas gain-of-function mutations cause a germline tumor phenotype. A genetic pathway has been established in which *glp-1* negatively regulates the genes *gld-1* and *gld-2*, whose inactivation also results in a germline tumor. Rather than general regulation of cell division, the *glp-1* pathway regulates the switch between mitotic and meiotic division in the germline. A failure to undergo this transition probably results in continued mitotic divisions and a tumorous phenotype. It is attractive to apply such genetic approaches more broadly to gain insight into important cell cycle decisions that are still little understood.

4.2. Maternal Product and the Embryonic Cell Cycle

The early divisions in *C. elegans* eggs are the subject of intense study for a number of reasons. The fertilized egg is relatively large (~50 m in length), which allows in vivo monitoring of centrosome duplication, chromosome segregation, and cell cleavage by time-lapse differential interference contrast microscopy. In addition, several early embryonic divisions are asymmetric, an important developmental variation of cell division that involves establishment of polarity, asymmetric segregation of components, and regulation of spindle forces. Furthermore, in contrast to early embryonic divisions in *Drosophila*, nuclear divisions are followed by cytoplasmic divisions, which resembles the situation in mammalian embryos and allows detailed studies of cytokinesis.

Maternal gene products deposited into the oocyte drive embryonic development. Consequently, homozygous cell cycle mutants obtained from heterozygous mothers usually complete embryogenesis, driven by the product of the single wild-type allele from the mother. Such animals usually display the first cell cycle defects during the L1 stage. However, for some genes, maternal product suffices until later larval stages, whereas for others, zygotic function is needed during late embryogenesis (*see* **Subheading 5.1.**). The important role of maternal product explains why several mutants identified in screens for animals with defects in the postembryonic cell lineages (*lin* mutants) defined genes involved in cell cycle control.

Temperature-sensitive mutations and RNAi offer opportunities to inactivate maternal as well as zygotic gene functions. For instance, homozygous *cdk-1(null)* mutants complete embryogenesis but do not undergo any larval divisions ***(9)***. In contrast, *cdk-1(RNAi)* embryos arrest at the one-cell stage and even fail to progress through M-phase of meiosis I. Thus, the results from RNAi and mutant studies are frequently additive, rather than redundant.

C. elegans embryonic divisions are not synchronous, making it difficult to conclude which embryonic divisions consist solely of S- and M-phases, and when G_2- and G_1-phases are introduced. Meiosis completes approx 20 min after fertilization and is followed by a number of mitotic division cycles that each complete in about 15 min. The time between divisions increases slowly over the next divisions. Importantly, even after the first division, the cell cycle timing of the daughters differs, with cells in the germ cell lineage being the slowest to divide. The germline founder cells are also the first to arrest division completely, at approx the 100-cell stage, 3 h into embryonic development. These cells appear to be the only cells that arrest with a G_2 DNA content; thus G_1-phase may not as yet be introduced at this stage or in this lineage. Just a few hours into embryonic development, cells in different lineages diverge greatly in cell cycle profiles. Certain cells continue rapid divisions, others divide after an extended interphase of 2 h and more, and yet other cells become postmitotic or only resume divisions during larval development, after an arrest of more than 18 h. Thus the regulation of cell cycle events and the introduction of G_1- and G_2-phases are determined by the cell lineage.

Although the progressive time periods between rounds of DNA synthesis strongly indicate introduction of G_1-phases ***(12)***, genes with essential roles in embryonic G_1 progression have not been identified. Notably, inactivation of *cyd-1* cyclin D or *cdk-4* Cdk-4/6 by mutation and RNAi results in G_1 arrest during larval development but allows completion of nearly all embryonic divisions ***(5,13)***. Only the final embryonic intestinal divisions do not take place in *cyd-1(0)* embryos, and the very last embryonic divisions, those that generate a hypodermal precursor cell (V5) and neuroblast (Q), are absent in *cdk-4(RNAi)* embryos. Thus, a true G_1-phase appears to be part of at least these late embryonic cell cycles. *cye-1* cyclin E RNAi embryos arrest at about the 100-cell stage, either in S-phase or at the G_1/S transition ***(14)***. Based on green fluorescent protein (GFP) reporter expression, *cyd-1* transcription initiates before the 300-cell stage. Therefore, we expect that various cell cycles in fact include a G_1-phase by this stage, possibly regulated through redundant mechanisms.

4.3. The Somatic Cell Cycle

In total, 51 somatic precursor cells continue to proliferate during larval development to generate the postembryonic cell lineages in the hermaphrodite. These cells appear to contain a 2n DNA content at the time of hatching and go through a DNA synthesis phase before initiating mitosis (***5,9***; our unpublished results). Moreover, they generally depend on the function of G_1/S and G_2/M control genes. Therefore, postembryonic "blast" cells and their descendents appear to follow canonical cell cycles in which the S- and M-phases are separated by Gap phases. However, the length of interphase varies greatly: divisions frequently follow each other within 1 h, but some cells remain quiescent for 20 h, before dividing again two larval stages later (**Fig. 1**, VPC). This latter type of cell cycle is particularly attractive for studies of G_1 control.

4.4. The Endoreduplication Cycle

Cells in two different tissues undergo typical endoreduplication cycles during larval development ***(15)***. Such cycles are characterized by a DNA synthesis phase that is not followed by M-phase, thus doubling the DNA ploidy with each additional cycle. Fourteen of the 20 intestinal cells undergo a final nuclear division at the end of the L1 stage (**Fig. 1**). Subsequently, all intestinal nuclei go through an endoreduplication cycle during each larval stage, which results in a 32n DNA content in adult animals.

A major part of the hypodermis (skin) is formed by fusion of individual cells. Cells that are added to this syncytium during larval development usually undergo a round of DNA synthesis just before fusion ***(15)***. Consequently the skin contains a mixture of nuclei with 2n and 4n DNA contents.

It has been proposed that endoreplication cycles may correspond to nuclear or cellular divisions in related species. Increases in ploidy correlate with increases in growth in other systems. Interestingly, endoreduplication cycles in the worm are limited to the two cell types that display the most significant growth during larval development. Therefore, these specialized cycles probably support cell growth by increasing the number of gene copies and, ultimately, protein products.

4.5. L1 Arrest and Dauer Formation

Despite the invariance of the cell lineage, external factors can greatly affect the timing of development and cell division. One factor affecting all stages is temperature. Development from zygote to adult takes about twice as long at 15℃ compared with 25℃, and larval development arrests between 6 and 10℃. Other external factors can trigger developmental arrest at two specific stages. Larvae that hatch in the absence of food will not initiate postembryonic growth and development and can live for several weeks as early L1 larvae. The addition of food (*E. coli* bacteria) immediately releases this arrest. Thus, feeding of arrested L1 animals can be used in the laboratory to obtain highly synchronized populations.

In addition, young larvae adopt an alternative larval program if the food supply becomes limiting or the population too dense during the first larval stage. This results in *dauer* larvae, which do not eat or grow, have thick protective cuticles, and are specialized for survival during harsh conditions. Importantly, it is currently not understood how the environmental factors that trigger or release L1 and dauer arrest connect to the cell cycle machinery.

5. Molecules Involved in the *C. elegans* Cell Cycle

This section describes results and conclusions from *C. elegans* cell cycle studies, focusing on the molecules discussed in other parts of this book. We apologize to those whose work could not be included or cited because of space limitations.

5.1. Cyclin-Dependent-Kinases

As in other metazoans, specific transitions in the *C. elegans* cell cycle are regulated by multiple distinct cyclin-dependent kinases (CDKs). These CDKs act in conjunction with cyclin partners that resemble members of the cyclin D, E, A, B, and B3 subfamilies in other organisms. In the first study to address the function of CDKs in *C. elegans*, the *ncc-1/cdk-1* gene was identified as the functional orthologue of mammalian Cdk1 ***(9)***. The CeCDK-1 kinase was shown to be essential for progression through meiotic

M-phase and entry into mitosis, but not for G_1- and S-phase. Endoreduplication cycles continued in *cdk-1* mutants and intestinal nuclei accumulated a 32n DNA content.

CDK-1 probably requires multiple different B-type cyclins for its mitotic functions (our unpublished results). Three typical cyclin B genes, *cyb-1*, *cyb-2.1*, and *cyb-2.2* are expressed in addition to a distinct member of the cyclin B3 subfamily, *cyb-3*. Whereas RNAi of *cyb-1* and *cyb-3* each results in embryonic lethality, simultaneous inactivation of both cyclins results in earlier arrest and mimics *cdk-1* loss of function. These results support the notion that different mitotic cyclins have functions that are partly distinct and partly overlapping.

The *C. elegans* genome contains a single gene encoding a Cdk4/Cdk6-related kinase, *cdk-4*, and only one D-type cyclin gene, *cyd-1*. In a reverse genetics study, these genes were found to be essential for the G_1/S transition during larval development **(13)**. In addition, both *cdk-4* and *cyd-1* were identified in a screen for genes essential for G_1 progression **(5)**. Similarity in phenotype and direct binding in vitro support the idea that *cyd-1*/*cdk-4* acts in a complex.

C. elegans is the first system in which an essential role for D-type kinases has been identified. This allowed researchers to address several interesting questions, most importantly: what critical substrates require phosphorylation/association by Cdk4/cyclin D in order to progress into S-phase? Based on mammalian studies, members of the retinoblastoma family (Rb) were obvious candidates. Indeed, null mutations in *lin-35*, the single *C. elegans* Rb family member, suppressed the growth and cell cycle arrest of *cyd-1* and *cdk-4* larvae. These results strongly support the hypothesis that D-type kinases act to overcome G1 inhibition by the pRb protein family. However, the rescue was incomplete, which indicated that *cyd-1* and *cdk-4* have critical functions in addition to *lin-35* Rb inactivation.

One candidate target is CKI-1, a Cdk inhibitor of the Cip/Kip family. Cell division in *cyd-1* and *cdk-4* mutants was restored in part by *cki-1* inactivation, and simultaneous inactivation of *lin-35* Rb and *cki-1* Cip/Kip resulted in a synergistic increase. D-type kinases have been suggested to inactivate Cip/Kip inhibitors through sequestration. This possibility is consistent with the results in *C. elegans* but not in *Drosophila* (see **Subheading 6.**). A third potential downstream target of *cyd-1*/*cdk-4* is defined by a recessive mutation that completely overcomes the *cyd-1* and *cdk-4* requirement when combined with *lin-35* inactivation (our preliminary results). Importantly, this mutation causes extra cell divisions in otherwise wild-type animals, which strongly indicates a role as a negative regulator of cell cycle entry. It will be of great interest to identify the molecular nature of this gene and to determine whether it acts with *cki-1* or in a parallel pathway.

It is currently not clear whether *C. elegans* has a true Cdk2 ortholog. The best candidate is K03E5.3, which shares 43% amino acid identity with human Cdk2 **(9)**. Inhibition of this gene by RNAi resulted in a highly variable cell cycle phenotype, with animals arresting during embryogenesis, during early or late larval development, or as sterile adults. Additional studies or mutations in the gene have not been described.

The *cye-1* cyclin E and *cya-1* cyclin A genes encode cyclins that preferentially bind Cdk2 in other systems. Surprisingly, *cye-1* deletion animals show normal development until late in the third larval stage ***(14)***. This modest phenotype apparently depends on long-lasting maternal function, as RNAi results in embryonic lethality at approximately the 100-cell stage. In contrast, *cya-1* RNAi does not produce an early phenotype, and defects in cell division only become apparent by late larval development.

Several other members of the Cdk superfamily are present in *C. elegans*, including a Cdk7/Mo15 ortholog for which a cell cycle role has been described ***(16)***. The combination of a temperature-sensitive allele and RNAi of *cdk-7* resulted in a one cell arrest similar to *cdk-1(RNAi)* embryos ***(17)***. In addition, partial inactivation of *cdk-7* interfered with transcription and phosphorylation of the RNA polymerase C-terminal domain (CTD). These data support dual activities of CDK-7 as both CDK-activating kinase (CAK) and CTD kinase.

Two other CDKs, *cdk-8* and *cdk-9*, are probably involved specifically in transcription regulation ***(16)***. CDK-5 is remarkably close to human Cdk5, sharing 74% identical at the amino acid level, but this kinase has not been implicated in cell cycle control in either system ***(9)***.

5.2. Inhibitors of Cyclin-Dependent Kinases

The sequence of the *C. elegans* genome revealed two members of the Cip/Kip family of CDK inhibitors (CKIs), most closely related to $p27^{Kip1}$ ***(18,19)***. Inactivation by RNAi showed that *cki-1* is critical for cell cycle arrest, whereas only minor effects have been attributed to *cki-2* RNAi ***(5,18 ,19)***. In *cki-1(RNAi)* larvae, postembryonic precursor cells fail to arrest in G_1 and express the S-phase marker *rnr::GFP* ***(5,18 ,19)***. In addition, such animals display extra cell divisions in a number of different tissues, including the skin and intestine. In a more recent study, deletions that include *cki-1* and *cki-2* were shown to result in a specific embryonic lethal phenotype, characterized by hyperplasia in multiple somatic lineages, apoptosis, and differentiation and morphogenesis defects ***(20)***. As these defects were rescued by a *cki-1* transgene and reproduced by *cki-1(RNAi)*, *cki-1* may also control embryonic divisions and directly or indirectly affect differentiation.

Not all divisions are controlled by *cki-1* Cip/Kip. This inhibitor appears to be rate-limiting for a subset of cells that normally exit the cycle or are in a prolonged G_1-phase. Thus, *cki-1* probably forms only one of multiple regulatory mechanisms that promote developmental exit from cell division. As mentioned above, genetic studies have shown that *cki-1* acts in parallel to *lin-35* Rb, which probably provides a second level of regulation. How these and other cell cycle regulators act in developmental control pathways is a topic of major interest.

5.3. The wee1 and cdc25 Families

In all eukaryotes studied, cell cycle CDKs contain critical threonine and/or tyrosine residues within the glycine-rich ATP binding loop (GXG**TY**G, corresponding to Thr14 and Tyr15 in fission yeast Cdc2). These residues are sites for inhibitory phosphorylation by the Wee1 and Myt1 kinases. Subsequent dephosphorylation by the Cdc25 family of dual specificity phosphatases leads to activation. *C. elegans* CDK-1 and K03E5.3 each contain the corresponding Thr and Tyr residues. CDK-4 has Ala and Tyr residues at these positions, similar to vertebrate Cdk-4/Cdk6.

Phosphorylation and dephosphorylation of these sites in the *C. elegans* kinases has not been demonstrated in vivo or in vitro. However, the *C. elegans* genome encodes three Wee1/Myt-1 kinases and four Cdc25-related phosphatases. Specific gain-of-function mutations have been identified with surprising lineage-specific effects. Two independent dominant mutations in the *cdc-25.1* gene cause hyperproliferation of intestinal cells ***(21,22)***. Gain-of-function alleles of *wee-1.3* Myt1 cause spermatogenesis defects, apparently owing to G_2/M arrest during male meiosis ***(23)***. However, loss-of-function alleles of *wee-1.3* obtained as intragenic suppressors of this phenotype cause embryonic lethality. This indicates a more general cell cycle role for this kinase, rather than a lineage-specific function. The same conclusion was drawn for *cdc-25.1*, based on proliferation defects in multiple lineages following RNAi.

5.4. The Rb/E2F Pathway for G_1 Control

The single *C. elegans* member of the retinoblastoma gene family, *lin-35* Rb, was originally identified because of its role in formation of the vulva [24]. As such, it acts in one of two redundant pathways that inhibit expression of the vulval cell fate. Simultaneous inactivation of a gene in each pathway causes additional cells to adopt a vulval fate, which results in animals with multiple vulvae. As this phenotype depends on simultaneous inactivation of two genes, it has been named the *synthetic multivulva* (synMuv) phenotype, and the genes involved have been grouped into class A and class B synMuv genes. Interestingly, the class B synMuv genes not only include *lin-35* Rb, but also other Rb pathway genes and components of the nucleosome remodeling and deacetylase (NURD) complex. For instance, *efl-1* E2F, *dpl-1* DP, *lin-53* RbAP46/48, and *hda-1* HDAC1 all act as synMuv B genes (*see* **ref. 25**). Thus, inhibition of the vulval cell fate apparently involves transcriptional repression and recruitment of Rb/E2F and NURD complexes.

The role of the *C. elegans* Rb pathway in cell cycle regulation was discovered more recently ***(5,25)***. Loss of function of *lin-35* Rb alone does not result in additional cell division or precocious S-phase entry ***(5)***. Instead, the *cki-1* Cip/Kip inhibitor appears to be rate-limiting and is sufficient to keep the cell cycle in check, even when *Rb* function is absent. However, as described above, *lin-35* Rb mutations partly overcome the cell cycle arrest of *cyd-1* and *cdk-4* mutants, and simultaneous inactivation of *lin-35* strongly enhances the number of extra cells in *cki-1(RNAi)* larvae ***(5)***. Therefore, *lin-35* Rb does indeed act as a negative regulator of G_1 progression, but this is evident only in sensitized genetic backgrounds.

The same methods were applied to examine a cell cycle role for other synMuv genes and *Rb* pathway components ***(25)***. G_1 control functions were identified for several other class B synMuv genes, but not for class A genes. Specifically, *efl-1* E2F negatively regulates cell cycle entry, whereas *dpl-1* DP appeared to act as both a positive and negative regulator. In addition, a negative G_1 regulatory function was identified for *lin-9* ALY, as well as *lin-15B* and *lin-36*, which encode novel proteins. These results further demonstrate the conserved cell cycle functions of Rb/E2F complexes and emphasize the potential for genetic identification of novel G_1 regulators in *C. elegans*.

5.5. The p53 Pathway

An exciting recent discovery is the identification of a *C. elegans p53*-like gene, *cep-1* ***(20,26)***. This gene is only distantly related to *TP53* and is equally close to the other family members, *p63* and *p73*. However, *cep-1* encodes a protein with the hallmark domains of mammalian *p53* and conservation of the residues most frequently mutated in human cancer. Furthermore, *cep-1* is required for radiation-induced apoptosis in the *C. elegans* germline, suggesting a conserved role in the DNA damage response. Other reported functions include contribution to proper meiotic chromosome segregation and survival under hypoxic conditions or during starvation ***(20)***. Similar to mouse and frog *p53*, *cep-1* appears to be highly expressed during embryogenesis. These levels likely are tightly controlled, as heat-shock–induced expression caused fully penetrant embryonic lethality. The conservation of a *p53* pathway in *C. elegans* makes this organism highly attractive for genetic dissection of the regulation and function of *p53*, possibly with major implications for human cancer.

5.6. SCF and APC

Protein degradation through ubiquitin-mediated processes plays an important role in the regulation of the *C. elegans* cell cycle, as has been shown in other eukaryotes. In two cell lineage mutants identified in early screens, *lin-19/cul-1* and *lin-23* mutants, postembryonic blast cells undergo excessive cell divisions ***(27,28)***. Both genes were found to encode components of the SCF ubiquitin/ligase complex composed of Skp1, cullin, F-box, and Rbx1/Roc1 proteins. In fact, *lin-19* was the founding member of the cullin family, and this gene has been renamed *cul-1* ***(28)***. LIN-23 is an F-box WD repeat protein, most similar to the β-TRCP F-box protein in *Drosophila* ***(27)***.

The cullin family comprises six different genes in *C. elegans*, as in humans, and cell cycle related functions have been reported for four of these. Deletion of *cul-2* results in G_1 arrest of germ precursor cells, whereas *cul-2(RNAi)* embryos show defects in chromosome condensation and segregation ***(19)***. Human CUL2 and the von Hippel-Lindau tumor suppressor protein (VHL) form part of an SCF-related ubiquitin ligase involved in degradation of hypoxia-inducible factor (HIF). Although *C. elegans* VHL-1 and HIF-1 are indeed part of a hypoxia-induced pathway, *cul-2* has not been implicated as yet. CUL-3 is specifically required for degradation of MEI-1, a subunit of a katanin complex that severs microtubules and is essential for meiotic chromosome segregation ***(29)***. Finally, *cul-4* was shown to be essential to prevent re-replication of DNA, probably by acting in the degradation of the replication licencing factor CDT-1 ***(30)***.

Ubiquitin-mediated degradation is also important in other transitions. A screen for temperature-sensitive embryonic lethal mutations revealed a large number of mutants that arrest in metaphase of meiosis I ***(31)***. These **m**etaphase to **a**naphase **t**ransition (*mat*) mutants were found to encode components of the anaphase-promoting complex (APC), an E3 ubiquitin ligase with M-phase and G1 functions. As homologs of spindle assembly checkpoint genes are also conserved, *C. elegans* provides a metazoan system in which the connection between spindle assembly and anaphase initiation can be explored genetically.

5.7. Aurora and Polo Kinases

Three different polo-related kinases (PLK) and two Aurora-/Ipl-1-related (AIR) kinases have been described in *C. elegans*. At least one of the polo kinases, PLK-1, is essential for progression through meiotic M-phase ***(32)***. Studies of AIR-1 Aurora-A and AIR-2 Aurora-B have provided significant insights into the functions of each of these kinase subfamilies. AIR-1 localizes to centrosomes and appears to be necessary for their maturation ***(33,34)***. AIR-2 shows distinct localizations to chromatin and the spindle midzone, probably reflecting its essential functions in chromosome segregation and cytokinesis (*see* **refs. *35*** and ***36*** and references therein). Several critical substrates and partners have been identified. AIR-2 phosphorylates histone H3, the meiotic cohesin REC-8, and the inner centromere protein ICP-1. Interactions with ICP-1 and the survivin homolog BIR-1 determine AIR-2 localization; and in return, localization of the kinesin motor protein ZEN-4/CeMKLP1 depends on interaction with AIR-2. The roles of these and other proteins in formation of the cytokinetic furrow and completion of cell division are topics of intense study.

6. Summary and Conclusions

Progress made over the past decade has clearly established cell cycle research in *C. elegans* and provided important insights. Pathways and mechanisms have been found to be surprisingly well conserved, with the Rb/E2F pathway as a good example. The *C. elegans* embryo has already become one of the leading models for studies of spindle function, chromosome segregation, and cytokinesis. The visibility of these processes in the early embryo and the efficiency of RNAi have both contributed to this success. Later embryonic and larval divisions have been studied to a lesser extent but also hold great promise. It is these divisions that can provide clues as to how cell division is coordinated with growth, differentiation, and tissue generation. In the future, genome-wide feeding RNAi in *C. elegans* will probably reveal novel cell cycle genes and molecular mechanisms. Such screens can even be modified to reveal suppressors and enhancers of cell cycle mutants.

Recent progress has been made in some areas, while others still need improvement. *C. elegans* cell-cycle research has progressed through its initial phase, and the availability of cell cycle mutants and reagents is expanding. Biochemical approaches have traditionally not been applied but are gaining in popularity. Two weaknesses of system are the current absence of a *C. elegans* tissue culture cell line, which prevents the possibility of analyzing synchronized cell populations, and the lack of methods for efficient homologous integration, which can be used to inactivate or modify gene functions in yeast and mammals.

The addition of another cell cycle system has obvious advantages. Each model has its strengths and drawbacks, and ultimately the question addressed determines what system is most suitable. Also, parallel approaches in multiple systems allow the field to take advantage of the strengths of each system and to circumvent its limitations. Cell cycle mechanisms tend to be highly conserved, but interesting differences have also been noted. For example, cyclin D kinases are essential for G_1 progression in *C.*

elegans but not in *Drosophila*, which may reveal information about their targets. Interestingly, worm CYD-1/CDK-4 binds to and counteracts the inhibitor CKI-1 Cip/Kip, a function conserved in mammals but not in *Drosophila*. Thus, the fly Cip/Kip inhibitor, *dacapo*, is probably regulated through alternative mechanisms, which reduces the importance of *Dm* cyclin D/Cdk4-6. This illustrates how small differences in the levels or properties of cell cycle regulators may determine which factors are rate-limiting, supporting studies in multiple model systems.

In summary, *C. elegans* is one of the most powerful genetic animal systems and has much to offer to the cell cycle field. Its popularity and relevance for cell cycle research will only become more apparent in the years to come.

Acknowledgments

We thank Ridgely Fisk Green and Dayalan Srinivasan for critically reading the manuscript and Mike Boxem for help with figures. S.v.d.H. is funded by grants from the NIH.

References

1. Brenner, S. (1974) The genetics of *Caenorhabditis elegans*. *Genetics* **77,** 71–94.
2. Sulston, J. E., et al. (1983) The embryonic cell lineage of the nematode *Caenorhabditis elegans*. *Dev. Biol.* **100,** 64–119.
3. Sulston, J. E. and Horvitz, H. R. (1977) Post-embryonic cell lineages of the nematode, Caenorhabditis elegans. *Dev. Biol.* **56,** 110–156.
4. OConnell, K. F., Leys, C. M., and White, J. G. (1998) A genetic screen for temperature-sensitive cell-division mutants of Caenorhabditis elegans. *Genetics* **149,** 1303–1321.
5. Boxem, M. and van den Heuvel, S. (2001) *lin-35* Rb and *cki-1* Cip/Kip cooperate in developmental regulation of G1 progression in *C. elegans*. *Development* **128,** 4349–4359.
6. Jansen, G., et al. (1997) Reverse genetics by chemical mutagenesis in *Caenorhabditis elegans*. *Nat. Genet.* **17,** 119–121.
7. Fire, A., Montgomery, M. K., Kostas, S. A., Driver, S. E., Mello, C. C. (1998) Potent and specific genetic interference by double-stranded RNA in *Caenorhabditis elegans*. *Nature* **391,** 806–811.
8. Kamath, R. S., Fraser, A. G., Dong, Y., Poulin, G., et al. (2003) Systematic functional analysis of the *Caenorhabditis elegans* genome using RNAi [Comment]. *Nature* **421,** 231–237.
9. Boxem, M., Srinivasan, D. G., and van den Heuvel, S. (1999) The *Caenorhabditis elegans* gene *ncc-1* encodes a cdc2-related kinase required for M phase in meiotic and mitotic cell divisions, but not for S phase. *Development* **126,** 2227–2239.
10. Riddle, D. L., et al. *C. elegans II*. 1997, Cold Spring Harbor Laboratory Press, Cold Spring Harbor, NY.
11. Wood, W. and the Community of *C. elegans* Researchers, eds. (1988) *The Nematode* Caenorhabditis elegans. Cold Spring Harbor Laboratory: Cold Spring Harbor Press, NY.
12. Edgar, L. G. and McGhee, J. D. (1988) DNA synthesis and the control of embryonic gene expression in *C. elegans*. *Cell* **53,** 589–599.
13. Park, M. and Krause, M. W. (1999) Regulation of postembryonic G_1 cell cycle progression in *Caenorhabditis elegans* by a cyclin D/CDK-like complex. *Development* **126,** 4849–4860.
14. Fay, D. S. and Han, M. (2000) Mutations in *cye-1*, a *Caenorhabditis elegans* cyclin E homolog, reveal coordination between cell-cycle control and vulval development. *Development* **127,** 4049–4060.

15. Hedgecock, E. M. and White, J. G. (1985) Polyploid tissues in the nematode *Caenorhabditis elegans*. *Dev. Biol.* **107,** 128–133.
16. Liu, J. and Kipreos, E. T. (2000) Evolution of cyclin-dependent kinases (CDKs) and CDK-activating kinases (CAKs): differential conservation of CAKs in yeast and metazoa. *Mol. Biol. Evol.* **17,** 1061–1074.
17. Wallenfang, M. R. and Seydoux, G. (2002) cdk-7 Is required for mRNA transcription and cell cycle progression in *Caenorhabditis elegans* embryos. *Proc. Natl. Acad. Sci. USA* **99,** 5527–5532.
18. Hong, Y., Roy, R., and Ambros, V. (1998) Developmental regulation of a cyclin-dependent kinase inhibitor controls postembryonic cell cycle progression in *Caenorhabditis elegans*. *Development* **125,** 3585–3597.
19. Feng, H., Zhong, W., Punkosdy, G., et al. (1999) CUL-2 is required for the G1-to-S-phase transition and mitotic chromosome condensation in *Caenorhabditis elegans*. *Nat. Cell Biol.* **1,** 486–492.
20. Derry, W. B., Putzke, A. P., and Rothman, J. H. (2001) *Caenorhabditis elegans* p53: role in apoptosis, meiosis, and stress resistance. *Science* **294,** 591–595.
21. Kostic, I. and Roy, R. (2002) Organ-specific cell division abnormalities caused by mutation in a general cell cycle regulator in *C. elegans*. *Development* **129,** 2155–2165.
22. Clucas, C., Cabello, J., Bussing, I., Schnabel, R., Johnstone, I. L. (2002) Oncogenic potential of a *C. elegans* cdc25 gene is demonstrated by a gain-of-function allele. *EMBO J.* **21,** 665–674.
23. Lamitina, S. T. and LHernault, S. W. (2002) Dominant mutations in the *Caenorhabditis elegans* Myt1 ortholog wee-1.3 reveal a novel domain that controls M-phase entry during spermatogenesis. *Development* **129,** 5009–5018.
24. Lu, X. and Horvitz, H. R. (1998) *lin-35* and *lin-53*, two genes that antagonize a *C. elegans* Ras pathway, encode proteins similar to Rb and its binding protein RbAp48. *Cell* **95,** 981–991.
25. Boxem, M. and van den Heuvel, S. (2002) *C. elegans* class B synthetic multivulva genes act in G(1) regulation. *Curr. Biol.* **12,** 906–911.
26. Schumacher, B., Hofmann, K., Boulton, S., Gartner, A. (2001) The *C. elegans* homolog of the p53 tumor suppressor is required for DNA damage-induced apoptosis. *Curr. Biol.* **11,** 1722–1727.
27. Kipreos, E. T., Gohel, S. P., and Hedgecock, E. M. (2000) The *C. elegans* F-box/WD-repeat protein LIN-23 functions to limit cell division during development. *Development* **127,** 5071–5082.
28. Kipreos, E. T., Lander, L. E., Wing, J. P., He, W. W., Hedgecock, E. M. (1996) *cul-1* is required for cell cycle exit in *C. elegans* and identifies a novel gene family. *Cell* **85,** 829–839.
29. Kurz, T., Pintard, L., Willis, J. H., et al. (2002) Cytoskeletal regulation by the Nedd8 ubiquitin-like protein modification pathway. *Science* **295,** 1294–1298.
30. Zhong, W., Feng, H., Santiago, F. E., Kipreos, E.T. (2003) CUL-4 ubiquitin ligase maintains genome stability by restraining DNA-replication licensing. *Nature* **423,** 885–889.
31. Golden, A., Sadler, P. L., Wallenfang, M. R., et al. (2000) Metaphase to anaphase (mat) transition-defective mutants in *Caenorhabditis elegans*. *J. Cell Biol.* **151,** 1469–1482.
32. Chase, D., Serafinas, C., Ashcroft, N., et al. (2000) The polo-like kinase PLK-1 is required for nuclear envelope breakdown and the completion of meiosis in *Caenorhabditis elegans*. *Genesis* **26,** 26–41.

33. Hannak, E., Kirkham, M., Hyman, A. A., Oegema, K. (2001) Aurora-A kinase is required for centrosome maturation in *Caenorhabditis elegans*. *J. Cell Biol.* **155,** 1109–1116.
34. Schumacher, J. M., Ashcroft, N., Donovan, P. J., Golden, A. (1998) A highly conserved centrosomal kinase, AIR-1, is required for accurate cell cycle progression and segregation of developmental factors in *Caenorhabditis elegans* embryos. *Development* **125,** 4391–4402.
35. Kaitna, S., Pasierbek, P., Jantsch, M., Loidl, J., Glotzer, M. (2002) The aurora B kinase AIR-2 regulates kinetochores during mitosis and is required for separation of homologous chromosomes during meiosis. *Curr. Biol.* **12,** 798–812.
36. Rogers, E., Bishop, J. D., Waddle, J. A., Schumacher, J. M., Lin, R. (2002) The aurora kinase AIR-2 functions in the release of chromosome cohesion in *Caenorhabditis elegans* meiosis. *J. Cell. Biol.* **157,** 219–229.

4

Developmental Control of Growth and Cell Cycle Progression in *Drosophila*

Lisa Swanhart, Jeremy Kupsco, and Robert J. Duronio

Summary

Drosophila melanogaster provides an outstanding experimental system to study the regulation of cell cycle progression during animal development. Sophisticated forward and reverse genetic techniques and the ability to observe detailed cell biological phenomena in vivo have allowed an unparalleled analysis of the cell cycle in the context of a whole animal. This chapter provides an overview of the diverse modes of cell cycle control that are utilized at different stages of *Drosophila* development.

Key Words

Drosophila; cell cycle; embryogenesis; oogenesis; endocycle; polyploidy; growth control; imaginal discs.

1. Introduction

Drosophila melanogaster has been a powerful experimental system in biology for more than 100 years. Many seminal discoveries have been made by studying this organism. *Drosophila* was originally used to construct the first genetic map, to prove that genes reside on chromosomes, to demonstrate that X-irradiation causes mutations, to position-clone a gene, and to rescue a mutant phenotype with a transgene ***(1)***. With a virtually complete genome sequence and technological innovations in both forward and reverse genetic methods ***(2–4)***, *Drosophila* researchers continue to make advances in the understanding of many cell biological processes, including progression through the cell cycle. *Drosophila* has been particularly useful for studying how control of the cell cycle is essential for, and coupled to, the dramatic morphological events that occur during development ***(5–8)***. One paradigm that has emerged from these studies is how the remarkable plasticity of the cell cycle is exploited to support the developmental strategies of the organism. Our goal is to illustrate this point as we

From: *Methods in Molecular Biology, vol. 296, Cell Cycle Control: Mechanisms and Protocols*
Edited by: T. Humphrey and G. Brooks © Humana Press Inc., Totowa, NJ

examine the variety of mechanisms used to control cell cycle progression during the *Drosophila* life cycle.

2. Three Modes of Cell Cycle Control During the *Drosophila* Life Cycle

Drosophila is a holometabolous insect, which means it undergoes complete metamorphosis in four morphologically distinct developmental stages: egg (embryo), larva, pupa, and adult. This entire life cycle, from egg to fertile adult, takes approximately 2 wk. A different mode of cell cycle control predominates at each stage of development. *Stereotyped* cell cycle control occurs during embryogenesis, when an invariant developmental program dictates a more or less identical series of cell cycle events that are not influenced by external growth conditions ***(9)*** (**Fig. 1**). The free-living larva displays *growth-regulated* cell cycles that support the development and proliferation of diploid imaginal cells that are the precursors of adult structures such as wings, eyes, and legs ***(10,11)***. The production of eggs by adult female flies relies on a *modified* cell cycle program during oogenesis that features developmentally regulated polyploidization and gene amplification ***(12,13)***. The following sections provide an overview of the types of molecular regulation that support these different modes of cell cycle control.

3. Embryonic Development

3.1. Syncytial Nuclear Division Cycles

The first 13 cycles of *Drosophila* embryogenesis are nuclear division cycles that occur in a common cytoplasm and result in approx 6000 cells for the blastoderm embryo. These syncytial cycles are very rapid and consist of only S-phase and mitosis without intervening G1 and G2. Zygotic transcription of most genes does not occur at this stage, and all the cell cycle components necessary to run these cycles are provided to the egg by the mother ***(14,15)***. This includes machinery that functions in all division cycles (e.g. cyclins A and B, stg^{cdc25}, cdc2), as well as a set of specialized machinery that functions only at this stage of development (e.g., plu, png, gnu, frühstart; *see* second paragraph below).

Cyclin synthesis and destruction are necessary for the entry into and exit from mitosis, respectively. Interestingly, when measured on a per embryo basis in the very earliest nuclear cycles (i.e., one to seven) the total pool of cyclin A and cyclin B is not destroyed during mitosis ***(16)***, rather, a localized pool of mitotic spindle-associated cyclin is destroyed to permit mitotic exit ***(17)***. In support of this model, the injection into the syncytial embryo of mRNA encoding a non-degradable mutant form of cyclin B arrests cells in mitosis ***(18)***. Cyclin destruction during mitosis requires an E3 ubiquitin ligase called the anaphase-promoting complex (APC). Cdc27 and Cdc16, two subunits of the APC, translocate to the nucleus at the beginning of mitosis but are not enriched on spindles ***(19)***. In contrast, Fzy/Cdc20, which binds to and presents cyclin B as a substrate to the APC, interacts with microtubules and is concentrated on spindles in vivo ***(20)***. This may contribute to the ubiquitination and destruction of the spindle-associated cyclin and consequent inactivation of Cdc2. Such local regulation of the cell cycle machinery provides nuclear autonomy in a common cytoplasm.

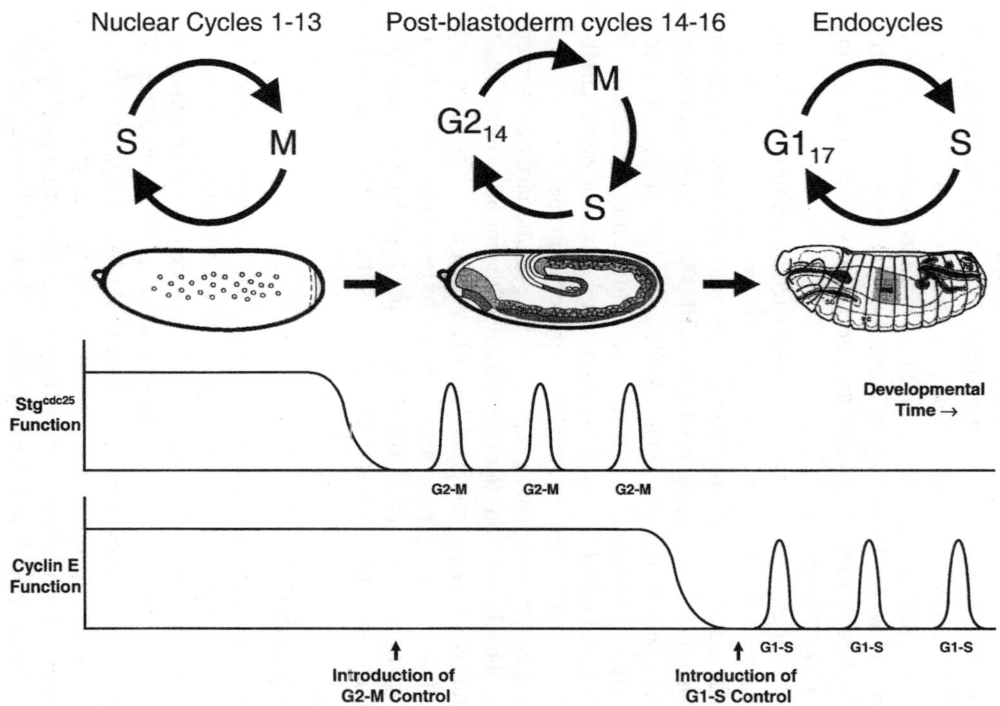

Fig. 1. Embryonic cell cycle control in *Drosophila*. During the S/M syncytial nuclear cycles, stg^{cdc25} and cyclin E provided by the mother are present continuously, as are all other essential cell cycle factors. The downregulation of stg^{cdc25} mRNA and protein at the maternal/zygotic transition causes the accumulation of inactive cyclin/cdc2 complexes and the appearance of G_2-phase in cycle 14 of the blastoderm embryo. During gastrulation the transcription of stg^{cdc25} is induced in late G_2 in response to developmental signals and triggers the entry into mitosis of cycles 14–16. Thus, the transition from continuous stg^{cdc25} expression to regulated stg^{cdc25} expression is the mechanism by which G_2/M control is introduced during *Drosophila* development. Likewise, the introduction of G_1/S control is a result of the transition from continuous to regulated cyclin E function. The downregulation of cyclin E/cdk2 activity is mediated by expression of the cdk inhibitor Dacapo and termination of *cyclin E* transcription by RBF/E2F complexes. Subsequent E2F-directed cyclin E transcription helps induce each subsequent endo S-phase.

The coordination of S-phase and mitosis in the nuclear cycles requires a specialized group of genes that appear to function only at this stage of development. Mothers that are homozygous for mutations in *pan gu* (*png*), *plutonium* (*plu*), or *giant nuclei* (*gnu*) produce eggs that do not develop and that contain highly polyploid nuclei, rather than an exponentially expanding complement of diploid nuclei ***(21–23)***. Polyploid nuclei are also found in unfertilized mutant eggs, indicating that these genes also act to suppress replication prior to the first zygotic nuclear cycle. The level of cyclin B and cdc2 kinase activity is reduced relative to wild type in these mutants, and increasing cyclin B gene dosage suppresses the polyploid phenotype ***(24)***. This suggests a failure to properly enter mitosis and/or prevent re-replication. *png* encodes an S/T kinase, *plu* encodes a 19-kDa ankyrin repeat protein, and *gnu* encodes a novel phosphoprotein ***(25–27)***. The mechanism by which these factors control cyclin levels in the early embryo is not known, although they physically and genetically interact and thus presumably function in a common pathway ***(26)***.

During cycles 9–13, the length of interphase increases each cycle, possibly owing to depletion of an essential replication factor. At this stage the coordination of S-phase and mitosis requires an additional mechanism. Here, a maternally provided replication checkpoint pathway ensures that mitosis does not begin until DNA replication is completed. *mei-41* and *grapes* encode the *Drosophila* ATR and Chk1 kinases, respectively, and function to extend interphase and prevent precocious entry into mitosis in cycles 12–13 ***(28–32)***. Maternal *mei-41* or *grapes* mutants display mitotic defects probably because of incompletely replicated chromosomes and/or improper chromosome condensation. The kinases encoded by these genes act in an evolutionarily conserved pathway that controls the phosphorylation state and activity of the mitotic kinase, cdc2. Consistent with this, embryos lacking maternal Dwee1, which mediates inhibitory Tyr15 phosphorylation of cdc2, display phenotypes similar to *grapes* and *mei-41* mutants ***(33)***. These genes are not required prior to nuclear cycle 11, perhaps because the rapidity of genome duplication in the earliest cycles obviates the need to delay entry into mitosis.

3.2. Introduction of G_2 Control

Dramatic developmental and cell cycle changes take place during interphase of cycle 14. Cellularization occurs as new plasma membrane is synthesized and moves down between each nucleus, which by this time have all migrated to the periphery of the egg. S-phase begins immediately after mitosis of nuclear cycle 13, and then all cells pause in the first G_2 phase of development during cycle 14. At this time many maternal mRNA and proteins are degraded, and the bulk of zygotic transcription commences. Hence, this event has been called the maternal/zygotic transition (MZT) and is functionally equivalent to the midblastula transition in well-studied models of vertebrate development such as *Xenopus laevis*, the African clawed frog.

G_2 appears because of destruction of the maternal pool of string (stg^{cdc25}) phosphatase, whose function is to remove the inhibitory phosphate from Tyr15 of cdc2 (**Fig. 1**). stg^{cdc25} is present continuously during the syncytial cycles and thereby maintains cdc2 in the active state. After maternal stg^{cdc25} is destroyed during cycle 13, inactive Tyr15-phosphorylated cdc2/cyclin A and B complexes accumulate, precluding entry into mitosis and causing a pause in G_2 of cycle 14 ***(16,34)***. Changing the maternal dose of stg^{cdc25} and twine, a paralogous phosphatase expressed only in the germline and early embryo ***(35,36)***, alters the total number of syncytial nuclear divisions, suggesting that the regulation of the phosphatases that act on cdc2 controls the timing of the MZT ***(34)***.

The occurrence of the MZT precisely during cycle 14 is dictated by the nuclear/cytoplasmic (N/C) ratio. This was best illustrated by analysis of haploid embryos, which first enter G_2 in cycle 15 rather than 14 ***(37)***. Activation of the *frühstart* gene appears to induce the MZT ***(38)***. High-level, transient *frühstart* gene expression is triggered by the N/C ratio at the beginning of cycle 14 and results in the appearance of G_2. Mutation of *frühstart* causes an extra syncytial nuclear cycle, and precocious expression of *frühstart* causes nuclear cycle arrest prior to interphase 14. *frühstart* encodes a short protein with no known homologs. Its molecular function is unknown, although it somehow participates in the downregulation of stg^{cdc25} RNA and/or protein ***(38,39)***.

3.3. Postblastoderm Cell Division Cycles

The G_2/M transition of embryonic cycle 14 ($G2_{14}$) is the first time zygotic gene expression is required for cell cycle progression. Gastrulation begins during G_2 of cycle 14, and different groups of cells referred to as mitotic domains (MDs) enter M-phase cycle 14 at different times, generating a stereotypic spatial–temporal pattern that coordinates mitosis with cell movements ***(40)***. stg^{cdc25} mutants arrest in G_2 cycle 14, indicating a requirement for zygotic expression of stg^{cdc25} and the consequent activation of cyclin/cdc2 complexes ***(41,42)***. stg^{cdc25} transcription coincides with the mitotic domains and is controlled by an extensive array of enhancers that integrates spatial and temporal patterning information through the activity of numerous transcription factors ***(43,44)***. Cell cycles 15 and 16 also lack a G_1 and are regulated at the G_2/M transition by *stg* transcription ***(43)***.

The regulation of stg^{cdc25} defines a paradigm of stereotypic cell cycle control during *Drosophila* development, whereby a major cell cycle transition is controlled by transcription of a single, key regulatory gene that contains a complex, modular upstream control region. This is an efficient mechanism in the embryo because essentially all the other components required for cell cycle progression are provided constitutively. In this way, the production of a single gene that responds directly to developmental information is sufficient to control cell cycle progression.

The postblastoderm embryonic divisions (i.e., cycles 14–16) have also been useful for studying intrinsic cell cycle controls. *Drosophila* cyclins A, B, and B3 cooperate to control progression through mitosis and function redundantly in some regards and uniquely in others. For instance, whereas cyclin B and B3 single mutants are viable, double mutants between the A and B cyclins display mitotic defects ***(45,46)***. Embryonic cyclin A does not bind cdk2 nor play a major role in S-phase ***(47)***. However, cyclin A mutant embryos enter a developmentally scheduled S-phase in tracheal precursor cells without completing mitosis 16 ***(48)***, providing early support for the evolutionarily conserved role of cyclin/cdc2 activity in suppressing re-replication within a cell division cycle. The analysis of mutant phenotypes during the postblastoderm divisions have also increased our understanding of the function of genes controlling cytokinesis (e.g., *pebble*; ***49–51***), the release of sister chromatid adhesion during anaphase (e.g., *pimples, three rows, fizzy*; ***52–57***), and centrosome duplication ***(58)***.

3.4. Introduction of G_1 Control

Cycles 1–16 lack a G_1-phase, and all components required for DNA replication are provided constitutively, from either maternal stores or ubiquitous expression during the postblastoderm cycles ***(59,60)***. Most cells enter G_1-phase for the first time in the 17th cell cycle ***(41)***. Subsequent entry into S-phase requires expression of cyclin E, since *cyclin E* mutants arrest in G_1 of cycle 17 ***(47)***. Maternal cyclin E function is presumably sufficient for S-phases 1–16. The introduction of G_1 control in cycle 17 is analogous to the introduction of G_2 control earlier in embryogenesis, in that the activity of an essential regulator of S-phase, cyclin E/cdk2, is rapidly downregulated.

Two mechanisms are used to downregulate cyclin E/cdk2: inhibition of cyclin E/cdk2 kinase activity, and termination of *cyclin E* transcription. Kinase inhibition is achieved by transcriptional activation of *dacapo* (*dap*), which encodes a p21/p27-type of cdk inhibitor ***(61,62)***. *dap* transcription is upregulated in late interphase of cycle 16 and, like stg^{cdc25} depends on a complex upstream control region that integrates spatiotemporal developmental information independently of cell cycle phase ***(63,64)***. *dap* mutant embryos fail to arrest in G_1 of cycle 17, and inappropriately enter a single additional cell division cycle before arresting in G_1 of cycle 18 ***(61,62)***.

Transcriptional downregulation of *cyclin E* helps maintain cell cycle arrest in G_1 of cycle 17. As in vertebrates, the transcriptional regulation of *cyclin E* and other S-phase genes is mediated by the pRB/E2F pathway ***(65,66)***. *Drosophila* contains two pRB homologs, RBF1 and RBF2, and two dimeric E2F transcription factors, E2F1/DP and E2F2/DP ***(67–71)***. Prior to cycle 17, E2F target genes are expressed continuously, and after the introduction of G_1 regulation, they are expressed coordinately with S-phase; this requires the function of E2F1/DP ***(72–74)***. E2F2 is not required during embryogenesis ***(75)***. E2F1/DP activity is controlled by RBF1. Mutant embryos lacking *Rbf1* gene function fail to downregulate the transcription of E2F targets during G_1 cycle 17, including *cyclin E* ***(76)***. Interestingly, *Rbf1* mutant epidermal cells enter G_1 cycle 17 on schedule, probably because of *dap* expression and the resulting inhibition of cyclin E/cdk2 kinase activity. However, they cannot maintain G_1 arrest, and some cells eventually enter an inappropriate S-phase of cycle 17 ***(76)***, probably because *dap* transcription is eventually downregulated as it is in wild type. Whereas RBF1 activity is mediated by phosphorylation as it is in mammals ***(77)***, the developmental mechanism by which RBF1 is activated during late cycle 16 to inhibit E2F is not known. Cyclin D/cdk4 is not the sole regulator of E2F ***(78)***, and the Janus kinase/signal transducer and activator of infection (JAK/STAT) pathway has recently been implicated in control of E2F activity in embryos ***(79)***.

3.5. Endocycles

Most cells remain arrested in G_1 of cycle 17 for the remainder of embryogenesis, although there are some notable exceptions. Cells in the central nervous system continue to proliferate via cell cycles regulated at G_2/M by *stg*[cdc25]. The cells in several tissues that will make up the larva proper (e.g., the midgut, hindgut, salivary glands) begin endocycles, a *modified* cell cycle in which polyploidy is achieved through G/S cycles that lack mitosis ***(7,80)***. With the exception of imaginal cells, which are diploid, larval growth is achieved via the endocycle rather than via cell proliferation. Cyclin E/cdk2 is required for DNA replication during endo S-phase ***(47,73,81)***, and it functions at least in part to load the minichromosome maintenance complex (MCM) replicative helicase onto chromatin ***(82)***. Interestingly, this activity appears to be conserved in mammalian cells ***(83)***. Like many other components of the replication machinery, cyclin E transcription in the endocycle occurs in coordination with S-phase and requires E2F1/DP ***(73,74,84)***. Bypass of this periodic transcriptional control either by forced, continuous expression of *cyclin E* or elimination of E2F activity blocks polyploidization ***(85–87)***. This suggests that cyclic activation/deactivation of cyclin E/cdk2 is an essential aspect of the repeated rounds of endo S-phase and that continuous cyclin E/cdk2 kinase activity prevents the assembly of pre-replication complexes (RCs).

The mitotic cyclins A, B, and B3 are not expressed during endocycles ***(46,88)***, and their downregulation is essential to make the transition from cell division cycle to endocycle. Mutation of *fizzy-related* (*fzr*), which encodes a cdh1-type adaptor molecule that presents cyclins to the APC E3 ubiquitin ligase complex, prevents destruction of cyclin A and B during interphase. This consequently inhibits endo S-phase by imposing a re-replication block ***(89)***. FZR activity is inhibited by RCA1, the fly homolog of mammalian Emi1 ***(90)***. *rca1* mutants cause premature degradation of cyclins in G_2 and a failure to enter mitosis, but still enter endocycles ***(90–92)***.

4. Cell Cycle Control in Imaginal Discs

4.1. Imaginal Disc Development

The imaginal discs of *Drosophila* provide an ideal system to study growth-regulated cell cycles. Unlike the growth-independent and G_2-regulated cell cycles of the early embryo, imaginal disc cells proliferate with cell division cycles that contain a G_1-phase and require growth. Imaginal disc are sacks of epithelial cells that differentiate into the structures of the adult head, thorax, and genitalia. Groups of 10–50 imaginal disc precursor cells arrest in G_1 of cycle 17 during embryogenesis. After larval hatching, these cells undergo a period of rapid proliferation as the larva feeds and grows. During this time the imaginal cells become specified by developmental signals, and in most cases undergo final differentiation in the pupae.

4.2. Growth Vs Cell Cycle Control in the Wing Imaginal Disc

Growth is defined as the accumulation of mass over time and can be measured on a level as small as the cell (cellular growth) or as large as the entire organism. When growth accompanies progress through the cell division cycle, proliferation occurs. However, growth and cell proliferation are not synonymous. Large neurons and endocycling polyploid cells provide good examples of when growth occurs without cell proliferation. Conversely, cell proliferation without growth is accompanied by a reduction in cell size, as occurs during early embryonic development in many species.

Forward genetic screens in *Drosophila* have revealed that the insulin receptor pathway (**Fig. 2**) is a key regulator of cellular and organismal growth ***(93–96)***. Loss-of-function mutations of the insulin receptor (Inr) ***(97)*** or the Inr receptor substrate chico ***(98)*** or loss of the downstream effectors PI3K ***(99)***, dAKT/dPKB ***(100)***, dS6K ***(101)***, dTOR ***(102)***, and Rheb ***(103–106)*** all result in a small cell phenotype owing to a reduction in growth without a concomitant decrease in proliferation. Loss of the negative regulator of phosphatidylinositol-3-like kinase (PI3K) signaling, phosphatase and tensin homolog (PTEN) ***(107–110)***, or loss of the negative regulators of dTOR, tuberous sclerosis 1 and 2 (Tsc1 and Tsc2/gigas) ***(111–113)***, leads to an increase in cellular growth without an apparent increase in proliferation, causing cells to become larger. Although the Inr pathway appears not to regulate proliferation directly, disruption of the pathways does cause changes in cell cycle phasing. Loss-of-function mutants of *PTEN*, *Tsc1*, and *gigas* (*Tsc2*), as well as overexpression of PI3K, cause a decrease in the length of G_1, but the rate of proliferation remains unchanged because longer S- and G_2-phases compensate for the shorter G_1. An increased level of cyclin E is observed in *Tsc1* or *Tsc2* mutant cells, and this is probably responsible for the shortened G_1-phase. Conversely, significant decreases in cyclin E levels and lengthening of G_1-phase occur in *dTOR* mutant cells, which compensate through a decrease in S and G_2 length ***(102,113)***.

Studies of the *Drosophila* homologs of the Myc and Ras oncoproteins have yielded similar results. Mitotic clonal analysis in wing imaginal discs has shown that *dmyc* mutant cells are smaller than wild type, whereas overexpression of *dmyc* leads to larger cell and clone size ***(114)***. Furthermore, *dmyc* overexpression shortens the length of G_1 without increasing proliferation because the cells compensate with longer S- and G_2-phases. This change in cell cycle phasing is associated with a posttranscriptional increase in cyclin E levels ***(115)***. Thus, as with the Inr pathway, it appears that growth and cell cycle progression can be regulated independently such that growth can occur without a concomitant increase in cellular proliferation. Expression of an activated form of Ras, Ras^{V12}, in wing discs results in an increase in cell size and growth, a shortened G_1-phase, and hyperplasia ***(116–118)***. Ras^{V12} induces an increase in dmyc protein levels and activates PI3K signaling, showing that Ras is able to affect growth by signaling through both these pathways ***(115)***.

Drosophila cyclin D and its associated kinase cdk4 are dispensable for cell cycle progression, but they stimulate growth ***(119,120)***. *cdk4* mutant flies are viable, but they are smaller than wild type and sterile. Overexpression of cyclin D/cdk4 in wing disc cells increases the rate of proliferation without affecting cell cycle phasing or cell size, whereas overexpression in postmitotic cells of the eye or endocycling salivary glands causes hypertrophy. Thus, unlike activation of the PI3K pathway, cyclin D/cdk4 stimulates growth with no changes in cell cycle phasing.

4.3. Growth and Cell Cycle Progression Can Be Uncoupled

Studies performed in the *Drosophila* wing disc demonstrate that cellular proliferation can be uncoupled from growth. Developmental patterning divides the wing imaginal disc into anterior and posterior compartments. When cyclin E or *stgcdc25* is over-expressed specifically in the posterior compartment, the durations of G_1- and G_2-

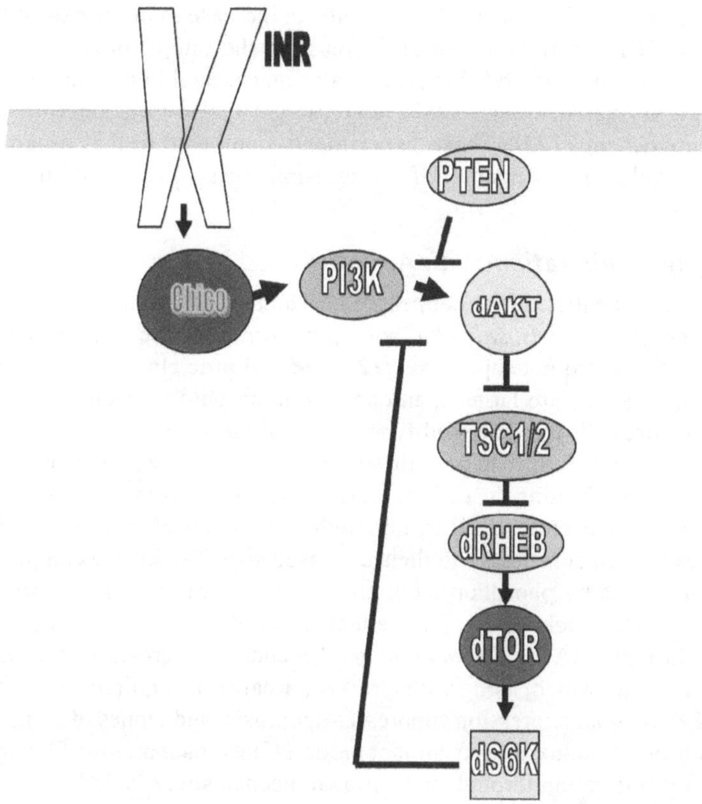

Fig. 2. Proposed model of Inr signaling pathway. The binding of growth factors to the insulin receptor (INR) leads to activation of the phosphatidylinositol-3 kinase (PI3K) by Chico. The negative regulator PTEN can inhibit the pathway by blocking the activity of PI3K. PI3K is required for the activation of dAKT, which negatively regulates Tsc1/2 by phosphorylation, allowing for activation of a small GTPase Rheb. Rheb activity leads to the stimulation of dTOR kinase and the subsequent activation of dS6K, which phosphorylates growth-related targets. S6K regulates PI3K through a negative feedback loop, although the mechanism for this is not well understood.

phase is reduced, respectively ***(121)***. However, in both situations the rate of cell division and the total number of cells in the posterior compartment remain unchanged because of cell cycle compensation. Cells that overexpress cyclin E compensate by extending S-phase, whereas cells that overexpress stg^{cdc25} compensate by extending G_1-phase. The mechanism by which this occurs is not known. In contrast, overexpression of E2F1/DP in the posterior compartment induces the transcription of both *cyclin E* and stg^{cdc25}, which shortens both G_1- and G_2-phases and causes an approximate doubling of the normal cell number. Interestingly, the overall size of the posterior compartment remains unchanged, although cells are much smaller than those in the anterior compartment. Thus, despite an increase in the rate of cell division in response to E2F1/DP, there is not an increase in the rate of growth. Conversely, wing imaginal discs in which *cdc2* is specifically inactivated in the anterior compartment grow to a normal size but contain fewer, larger cells in the anterior versus the posterior compartment ***(122)***. These experiments demonstrate that organ size and growth are controlled independently of cell division during *Drosophila* imaginal disc development.

4.4. Balancing Proliferation With Apoptosis

Several recently identified genes appear to coordinate cell proliferation, growth, and apoptosis. Three genes, *warts*, *salvador*, and *hippo* form a kinase complex that appears to restrict growth and promote apoptosis ***(123–128)***. Mitotic clones of cells mutant for any of these three genes are larger than corresponding wild-type clones. The mutant clones contain more cells than the wild-type clones, although the cells are the same size as wild type. Therefore, perturbation of these genes increases the rate of growth and cell proliferation without affecting the cell cycle. How this kinase complex regulates growth and proliferation is currently unclear, although the mutant clones have a decreased apoptotic index, which contributes to their increased size. The kinase complex appears to stimulate apoptosis by phosphorylation and inhibition of DIAP1, a *Drosophila* inhibitor of apoptosis, which allows for the activation of proapoptotic caspases. Other pathways act through DIAP1. The *bantam* gene encodes a micro-RNA that suppresses the expression of Hid, which, along with Grim and Reaper, is required for the inhibition of DIAP1 ***(129)***. *bantam* expression suppresses apoptosis, and clones of *bantam* mutant cells grow poorly. In addition to its antiapoptotic effects, bantam also functions to increase rates of proliferation through an unknown mechanism ***(130,131)***.

4.5. Growth-Regulated and Stereotyped Cell Cycle Control in the Eye Imaginal Disc

The *Drosophila* eye imaginal disc provides another excellent system to study cell cycle control during organ development ***(132)***. During the first two larval instars the eye discs grow via cell proliferation as described above in **Subheading 4.2.** for wing discs. In the third larval instar, a wave of differentiation sweeps across the disc in a posterior-to-anterior fashion. This wave is accompanied by an apical cell constriction that causes a visible indentation in the epithelial cell layer called the morphogenetic furrow (MF). Undifferentiated cells anterior to the MF proliferate asynchronously, whereas cells posterior to the furrow differentiate. Cells within the MF arrest in G_1. During this G_1 arrest, groups of five photoreceptor cells (called *preclusters*) permanently exit the cell cycle and begin to differentiate. The cells that are not included in these preclusters enter a synchronous, final round of cell division, referred to as the second mitotic wave, just posterior to the furrow. The resulting pool of cells differentiates into a variety of additional cell types including bristle cells, cone cells, and pigment cells ***(133)***.

The initiation and progression of the MF are regulated positively by Hedgehog (Hh) and Decapentaplegic (Dpp) signaling, and negatively by Wingless (Wg) signaling ***(134,135)***. Hh is secreted from the differentiating neurons posterior to the MF. Mutation of *smoothened*, which encodes the Hh receptor, blocks entry into the second mitotic wave, whereas overexpression of the transcriptional effector of Hh signaling, Cubitius interruptus (Ci), in the G_1 cells of the MF causes precocious entry into S-phase ***(136)***. Hh induces transcription of *cyclin E* in the MF, and chromatin immunoprecipitation assays indicate that Ci binds directly to the *cyclin E* promoter. This strongly suggests that Hh is responsible for the induction of S-phase in the second mitotic wave by the induction of *cyclin E* ***(136)***. This provides an explanation for why E2F-directed transcription is not required for *cyclin E* expression in the MF ***(137)***, unlike in the embryonic endocycles as described above in **Subheading 3.5.** Moreover, expression of a mutant RBF1 that is not inhibited by cdk phosphorylation and that constitutively represses E2F does not prevent entry into S-phase of the second mitotic wave ***(77)***. These data are consistent with the observation that *cyclin E*, like *stg*^{cdc25} and *dap*, contains a complex upstream control region containing modular cell type-specific enhancers that control its response to different developmental signals ***(138)***. Epidermal growth factor receptor (EGFR) signaling also controls cell cycle progression in the eye disc. Cells in EGFR mutant clones arrest in G_2 of the second mitotic wave, whereas overexpression of the EGFR drives cells into mitosis ***(139)***.

Several genes participate in the G_1 arrest of the MF. Mutation of *roughex* (*rux*) causes precocious entry into S-phase in the MF ***(140)***. *rux* encodes a novel protein that binds to and inhibits cyclin A/cdk complexes ***(141,142)***. In the MF Rux prevents inappropriate cyclin A-dependent kinase activity, which can drive cells into S-phase ***(143)***. Interestingly, the MF is the only situation thus far described in which *rux* is required to maintain G_1 arrest, although it can function elsewhere (e.g., the embryo ***[143]***). Evidence also suggests that Dpp signaling is required for G_1 arrest. Clones of cells in the MF that are mutant for the Dpp receptor *thickveins* continue to cycle asynchronously as if they were anterior to the furrow, whereas overexpression of Dpp transiently inhibits S-phase ***(144)***. Dpp-induced arrest appears to occur independently of both Rux and Dap (i.e., p21/p27 ***[144]***). Thus, multiple mechanisms are used to establish G_1 arrest in the MF.

5. Oogenesis

Oogenesis offers a unique opportunity to determine the mechanisms by which the canonical cell cycle is modified to achieve a specific developmental outcome ***(12,13)***. Each ovary consists of approx 16 ovarioles, which are essentially egg assembly lines that contain all developmental stages of oogenesis (**Fig. 3**). At the most anterior region of each ovariole is the germarium, a specialized structure that contains both germ line (GSCs) and somatic stem cells (SSCs). Two to three GSCs are located at the anterior tip of the germarium (region 1), where they are in close contact with specialized, postmitotic somatic cells called the terminal filament cells and the cap cells ***(145)***. GSCs divide asymmetrically to produce a daughter stem cell and a cystoblast ***(146)***. The cystoblast undergoes four synchronous mitotic divisions with incomplete cytokinesis to produce a 16-cell cyst. The cytoplasm of these 16 cystocyte cells remains interconnected by an actin-rich structure known as the ring canal. Before exiting the

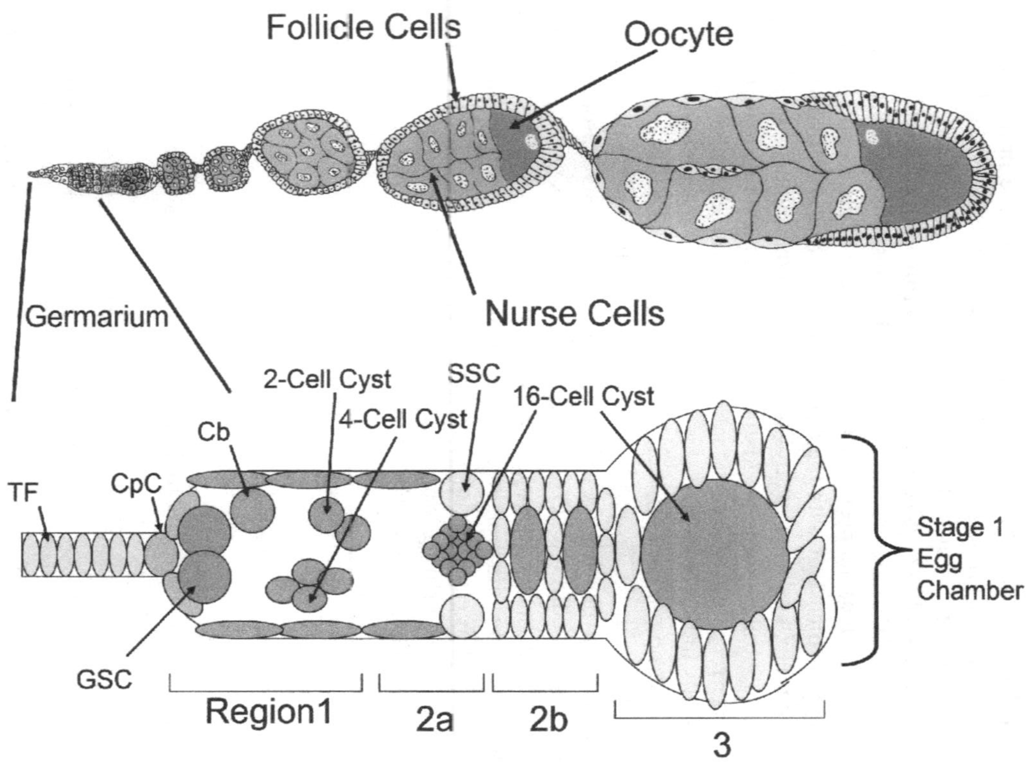

Fig. 3. *Drosophila* oogenesis. Schematic drawing of an ovariole is shown at the top. Each ovariole represents an egg chamber assembly line, with egg chambers of increasing maturity located from anterior to posterior (left to right as shown) The germarium, located at the anterior tip of the ovariole, is diagrammed at the bottom. TF, terminal filament cells; CpC, cap cells; GSC, germline stem cell; SSC, somatic stem cell; Cb, cystoblast; SSC, somatic stem cell.

germarium, the cyst is encapsulated by the surrounding follicle cells to create an egg chamber (region 3). Follicle cells are produced by two to three somatic stem cells that are located in the region 2a–2b of the germarium ***(147)***. Subsequent stages of egg chamber development are marked by growth and patterning of the oocyte.

5.1. Control of Germline Stem Cell Proliferation

The proliferation of the GSCs is controlled by genes that act within the germline and by intercellular signaling between the soma and the germ line. *piwi*, which encodes an evolutionarily conserved protein of the Argonaute family ***(148)***, functions in both processes. *piwi* is expressed in the GSCs and is required for their self-renewal ***(146,148–152)***. *piwi* is also expressed in the cap and terminal filament cells, which are required for GSC maintenance. Signals from these cells creates a microenvironment, or "niche," that is necessary for the asymmetric division of GSCs, as well as for the coordinated division of the SSCs ***(145,153,154)***.

fs(1)Yb (Yb) is a 117-kDa protein that has limited homology to the p68 family of RNA helicases ***(152)***. *Yb* is expressed in the terminal filament cells and the cap cells and is thought to control the expression of *piwi* in these cell types ***(155)***. In both *Yb* and *piwi* mutants, GSCs cannot maintain their fate, resulting in ovarioles that contain only two three egg chambers and a small number of ill-fated germ cells ***(146,151,152,155)***. *dpp* encodes a transforming growth factor-β-type morphogen that is expressed in the cap cells as well as the inner sheath cells ***(153)***. Mutation of *dpp* slows GSC cell division and accelerates stem cell loss ***(156)***.

To ensure the proper development of an egg chamber, the GSC divisions must be coordinated with the divisions of the SSCs. Many conserved developmental signaling pathways are involved in the control of the SSCs, including both the *hh* and *wg* pathways. *hh* and *wg* are expressed in the terminal filament cells and the cap cells ***(155,157)***. In *hh* mutants, fewer somatic cells are present, and encapsulation of the germline cysts fails to occur ***(157)***. The expression of *hh* is mediated by *Yb*, and in *Yb* mutants somatic defects include germarium–egg chamber fusion, improper partitioning of germline cysts by follicle cells, and missing interfollicular stalk cells, which normally act to separate individual egg chambers ***(158)***. Mutation of *disheveled* (*dsh*) and *armadillo* (*arm*), transducers of the *wg* signal, result in the loss of SSCs, whereas mutation of wg pathway antagonists *shaggy* (*shg*) and *Axin* results in overproliferation of the follicle cells ***(159)***.

5.2. Regulation of Cystoblast Cell Cycles

One of the two GSC daughter cells becomes a cystoblast. (The other remains a GSC.) This cell is the progenitor of the 15 nurse cells and single oocyte of the egg chamber. The *bag of marbles* (*bam*) and the *arrest* (*aret*) genes are required for cystoblast differentiation. In *bam* mutant germ cells, the cystoblast fate is not adopted, and the ovary fills up with an excess of mitotically active stem cells ***(160–162)***. The *aret* gene encodes a translational repressor and is required for *bam* function such that in *aret* mutants the cystoblast does not develop even though BAM is properly expressed ***(163,164)***.

The switch from asymmetric division of the GSCs to the mitotic divisions of the cystoblast also requires *bam* function. Threshold levels of this gene are necessary to preserve the four mitotic divisions of the cystoblast. *Encore* (*enc*) is required to maintain this threshold level of *bam*. In *enc* mutants, the expression domain of *bam* is expanded, resulting in one extra round of mitosis ***(165)***. The *aret* gene is also required for the controlled divisions of the cystoblast. In aret mutants, cystocytes undergo more than four rounds of mitosis, and these mitoses are marked by complete cytokinesis ***(163)***. These *aret* mutant phenotypes are reminiscent of those phenotypes seen in *ovarian tumor* (*otu*) mutants, which produce tumorous egg chambers ***(166)***.

Following completion of the mitotic divisions, all 16 cystocyte cells enter a long S-phase that represents the premeiotic S-phase of the future oocyte ***(167)***. Upon determination, the oocyte will arrest in prophase of meiosis I; the remaining 15 cells will adopt a nurse cell fate and begin to endocycle. Oocyte cell cycle arrest requires the cyclin E/Cdk2 inhibitor encoded by the *dacapo* gene (Dap), which is expressed at high levels in this cell. Without Dap, the oocyte enters the endocycle and becomes polyploid ***(168)***. Thus, the inhibition of cyclin E/Cdk2 kinase is necessary to maintain oocyte fate.

5.3. Nurse Cell Endocycles

To support growth of the oocyte, nurse cells become highly polyploid (1024C). Initially these endocycles are marked by complete replication of the entire genome (both euchromatin and heterochromatin), and the sister chromatids are held in close association to generate polytene chromosomes. When nurse cell ploidy reaches 32C, the chromosomes undergo dramatic structural rearrangements that ultimately result in dispersal of the chromatin. The nurse cells are now considered polyploid and at this point undergo a change in endo S-control such that only euchromatin and not heterochromatin is replicated ***(7,169)***. Near the end of oogenesis, these polyploid nurse cells transmit their cytoplasmic contents to the oocyte and then commit apoptosis. These two processes are intimately coupled via an incompletely understood process that requires regulation of f-actin bundling, caspases, and the E2F1/DP transcription factor ***(170–174)***.

As with endocycles at other stages of development ***(48,73,87)***, the nurse cell endocycles are controlled by cyclin E/cdk2 ***(81)***. Cyclin E protein accumulation oscillates in the nurse cell endocycle and is highest during S-phase. Dap levels also oscillate in the nurse cells, and *dap* expression requires cyclin E function ***(175)***. This is probably a major aspect of controlling oscillation of cyclin E/cdk2 kinase activity during the endocycle. The induction of cyclin E triggers origin firing as well as the expression of its own inhibitor. This in turn depletes cyclin E/cdk2 kinase activity, allowing reassembly of pre-RC complexes and re-replication in the next endo S-phase ***(7)***. The underreplication of heterochromatin also depends on downregulation of cyclin E expression prior to the end of endo S-phase ***(81)***.

A trademark of the endocycle is the absence of mitosis. Neither of the mitotic cyclins A or B is present in endocycling nurse cells ***(12)***, and their absence depends on the action of the APC E3 ubiquitin ligase. *morula* encodes the cullin-like APC2 subunit, and hypomorphic alleles of *morula* cause female sterility ***(176,177)***. *morula* mutant nurse cells inappropriately accumulate cyclin B and display aspects of mitosis, including chromatin condensation and polymerization of microtubules ***(176,177)***. This indicates that the continuous destruction of mitotic cyclins is necessary to suppress mitotic functions and promote the nurse cell endocycle.

5.4 Cell Cycle Control in Somatic Follicle Cells

The somatic follicle cells that surround the germ line cyst undergo a number of essential cell cycle changes during oogenesis ***(12,178)***. After encapsulation of the 16-cell cyst, five rounds of follicle cell mitosis occurs ***(179)***. The follicle cells subsequently begin to endocycle. The transition from mitotic cycle to endocycle requires Notch signaling between the germline and soma. At the time of the switch, the Notch ligand Delta is upregulated in the germ line and the Notch receptor is concentrated in the follicle cells ***(180)***. Both mutations of *Delta* in the germline and mutations of *Notch* in the follicle cells result in a failure to terminate follicle cell mitotic divisions and prevent the transition to polyploidy ***(180,181)***. As in the early embryo, the mechanism by which Notch signaling terminates mitosis is via transcriptional downregulation of stg^{cdc25} ***(181)***.

Follicle cells undergo three endocycles and reach a ploidy of 16C, at which time genomic DNA replication ceases. Subsequently, several loci, including those on chromosomes 1 and 3 that encode chorion proteins, begin gene amplification ***(13,178,182)***. Polyploidy and chorion gene amplification provide the follicle cells with the biosynthetic capacity needed to generate the eggshell. Gene amplification is a developmentally controlled process that requires the specific, repeated firing of the chorion origins, ACE3 and ACE1, at the same time that virtually all other genomic origins of replication are inactivated ***(13,183,184)***. Many of the players that participate in and that regulate replication during the mitotic cycle play an important role in gene amplification. Female sterile mutations in genes encoding pre-RC components or replication factors such as Mcm6, $chiffon^{DBF4}$, and Orc2 prevent chorion gene amplification and cause the production of eggs with thin shells ***(178,185–188)***. Cyclin E/Cdk2 is also necessary for chorion gene amplification ***(178)***.

Members of the E2F/DP family of transcription factors play a critical role in chorion gene amplification. Transcriptional repression of pre-RC components by the E2F2/DP/RBF1 and E2F1/DP/RBF1 complexes contributes to the suppression of genomic replication ***(75,189)***, and overexpression of the E2F-target Orc1 in follicle cells results in genomic replication at the expense of gene amplification ***(190)***. In addition, RBF/E2F complexes play a direct role in controlling chorion gene amplification that is independent of transcription. Chromatin immunoprecipitation analyses demonstrate that RBF1, E2F1, and DP proteins bind in close proximity to chorion origins ***(191)***. Mutation of *E2f1* or *Dp* reduces chorion gene amplification and disrupts the localization of origin recognition complex (ORC) components to sites of gene amplification ***(192)***. Conversely, mutation of Rbf1 causes overamplification ***(191)***. Moreover, Rbf1 and E2f1 coimmunoprecipitate with Orc2 from ovary extracts. The precise mechanism by which RBF/E2F complexes control gene amplification is not known, although they may affect local chromatin structure ***(193)***.

6. Conclusions

Drosophila uses a variety of modes of control to coordinate cell cycle progression with growth and morphogenesis during development. One of the interesting features of these different modes of control is that they all rely on a common set of regulatory molecules that are highly conserved throughout evolution. The powerful genetic and cell biological methods available to *Drosophila* promises future insight into how developmental programs manipulate such regulators to achieve precise control of cell cycle progression.

References

1. Rubin, G. M. and Lewis, E. B. (2000) A brief history of *Drosophila*'s contributions to genome research. *Science* **287,** 2216–2218.
2. Adams, M. D., Celniker, S. E., Holt, R. A., et al. (2000) The genome sequence of *Drosophila melanogaster*. *Science* **287,** 2185–2195.
3. Adams, M. D. and Sekelsky, J. J. (2002) From sequence to phenotype: reverse genetics in *Drosophila melanogaster*. *Nat. Rev. Genet.* **3,** 189–198.
4. St. Johnston, D. (2002) The art and design of genetic screens: *Drosophila melanogaster*. *Nat. Rev. Genet.* **3,** 176–188.
5. Edgar, B. A. and Lehner, C. F. (1996) Developmental control of cell cycle regulators: a fly's perspective. *Science* **274,** 1646–1652.
6. Follette, P. J. and OFarrell, P. H. (1997) Cdks and the *Drosophila* cell cycle. *Curr. Opin. Genet. Dev.* **7,** 17–22.
7. Edgar, B. A. and Orr-Weaver, T. L. (2001) Endoreplication cell cycles: more for less. *Cell* **105,** 297–306.
8. Vidwans, S. J. and Su, T. T. (2001) Cycling through development in *Drosophila* and other metazoa. *Nat. Cell Biol.* **3,** E35–E39.
9. Foe, V. E., Odell, G. M., and Edgar, B. A. (1993) Mitosis and morphogenesis in the *Drosophila* embryo: Point and counterpoint. In *The Development of Drosophila melanogaster*, Vol. 1 (Bate, M. and Martinez Arias, A., eds.) Cold Spring Harbor Laboratory Press, Cold Spring Harbor, NY, pp. 149–300.
10. Cohen, S. M. (1993) Imaginal disc development. In *The Development of Drosophila melanogaster*, Vol. 2, (Bate, M. and Martinez Arias, A., eds.) Cold Spring Harbor Laboratory Press, Cold Spring Harbor, NY, pp. 747–842.
11. Serrano, N. and OFarrell, P. H. (1997) Limb morphogenesis: connections between patterning and growth. *Curr. Biol.* **7,** R186–195.
12. Royzman, I. and Orr-Weaver, T. L. (1998) S phase and differential DNA replication during *Drosophila* oogenesis. *Genes Cells* **3,** 767–776.
13. Calvi, B. R. and Spradling, A. C. (1999) Chorion gene amplification in *Drosophila*: A model for metazoan origins of DNA replication and S-phase control. *Methods* **18,** 407–417.
14. Wieschaus, E. and Sweeton, D. (1988) Requirements for X-linked zygotic gene activity during cellularization of early *Drosophila* embryos. *Development* **104,** 483–493.
15. Merrill, P. T., Sweeton, D., and Wieschaus, E. (1988) Requirements for autosomal gene activity during precellular stages of *Drosophila melanogaster*. *Development* **104,** 495–509.
16. Edgar, B. A., Sprenger, F., Duronio, R. J., Leopold, P., and OFarrell, P. H. (1994) Distinct molecular mechanisms regulate cell cycle timing at successive stages of *Drosophila* embryogenesis. *Genes Dev.* **8,** 440–452.

17. Huang, J. and Raff, J. W. (1999) The disappearance of cyclin B at the end of mitosis is regulated spatially in *Drosophila* cells. *EMBO J.* **18,** 2184–2195.
18. Su, T. T., Sprenger, F., DiGregorio, P. J., Campbell, S. D., and OFarrell, P. H. (1998) Exit from mitosis in *Drosophila* syncytial embryos requires proteolysis and cyclin degradation, and is associated with localized dephosphorylation. *Genes Dev.* **12,** 1495–1503.
19. Huang, J. Y. and Raff, J. W. (2002) The dynamic localisation of the *Drosophila* APC/C: evidence for the existence of multiple complexes that perform distinct functions and are differentially localised. *J. Cell Sci.* **115,** 2847–2856.
20. Raff, J. W., Jeffers, K., and Huang, J. Y. (2002) The roles of Fzy/Cdc20 and Fzr/Cdh1 in regulating the destruction of cyclin B in space and time. *J. Cell Biol.* **157,** 1139–1149.
21. Shamanski, F. L. and Orr-Weaver, T. L. (1991) The *Drosophila* plutonium and pan gu genes regulate entry into S phase at fertilization. *Cell* **66,** 1289–1300.
22. Freeman, M., Nusslein-Volhard, C., and Glover, D. M. (1986) The dissociation of nuclear and centrosomal division in gnu, a mutation causing giant nuclei in *Drosophila*. *Cell* **46,** 457–468.
23. Freeman, M. and Glover, D. M. (1987) The gnu mutation of *Drosophila* causes inappropriate DNA synthesis in unfertilized and fertilized eggs. *Genes Dev.* **1,** 924–930.
24. Lee, L. A., Elfring, L. K., Bosco, G., and Orr-Weaver, T. L. (2001) A genetic screen for suppressors and enhancers of the *Drosophila* PAN GU cell cycle kinase identifies cyclin B as a target. *Genetics* **158,** 1545–1556.
25. Axton, J. M., Shamanski, F. L., Young, L. M., Henderson, D. S., Boyd, J. B., and Orr-Weaver, T. L. (1994) The inhibitor of DNA replication encoded by the *Drosophila* gene plutonium is a small, ankyrin repeat protein. *EMBO J.* **13,** 462–470.
26. Fenger, D. D., Carminati, J. L., Burney-Sigman, D. L., et al. (2000) PAN GU: a protein kinase that inhibits S phase and promotes mitosis in early *Drosophila* development. *Development* **127,** 4763–4774.
27. Renault, A. D., Zhang, X. H., Alphey, L. S., et al. (2003) giant nuclei is essential in the cell cycle transition from meiosis to mitosis. *Development* **130,** 2997–3005.
28. Fogarty, P., Campbell, S. D., Abu-Shumays, R., et al. (1997) The *Drosophila* grapes gene is related to checkpoint gene chk1/rad27 and is required for late syncytial division fidelity. *Curr. Biol.* **7,** 418–426.
29. Sibon, O. C., Stevenson, V. A., and Theurkauf, W. E. (1997) DNA-replication checkpoint control at the *Drosophila* midblastula transition. *Nature* **388,** 93–97.
30. Sibon, O. C., Laurencon, A., Hawley, R., and Theurkauf, W. E. (1999) The *Drosophila* ATM homologue Mei-41 has an essential checkpoint function at the midblastula transition. *Curr. Biol.* **9,** 302–312.
31. Su, J. Y., Rempel, R. E., Erikson, E., and Maller, J. L. (1995) Cloning and characterization of the Xenopus cyclin-dependent kinase inhibitor p27XIC1. *Proc. Natl. Acad. Sci. USA* **92,** 10187–10191.
32. Yu, K. R., Saint, R. B., and Sullivan, W. (2000) The Grapes checkpoint coordinates nuclear envelope breakdown and chromosome condensation. *Nat. Cell Biol.* **2,** 609–615.
33. Price, D., Rabinovitch, S., OFarrell, P. H., and Campbell, S. D. (2000) *Drosophila* wee1 has an essential role in the nuclear divisions of early embryogenesis. *Genetics* **155,** 159–166.
34. Edgar, B. A. and Datar, S. A. (1996) Zygotic degradation of two maternal Cdc25 mRNAs terminates *Drosophila's* early cell cycle program. *Genes Dev.* **10,** 1966–1977.
35. Courtot, C., Fankhauser, C., Simanis, V., and Lehner, C. F. (1992) The *Drosophila* cdc25 homolog twine is required for meiosis. *Development* **116,** 405–416.

36. Alphey, L., Jimenez, J., White, C. H., Dawson, I., Nurse, P., and Glover, D. M. (1992) *twine*, a cdc25 homolog that functions in the male and female germline of *Drosophila*. *Cell* **69,** 977–988.
37. Edgar, B. A., Kiehle, C. P., and Schubiger, G. (1986) Cell cycle control by the nucleo-cytoplasmic ratio in early *Drosophila* development. *Cell* **44,** 365–372.
38. Grosshans, J., Muller, H. A., and Wieschaus, E. (2003) Control of cleavage cycles in *Drosophila* embryos by fruhstart. *Dev. Cell* **5,** 285–294.
39. Grosshans, J. and Wieschaus, E. (2000) A genetic link between morphogenesis and cell division during formation of the ventral furrow in *Drosophila*. *Cell* **101,** 523–531.
40. Foe, V. E. and Odell, G. M. (1989) Mitotic domains partition fly embryos, reflecting early cell biological consequences of determination in progress. *Am. Zool.* **29,** 617–652.
41. Edgar, B. A. and OFarrell, P. H. (1989) Genetic control of cell division patterns in the *Drosophila* embryo. *Cell* **57,** 177–187.
42. Edgar, B. A. and OFarrell, P. H. (1990) The three postblastoderm cell cycles of *Drosophila* embryogenesis are regulated in G2 by string. *Cell* **62,** 469–480.
43. Edgar, B. A., Lehman, D. A., and OFarrell, P. H. (1994) Transcriptional regulation of string (cdc25): a link between developmental programming and the cell cycle. *Development* **120,** 3131–3143.
44. Lehman, D. A., Patterson, B., Johnston, L. A., et al. (1999) Cis–regulatory elements of the mitotic regulator, string/Cdc25. *Development* **126,** 1793–1803.
45. Knoblich, J. A. and Lehner, C. F. (1993) Synergistic action of *Drosophila* cyclins A and B during the G2–M transition. *EMBO J.* **12,** 65–74.
46. Jacobs, H. W., Knoblich, J. A., and Lehner, C. F. (1998) *Drosophila* cyclin B3 is required for female fertility and is dispensable for mitosis like cyclin B. *Genes Dev.* **12,** 3741–3751.
47. Knoblich, J. A., Sauer, K., Jones, L., Richardson, H., Saint, R., and Lehner, C. F. (1994) Cyclin E controls S phase progression and its down-regulation during *Drosophila* embryogenesis is required for the arrest of cell proliferation. *Cell* **77,** 107–120.
48. Sauer, K., Knoblich, J. A., Richardson, H., and Lehner, C. F. (1995) Distinct modes of cyclin E/cdc2c kinase regulation and S-phase control in mitotic and endoreduplication cycles of *Drosophila* embryogenesis. *Genes Dev.* **9,** 1327–1339.
49. Lehner, C. F. (1992) The pebble gene is required for cytokinesis in *Drosophila*. *J. Cell Sci.* **103,** 1021–1030.
50. Hime, G. and Saint, R. (1992) Zygotic expression of the pebble locus is required for cytokinesis during postblastoderm mitoses of *Drosophila*. *Development* **114,** 165–171.
51. Echard, A. and OFarrell, P. H. (2003) The degradation of two mitotic cyclins contributes to the timing of cytokinesis. *Curr. Biol.* **13,** 373–383.
52. Sigrist, S., Jacobs, H., Stratmann, R., and Lehner, C. F. (1995) Exit from mitosis is regulated by *Drosophila* fizzy and the sequential destruction of cyclins A, B and B3. *EMBO J.* **14,** 4827–4838.
53. D'Andrea, R. J., Stratmann, R., Lehner, C. F., John, U. P., and Saint, R. (1993) The three rows gene of *Drosophila melanogaster* encodes a novel protein that is required for chromosome disjunction during mitosis. *Mol. Biol. Cell* **4,** 1161–1174.
54. Leismann, O., Herzig, A., Heidmann, S., and Lehner, C. F. (2000) Degradation of *Drosophila* PIM regulates sister chromatid separation during mitosis. *Genes Dev.* **14,** 2192–2205.
55. Stratmann, R. and Lehner, C. F. (1996) Separation of sister chromatids in mitosis requires the *Drosophila* pimples product, a protein degraded after the metaphase/anaphase transition. *Cell* **84,** 25–35.

56. Vidwans, S. J., DiGregorio, P. J., Shermoen, A. W., et al. (2002) Sister chromatids fail to separate during an induced endoreplication cycle in *Drosophila* embryos. *Curr. Biol.* **12,** 829–833.
57. Parry, D. H., Hickson, G. R., and OFarrell, P. H. (2003) Cyclin B destruction triggers changes in kinetochore behavior essential for successful anaphase. *Curr. Biol.* **13,** 647–653.
58. Vidwans, S. J., Wong, M. L., and OFarrell, P. H. (1999) Mitotic regulators govern progress through steps in the centrosome duplication cycle. *J. Cell Biol.* **147,** 1371–1378.
59. Duronio, R. J. and OFarrell, P. H. (1994) Developmental control of a G1-S transcriptional program in *Drosophila. Development* **120,** 1503–1515.
60. Richardson, H. E., OKeefe, L. V., Reed, S. I., and Saint, R. (1993) A *Drosophila* G1-specific cyclin E homolog exhibits different modes of expression during embryogenesis. *Development* **119,** 673–690.
61. de Nooij, J. C., Letendre, M. A., and Hariharan, I. K. (1996) A cyclin-dependent kinase inhibitor, Dacapo, is necessary for timely exit from the cell cycle during *Drosophila* embryogenesis. *Cell* **87,** 1237–1247.
62. Lane, M. E., Sauer, K., Wallace, K., Jan, Y. N., Lehner, C. F., and Vaessin, H. (1996) Dacapo, a cyclin-dependent kinase inhibitor, stops cell proliferation during *Drosophila* development. *Cell* **87,** 1225–1235.
63. Liu, T. H., Li, L., and Vaessin, H. (2002) Transcription of the *Drosophila* CKI gene dacapo is regulated by a modular array of cis-regulatory sequences. *Mech. Dev.* **112,** 25–36.
64. Meyer, C. A., Kramer, I., Dittrich, R., Marzodko, S., Emmerich, J., and Lehner, C. F. (2002) *Drosophila* p27Dacapo expression during embryogenesis is controlled by a complex regulatory region independent of cell cycle progression. *Development* **129,** 319–328.
65. Dyson, N. (1998) The regulation of E2F by pRB-family proteins. *Genes Dev.* **12,** 2245–2262.
66. DeGregori, J. (2002) The genetics of the E2F family of transcription factors: shared functions and unique roles. *Biochem. Biophys. Acta* **1602,** 131–150.
67. Dynlacht, B. D., Brook, A., Dembski, M., Yenush, L., and Dyson, N. (1994) DNA-binding and trans-activation properties of *Drosophila* E2F and DP proteins. *Proc. Natl. Acad. Sci. USA* **91,** 6359–6363.
68. Ohtani, K. and Nevins, J. R. (1994) Functional properties of a *Drosophila* homolog of the E2F1 gene. *Mol. Cell. Biol.* **14,** 1603–1612.
69. Sawado, T., Yamaguchi, M., Nishimoto, Y., Ohno, K., Sakaguchi, K., and Matsukage, A. (1998) dE2F2, a novel E2F-family transcription factor in *Drosophila melanogaster. Biochem. Biophys. Res. Commun.* **251,** 409–415.
70. Du, W., Vidal, M., Xie, J. E., and Dyson, N. (1996) RBF, a novel RB-related gene that regulates E2F activity and interacts with cyclin E in *Drosophila. Genes Dev.* **10,** 1206–1218.
71. Stevaux, O., Dimova, D., Frolov, M. V., Taylor-Harding, B., Morris, E., and Dyson, N. (2002) Distinct mechanisms of E2F regulation by *Drosophila* RBF1 and RBF2. *EMBO J.* **21,** 4927–4937.
72. Duronio, R. J., OFarrell, P. H., Xie, J. E., Brook, A., and Dyson, N. (1995) The transcription factor E2F is required for S phase during *Drosophila* embryogenesis. *Genes Dev.* **9,** 1445–1455.
73. Duronio, R. J. and OFarrell, P. H. (1995) Developmental control of the G1 to S transition in *Drosophila*: cyclin E is a limiting downstream target of E2F. *Genes Dev.* **9,** 1456–1468.
74. Royzman, I., Whittaker, A. J., and Orr-Weaver, T. L. (1997) Mutations in *Drosophila* DP and E2F distinguish G1-S progression from an associated transcriptional program. *Genes Dev.* **11,** 1999–2011.

75. Cayirlioglu, P., Ward, W. O., Silver Key, S. C., and Duronio, R. J. (2003) Transcriptional repressor functions of *Drosophila* E2F1 and E2F2 cooperate to inhibit genomic DNA synthesis in ovarian follicle cells. *Mol. Cell. Biol.* **23,** 2123–2134.
76. Du, W. and Dyson, N. (1999) The role of RBF in the introduction of G1 regulation during *Drosophila* embryogenesis. *EMBO J.* **18,** 916–925.
77. Xin, S., Weng, L., Xu, J., and Du, W. (2002) The role of RBF in developmentally regulated cell proliferation in the eye disc and in Cyclin D/Cdk4 induced cellular growth. *Development* **129,** 1345–1356.
78. Meyer, C. A., Jacobs, H. W., and Lehner, C. F. (2002) Cyclin D-cdk4 is not a master regulator of cell multiplication in *Drosophila* embryos. *Curr. Biol.* **12,** 661–666.
79. Chen, X., Oh, S. W., Zheng, Z., Chen, H. W., Shin, H. H., and Hou, S. X. (2003) Cyclin D-Cdk4 and cyclin E-Cdk2 regulate the Jak/STAT signal transduction pathway in *Drosophila. Dev. Cell* **4,** 179–190.
80. Smith, A. V. and Orr-Weaver, T. L. (1991) The regulation of the cell cycle during *Drosophila* embryogenesis: the transition to polyteny. *Development* **112,** 997–1008.
81. Lilly, M. A. and Spradling, A. C. (1996) The *Drosophila* endocycle is controlled by cyclin E and lacks a checkpoint ensuring S-phase completion. *Genes Dev.* **10,** 2514–2526.
82. Su, T. T. and OFarrell, P. H. (1998) Chromosome association of minichromosome maintenance proteins in *Drosophila* endoreplication cycles. *J. Cell Biol.* **140,** 451–460.
83. Geng, Y., Yu, Q., Sicinska, E., et al. (2003) Cyclin E ablation in the mouse. *Cell* **114,** 431–443.
84. Duronio, R. J., Bonnette, P. C., and OFarrell, P. H. (1998) Mutations of the *Drosophila* dDP, dE2F, and cyclin E genes reveal distinct roles for the E2F-DP transcription factor and cyclin E during the G1-S transition. *Mol. Cell. Biol.* **18,** 141–151.
85. Follette, P. J., Duronio, R. J., and OFarrell, P. H. (1998) Fluctuations in cyclin E levels are required for multiple rounds of endocycle S phase in *Drosophila. Curr. Biol.* **8,** 235–238.
86. Weiss, A., Herzig, A., Jacobs, H., and Lehner, C. F. (1998) Continuous cyclin E expression inhibits progression through endoreduplication cycles in *Drosophila. Curr. Biol.* **8,** 239–242.
87. Weng, L., Zhu, C., Xu, J., and Du, W. (2003) Critical role of active repression by E2F and Rb proteins in endoreplication during *Drosophila* development. *EMBO J.* **22,** 3865–3875.
88. Lehner, C. F. and OFarrell, P. H. (1990) The roles of *Drosophila* cyclins A and B in mitotic control. *Cell* **61,** 535–547.
89. Sigrist, S. J. and Lehner, C. F. (1997) *Drosophila* fizzy-related down-regulates mitotic cyclins and is required for cell proliferation arrest and entry into endocycles. *Cell* **90,** 671–681.
90. Reimann, J. D., Freed, E., Hsu, J. Y., Kramer, E. R., Peters, J. M., and Jackson, P. K. (2001) Emi1 is a mitotic regulator that interacts with Cdc20 and inhibits the anaphase promoting complex. *Cell* **105,** 645–655.
91. Grosskortenhaus, R. and Sprenger, F. (2002) Rca1 inhibits APC-Cdh1(Fzr) and is required to prevent cyclin degradation in G2. *Dev. Cell* **2,** 29–40.
92. Dong, X., Zavitz, K. H., Thomas, B. J., Lin, M., Campbell, S., and Zipursky, S. L. (1997) Control of G1 in the developing *Drosophila* eye: rca1 regulates Cyclin A. *Genes Dev.* **11,** 94–105.
93. Edgar, B. A. (1999) From small flies come big discoveries about size control. *Nat. Cell Biol.* **1,** E191–E193.
94. Stocker, H., and Hafen, E. (2000) Genetic control of cell size. *Curr. Opin. Genet. Dev.* **10,** 529–535.

95. Johnston, L. A. and Gallant, P. (2002) Control of growth and organ size in *Drosophila. Bioessays* **24,** 54–64.
96. Saucedo, L. J., and Edgar, B. A. (2002) Why size matters: altering cell size. *Curr. Opin. Genet. Dev.* **12,** 565–571.
97. Chen, C., Jack, J., and Garofalo, R. S. (1996) The *Drosophila* insulin receptor is required for normal growth. *Endocrinology* **137,** 846–856.
98. Bohni, R., Riesgo-Escovar, J., Oldham, S., et al. (1999) Autonomous control of cell and organ size by CHICO, a *Drosophila* homolog of vertebrate IRS1-4. *Cell* **97,** 865–875.
99. Weinkove, D., Neufeld, T. P., Twardzik, T., Waterfield, M. D., and Leevers, S. J. (1999) Regulation of imaginal disc cell size, cell number and organ size by *Drosophila* class I(A) phosphoinositide 3-kinase and its adaptor. *Curr. Biol.* **9,** 1019–1029.
100. Verdu, J., Buratovich, M. A., Wilder, E. L., and Birnbaum, M. J. (1999) Cell-autonomous regulation of cell and organ growth in *Drosophila* by Akt/PKB. *Nat. Cell Biol.* **1,** 500–506.
101. Montagne, J., Stewart, M. J., Stocker, H., Hafen, E., Kozma, S. C., and Thomas, G. (1999) *Drosophila* S6 kinase: a regulator of cell size. *Science* **285,** 2126–2129.
102. Zhang, H., Stallock, J. P., Ng, J. C., Reinhard, C., and Neufeld, T. P. (2000) Regulation of cellular growth by the *Drosophila* target of rapamycin dTOR. *Genes Dev.* **14,** 2712–2724.
103. Zhang, Y., Gao, X., Saucedo, L. J., Ru, B., Edgar, B. A., and Pan, D. (2003) Rheb is a direct target of the tuberous sclerosis tumour suppressor proteins. *Nat. Cell Biol.* **5,** 578–581.
104. Saucedo, L. J., Gao, X., Chiarelli, D. A., Li, L., Pan, D., and Edgar, B. A. (2003) Rheb promotes cell growth as a component of the insulin/TOR signalling network. *Nat. Cell Biol.* **5,** 566–571.
105. Stocker, H., Radimerski, T., Schindelholz, B., et al. (2003) Rheb is an essential regulator of S6K in controlling cell growth in *Drosophila. Nat. Cell Biol.* **5,** 559–565.
106. Patel, P. H., Thapar, N., Guo, L., et al. (2003) *Drosophila* Rheb GTPase is required for cell cycle progression and cell growth. *J. Cell Sci.* **116,** 3601–3610.
107. Gao, X., Neufeld, T. P., and Pan, D. (2000) *Drosophila* PTEN regulates cell growth and proliferation through PI3K-dependent and -independent pathways. *Dev. Biol.* **221,** 404–418.
108. Scanga, S. E., Ruel, L., Binari, R. C., et al. (2000) The conserved PI3K/PTEN/Akt signaling pathway regulates both cell size and survival in *Drosophila. Oncogene* **19,** 3971–3977.
109. Goberdhan, D. C., Paricio, N., Goodman, E. C., Mlodzik, M., and Wilson, C. (1999) *Drosophila* tumor suppressor PTEN controls cell size and number by antagonizing the Chico/PI3-kinase signaling pathway. *Genes Dev.* **13,** 3244–3258.
110. Huang, H., Potter, C. J., Tao, W., et al. (1999) PTEN affects cell size, cell proliferation and apoptosis during *Drosophila* eye development. *Development* **126,** 5365–5372.
111. Gao, X. and Pan, D. (2001) TSC1 and TSC2 tumor suppressors antagonize insulin signaling in cell growth. *Genes Dev.* **15,** 1383–1392.
112. Gao, X., Zhang, Y., Arrazola, P., et al. (2002) Tsc tumour suppressor proteins antagonize amino-acid-TOR signalling. *Nat. Cell Biol.* **4,** 699–704.
113. Tapon, N., Ito, N., Dickson, B. J., Treisman, J. E., and Hariharan, I. K. (2001) The *Drosophila* tuberous sclerosis complex gene homologs restrict cell growth and cell proliferation. *Cell* **105,** 345–355.
114. Johnston, L. A., Prober, D. A., Edgar, B. A., Eisenman, R. N., and Gallant, P. (1999) *Drosophila* myc regulates cellular growth during development. *Cell* **98,** 779–790.

115. Prober, D. A. and Edgar, B. A. (2002) Interactions between Ras1, dMyc, and dPI3K signaling in the developing *Drosophila* wing. *Genes Dev.* **16,** 2286–2299.
116. Karim, F. D. and Rubin, G. M. (1998) Ectopic expression of activated Ras1 induces hyperplastic growth and increased cell death in *Drosophila* imaginal tissues. *Development* **125,** 1–9.
117. Prober, D. A. and Edgar, B. A. (2000) Ras1 promotes cellular growth in the *Drosophila* wing. *Cell* **100,** 435–446.
118. Prober, D. A. and Edgar, B. A. (2001) Growth regulation by oncogenesnew insights from model organisms. *Curr. Opin. Genet. Dev.* **11,** 19–26.
119. Datar, S. A., Jacobs, H. W., de la Cruz, A. F., Lehner, C. F., and Edgar, B. A. (2000) The *Drosophila* cyclin D-Cdk4 complex promotes cellular growth. *EMBO J.* **19,** 4543–4554.
120. Meyer, C. A., Jacobs, H. W., Datar, S. A., Du, W., Edgar, B. A., and Lehner, C. F. (2000) *Drosophila* cdk4 is required for normal growth and is dispensable for cell cycle progression. *EMBO J.* **19,** 4533–4542.
121. Neufeld, T. P., de la Cruz, A. F., Johnston, L. A., and Edgar, B. A. (1998) Coordination of growth and cell division in the *Drosophila* wing. *Cell* **93,** 1183–1193.
122. Weigmann, K., Cohen, S. M., and Lehner, C. F. (1997) Cell cycle progression, growth and patterning in imaginal discs despite inhibition of cell division after inactivation of *Drosophila* Cdc2 kinase. *Development* **124,** 3555–3563.
123. Justice, R. W., Zilian, O., Woods, D. F., Noll, M., and Bryant, P. J. (1995) The *Drosophila* tumor suppressor gene warts encodes a homolog of human myotonic dystrophy kinase and is required for the control of cell shape and proliferation. *Genes Dev.* **9,** 534–546.
124. Xu, T., Wang, W., Zhang, S., Stewart, R. A., and Yu, W. (1995) Identifying tumor suppressors in genetic mosaics: the *Drosophila* lats gene encodes a putative protein kinase. *Development* **121,** 1053–1063.
125. Kango-Singh, M., Nolo, R., Tao, C., et al. (2002) Shar-pei mediates cell proliferation arrest during imaginal disc growth in *Drosophila. Development* **129,** 5719–5730.
126. Tapon, N., Harvey, K. F., Bell, D. W., et al. (2002) *salvador* Promotes both cell cycle exit and apoptosis in *Drosophila* and is mutated in human cancer cell lines. *Cell* **110,** 467–478.
127. Harvey, K. F., Pfleger, C. M., and Hariharan, I. K. (2003) The *Drosophila* Mst ortholog, hippo, restricts growth and cell proliferation and promotes apoptosis. *Cell* **114,** 457–467.
128. Wu, S., Huang, J., Dong, J., and Pan, D. (2003) *hippo* encodes a Ste-20 family protein kinase that restricts cell proliferation and promotes apoptosis in conjunction with *salvador* and *warts*. *Cell* **114,** 445–456.
129. Yoo, S. J., Huh, J. R., Muro, I., et al. (2002) Hid, Rpr and Grim negatively regulate DIAP1 levels through distinct mechanisms. *Nat. Cell Biol.* **4,** 416–424.
130. Hipfner, D. R., Weigmann, K., and Cohen, S. M. (2002) The bantam gene regulates *Drosophila* growth. *Genetics* **161,** 1527–1537.
131. Brennecke, J., Hipfner, D. R., Stark, A., Russell, R. B., and Cohen, S. M. (2003) *bantam* encodes a developmentally regulated microRNA that controls cell proliferation and regulates the proapoptotic gene *hid* in *Drosophila. Cell* **113,** 25–36.
132. Baker, N. E. (2001) Cell proliferation, survival, and death in the *Drosophila* eye. *Semin. Cell Dev. Biol.* **12,** 499–507.
133. de Nooij, J. C. and Hariharan, I. K. (1995) Uncoupling cell fate determination from patterned cell division in the *Drosophila* eye. *Science* **270,** 983–985.

134. Ma, C. and Moses, K. (1995) Wingless and patched are negative regulators of the morphogenetic furrow and can affect tissue polarity in the developing *Drosophila* compound eye. *Development* **121,** 2279–2289.
135. Treisman, J. E. and Rubin, G. M. (1995) wingless inhibits morphogenetic furrow movement in the Drosophila eye disc. *Development* **121,** 3519–3527.
136. Duman-Scheel, M., Weng, L., Xin, S., and Du, W. (2002) Hedgehog regulates cell growth and proliferation by inducing cyclin D and cyclin E. *Nature* **417,** 299–304.
137. Frolov, M. V., Huen, D. S., Stevaux, O., et al. (2001) Functional antagonism between E2F family members. *Genes Dev.* **15,** 2146–2160.
138. Jones, L., Richardson, H., and Saint, R. (2000) Tissue-specific regulation of cyclin E transcription during *Drosophila melanogaster* embryogenesis. *Development* **127,** 4619–4630.
139. Baker, N. E. and Yu, S. Y. (2001) The EGF receptor defines domains of cell cycle progression and survival to regulate cell number in the developing *Drosophila* eye. *Cell* **104,** 699–708.
140. Thomas, B. J., Gunning, D. A., Cho, J., and Zipursky, L. (1994) Cell cycle progression in the developing *Drosophila* eye: roughex encodes a novel protein required for the establishment of G1. *Cell* **77,** 1003–1014.
141. Avedisov, S. N., Krasnoselskaya, I., Mortin, M., and Thomas, B. J. (2000) Roughex mediates G(1) arrest through a physical association with cyclin A. Mol. *Cell. Biol.* **20,** 8220–8229.
142. Foley, E., OFarrell, P. H., and Sprenger, F. (1999) Rux is a cyclin-dependent kinase inhibitor (CKI) specific for mitotic cyclin-Cdk complexes. *Curr. Biol.* **9,** 1392–1402.
143. Sprenger, F., Yakubovich, N., and OFarrell, P. H. (1997) S-phase function of *Drosophila* cyclin A and its downregulation in G1 phase. *Curr. Biol.* **7,** 488–499.
144. Horsfield, J., Penton, A., Secombe, J., Hoffman, F. M., and Richardson, H. (1998) *decapentaplegic* is required for arrest in G1 phase during *Drosophila* eye development. *Development* **125,** 5069–5078.
145. Jones, L. (2001) Stem cells: so whats in a niche? Curr. Biol. 11, R484–R486.
146. Lin, H., and Spradling, A. C. (1997) A novel group of *pumilio* mutations affects the asymmetric division of germline stem cells in the *Drosophila* ovary. *Development* **124,** 2463–2476.
147. Margolis, J. and Spradling, A. (1995) Identification and behavior of epithelial stem cells in the *Drosophila* ovary. *Development* **121,** 3797–3807.
148. Cox, D. N., Chao, A., and Lin, H. (2000) *piwi* encodes a nucleoplasmic factor whose activity modulates the number and division rate of germline stem cells. *Development* **127,** 503–514.
149. Forbes, A. and Lehmann, R. (1998) *Nanos* and *Pumilio* have critical roles in the development and function of *Drosophila* germline stem cells. *Development* **125,** 679–690.
150. Parisi, M. and Lin, H. (1999) The *Drosophila pumilio* gene encodes two functional protein isoforms that play multiple roles in germline development, gonadogenesis, oogenesis and embryogenesis. *Genetics* **153,** 235–250.
151. Cox, D. N., Chao, A., Baker, J., Chang, L., Qiao, D., and Lin, H. (1998) A novel class of evolutionarily conserved genes defined by piwi are essential for stem cell self-renewal. *Genes Dev.* **12,** 3715–3727.

152. King, F. J. and Lin, H. (1999) Somatic signaling mediated by fs(1)Yb is essential for germline stem cell maintenance during *Drosophila* oogenesis. *Development* **126,** 1833–1844.
153. Xie, T., and Spradling, A. C. (2000) A niche maintaining germ line stem cells in the *Drosophila* ovary. *Science* **290,** 328–330.
154. Song, X., Zhu, C. H., Doan, C., and Xie, T. (2002) Germline stem cells anchored by adherens junctions in the *Drosophila* ovary niches. *Science* **296,** 1855–1857.
155. King, F. J., Szakmary, A., Cox, D. N., and Lin, H. (2001) *Yb* modulates the divisions of both germline and somatic stem cells through *piwi*- and *hh*-mediated mechanisms in the *Drosophila* ovary. *Mol. Cell* **7,** 497–508.
156. Xie, T. and Spradling, A. C. (1998) decapentaplegic is essential for the maintenance and division of germline stem cells in the *Drosophila* ovary. *Cell* **94,** 251–260.
157. Forbes, A. J., Lin, H., Ingham, P. W., and Spradling, A. C. (1996) hedgehog is required for the proliferation and specification of ovarian somatic cells prior to egg chamber formation in *Drosophila. Development* **122,** 1125–1135.
158. King, F. J., Szakmary, A., Cox, D. N., and Lin, H. (2001) *Yb* modulates the divisions of both germline and somatic stem cells through *piwi*- and *hh*-mediated mechanisms in the *Drosophila* ovary. *Mol. Cell* **7,** 497–508.
159. Song, X. and Xie, T. (2003) Wingless signaling regulates the maintenance of ovarian somatic stem cells in *Drosophila. Development* **130,** 3259–3268.
160. McKearin, D. M. and Spradling, A. C. (1990) *bag-of-marbles*: a *Drosophila* gene required to initiate both male and female gametogenesis. *Genes Dev.* **4,** 2242–2251.
161. McKearin, D., and Ohlstein, B. (1995) A role for the *Drosophila* bag-of-marbles protein in the differentiation of cystoblasts from germline stem cells. *Development* **121,** 2937–2947.
162. Ohlstein, B. and McKearin, D. (1997) Ectopic expression of the *Drosophila* Bam protein eliminates oogenic germline stem cells. *Development* **124,** 3651–3662.
163. Parisi, M. J., Deng, W., Wang, Z., and Lin, H. (2001) The arrest gene is required for germline cyst formation during *Drosophila* oogenesis. *Genesis* **29,** 196–209.
164. Webster, P. J., Liang, L., Berg, C. A., Lasko, P., and Macdonald, P. M. (1997) Translational repressor bruno plays multiple roles in development and is widely conserved. *Genes Dev.* **11,** 2510–2521.
165. Hawkins, N. C., Thorpe, J., and Schupbach, T. (1996) *Encore*, a gene required for the regulation of germ line mitosis and oocyte differentiation during *Drosophila* oogenesis. *Development* **122,** 281–290.
166. King, R. C., and Storto, P. D. (1988) The role of the *otu* gene in *Drosophila* oogenesis. *Bioessays* **8,** 18–24.
167. Spradling, A. C. (1993) Developmental genetics of oogenesis. In The Development of *Drosophila melanogaster*, Vol. 1 (Bate, M. and Martinez Arias, A., eds.) Cold Spring Harbor Laboratory Press, Cold Spring Harbor, NY, pp. 1–70.
168. Hong, A., Lee-Kong, S., Iida, T., Sugimura, I., and Lilly, M. A. (2003) The p27cip/kip ortholog dacapo maintains the *Drosophila* oocyte in prophase of meiosis I. *Development* **130,** 1235–1242.
169. Dej, K. J. and Spradling, A. C. (1999) The endocycle controls nurse cell polytene chromosome structure during *Drosophila* oogenesis. *Development* **126,** 293–303.
170. McCall, K. and Steller, H. (1998) Requirement for DCP-1 caspase during *Drosophila* oogenesis. *Science* **279,** 230–234.
171. Foley, K. and Cooley, L. (1998) Apoptosis in late stage *Drosophila* nurse cells does not require genes within the H99 deficiency. *Development* **125,** 1075–1082.

172. Matova, N., Mahajan-Miklos, S., Mooseker, M. S., and Cooley, L. (1999) *Drosophila* Quail, a villin-related protein, bundles actin filaments in apoptotic nurse cells. *Development* **126,** 5645–5657.
173. Myster, D. L., Bonnette, P. C., and Duronio, R. J. (2000) A role for the DP subunit of the E2F transcription factor in axis determination during *Drosophila* oogenesis. *Development* **127,** 3249–3261.
174. Royzman, I., Hayashi-Hagihara, A., Dej, K. J., Bosco, G., Lee, J. Y., and Orr-Weaver, T. L. (2002) The E2F cell cycle regulator is required for *Drosophila* nurse cell DNA replication and apoptosis. *Mech. Dev.* **119,** 225–237.
175. de Nooij, J. C., Graber, K. H., and Hariharan, I. K. (2000) Expression of the cyclin-dependent kinase inhibitor Dacapo is regulated by cyclin E. *Mech. Dev.* **97,** 73–83.
176. Reed, B. H. and Orr-Weaver, T. L. (1997) The Drosophila gene morula inhibits mitotic functions in the endo cell cycle and the mitotic cell cycle. *Development* **124,** 3543–3553.
177. Kashevsky, H., Wallace, J. A., Reed, B. H., Lai, C., Hayashi-Hagihara, A., and Orr-Weaver, T. L. (2002) The anaphase promoting complex/cyclosome is required during development for modified cell cycles. *Proc. Natl. Acad. Sci. USA* **99,** 11217–11222.
178. Calvi, B. R., Lilly, M. A., and Spradling, A. C. (1998) Cell cycle control of chorion gene amplification. *Genes Dev.* **12,** 734–744.
179. Roth, S. (2001) *Drosophila* oogenesis: coordinating germ line and soma. *Curr. Biol.* **11,** R779–R781.
180. Lopez-Schier, H., and St. Johnston, D. (2001) Delta signaling from the germ line controls the proliferation and differentiation of the somatic follicle cells during *Drosophila* oogenesis. *Genes Dev.* **15,** 1393–1405.
181. Deng, W. M., Althauser, C., and Ruohola-Baker, H. (2001) Notch-Delta signaling induces a transition from mitotic cell cycle to endocycle in *Drosophila* follicle cells. *Development* **128,** 4737–4746.
182. Spradling, A. C. (1999) ORC binding, gene amplification, and the nature of metazoan replication origins. *Genes Dev.* **13,** 2619–2623.
183. Tzolovsky, G., Deng, W. M., Schlitt, T., and Bownes, M. (1999) The function of the broad-complex during *Drosophila melanogaster* oogenesis. *Genetics* **153,** 1371–1383.
184. Lu, L., Zhang, H., and Tower, J. (2001) Functionally distinct, sequence-specific replicator and origin elements are required for *Drosophila* chorion gene amplification. *Genes Dev.* **15,** 134–146.
185. Landis, G., Kelley, R., Spradling, A. C., and Tower, J. (1997) The k43 gene, required for chorion gene amplification and diploid cell chromosome replication, encodes the *Drosophila* homolog of yeast origin recognition complex subunit 2. *Proc. Natl. Acad. Sci. USA* **94,** 3888–3892.
186. Landis, G. and Tower, J. (1999) The *Drosophila* chiffon gene is required for chorion gene amplification, and is related to the yeast Dbf4 regulator of DNA replication and cell cycle. *Development* **126,** 4281–4293.
187. Austin, R. J., Orr-Weaver, T. L., and Bell, S. P. (1999) *Drosophila* ORC specifically binds to ACE3, an origin of DNA replication control element. *Genes Dev.* **13,** 2639–2649.
188. Schwed, G., May, N., Pechersky, Y., and Calvi, B. R. (2002) Drosophila minichromosome maintenance 6 is required for chorion gene amplification and genomic replication. *Mol. Biol. Cell* **13,** 607–620.
189. Cayirlioglu, P., Bonnette, P. C., Dickson, M. R., and Duronio, R. J. (2001) *Drosophila* E2f2 promotes the conversion from genomic DNA replication to gene amplification in ovarian follicle cells. *Development* **128,** 5085–5098.

190. Asano, M. and Wharton, R. P. (1999) E2F mediates developmental and cell cycle regulation of ORC1 in *Drosophila*. *EMBO J.* **18,** 2435–2448.
191. Bosco, G., Du, W., and Orr-Weaver, T. L. (2001) DNA replication control through interaction of E2F-RB and the origin recognition complex. *Nat. Cell Biol.* **3,** 289–295.
192. Royzman, I., Austin, R. J., Bosco, G., Bell, S. P., and Orr-Weaver, T. L. (1999) ORC localization in *Drosophila* follicle cells and the effects of mutations in dE2F and dDP. *Genes Dev.* **13,** 827–840.
193. Cayirlioglu, P. and Duronio, R. J. (2001) Cell cycle: Flies teach an old dogma new tricks. *Curr. Biol.* **11,** R178–R181.

5

The *Xenopus* Cell Cycle

An Overview

Anna Philpott and P. Renee Yew

Summary

Oocytes, eggs, and embryos from the frog *Xenopus laevis* have been an important model system for studying cell cycle regulation for several decades. First, progression through meiosis in the oocyte has been extensively investigated. Oocyte maturation has been shown to involve complex networks of signal transduction pathways, culminating in the cyclic activation and inactivation of maturation promoting factor (MPF), which is composed of cyclin B and cdc2. After fertilization, the early embryo undergoes rapid simplified cell cycles, which have been recapitulated in cell-free extracts of *Xenopus* eggs. Experimental manipulation of these extracts has given a wealth of biochemical information about the cell cycle, particularly concerning DNA replication and mitosis. Finally, cells of older embryos adopt a more somatic-type cell cycle and have been used to study the balance between cell cycle and differentiation during development.

Key Words

Xenopus; cell cycle; oocyte; egg, embryo; DNA replication; initiation; meiosis; mitosis; midblastula transition; MPF; cdk; cyclin; APC/C; ubiquitin; proteolysis.

1. Introduction

Xenopus eggs contain vast stockpiles of proteins that enable them to undergo 12 rounds of cell division before the onset of zygotic transcription. Much work looking at control of S- and M-phases has been performed in cell-free extracts of *Xenopus* eggs. After fertilization, *Xenopus* eggs undergo 12 synchronous rounds of cell division in around 8 h. These are specialized cell cycles occurring in the absence of transcription *(1)*. Therefore, all the RNAs and most of the proteins required for these early rounds of

From: *Methods in Molecular Biology, vol. 296, Cell Cycle Control: Mechanisms and Protocols*
Edited by: T. Humphrey and G. Brooks © Humana Press Inc., Totowa, NJ

division have been stockpiled in the unfertilized eggs, and cyclical translation and degradation of cyclin B alone are sufficient to drive these early cell cycles *(2,3)*.

Many of the biochemical and cell biological processes associated with the cell cycle, such as DNA replication, cyclin-dependent kinase (cdk) activation, nuclear envelope breakdown and assembly, mitotic spindle formation, and chromosome condensation and decondensation, will occur in extracts of eggs produced by centrifugation. These egg extracts have several advantages as a model system; large volumes (milliliters) of extract can be produced from the eggs of a single frog, and the cell-free nature of these extracts means that proteins can be added easily and also removed by immunodepletion. Extracts can support accurate, once-per-cell-cycle, replication of exogenous DNA templates, usually sperm nuclei *(4)*. In addition, extracts have been prepared that undergo multiple rounds of mitosis, supporting processes including nuclear envelope breakdown, chromosome condensation, and spindle formation *(3)*. Studies using egg extracts have given us a great wealth of information about the cell cycle that is applicable to a wide range of organisms.

Regulation of the cell cycle varies through development from oocyte formation and maturation through early cleavage stages of the embryonic blastula to development of the tadpole and probably beyond. This review covers our understanding of cell cycle regulation at different stages of *Xenopus* embryonic development.

2. Oocyte Maturation in *Xenopus*

Fully grown *Xenopus* oocytes are arrested at the first meiotic metaphase in a G_2-like state until progesterone triggers meiotic maturation. In summary, the maturing oocyte passes through meiosis I, undergoes germinal vesicle breakdown, emitting the first polar body, and then enters meiosis II without an intervening interphase. The oocyte then arrests at metaphase of the second meiotic division until fertilization (reviewed in **refs. *5*** and ***6***). Regulation of the G_2/M transition is brought about by the activity of maturation-promoting factor (MPF), which is composed of cdc2 and B-type cyclins *(7–11)*. MPF activity is upregulated on progression through meiosis I and is inactivated between meiosis I and II, but then rises again at meiosis II, where it is kept in an active state by cytostatic factor (CSF; for detailed reviews, *see* **refs. *5*, *6***, and ***12***). For an overview, *see* **Fig. 1**.

Arrest at meiotic metaphase I of the fully grown oocyte is maintained by inhibitory phosphorylation of cdc2 on Thre14 and Tyr15 *(13–15)*. Although pathways regulating this cell cycle arrest are still not fully understood, it is known that inhibitory phosphorylation of cdc2 is brought about by myt1 and removed by activated cdc25C *(16–18)*.

Progesterone treatment triggers oocyte maturation, allowing passage through meiotic metaphase I, and leads to a rapid lowering of cyclic adenosine monophosphate (cAMP) and hence protein kinase A (PKA) activity (for a comprehensive review, *see* **refs. *6***). Interestingly, PKA and the checkpoint protein chk1 are thought to inhibit progesterone-mediated maturation, possibly via cdc25C phosphorylation or other unidentified mechanisms, which maintain cdc2 in its inhibited state *(19)*. New protein synthesis is required for the progression through meiosis I, and one of the most important new proteins to be synthesized in response to progesterone is Mos, an activator of

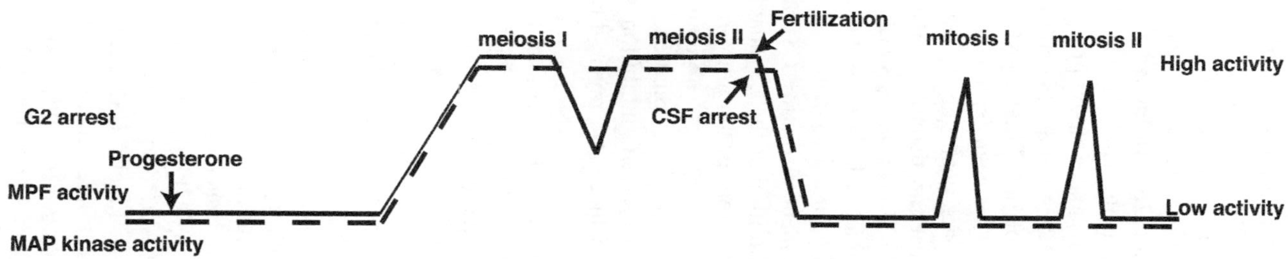

Fig. 1. A schematic diagram of oocyte maturation and fertilization. In response to progesterone, cyclical changes in MPF activity (solid line) and mitogen-activated protein (MAP) kinase activity (dotted line) drive fully grown oocytes, which have arrested in a G_2-like state, through meiosis I and II. Mature oocytes arrest in meiotic metaphase II with high levels of maturation promoting factor (MPF) maintained by the action of cytostatic factor (CSF). CSF activity is destroyed by a wave of calcium released from intracellular stores at fertilization, allowing the embryo to undergo mitotic cell division.

the MAP kinase pathway *(20–22)*. This synthesis results from cytoplasmic polyadenylation of Mos messenger RNA, which promotes translation *(23,24)* as well as stabilization of Mos protein (for a review, *see* **ref. 25**). Mos activity then remains high and is required for suppression of DNA synthesis between meiosis I and II as well as reactivation and stabilization of MPF in meiosis II in the form of CSF.

Mos activates mitogen-activated protein (MAP)/Erk1 kinase in oocytes by acting as a MAP kinase kinase kinase (MAPKKK) *(26–28)*. Indeed, the role of Mos in oocyte maturation may be explained by its ability to ultimately activate MAP kinase, as blocking MAP kinase activity under some circumstances blocks progesterone-induced maturation, whereas constitutively active MAP kinase induces maturation in the absence of progesterone *(29,30)*. However, it should be noted that oocyte maturation may occur in response to progesterone even when Mos has been depleted by antisense morpholino oligonucleotides. MAP Kinase may feed back positively by stimulating Mos messenger RNA polyadenylation *(31)*. MAP kinase can also inhibit Mos protein degradation, at least in part through phosphorylation of Mos at serine 3 *(32,33)*.

A crucial downstream target for Mos-induced MAP kinase during oocyte maturation is the kinase p90rsk, which plays important roles in the meiotic cell cycle. In particular, p90rsk is an important component mediating CSF arrest at metaphase II *(34,35)*. Activation of p90rsk leads to inhibition of the E3, anaphase-promoting complex or cyclosome (APC/C), which is required for ubiquitin-mediated proteolysis of cyclin B and so progression through mitosis *(36)* (*see* **Subheading 5.** below). Recent evidence has shown that Bub1, a component of the spindle assembly checkpoint, is activated by p90rsk, and this may mediate APC/C inhibition *(37)*. Phosphorylated/active p90rsk can also phosphorylate and inactivate myt1, directly promoting cdc2 activity *(38)*.

Activation of cdc2 kinase, in the form of MPF, is crucial for progression through meiosis and is an integration point for several pathways involved in oocyte maturation. Cdc2 kinase activation is tightly temporally controlled in *Xenopus* oocytes. As described above, p90rsk may inactivate the inhibitory kinase myt1 *(38)*, whereas PKA and chk1 may regulate CDC25C *(19)*, both of which ultimately regulate cdc2 kinase. In addition, a polo-like kinase, Plx1, can bind to cdc25 and phosphorylate and activate it in vitro *(39)*. It has also been suggested that direct phosphorylation of cyclin B by MAP kinase may influence its nuclear localization and hence the activity of cdc2 kinase (for reviews, *see* **refs. 6** and **40**). Blocking cdc2 kinase activity inhibits progesterone-induced oocyte maturation *(41)*, whereas injection of cyclin/cdc2 complexes alone promotes maturation *(11)*. On exit from meiosis I, cyclin B is degraded to around 50% of its interphase level, partially inactivating MPF, but chromatin remains condensed, and DNA replication does not occur. However, this is followed by resynthesis of cyclin B, which occurs on entry into meiosis II. Finally CSF activity is then established, maintaining high MPF cdc2 kinase levels until fertilization (**42**; for review, *see* **ref. 6**).

After oocyte maturation, unfertilized eggs arrest in meiotic metaphase II with high levels of MPF stabilized in the form of CSF. As described above, Mos is a component of CSF activity. However, recently it has been shown that a further component of CSF, emi1, may be primarily responsible for stabilization of cyclin B in metaphase-arrested eggs, in a MAP kinase-independent manner, by inhibiting the APC/C, which targets cyclin B for ubiquitin-mediated proteolysis *(43)*.

Fertilization results in a wave of calcium influx from intracellular stores, which destroys the CSF activity along with Mos protein and results in activation of cdc2 kinase *(44)*. However, the timing of Mos degradation with respect to MPF inactivation indicates that Mos degradation does not direct MPF loss. Instead, it seems likely that cdc2 inactivation is brought about primarily by disruption of emi1 association with the APC/C activator cdc20 *(43,45)*. Then activated APC/C degrades cyclin B and allows entry of the fertilized egg into interphase.

3. Regulation of S-Phase

After release from CSF arrest at fertilisation, both egg and sperm chromosomes decondense *(46)*, nuclear envelopes reform *(47)*, and DNA replication commences *(4)*. The large stores of replication proteins and the ability to add and deplete proteins has made these egg extracts a very valuable system for elucidating biochemical mechanisms of DNA replication control, as outlined below. Some of the findings from frog eggs have now been recapitulated in extracts from mammalian cells *(48)*.

S-phase is extremely rapid in early embryonic cell cycles, which last only 20–30 min. To allow all the DNA to be duplicated so quickly, origins of replication are close together and do not seem to require specific DNA sequences *(49)*. However, origins of replication are defined by the binding of origin recognition complex (ORC) proteins. The ORC proteins provide a binding site for so-called replication licensing proteins including those of the minichromosome maintenance complex (MCMs; *see* **Subheading 4.** below), which allow or 'licence' an origin to initiate. It is the cell cycle-regulated binding of pre-replication complex (pre-RC) proteins to an origin that ensures that each piece of DNA will be replicated only once per cell cycle (for review, *see* **ref. *50***).

4. The Temporal Events of DNA Replication Initiation

The steps of DNA replication initiation are evolutionarily conserved from budding yeast to humans and require the assembly of the pre-replication complex or pre-RC on DNA (**Fig. 2**). The pre-RC is comprised of the origin recognition complex (ORC1-6), Cdc6, Cdt1, and the mini-chromosome maintenance (MCM2–7) proteins *(51–65)*. The next step of replication initiation requires the activities of the cdc7/Dbf4 kinase, the cdk2/cyclin E kinase, and MCM10, resulting in the recruitment of Cdc45 to the pre-RC to form the pre-initiation complex (pre-IC) *(51,53,54,60,65–71)*. Origin unwinding, mediated by the putative MCM2–7 helicase, is followed by the recruitment of replication protein A (RPA) and DNA polymerase α-primase *(65,72,73)*. No requirement for cyclin A in the onset of S-phase in *Xenopus* eggs has been observed, but cdk2/cyclin A activity is required for the start of DNA replication in mammals and most likely plays a role in S-phase onset during the *Xenopus* somatic cell cycle *(74–76)*. Replication factor C (RFC) and proliferating cell nuclear antigen (PCNA) mediate DNA polymerase switching, the last step of replication initiation *(73,77–81)*. Processive DNA synthesis, or elongation, is mediated by PCNA and DNA polymerase δ/ε (reviewed in **refs. *73,78–80*,** and ***82***). Finally, DNA that has replicated once must

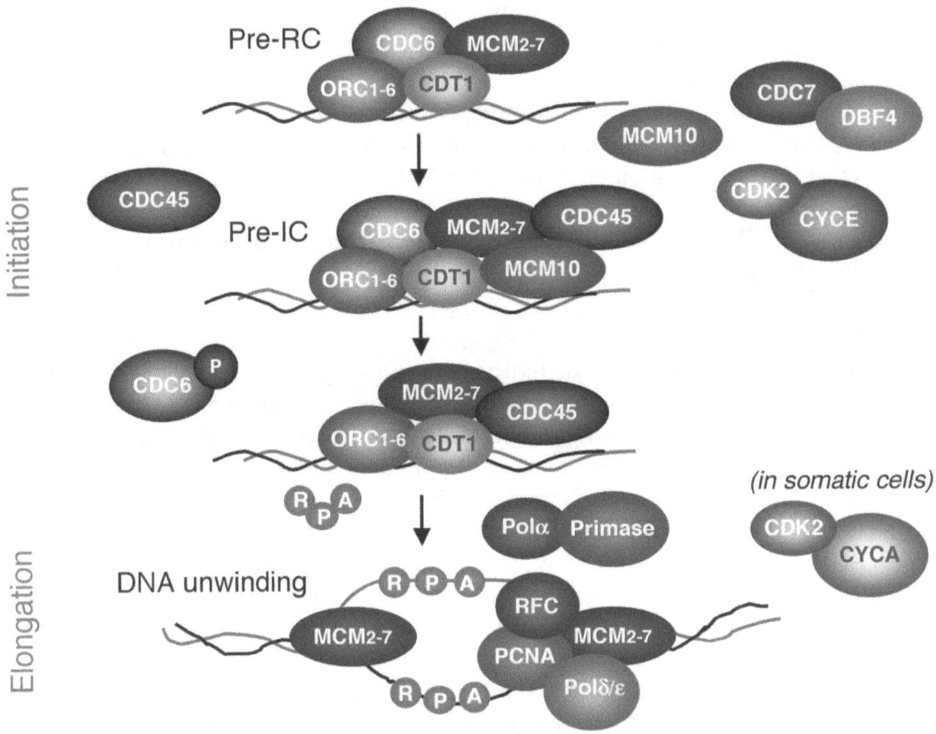

Fig. 2. The molecular events of DNA replication initiation. During the onset of S-phase in *Xenopus*, the origin recognition complex (ORC), cdc6, cdt1, and the mini-chromosome maintenance (MCM) complex form the pre-replication complex (pre-RC). The activities of the S-phase kinases, cdc7/dbf4 and cdk2/cyclin E (cycE), as well as Mcm10, mediate the tight binding of cdc45 to the origin, resulting in the formation of the pre-initiation complex (pre-IC). Just prior to DNA unwinding, the pre-IC undergoes specific changes that are not well characterized in *Xenopus* but that may include the phosphorylation and release of cdc6 as well as the post-translational modification of other pre-RC components. These alterations ensure that DNA replication occurs only once during each cell cycle. DNA unwinding is mediated by the putative MCM2-7 helicase and stabilized by the single-strand DNA binding protein, replication protein A (RPA), followed by primer synthesis by DNA polymerase α (Polα)/primase. DNA polymerase switching is mediated by replication factor C (RFC) and proliferating cell nuclear antigen (PCNA), resulting in processive DNA synthesis or elongation by DNA polymerase δ or ε. Although not highly expressed in *Xenopus* eggs, the cdk2/cyclin A2 (cycA) kinase is likely to play an important role in the onset of S-phase in the *Xenopus* somatic cell cycle.

be prevented from doing so again until mitosis is completed, and this is brought about by inactivation of the licensing system prior to the onset of S-phase. This may be brought about partly by exclusion of cdc6 from the nucleus after licensing has occurred and partly by the action of geminin, a cdt1 binding protein and inhibitor of MCM2–7 recruitment. Geminin accumulates during G_1- and S-phases to prevent relicensing but becomes abruptly degraded upon exit from mitosis ***(83,84)***.

5. Entry Into Mitosis: Activation of Cdc2/Cyclin B

After DNA replication is completed, cells of early *Xenopus* embryos enter directly into mitosis, apart from the first cell cycle after fertilization which may contain a G_2-like period ***(85)***. Entry into mitosis requires the activation of the cdc2 kinase. This activation involves the synthesis, accumulation, and binding of cyclin B and the proper phosphorylation status of cdc2. During the G_2-phase of the *Xenopus* embryonic cell cycle, cyclin B is synthesized and reaches a threshold level just before entry into M-phase ***(86)***. Once synthesized, cyclin B binds to cdc2 and, based on the crystal structure of human cdk2-cyclin A, most likely alters the conformation of the cdc2 active-site cleft by reorienting the T-loop region of cdc2 ***(87)***. Cdc2 itself is also regulated by a complicated pattern of phosphorylation and dephosphorylation. The activation of cdc2 is mediated in part by the phosphorylation of Threonine 161 by cdk-activating kinase (CAK) ***(88,89)***. Prior to the onset of mitosis, the cdc2/cyclin B complex is kept in an inactive state by the phosphorylation of cdc2 at Tyr15 by Wee1/Mik1/Myt1 and Thr14 by Myt1 ***(16,90)***. The onset of mitosis is triggered by the dephosphorylation of Tyr15 and Thr14 on cdc2 by the dual-specificity phosphatase cdc25 ***(17,18,91,92)***. The activity of cdc25 is further enhanced by phosphorylation by cdc2/cyclin B, creating a feedback loop that results in the rapid activation of cdc2/cyclin B and the entry into mitosis ***(93–95)***.

Cdc25 plays an important role in the DNA replication checkpoint, which ensures that DNA replication is completed before the start of mitosis ***(96)***. In the presence of unreplicated DNA, the checkpoint inducer protein, Chk1, phosphorylates cdc25 at Serine 287 ***(97)***. This phosphorylation of cdc25 confers the binding to 14-3-3 proteins, which negatively regulate the mitotic inducing activity of cdc25 ***(97–99)***. Once cdc2/cyclin B is activated, it targets multiple substrates for phosphorylation, which in turn triggers the mitotic events of nuclear envelope breakdown, chromosome condensation, and spindle assembly ***(100–102)***.

6. Ubiquitin-Mediated Proteolysis

The levels of key cell cycle proteins are regulated by ubiquitin-mediated proteolysis. This is particularly well understood for the exit of mitosis, which is triggered by the ubiquitin-mediated proteolysis of key substrates including cyclin B ***(103)***, the controlled degradation of which is described in detail below. Ubiquitin is a highly conserved 76-amino acid protein, which, when covalently attached to lysine residues of proteins, targets them for proteolysis. This pathway requires the activities of E1 (or UBA~~ub~~iquitin-activating enzyme), E2 (or UBC~~ub~~iquitin-conjugating enzyme), and E3 (or UBR~~ub~~iquitin recognition factor or ubiquitin protein ligase) (reviewed in **refs. *104–106***). In an ATP-dependent process, the E1 enzyme covalently attaches the C-terminus of ubiquitin to an internal cysteine residue via a thioester bond. The activated ubiquitin is then transferred to a specific cysteine residue on the E2 enzyme. The E2 enzyme then mediates the transfer of ubiquitin to protein substrates via isopeptide bonds at specific internal lysine residues, most often in association with an E3 accessory enzyme. The E3 component primarily functions to confer substrate specificity, typically by recruiting both the E2 and the substrate into close proximity for ubiquitin transfer. Together these enzymes mediate the polyubiquitination of substrates, a molecular signal, which generally targets proteins for proteolysis by the 26S proteasome.

7. Regulation of APC/C Function During the Progression Through Mitosis

The progression through mitosis in *Xenopus* is regulated temporally by the stepwise ubiquitin-mediated proteolysis of mitotic substrates including cyclin B by the APC/C, which has been extensively analyzed biochemically in extracts from *Xenopus* eggs (**Fig. 3**; ***83,107–109***). APC/C is comprised of at least 10 subunits in mammals and *Xenopus* and is directed toward distinct mitotic substrate classes through the binding of the substoichiometric APC/C activators, cdc20 and cdh1 ***(110,111)***. Cdc20 and cdh1 are WD40 repeat-containing proteins that are necessary and limiting substrate-specific activators of APC/C (reviewed in **ref. *112***). Generally, APC/C^{Cdc20} targets substrates containing the structural motif called the destruction (D)-box (a conserved nine-amino acid element), whereas APC/C^{Cdh1} targets substrates containing a D-box, the structural motif called a KEN (lysine, glutamic acid, and asparagine) box, or a combination of both elements ***(113)***. In the *Xenopus* embryo, Cdc20 plays an essential role in cyclin proteolysis, whereas cdh1 appears to be dispensable in the embryo and may not be expressed until establishment of the G_1-phase of the somatic cell cycle, as observed in mammalian cells ***(114)***. APC/C^{Cdc20} is activated by cdc2/cyclin B phosphorylation and mediates the ubiquitination and degradation of securin, which promotes sister chromatid separation at the metaphase-to-anaphase transition ***(108)***. During chromosome segregation, the activity of APC/C is negatively regulated by the spindle assembly checkpoint protein, Mad2 ***(115)***. Although not yet shown in *Xenopus* or other vertebrates, APC/C^{cdc20} in budding yeast also mediates activation of the cdc14 phosphatase, which in turn dephosphorylates Cdh1 and activates APC/C^{Cdh1} ***(112)***. In the *Xenopus* embryo, APC/C^{cdc20} alone appears to be sufficient to mediate the ubiquitination of cyclins and securin, resulting in the progression through and exit from mitosis ***(108,114)***. However, in budding yeast and somatic cells, the exit from mitosis is triggered by the proteolysis of cyclin B by APC/C^{cdh1} ***(112,114)***. Upon the exit from mitosis, APC/C^{cdh1} also mediates the degradation of geminin, a cdt1 binding protein and an inhibitor of pre-RC formation ***(63,64,83)***. It is likely that the inactivation of APC/C^{cdh1} at the G_1- to S-phase boundary allows cyclin A to accumulate in order to mediate S-phase onset in mammalian cells (and most likely in *Xenopus* somatic cells), whereas the accumulation of geminin inhibits the function of the replication protein cdt1 and aids in preventing the reinitiation of DNA replication.

8. Cyclin A Vs Cyclin B in *Xenopus*

Prior to the midblastula transition (MBT), both cyclins A and B bind to and activate cdc2, whereas a zygotic form of cyclin A expressed after the MBT, cyclin A2, also binds to and activates cdk2 ***(116)***. In *Xenopus*, the destruction of both cyclins A and B is dependent on an intact D-box, but the D-box of cyclin A1 cannot functionally replace the D-box of cyclin B, suggesting that A- and B-type cyclins are differentially regulated by the APC/C ***(117)***. The degradation of both cyclins A and B is required for the exit from mitosis in early *Xenopus* embryos; in clam oocytes and *Drosophila* embryos, the

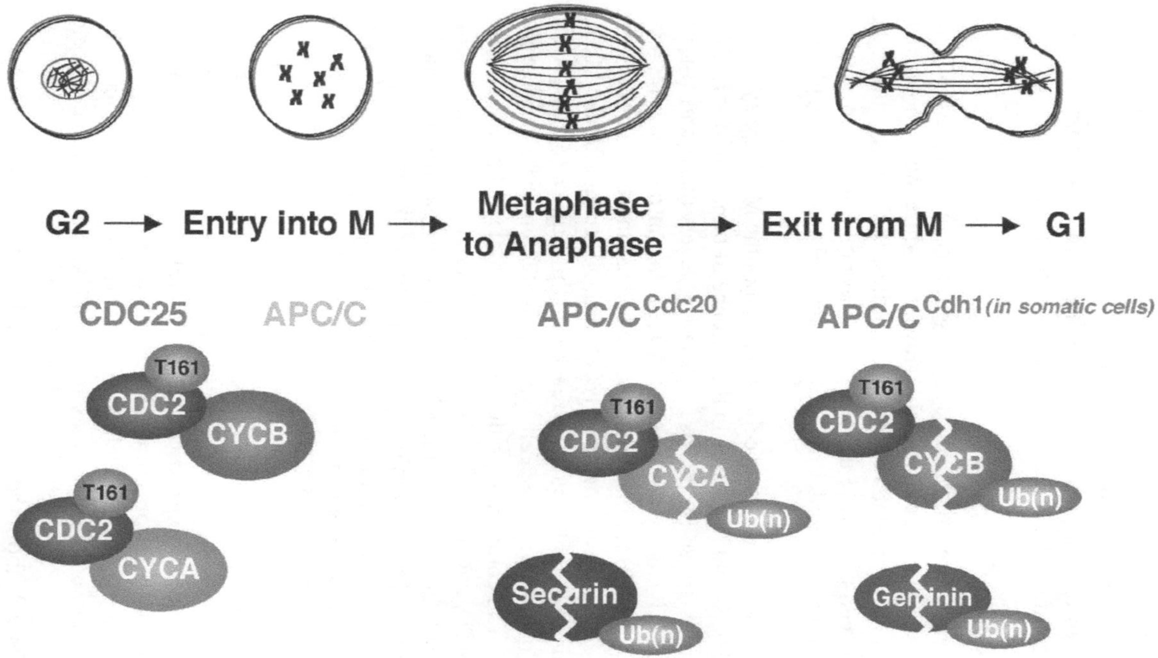

Fig. 3. The events driving mitotic progression. During the entry into mitosis when the E3 called anaphase promoting complex/cyclosome (APC/C) is inactive, cyclin B (CycB) and A1 (CycA) accumulate during a pseudo G_2-phase and complex to cdc2. The activation of cdc2/cyclin is triggered by the phosphorylation of cdc2 at Threonine 161 (T161) and the dephosphorylation of cdc2 by cdc25, resulting in the entry into mitosis. During the metaphase-to-anaphase transition, APC/C is activated by cdc20 and targets Securin for polyubiquitination (Ub[n]) and proteolysis, thereby allowing sister chromatid separation. The stepwise ubiquitination and proteolysis of cyclin A, geminin, and cyclin B by APC/C^{cdc20} in the *Xenopus* embryo trigger the exit from mitosis, although during the *Xenopus* somatic cell cycle, the degradation of mitotic cyclins and geminin are probably mediated by APC/C^{cdh1}. Following mitotic exit, cells enter a pseudo G_1-phase in which the proteolysis of geminin allows the formation of the pre-RC.

proteolysis of cyclin A precedes the proteolysis of cyclin B ***(118,119)***. Cdc2/cyclin B and cdc2/cyclin A appear to have different, nonoverlapping functions during the early embryonic *Xenopus* cell cycle, and they most likely differ in their substrate specificity. Whereas antisense oligonucleotides directed against cyclin B prevent the entry into mitosis, antisense oligonucleotides directed against cyclin A1 result in the premature activation of cdc2/cyclin B kinase activity and the premature onset of mitosis in *Xenopus* egg extracts ***(120)***. The treated extracts replicate DNA when induced into interphase, although these extracts enter mitosis before the completion of S-phase ***(120)***.

9. Cell Cycle and Development

Most of the information above describes cell cycle regulation in *Xenopus* oocytes and eggs, which have been investigated extensively. However, while much molecular and biochemical insight has been obtained from these studies, the fact remains that meiosis and early embryonic cell cycles are unusual. *Xenopus* embryos are proving to be an increasingly interesting in vivo model for studying normal cell cycle regulation during development.

As described above, the first 12 embryonic cell cycles before the MBT are synchronous and consist of alternating S- and M-phases in the absence of transcription ***(1,121)***. These cell cycles are primarily driven by the rise and fall in levels of A- and B-type cyclins bound to cdc2, which drive entry into mitosis ***(3,122)***. In contrast, cyclin E levels are fairly uniform prior to the MBT, although associated cdk2 kinase levels do oscillate around twofold ***(123,124)***. However, drastic cell cycle changes occur after the MBT, triggered by a critical nucleus-to-cytoplasmic ratio reached at this time ***(1,85,121)***. After the MBT, cells lose their synchrony, and the cell cycle gradually lengthens to incorporate a G_1- and G_2-phase ***(85)***. Concomitant with the MBT, maternal stockpiles of cyclin E are degraded, to be replaced by zygotic cyclin E ***(123,124)***. Moreover, maternal cyclin A1, which complexes to cdc2, becomes degraded at the onset of gastrulation and is replaced by cyclin A2 ***(116)***. Cyclin A2 complexes increasingly to cdk2 as development progresses, when it appears to play a role much more similar to mammalian cyclin A ***(116)***. Moreover, cells in *Xenopus* embryos do not acquire the ability to undergo apoptosis until midgastrulation stages ***(125,126)***.

Xenopus has a single identified cdk inhibitor of the cip/kip family named xicl ***(127)***. (Kix1 is thought to represent a pseudotetraploid allele rather than a distinct gene ***[128]***). Although several studies have investigated the degradation of Xicl in *Xenopus* egg ***(129,130)***, in fact, Xicl only becomes expressed strongly after the MBT. This supports the generally held view that cdk inhibitors control G_1- and S-phase progression. *Xenopus* embryos post MBT do express D-type cyclins ***(131)*** as well as the retinoblastoma protein and E2Fs ***(132–134)***, although as yet, they have not been shown to play a role in cell cycle regulation.

Although we now know much about the molecular mechanisms of cell cycle regulation, little is known about the coordination of cell cycling and differentiation events during development. However, recent experiments in *Xenopus* embryos have begun to shed some light on this area. Interestingly, cell cycle regulators such as cyclins, cdks, and Xicl are not expressed uniformly at the RNA level in dividing tissues, but instead are found in tissue-specific patterns which differ between cell cycle components ***(131)***. This might indicate potential additional functions beyond simple cell cycle regulation. This appears to be true at least for the cdk inhibitor Xicl. Overexpression of Xicl in the developing *Xenopus* retina results in neuroblasts adopting a glial over a neural fate ***(135)***. Moreover, this activity resides in the N-terminus of the molecule and is distinct from Xicl's ability to arrest the cell cycle or inhibit overall cdk kinase activity. It has subsequently been shown that the same region of Xicl is required for differentiation of both nerve and muscle in the early embryo, over and above its ability to arrest the cell cycle ***(136,137)***.

Therefore, it seems that the roles of cell cycle regulators in control of differentiation and development go beyond their traditional functions (for reviews, *see* **refs. *138*** and ***139***). Other intriguing studies have shown that the cell cycle regulators E2F and geminin also play roles in embryonic patterning and differentiation, probably distinct from their ability to regulate the cell cycle ***(134,140)***, although their real mode of action is poorly understood. In a broader sense, it is still not clear how the cell cycle influences differentiation and vice versa. This will clearly be an important area in the future, and *Xenopus* is a good model system for these studies.

10. Conclusions

The simplified cell cycle systems of oocyte maturation and *Xenopus* eggs have provided much crucial information for elucidating molecular mechanisms of cell cycle control. Indeed, the almost unique advantages of *Xenopus*, namely, the ability to perform easy biochemistry including protein addition and immunodepletion from cell-free extracts that recapitulate basic cell cycle events, have put it at the forefront of cell cycle research in many areas. Pleasingly, much of the information we have gained from studying *Xenopus* embryos has been found to be applicable to mammalian cells. Future uses are likely to extend into study of more complex cell cycles of the older embryo and investigation of the links between cell division and differentiation in vivo. Indeed, *Xenopus*, one of the oldest systems for studying cell cycle regulation, is far from outliving its usefulness.

Acknowledgments

We would like to thank Anna Git for helpful reading of the manuscript. A.P. is supported by the British Heart Foundation (grant PG/03/068). P.R.Y is supported by the National Institutes of Health (grant GM066226-01), the U.S. Department of Defense (grants DAMD17-01-11-0589 and DAMD17-01-1-0415), and the National Science Foundation (grant MCB 9982543).

References

1. Newport, J. and Kirschner, M. (1982) A major developmental transition in early *Xenopus* embryos: II. Control of the onset of transcription. *Cell* **30,** 687–696.
2. Murray, A. W., Solomon, M. J., and Kirschner, M. W. (1989) The role of cyclin synthesis and degradation in the control of maturation promoting factor activity. *Nature* **339,** 280–286.

3. Murray, A. W. and Kirschner, M. W. (1989) Cyclin synthesis drives the early embryonic cell cycle. *Nature* **339,** 275–280.
4. Blow, J. J. and Laskey, R. A. (1986) Initiation of DNA replication in nuclei and purified DNA by a cell-free extract of *Xenopus* eggs. *Cell* **47,** 577–587.
5. Nebreda, A. R. and Ferby, I. (2000) Regulation of the meiotic cell cycle in oocytes. *Curr. Opin. Cell. Biol.* **12,** 666–675.
6. Tunquist, B. J. and Maller, J. L. (2003) Under arrest: cytostatic factor (CSF)-mediated metaphase arrest in vertebrate eggs. *Genes Dev.* **17,** 683–710.
7. Masui, Y. and Markert, C. L. (1971) Cytoplasmic control of nuclear behavior during meiotic maturation of frog oocytes. *J. Exp. Zool.* **177,** 129–145.
8. Gautier, J., Norbury, C., Lohka, M., Nurse, P., and Maller, J. (1988) Purified maturation-promoting factor contains the product of a *Xenopus* homolog of the fission yeast cell cycle control gene cdc2+. *Cell* **54,** 433–439.
9. Gautier, J. and Maller, J. L. (1991) Cyclin B in *Xenopus* oocytes: implications for the mechanism of pre-MPF activation. *EMBO J.* **10,** 177–182.
10. Dunphy, W. G., Brizuela, L., Beach, D., and Newport, J. (1988) The *Xenopus* cdc2 protein is a component of MPF, a cytoplasmic regulator of mitosis. *Cell* **54,** 423–431.
11. Lohka, M. J., Hayes, M. K., and Maller, J. L. (1988) Purification of maturation-promoting factor, an intracellular regulator of early mitotic events. *Proc. Natl. Acad. Sci. USA* **85,** 3009–3013.
12. Ferrell, J. E., Jr. (1999) *Xenopus* oocyte maturation: new lessons from a good egg. *Bioessays* **21,** 833–842.
13. Ferrell, J. E., Jr., Wu, M., Gerhart, J. C., and Martin, G. S. (1991) Cell cycle tyrosine phosphorylation of p34cdc2 and a microtubule-associated protein kinase homolog in *Xenopus* oocytes and eggs. *Mol. Cell Biol.* **11,** 1965–1971.
14. Jessus, C., Rime, H., Haccard, O., et al. (1991) Tyrosine phosphorylation of p34cdc2 and p42 during meiotic maturation of *Xenopus* oocyte. Antagonistic action of okadaic acid and 6-DMAP. *Development* **111,** 813–820.
15. Posada, J., Sanghera, J., Pelech, S., Aebersold, R., and Cooper, J. A. (1991) Tyrosine phosphorylation and activation of homologous protein kinases during oocyte maturation and mitogenic activation of fibroblasts. *Mol. Cell. Biol.* **11,** 2517–2528.
16. Mueller, P. R., Coleman, T. R., Kumagai, A., and Dunphy, W. G. (1995) Myt1: a membrane-associated inhibitory kinase that phosphorylates Cdc2 on both threonine-14 and tyrosine-15. *Science* **270,** 86–90.
17. Strausfeld, U., Labbe, J. C., Fesquet, D., et al. (1991) Dephosphorylation and activation of a p34cdc2/cyclin B complex in vitro by human CDC25 protein. *Nature* **351,** 242–245.
18. Gautier, J., Solomon, M. J., Booher, R. N., Bazan, J. F., and Kirschner, M. W. (1991) cdc25 is a specific tyrosine phosphatase that directly activates p34cdc2. *Cell* **67,** 197–211.
19. Nakajo, N., Oe, T., Uto, K., and Sagata, N. (1999) Involvement of Chk1 kinase in prophase I arrest of *Xenopus* oocytes. *Dev. Biol.* **207,** 432–444.
20. Sagata, N., Daar, I., Oskarsson, M., Showalter, S. D., and Vande Woude, G. F. (1989) The product of the mos proto-oncogene as a candidate "initiator" for oocyte maturation. *Science* **245,** 643–646.
21. Sagata, N., Oskarsson, M., Copeland, T., Brumbaugh, J., and Vande Woude, G. F. (1988) Function of c-mos proto-oncogene product in meiotic maturation in *Xenopus* oocytes. *Nature* **335,** 519–525.
22. Yew, N., Mellini, M. L., and Vande Woude, G. F. (1992) Meiotic initiation by the mos protein in Xenopus. *Nature* **355,** 649–652.

23. Gebauer, F., Xu, W., Cooper, G. M. & Richter, J. D. (1994) Translational control by cytoplasmic polyadenylation of c-mos mRNA is necessary for oocyte maturation in the mouse. *EMBO J.* **13,** 5712.
24. Sheets, M. D., Wu, M., and Wickens, M. (1995) Polyadenylation of c-mos mRNA as a control point in *Xenopus* meiotic maturation. *Nature 374,* 511–516.
25. Sagata, N. (1997) What does Mos do in oocytes and somatic cells? *Bioessays* **19,** 13–21.
26. Posada, J., Yew, N., Ahn, N. G., Vande Woude, G. F., and Cooper, J. A. (1993) Mos stimulates MAP kinase in *Xenopus* oocytes and activates a MAP kinase kinase in vitro. *Mol. Cell. Biol.* **13,** 2546–2553.
27. Nebreda, A. R. and Hunt, T. (1993) The c-mos proto-oncogene protein kinase turns on and maintains the activity of MAP kinase, but not MPF, in cell-free extracts of *Xenopus* oocytes and eggs. *EMBO J.* **12,** 1979–1986.
28. Shibuya, E. K. and Ruderman, J. V. (1993) Mos induces the in vitro activation of mitogen-activated protein kinases in lysates of frog oocytes and mammalian somatic cells. *Mol. Biol. Cell.* **4,** 781–790.
29. Gotoh, Y., Masuyama, N., Dell, K., Shirakabe, K., and Nishida, E. (1995) Initiation of *Xenopus* oocyte maturation by activation of the mitogen-activated protein kinase cascade. *J. Biol. Chem.* **270,** 25898–25904.
30. Haccard, O., Lewellyn, A., Hartley, R. S., Erikson, E., and Maller, J. L. (1995) Induction of Xenopus oocyte meiotic maturation by MAP kinase. *Dev. Biol.* **168,** 677–682.
31. Howard, E. L., Charlesworth, A., Welk, J., and MacNicol, A. M. (1999) The mitogen-activated protein kinase signaling pathway stimulates mos mRNA cytoplasmic polyadenylation during *Xenopus* oocyte maturation. *Mol. Cell. Biol.* **19,** 1990–1999.
32. Matten, W. T., Copeland, T. D., Ahn, N. G., and Vande Woude, G. F. (1996) Positive feedback between MAP kinase and Mos during *Xenopus* oocyte maturation. *Dev. Biol.* **179,** 485–492.
33. Nishizawa, M., Furuno, N., Okazaki, K., Tanaka, H., Ogawa, Y., and Sagata, N. (1993) Degradation of Mos by the N-terminal proline (Pro2)-dependent ubiquitin pathway on fertilization of *Xenopus* eggs: possible significance of natural selection for Pro2 in Mos. *EMBO J.* **12,** 4021–4027.
34. Bhatt, R. R. and Ferrell, J. E., Jr. (1999) The protein kinase p90 rsk as an essential mediator of cytostatic factor activity. *Science* **286,** 1362–1365.
35. Gross, S. D., Schwab, M. S., Lewellyn, A. L., and Maller, J. L. (1999) Induction of metaphase arrest in cleaving *Xenopus* embryos by the protein kinase p90Rsk. *Science* **286,** 1365–1367.
36. Gross, S. D., Schwab, M. S., Taieb, F. E., Lewellyn, A. L., Qian, Y. W., and Maller, J. L. (2000) The critical role of the MAP kinase pathway in meiosis II in *Xenopus* oocytes is mediated by p90(Rsk) *Curr. Biol.* **10,** 430–438.
37. Schwab, M. S., Roberts, B. T., Gross, S. D., et al. (2001) Bub1 is activated by the protein kinase p90(Rsk) during *Xenopus* oocyte maturation. *Curr. Biol.* **11,** 141–150.
38. Palmer, A., Gavin, A. C., and Nebreda, A. R. (1998) A link between MAP kinase and p34(cdc2)/cyclin B during oocyte maturation: p90(rsk) phosphorylates and inactivates the p34(cdc2) inhibitory kinase Myt1. *EMBO J.* **17,** 5037–5047.
39. Kumagai, A. and Dunphy, W. G. (1996) Purification and molecular cloning of Plx1, a Cdc25-regulatory kinase from *Xenopus* egg extracts. *Science* **273,** 1377–1380.

40. Takizawa, C. G. and Morgan, D. O. (2000) Control of mitosis by changes in the subcellular location of cyclin-B1-Cdk1 and Cdc25C. *Curr. Opin. Cell. Biol.* **12,** 658–665.
41. Nebreda, A. R., Gannon, J. V., and Hunt, T. (1995) Newly synthesized protein(s) must associate with p34cdc2 to activate MAP kinase and MPF during progesterone-induced maturation of *Xenopus* oocytes. *EMBO J.* **14,** 5597–5607.
42. Taieb, F. E., Gross, S. D., Lewellyn, A. L., and Maller, J. L. (2001) Activation of the anaphase-promoting complex and degradation of cyclin B is not required for progression from meiosis I to II in *Xenopus* oocytes. *Curr. Biol.* **11,** 508–513.
43. Reimann, J. D. and Jackson, P. K. (2002) Emi1 is required for cytostatic factor arrest in vertebrate eggs. *Nature* **416,** 850–854.
44. Watanabe, N., Vande Woude, G. F., Ikawa, Y., and Sagata, N. (1989) Specific proteolysis of the c-mos proto-oncogene product by calpain on fertilization of *Xenopus* eggs. *Nature* **342,** 505–511.
45. Reimann, J. D., Freed, E., Hsu, J. Y., Kramer, E. R., Peters, J. M., and Jackson, P. K. (2001) Emi1 is a mitotic regulator that interacts with Cdc20 and inhibits the anaphase promoting complex. *Cell* **105,** 645–655.
46. Philpott, A., Leno, G. H., and Laskey, R. A. (1991) Sperm decondensation in *Xenopus* egg cytoplasm is mediated by nucleoplasmin. *Cell* **65,** 569–578.
47. Lohka, M. J. and Masui, Y. (1984) Roles of cytosol and cytoplasmic particles in nuclear envelope assembly and sperm pronuclear formation in cell-free preparations from amphibian eggs. *J. Cell. Biol.* **98,** 1222–1230.
48. Coverley, D., Laman, H., and Laskey, R. A. (2002) Distinct roles for cyclins E and A during DNA replication complex assembly and activation. *Nat. Cell. Biol.* **4,** 523–528.
49. Laskey, R. A. and Harland, R. M. (1980) Regulated replication of DNA microinjected into eggs of *Xenopus laevis*. *Cell* **21,** 761–771.
50. Tada, S. and Blow, J. J. (1998) The replication licensing system. *Biol. Chem.* **379,** 941–949.
51. Coverley, D. and Laskey, R. A. (1994) Regulation of eukaryotic DNA replication. *Annu. Rev. Biochem.* **63,** 745–776.
52. Rowles, A., Chong, J. P., Brown, L., Howell, M., Evan, G. I. & Blow, J. J. (1996) Interaction between the origin recognition complex and the replication licensing system in *Xenopus*. *Cell* **87,** 287–296.
53. Diffley, J. F. (1998) Replication conrol: choreographing replication origins. *Curr. Biol.* **8,** R771–R773.
54. Newlon, C. S. (1997) Putting it all together: building a prereplicative complex. *Cell* **91,** 717–720.
55. Carpenter, P. B., Mueller, P. R., and Dunphy, W. G. (1996) Role for a *Xenopus* Orc2-related protein in controlling DNA replication. *Nature* **379,** 357–360.
56. Coleman, T. R., Carpenter, P. B., and Dunphy, W. G. (1996) The *Xenopus* Cdc6 protein is essential for the initiation of a single round of DNA replication in cell-free extracts. *Cell* **87,** 53–63.
57. Kubota, Y., Mimura, S., Nishimoto, S., Takisawa, H., and Nojima, H. (1995) Identification of the yeast MCM3-related protein as a component of *Xenopus* DNA replication licensing factor. *Cell* **81,** 601–609.
58. Chong, J. P., Mahbubani, H. M., Khoo, C. Y., and Blow, J. J. (1995) Purification of an MCM-containing complex as a component of the DNA replication licensing system. *Nature* **375,** 418–421.
59. Madine, M. A., Khoo, C. Y., Mills, A. D., and Laskey, R. A. (1995) MCM3 complex required for cell cycle regulation of DNA replication in vertebrate cells. *Nature* **375,** 421–424.

60. Leatherwood, J. (1998) Emerging mechanisms of eukaryotic DNA replication initiation. *Curr. Opin. Cell. Biol.* **10,** 742–748.
61. Romanowski, P., Madine, M. A., and Laskey, R. A. (1996) XMCM7, a novel member of the *Xenopus* MCM family, interacts with XMCM3 and colocalizes with it throughout replication. *Proc. Natl. Acad. Sci. USA* **93,** 10189–10194.
62. Romanowski, P., Madine, M. A., Rowles, A., Blow, J. J., and Laskey, R. A. (1996) The *Xenopus* origin recognition complex is essential for DNA replication and MCM binding to chromatin. *Curr. Biol.* **6,** 1416–1425.
63. Tada, S., Li, A., Maiorano, D., Mechali, M., and Blow, J. J. (2001) Repression of origin assembly in metaphase depends on inhibition of RLF- B/Cdt1 by geminin. *Nat. Cell. Biol.* **3,** 107–113.
64. Wohlschlegel, J. A., Dwyer, B. T., Dhar, S. K., Cvetic, C., Walter, J. C., and Dutta, A. (2000) Inhibition of eukaryotic DNA replication by geminin binding to Cdt1. *Science* **290,** 2309–2312.
65. Lei, M. and Tye, B. K. (2001) Initiating DNA synthesis: from recruiting to activating the MCM complex. *J. Cell. Sci.* **114,** 1447–1454.
66. Jackson, P. K., Chevalier, S., Philippe, M., and Kirschner, M. W. (1995) Early events in DNA replication require cyclin E and are blocked by p21CIP1. *J. Cell. Biol.* **130,** 755–769.
67. Walter, J. and Newport, J. (2000) Initiation of eukaryotic DNA replication: origin unwinding and sequential chromatin association of Cdc45, RPA, and DNA polymerase alpha. *Mol. Cell.* **5,** 617–627.
68. Sclafani, R. A. (1998) Chromosomes in the Rocky Mountains. Yeast chromosome structure, replication and segregation, Snowmass, CO, USA, 8–13 August 1998. *Trends Genet.* **14,** 441–442.
69. Jackson, A. L., Pahl, P. M., Harrison, K., Rosamond, J., and Sclafani, R. A. (1993) Cell cycle regulation of the yeast Cdc7 protein kinase by association with the Dbf4 protein. *Mol. Cell. Biol.* **13,** 2899–908.
70. Mimura, S. and Takisawa, H. (1998) *Xenopus* Cdc45-dependent loading of DNA polymerase alpha onto chromatin under the control of S-phase Cdk. *EMBO J.* **17,** 5699–707.
71. Wohlschlegel, J. A., Dhar, S. K., Prokhorova, T. A., Dutta, A., and Walter, J. C. (2002) *Xenopus* Mcm10 binds to origins of DNA replication after Mcm2-7 and stimulates origin binding of Cdc45. *Mol. Cell.* **9,** 233–240.
72. Walter, J. C. (2000) Evidence for sequential action of cdc7 and cdk2 protein kinases during initiation of DNA replication in *Xenopus* egg extracts. *J. Biol. Chem.* **275,** 39773–39778.
73. Hubscher, U., Maga, G., and Spadari, S. (2002) Eukaryotic DNA polymerases. *Annu. Rev. Biochem.* **71,** 133–163.
74. Pagano, M., Pepperkok, R., Verde, F., Ansorge, W., and Draetta, G. (1992) Cyclin A is required at two points in the human cell cycle. *EMBO J.* **11,** 961–971.
75. Girard, F., Strausfeld, U., Fernandez, A., and Lamb, N. J. (1991) Cyclin A is required for the onset of DNA replication in mammalian fibroblasts. *Cell* **67,** 1169–1179.
76. Fotedar, A., Cannella, D., Fitzgerald, P., et al. (1996) Role for cyclin A-dependent kinase in DNA replication in human S phase cell extracts. *J. Biol. Chem.* **271,** 31627–31637.
77. Waga, S. and Stillman, B. (1998) The DNA replication fork in eukaryotic cells. *Annu. Rev. Biochem.* **67,** 721–751.
78. Bogan, J. A., Natale, D. A., and Depamphilis, M. L. (2000) Initiation of eukaryotic DNA replication: conservative or liberal? *J. Cell. Physiol.* **184,** 139–150.

79. DePamphilis, M. L. (2000) Review: nuclear structure and DNA replication. *J. Struct. Biol.* **129,** 186–197.
80. Fujita, M. (1999) Cell cycle regulation of DNA replication initiation proteins in mammalian cells. *Front. Biosci.* **4,** D816–D823.
81. Waga, S. and Stillman, B. (1994) Anatomy of a DNA replication fork revealed by reconstitution of SV40 DNA replication in vitro. *Nature* **369,** 207–212.
82. Blow, J. J. and Tada, S. (2000) Cell cycle. A new check on issuing the licence. [letter; comment]. *Nature* **404,** 560–561.
83. McGarry, T. J. and Kirschner, M. W. (1998) Geminin, an inhibitor of DNA replication, is degraded during mitosis. *Cell* **93,** 1043–1053.
84. Madine, M. & Laskey, R. (2001) Geminin bans replication licence. *Nat. Cell. Biol.* **3,** E49–E50.
85. Graham, C. F. and Morgan, R. W. (1966) Changes in the cell cycle during early amphibian development. *Dev. Biol.* **14,** 436–460.
86. King, R. W., Jackson, P. K., and Kirschner, M. W. (1994) Mitosis in transition [see comments]. *Cell* **79,** 563–571.
87. Russo, A. A., Jeffrey, P. D., Patten, A. K., Massague, J., and Pavletich, N. P. (1996) Crystal structure of the p27Kip1 cyclin-dependent-kinase inhibitor bound to the cyclin A-Cdk2 complex [see comments]. *Nature* **382,** 325–331.
88. Fesquet, D., Labbe, J. C., Derancourt, J., et al. (1993) The Mo15 gene encodes the catalytic subunit of a protein-kinase that activates Cdc2 and other cyclin-dependent kinases (Cdks) through phosphorylation of thr161 and its homologs. *EMBO J.* **12,** 3111–3121.
89. Solomon, M. J. (1994) The function(s) of CAK, the p34cdc2-activating kinase. *Trends Biochem. Sci.* **19,** 496–500.
90. Mueller, P. R., Coleman, T. R., and Dunphy, W. G. (1995) Cell cycle regulation of a *Xenopus* Wee1-like kinase. *Mol. Biol. Cell.* **6,** 119–134.
91. Kumagai, A. and Dunphy, W. G. (1991) The cdc25 protein controls tyrosine dephosphorylation of the cdc2 protein in a cell-free system. *Cell* **64,** 903–914.
92. Dunphy, W. G. and Kumagai, A. (1991) The cdc25 protein contains an intrinsic phosphatase activity. *Cell* **67,** 189–196.
93. Galaktionov, K. and Beach, D. (1991) Specific activation of cdc25 tyrosine phosphatases by B-type cyclins: evidence for multiple roles of mitotic cyclins. *Cell* **67,** 1181–1194.
94. Kumagai, A. and Dunphy, W. G. (1992) Regulation of the cdc25 protein during the cell cycle in *Xenopus* extracts. *Cell* **70,** 139–151.
95. Millar, J. B. and Russell, P. (1992) The cdc25 M-phase inducer: an unconventional protein phosphatase. *Cell* 68, 407–410.
96. Dasso, M. & Newport, J. W. (1990) Completion of DNA replication is monitored by a feedback system that controls the initiation of mitosis in vitro: studies in *Xenopus*. *Cell* **61,** 811–823.
97. Kumagai, A., Guo, Z., Emami, K. H., Wang, S. X., and Dunphy, W. G. (1998) The *Xenopus* Chk1 protein kinase mediates a caffeine-sensitive pathway of checkpoint control in cell-free extracts. *J. Cell. Biol.* **142,** 1559–1569.
98. Kumagai, A. and Dunphy, W. G. (1999) Binding of 14-3-3 proteins and nuclear export control the intracellular localization of the mitotic inducer Cdc25. *Genes Dev.* **13,** 1067–1072.
99. Kumagai, A., Yakowec, P. S., and Dunphy, W. G. (1998) 14-3-3 proteins act as negative regulators of the mitotic inducer Cdc25 in *Xenopus* egg extracts. *Mol. Biol. Cell.* **9,** 345–354.

100. Nigg, E. A., Blangy, A. & Lane, H. A. (1996) Dynamic changes in nuclear architecture during mitosis: on the role of protein phosphorylation in spindle assembly and chromosome segregation. *Exp. Cell. Res.* **229,** 174–180.
101. Nigg, E. A. (1991) The substrates of the cdc2 kinase. *Semin. Cell. Biol.* **2,** 261–270.
102. Stukenberg, P. T., Lustig, K. D., McGarry, T. J., King, R. W., Kuang, J., and Kirschner, M. W. (1997) Systematic identification of mitotic phosphoproteins. *Curr. Biol.* **7,** 338–348.
103. King, R. W., Deshaies, R. J., Peters, J. M., and Kirschner, M. W. (1996) How proteolysis drives the cell cycle. *Science* **274,** 1652–1659.
104. Hershko, A. and Ciechanover, A. (1998) The ubiquitin system. *Annu. Rev. Biochem.* **67,** 425–479.
105. Ciechanover, A., Orian, A., and Schwartz, A. L. (2000) Ubiquitin-mediated proteolysis: biological regulation via destruction. *Bioessays* **22,** 442–451.
106. Kornitzer, D. and Ciechanover, A. (2000) Modes of regulation of ubiquitin-mediated protein degradation. *J. Cell. Physiol.* **182,** 1–111.
107. King, R. W., Peters, J. M., Tugendreich, S., Rolfe, M., Hieter, P., and Kirschner, M. W. (1995) A 20S complex containing CDC27 and CDC16 catalyzes the mitosis-specific conjugation of ubiquitin to cyclin B. *Cell* **81,** 279–288.
108. Zou, H., McGarry, T. J., Bernal, T., and Kirschner, M. W. (1999) Identification of a vertebrate sister-chromatid separation inhibitor involved in transformation and tumorigenesis. *Science* **285,** 418–422.
109. Holloway, S. L., Glotzer, M., King, R. W., and Murray, A. W. (1993) Anaphase is initiated by proteolysis rather than by the inactivation of maturation-promoting factor. *Cell* **73,** 1393–1402.
110. Yu, H., Peters, J. M., King, R. W., Page, A. M., Hieter, P., and Kirschner, M. W. (1998) Identification of a cullin homology region in a subunit of the anaphase-promoting complex. *Science* **279,** 1219–1222.
111. Peters, J. M. (1999) Subunits and substrates of the anaphase-promoting complex. *Exp. Cell. Res.* **248,** 339–349.
112. Zachariae, W. and Nasmyth, K. (1999) Whose end is destruction: cell division and the anaphase-promoting complex. *Genes Dev.* **13,** 2039–2058.
113. Pfleger, C. M. and Kirschner, M. W. (2000) The KEN box: an APC recognition signal distinct from the D box targeted by Cdh1. *Genes Dev.* **14,** 655–665.
114. Fang, G., Yu, H., and Kirschner, M. W. (1998) Direct binding of CDC20 protein family members activates the anaphase-promoting complex in mitosis and G1. *Mol. Cell.* **2,** 163–171.
115. Chen, R. H., Waters, J. C., Salmon, E. D., and Murray, A. W. (1996) Association of spindle assembly checkpoint component XMAD2 with unattached kinetochores. *Science* **274,** 242–246.
116. Howe, J. A., Howell, M., Hunt, T., and Newport, J. W. (1995) Identification of a developmental timer regulating the stability of embryonic cyclin A and a new somatic A-type cyclin at gastrulation. *Genes Dev.* **9,** 1164–1176.
117. King, R. W., Glotzer, M., and Kirschner, M. W. (1996) Mutagenic analysis of the destruction signal of mitotic cyclins and structural characterization of ubiquitinated intermediates. *Mol. Biol. Cell.* **7,** 1343–1357.
118. Sudakin, V., Ganoth, D., Dahan, A., et al. (1995) The cyclosome, a large complex containing cyclin-selective ubiquitin ligase activity, targets cyclins for destruction at the end of mitosis. *Mol. Biol. Cell.* **6,** 185–197.
119. Hunt, T., Luca, F. C., and Ruderman, J. V. (1992) The requirements for protein synthesis and degradation, and the control of destruction of cyclins A and B in the meiotic and mitotic cell cycles of the clam embryo. *J. Cell. Biol.* **116,** 707–724.

120. Walker, D. H. and Maller, J. L. (1991) Role for cyclin A in the dependence of mitosis on completion of DNA replication. *Nature* **354,** 314–317.
121. Newport, J. & Kirschner, M. (1982) A major developmental transition in early *Xenopus* embryos: I. Characterization and timing of cellular changes at the midblastula stage. *Cell* **30,** 675–686.
122. Clarke, P. R., Leiss, D., Pagano, M., and Karsenti, E. (1992) Cyclin A- and cyclin B-dependent protein kinases are regulated by different mechanisms in *Xenopus* egg extracts. *EMBO J.* **11,** 1751–1761.
123. Rempel, R. E., Sleight, S. B., and Maller, J. L. (1995) Maternal *Xenopus* Cdk2-cyclin E complexes function during meiotic and early embryonic cell cycles that lack a G1 phase. *J. Biol. Chem.* **270,** 6843–6855.
124. Hartley, R. S., Rempel, R. E., and Maller, J. L. (1996) In vivo regulation of the early embryonic cell cycle in *Xenopus*. *Dev. Biol.* **173,** 408–419.
125. Hensey, C. and Gautier, J. (1997) A developmental timer that regulates apoptosis at the onset of gastrulation. *Mech. Dev.* **69,** 183–195.
126. Stack, J. H. and Newport, J. W. (1997) Developmentally regulated activation of apoptosis early in *Xenopus* gastrulation results in cyclin A degradation during interphase of the cell cycle. *Development* **124,** 3185–3195.
127. Su, J. Y., Rempel, R. E., Erikson, E., and Maller, J. L. (1995) Cloning and characterization of the *Xenopus* cyclin-dependent kinase inhibitor p27XIC1. *Proc. Natl. Acad. Sci. USA* **92,** 10187–10191.
128. Shou, W. and Dunphy, W. G. (1996) Cell cycle control by *Xenopus* p28Kix1, a developmentally regulated inhibitor of cyclin-dependent kinases. *Mol. Biol. Cell.* **7,** 457–469.
129. Chuang, L. C. and Yew, P. R. (2001) Regulation of nuclear transport and degradation of the *Xenopus* cyclin-dependent kinase inhibitor, p27Xic1. *J. Biol. Chem.* **276,** 1610–1617.
130. You, Z., Harvey, K., Kong, L., and Newport, J. (2002) Xic1 degradation in *Xenopus* egg extracts is coupled to initiation of DNA replication. *Genes Dev.* **16,** 1182–1194.
131. Vernon, A. E. and Philpott, A. (2003) The developmental expression of cell cycle regulators in *Xenopus laevis*. *Gene. Expr. Patterns* **3,** 179–192.
132. Destree, O. H., Lam, K. T., Peterson-Maduro, L. J., et al. (1992) Structure and expression of the Xenopus retinoblastoma gene. *Dev. Biol.* **153,** 141–149.
133. Philpott, A. and Friend, S. H. (1994) E2F and its developmental regulation in Xenopus laevis. *Mol. Cell. Biol.* **14,** 5000–5009.
134. Suzuki, A. and Hemmati-Brivanlou, A. (2000) Xenopus embryonic E2F is required for the formation of ventral and posterior cell fates during early embryogenesis. *Mol. Cell.* **5,** 217–229.
135. Ohnuma, S., Philpott, A., Wang, K., Holt, C. E., and Harris, W. A. (1999) p27Xic1, a Cdk inhibitor, promotes the determination of glial cells in *Xenopus* retina. *Cell* **99,** 499–510.
136. Vernon, A. E. and Philpott, A. (2003) A single cdk inhibitor, p27Xic1, functions beyond cell cycle regulation to promote muscle differentiation in *Xenopus*. *Development* **130,** 71–83.
137. Vernon, A. E., Devine, C., and Philpott, A. (2003) The cdk inhibitor p27Xic1 is required for differentiation of primary neurones in *Xenopus*. *Development* **130,** 85–92.
138. Cremisi, F., Philpott, A., and Ohnuma, S. (2003) Cell cycle and cell fate interactions in neural development. *Curr. Opin. Neurobiol.* **13,** 26–33.
139. Ohnuma, S., Philpott, A., and Harris, W. A. (2001) Cell cycle and cell fate in the nervous system. *Curr. Opin. Neurobiol.* **11,** 66–73.
140. Kroll, K. L., Salic, A. N., Evans, L. M., and Kirschner, M. W. (1998) Geminin, a neuralizing molecule that demarcates the future neural plate at the onset of gastrulation. *Development* **125,** 3247–3258.

6

The Mammalian Cell Cycle

An Overview

Jane V. Harper and Gavin Brooks

Summary

In recent years, we have witnessed major advances in our understanding of the mammalian cell cycle and how it is regulated. Normal mammalian cellular proliferation is tightly regulated at each phase of the cell cycle by the activation and deactivation of a series of proteins that constitute the cell cycle machinery. This review article describes the various phases of the mammalian cell cycle and focuses on the cell cycle regulatory molecules that act at each stage to ensure normal cellular progression.

Key Words

14-3-3; anaphase-promoting complex; CDC25; cyclins; cyclin-dependent kinases (Cdks); Cdk inhibitors; cytokinesis; DNA replication; E2F transcription factors; endoreduplication; MAP kinase; pocket proteins; SCF ubiquitin ligases.

1. Introduction

Cell division in mammalian cells is similar to that in other eukaryotes in that it represents an evolutionarily conserved process involving an ordered and tightly controlled series of molecular events. In mammals, the cell cycle consists of five distinct phases: three gap phases—G_0, in which cells remain in a quiescent or resting state, and G_1 and G_2, during which RNA synthesis and protein synthesis occur; S-phase during which DNA is replicated; and M-phase, in which cells undergo mitosis and cytokinesis (**Fig. 1**). G_0, G_1, S, and G_2 are referred to collectively as interphase (i.e., between mitoses). Some cells in the body remain quiescent for their whole lifetime and do not undergo cell division; however, stimulation of the cell by external factors such as

From: *Methods in Molecular Biology, vol. 296, Cell Cycle Control: Mechanisms and Protocols*
Edited by: T. Humphrey and G. Brooks © Humana Press Inc., Totowa, NJ

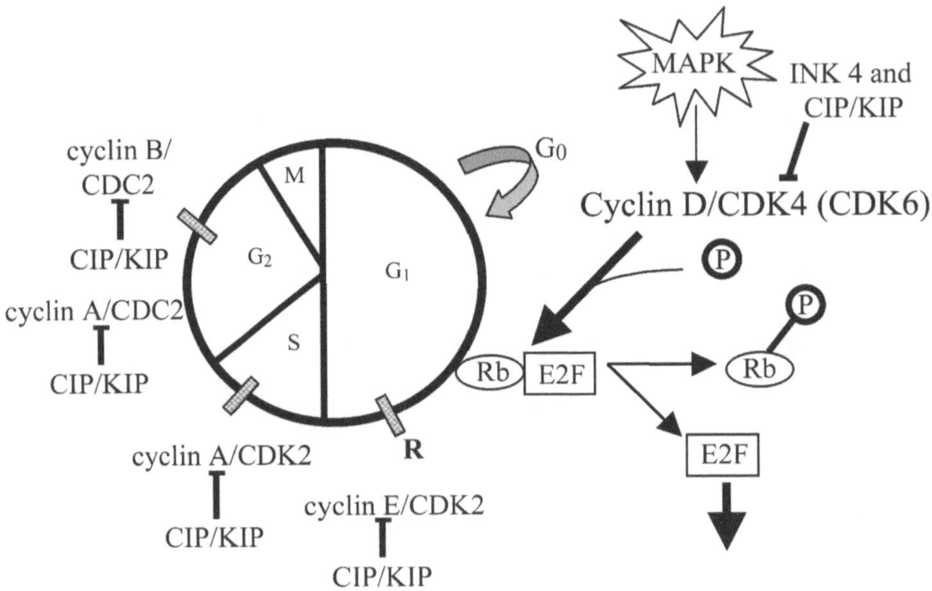

Fig. 1. The five distinct phases of the cell cycle are each controlled by specific cyclin/CDK complexes. The cyclin/CDK complexes in turn are negatively regulated by CIP/KIP and INK4 CDKI family members. E2F transcription factors function at the restriction point (R), leading to the activation of genes essential for DNA synthesis and cell cycle progression. E2F complexed with hypophosphorylated Rb cannot activate transcription. Hyperphosphorylation of Rb causes dissociation from E2F. Cell cycle checkpoints are shown as shaded bars. ⊣, inhibition step; →, activation step; Cdk, cyclin-dependent kinase; CIP, Cdk-interacting protein; INK4, inhibitior of Cdk4; KIP, kinase inhibitor protein; MAPK, mitogen-activated protein kinase.

mitogens causes these quiescent cells to reenter the cell cycle and undergo division. Binding of a growth factor molecule to its cell surface receptor can stimulate a number of signaling pathways, an example of which is the Ras-dependent mitogen-activated protein kinase (MAPK) pathway, which plays a major role in entry into G_1, as discussed in more detail **Subheading 2.1.** Once cells enter G_1, synthesis of the mRNAs and proteins necessary for DNA synthesis occurs, allowing cells to enter S-phase.

The mammalian cell cycle consists of a number of checkpoints that exist to ensure normal cell cycle progression. The primary checkpoint acts late in G_1 and is known as the restriction (R) point (**Fig. 1**). Once cells have passed this point, they normally are committed to a round of cell division. Other checkpoints exist in S-phase to activate DNA repair mechanisms if necessary and at the G_2/M transition to ensure that cells have fully replicated their DNA and that it is undamaged before they enter mitosis. Finally, there are checkpoint control mechanisms within mitosis to ensure that conditions remain suitable for the cell to complete cell division (cytokinesis).

The length of time for a mammalian cell to progress around the cell cycle and undergo division varies depending on the cell type but on average it takes approx 24 h.

Cell cycle time varies in different cell types as a consequence of differences in the time spent between cytokinesis and the restriction point (i.e., G_1). The time taken for a cell to pass from S-phase into M is extremely constant between cells and typically is in the region of approx 6 h for S-phase, 4 h for G_2 and 1–2 h for mitosis and cytokinesis ***(1)***.

Progression through each phase of the cell cycle is under the strict control of various cell cycle molecules, e.g., cyclins, cyclin-dependent kinases (Cdks), and Cdk inhibitors (CDKIs), which themselves are regulated by phosphorylation and dephosphorylation events ***(1)***. The Cdks play a crucial role in regulating cell cycle events once complexed with a cyclin regulatory subunit. Cyclin levels vary dramatically through the cell cycle as a consequence of changes in transcription and ubiquitin-mediated degradation (for review *see* **refs.** ***2*** and ***3***). Cdk activity is negatively regulated by association with various CDKIs. Specific cyclin/Cdk complexes become activated and thereby modulate a distinct phase(s) of the cell cycle (**Fig. 1**). For example, cyclin D/Cdk4(Cdk6) complexes initiate progression through G_1 by phosphorylating substrates, such as members of the retinoblastoma (Rb) family of pocket proteins (*see* **Subheading 3.1.**), that eventually lead to the activation of transcription of genes necessary for DNA synthesis and subsequent cell cycle progression; the cyclin E/Cdk2 complex is important in the G_1/S transition, where levels peak at the restriction point ***(4,5)***; cyclin A/Cdk2 is important during S-phase progression; and cyclin A/CDC2 (also known as Cdk1) and cyclin B/CDC2 are important for progression through G_2 and M. The regulation of cyclin synthesis and degradation, in addition to Cdk activity levels, are tightly controlled and is key to ordered progression through the mammalian cell cycle. The following sections will give an overview of the various stages of the mammalian cell cycle and the molecules that regulate progression through each stage of the cycle.

1.1. Cyclins and Cyclin-Dependent Kinases

As for other eukaryotic cells, the mammalian cell cycle is regulated by the sequential formation, activation, and inactivation of a series of cell cycle regulatory molecules that include the cyclins (regulatory subunits) and the Cdks (catalytic kinase subunits) (*2,3 ,6*; *see* also Chap. 16, this volume). Different cyclins bind specifically to different Cdks to form distinct complexes at specific phases of the cell cycle and thereby drive the cell from one phase to another. The cyclins are a family of proteins, which, as their name suggests, are synthesized and destroyed during each cell cycle. To date, eight cyclins have been described that directly affect cell cycle progression: cyclins A1 and A2, B1, -2, and –3, C, D1, -2, and –3, E1 and -2, F, G1 and G2, and H, which all share an approx 150-amino acid region of homology called the *cyclin box* that binds to the N-terminal end of specific Cdks (review, *see* **ref.** ***2***). Most cyclin mRNAs and proteins show a dramatic fluctuation in their expression during the cell cycle. For example, the expressions of cyclins A and B accumulate transiently at the onset of S-phase and in late G2, respectively, followed rapidly by their degradation via the ubiquitin–proteosome pathway, whereas cyclin D1 levels rise in G_1 and remain elevated until mitosis ***(3,6)***. In contrast, expression of the various Cdk molecules remains relatively constant throughout the cell cycle.

Little information is currently available regarding the recently described cyclins F and G, whereas cyclin H has been shown to form complexes specifically with Cdk7 to produce an enzyme known as Cdk-activating kinase (CAK) that is involved in the activation of CDC2 and Cdk2 kinases by phosphorylating Thr160 and Thr161, respectively *(7)*. Cyclin H/Cdk7 can also form a tertiary complex with the protein ménage-à-trois-1 (MAT-1), when it modulates gene transcription via RNA polymerase II activity *(8)*. Another cyclin, cyclin T, has recently been reported in the literature, although it does not appear to be involved with cell cycle progression directly. Cyclin T pairs with Cdk9 and is involved in various cellular processes, including basal transcription, signal transduction, and differentiation (reviewed in **refs.** ***9*** and ***10***).

The Cdks are a family of serine/threonine protein kinases that bind to, and are activated by, specific cyclins. To date, at least nine Cdks have been described: CDC2 (Cdk1), Cdk2, Cdk3, Cdk4, Cdk5, Cdk6, Cdk7, Cdk8, and CDK9. Cdks 4, 5, and 6 complex mainly with the cyclin D family and function during the G_0/G_1-phases of the cycle; Cdk2 can also bind with members of the cyclin D family but more commonly associates with cyclins A and E and functions during the G_1- phase and during the G_1/S transition. As just mentioned, Cdk7 is found in association with cyclin H and is able to phosphorylate either CDC2, Cdk2, or the C-terminal domain of the largest subunit of RNA polymerase II, in addition to the TATA box binding protein or TFIIE *(7)*. CDC2 is the mitotic Cdk and forms complexes with cyclins A and B that function in the G_2- and M-phases of the cell cycle. Cdk8 pairs with cyclin C and is found in a large multiprotein complex with RNA polymerase II, where it is thought to control RNA polymerase II function (reviewed in **ref.** ***11***). Finally, Cdk9 is a serine/threonine kinase related to CDC2 that pairs with T-type cyclins. The activity of the cyclin T/Cdk9 complex is not cell cycle regulated but is involved in many processes such as differentiation and basal transcription (reviewed in **refs.** ***9***, ***10***, and ***12***).

As stated above, specific Cdks bind to specific cyclins to form an active complex that integrates signals from extracellular molecules and controls progression through the cell cycle. The Cdk subunit on its own has no detectable kinase activity and requires sequential activation by cyclin binding and subsequent phosphorylation by CAK and dephosphorylation by CDC25 protein phosphatase (*see* **Subheading 6.1.**). This activation process occurs in a two-step manner, as follows:

1. Binding of the cyclin to the Cdk confers partial activity to the kinase. Cyclin binding causes a conformational change in the Cdk molecule, thereby bringing together specific residues involved in orienting ATP phosphate atoms ready for catalysis within the catalytic cleft. These conformational changes also set the stage for subsequent phosphorylation and full activation.
2. Phosphorylation of the cyclin/Cdk complex is performed by CAK which increases Cdk activity approximately 100-fold (13). Phosphorylation occurs on a conserved threonine residue within the T-loop region of the Cdk (Thr160 in CDC2 and Thr161 in other Cdks). Cyclin binding moves the T-loop to expose the phosphorylation site, allowing full activation of the Cdk.

Once activated, the various cyclin/Cdk complexes phosphorylate a number of specific substrates involved in cell cycle progression. Such substrates include the Rb pocket proteins, lamins, and histones. Evidence exists to suggest that cyclins may be involved in determining the substrate specificity of Cdks (reviewed in ref. 14). For example, cyclin A/Cdk2 and cyclin A/CDC2, but not cyclin B/CDC2, can phosphorylate p107, showing regulation of substrate specificity between kinases complexed with cyclins A and B (15). The E2F-1/DP-1 heterodimer is not a substrate for the active cyclin D-dependent kinases but is efficiently phosphorylated by cyclin B-dependent kinases (16). Interestingly, whereas phosphorylation of E2F-1/DP-1 by cyclin B-dependent kinases does not result in a loss of DNA binding activity, phosphorylation of this same heterodimer by cyclin A-dependent kinases does lead to loss of DNA binding (16). Thus, different Cdk complexes can exert contrasting effects on a common substrate depending on the complexed cyclin. The regulation of Cdks themselves by other molecules can also differ depending upon the bound cyclin. Thus, cyclin A/CDC2 complexes do not require activation by CDC25 phosphatase, whereas cyclin B/CDC2 complexes do ***(17)***.

1.2. Cyclin-Dependent Kinase Inhibitors

The cyclins and Cdks often are referred to as positive regulators of the eukaryotic cell cycle. A family of negative regulators also exists, the CDKIs ***(2,18–20)***. The CDKIs comprise two structurally distinct families, the INK4 (inhibitor of Cdk4) and CIP (Cdk-interacting protein)/KIP (kinase inhibitor protein) families (reviewed in **ref.** ***21***). The INK4 family includes p14, p15 (INK4B), p16 (INK4A), p18 (INK4C), and p19 (INK4D), which specifically inhibit G_1 cyclin/Cdk complexes (cyclin D/CDK4 and cyclin D/CDK6) and are involved in G_1-phase control. The CIP/KIP family includes p21 (CIP1/WAF1/SDI1), p27 (KIP1), and p57 (KIP2), which are 38–44% identical in the first 70-amino acid region of their amino terminia ~~region~~ region that is involved in cyclin binding and kinase inhibitory function ***(19,20 ,22)***. The CIP/KIP family displays a broader specificity than the INK4 family, since members interact with, and inhibit the kinase activities of, cyclin E/Cdk2, cyclin D/Cdk4, cyclin D/Cdk6, cyclin A/Cdk2, and cyclin B/CDC2 complexes and also function throughout the cell cycle ***(19)***. Members of the two CDKI families inhibit Cdk activity by distinct mechanisms. Thus, CIP/KIP family members bind to, and inhibit the activity of, the entire cyclin/Cdk complex ***(23)***, whereas INK4 family members block cyclin binding indirectly by causing allosteric changes in the Cdk and hence altering the cyclin binding site; they also act by distorting the ATP binding site that leads to reduced affinity for ATP ***(24)***.

In the case of p21, this CDKI has been shown to exist in both active and inactive cyclin/Cdk complexes, and it has been suggested that the stoichiometry of p21 binding to the cyclin/Cdk complex controls activation/inhibition of the complex ***(25)***. In support of this hypothesis, Zhang and colleagues ***(25)*** demonstrated that p21 exists both in catalytically active and inactive cyclin/Cdk complexes and that the addition of subsaturating concentrations of p21 to cyclin A/CDK2 complexes resulted in a progressive increase in Cdk2 activity, suggesting that low concentrations of p21 might function as a cyclin/Cdk assembly factor, whereas the binding of more than one p21 molecule is required to inhibit Cdk2 activity.

The tumour suppressor protein p53 also plays an important role in cell cycle arrest at the G_1 and G_2 checkpoints subsequent to inducing apoptosis ***(26–28)***. The p53 protein has a central sequence-specific DNA binding domain and a transcriptional activation domain at its amino-terminus; in response to DNA damage, it can induce the transcription of the CDKI p21, which inhibits the activation of various G_1 cyclin/Cdk complexes ***(22,27)***.

Antiproliferative signals such as contact inhibition, senescence ***(29)***, extracellular antimitogenic factors ***(30)***, and cell cycle checkpoints, such as p53 ***(31)***, induce expression of p27, p16, p15, and p21, respectively. The role of cell cycle molecules in regulating proliferation is highlighted by the fact that a number of these molecules are found to be mutated or deregulated in numerous tumors. Indeed, it has been suggested that most human tumours result from a mutation or deletion in one or more cell cycle regulators, especially those that control G_1/S progression. For example, *p16* is mutated in approximately one-third of all human cancers ***(24,32,33)***, and *p53* is the most frequently mutated gene identified in human tumors ***(34)***. Also, many types of tumors show low expression levels of *p27*, which is associated with a poor prognosis ***(35)***, and cyclin D1 ***(23,36)*** and B1 ***(37)*** are often found at increased levels in breast cancer. Furthermore, there is a direct correlation between inactivation of p53 function and cyclin B1 overexpression in many tumors ***(37)***, although no direct interaction between these two molecules has been shown. However, as a direct consequence of this correlation, it has been proposed that cyclin B1 could be used as a tumor antigen and a cancer vaccine in some instances ***(38)***. Cdks have also been found to be deregulated in some tumors; for example, Cdk4 is mutated in melanoma ***(39,40)***, and Cdk2 expression is increased in some breast cancer cells ***(41)***. Indeed, targeting Cdk2 expression with antisense oligonucleotides and RNA interference technologies reduces cellular proliferation in breast tumor cells ***(41)***.

2. The G_0/G_1 Transition

The mammalian cell cycle is influenced by external signals during the G_0- and G_1-phases. The mitogen-activated protein kinase (MAPK) cascade is one of the most ubiquitous signal transduction pathways; it regulates several biological processes including progression of the cell cycle. The MAPK cascade consists of three evolutionarily conserved protein kinases, i.e., MAPK kinase kinase (MAPKKK), MAPK kinase (MAPKK), and MAPK, which are activated sequentially in a Ras-dependent manner (reviewed in **ref.** ***42***).

The MAPK cascade influences cellular proliferation by targeting the cyclin D-dependent kinases ***(43–45)***. Evidence for this comes from the fact that cells that proliferate in the absence of mitogens, for example during embryogenesis, have very little cyclin D-dependent kinase activity ***(46)***.

2.1. Role of MAPK in G_1 Cell Cycle Progression

The activation of cyclin D/Cdk4 and cyclin D/Cdk6 complexes is essential for passage through the G_1-phase, and they exert their regulation on cell cycle progression by phosphorylating Rb pocket proteins. The Rb pocket protein family serves to repress the activity of the E2F transcription factors that themselves are essential for transcription of genes necessary for entry into S-phase (discussed in more detail in **Subheading 3.**). The Ras/MAPK pathway has been shown to control cyclin D gene expression directly. This is mediated primarily by MAPK, which controls activation of the activation protein-1 (AP-1) and ETS transcription factors that then transactivate the cyclin D promoter that contains specific binding sites for both AP-1 and ETS ***(43,47)***. Furthermore, expression studies using direct inhibitors of cyclin D/Cdk4(Cdk6) complexes (e.g., p21) inhibits Ras-induced proliferation ***(48)***. These data demonstrate that MAPK directly regulates cyclin D expression and, consequently, Cdk4 and Cdk6 activities.

The Ras/MAPK pathway also has been shown to regulate Cdks posttranscriptionally by affecting their assembly and catalytic activities. Although the primary role of p21 and p27 is to regulate the activity of Cdks negatively, they are also involved in the assembly of cyclin D/Cdk4(Cdk6) complexes during early G_1 (***49,50***; *see* **Subheading 1.2.**). The Ras/MAPK pathway has been shown to regulate directly the synthesis of the CIP/KIP family of inhibitors, and it has been demonstrated that growth factor stimulation of quiescent cells causes cell cycle reentry and transient expression of p21 that was dependent on MAPK activity ***(51)***.

Entry into S-phase is partly dependent on proteolytic degradation of p27, and this, in turn, has been shown to be dependent on MAPK activity ***(52,53)***. These investigators also observed that expression of Ras resulted in decreased p27 protein levels and an increase in E2F-dependent transcriptional activity ***(53)***.

Taken together, these data provide evidence for a role for the Ras/MAPK pathway in controlling G_1/S progression in mammalian cells by a number of mechanisms, including: (1) induction of cyclin D expression and subsequent release of E2F transcription factors following phosphorylation of Rb pocket proteins by cyclin D-dependent kinases; (2) assembly of cyclin A/Cdk2 and cyclin E/Cdk2 complexes by increasing levels of the CDKIs involved in cyclin/Cdk assembly; and (3) decreasing p27 levels.

More recently, a role for MAPK in regulating the G_2/M transition has been suggested. Thus, it has been shown that ionizing radiation can activate the MAPK pathway ***(54,55)*** and cells expressing a dominant-negative MAPKK are unable to recover from radiation-induced G_2/M arrest ***(56)***. Additionally, treatment of cells with a MAPK inhibitor induces G_2/M arrest concomitant with a reduction in cyclin B/CDC2 activity ***(57)***. These data suggest that the Ras/MAPK pathway plays a regulatory role at many points during the mammalian cell cycle.

The data discussed above demonstrate regulation of the cell cycle by the MAPK extracellular mitogenic signaling pathway. If the activity of the MAPK pathway were maintained at an abnormally high level, then this could lead to cellular transformation and tumorigenesis. Indeed, cells have developed a safety mechanism in order to counteract this possibility, as shown by the fact that expression of oncogenic Ras or constitutively active MAPKK causes cell cycle arrest with high levels of p21, which is expressed in a p53-dependent manner ***(58,59)***.

3. The G_1/S Transition

One of the most extensively studied substrates of the cyclin/Cdks is the retinoblastoma (Rb) family of pocket proteins. The phosphorylation status of the Rb pocket proteins plays a major role in controlling E2F transcriptional activity and subsequent S-phase entry by modulating passage through the restriction (R) point in late G_1 as discussed in the following sections.

3.1. The Retinoblastoma Pocket Protein Family

The Rb family of pocket proteins comprises a group of tumor suppressor proteins consisting of three members; pRb, p107, and p130. As their name suggests, these proteins contain a pocket region that binds cellular targets. This region also is capable of binding a number of viral oncoproteins such as the adenovirus E1A protein, SV40 large T antigen, and the human papillomavirus 16 E7 protein ***(60)***, demonstrating one mechanism by which tumor viruses can interfere with cell cycle progression in mammalian cells. In addition to phosphorylation events, the functions of different Rb family members are also regulated by changes in expression. During G_1, the Rb pocket proteins are found in a hypophosphorylated state in which they bind to members of the E2F transcription factor family (*see* **Subheading 3.2.** below). As cells progress through the cell cycle, these proteins become hyperphosphorylated as a result of phosphorylation by cyclin D/Cdk4(Cdk6) and cyclin E/Cdk2 complexes. Each family member also displays differential expression throughout the cell cycle. Thus, pRb is expressed throughout the cell cycle but is hyperphosphorylated and therefore inactivated in late G_1, although by mitosis it becomes dephosphorylated; p130 is highly expressed in G_0, whereas levels diminish as cells progress into S-phase, consistent with a role for p130 in maintaining quiescence (reviewed in **refs. *61***); and p107 shows a reciprocal expression pattern to p130 such that low levels are found in G_0, which then increase as cells progress through G_1 into S.

The importance of the Rb family of tumor suppressor proteins in controlling the restriction point is demonstrated by the fact that they are targets of deregulation in most types of human cancer ***(23,28,62)***; indeed, pRb has been reported to be mutated in approx 30% of all human cancers (reviewed in **ref. *63***).

The different actions of Rb pocket proteins with respect to E2F regulation was demonstrated in a study by Hurford et al. ***(64)***. These authors showed that pRb has distinct functions from p107 and p130. They also demonstrated that p107 and p130 functions overlap, since, in cells lacking p107 or p130, there were no changes in E2F-regulated transcription. However, in cells lacking both p107 and p130, or lacking pRb alone, an increase in E2F-regulated transcription was observed ***(64)***.

3.2. The E2F Transcription Factors

Another family of molecules that regulates the G_1/S transition is that comprising the E2F transcription factors. To date, seven E2F members have been described (E2Fs 1–7; ***65***), and these molecules exist as heterodimers paired with a DP subunit (**Fig. 2**). Two mammalian DP genes have so far been identified (DP-1 and -2) ***(66)***. E2F and DP proteins contain highly conserved DNA-binding and dimerization domains (**Fig. 2**). The E2F and DP proteins activate transcription in a synergistic manner, and DP proteins appear to act indirectly by enhancing the activity of E2F ***(67)***.

The Rb pocket proteins bind to, and sterically hinder transcriptional activity of, the E2F/DP complex, thereby enabling the E2F transcription factors to act as repressors of

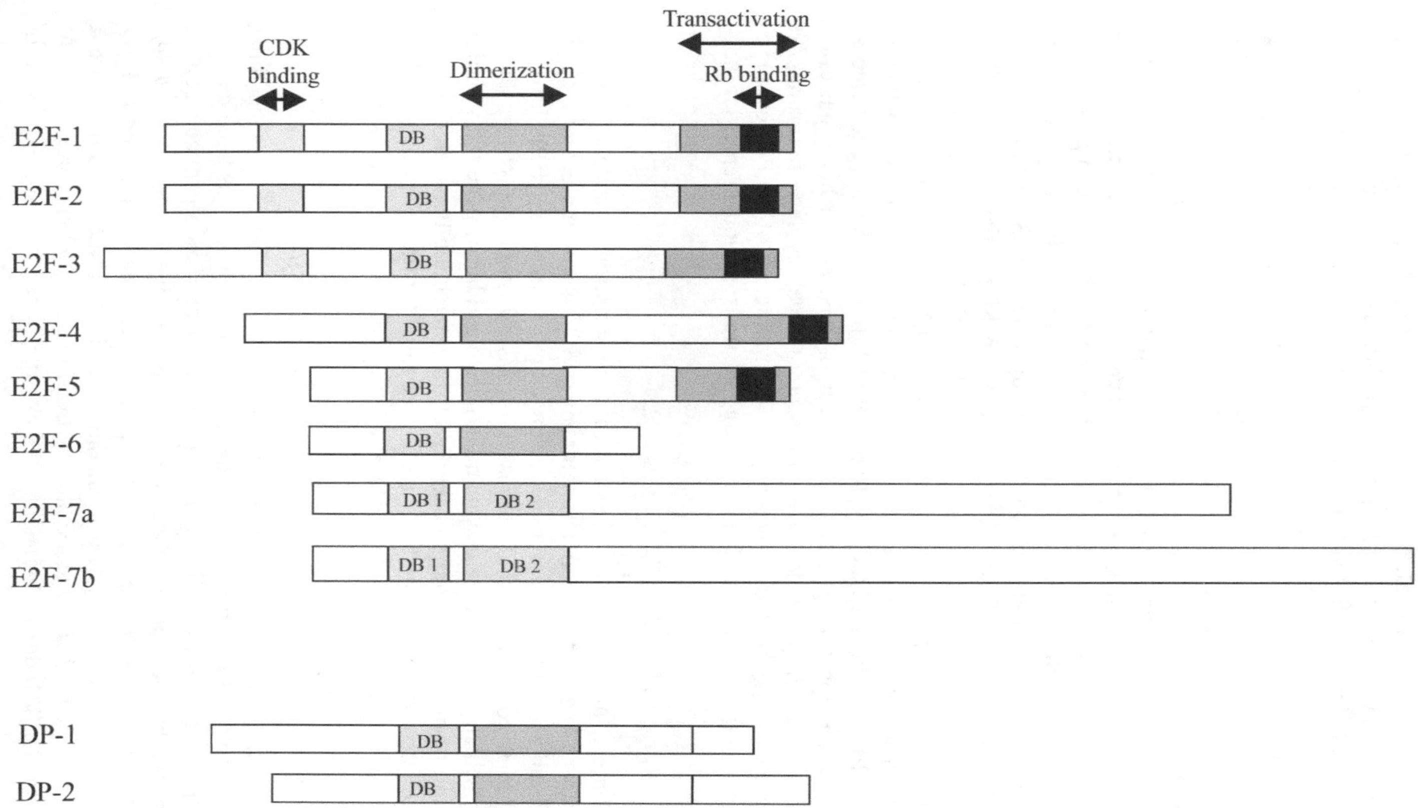

Fig. 2. E2F and DP conserved domain structures. DB represents the DNA binding domain of the E2F and DP family members. E2F7-a and -b contain two domains with high homology to the DNA binding domains of the E2F proteins (DB1 and DB2). The retinoblastoma (Rb) binding domain is located in the transactivation domain of the E2F proteins. Cdk, cyclin-dependent kinase.

gene transcription (reviewed in **ref. *65***). Phosphorylation of the pocket protein component of the E2F/pocket protein complex by cyclin D/Cdk4(Cdk6) complexes in the G_1-phase of the cycle leads to dissociation of the phosphorylated pocket protein and E2F, followed by E2F-mediated transactivation of promoters of genes necessary for S-phase progression, e.g., dihydrofolate reductase (*DHFR*), *cyclin E*, and *cyclin A* (**Fig. 1; *68–70***). With the exception of E2F-7, all members of the seven-member E2F family require heterodimerization with a DP subunit (DP-1 or DP-2) for full activity and share strong homology in their heterodimerization and DNA binding domains, their marked box, and, with the exception of E2F-6, a transactivation domain and a pocket protein binding region that resides within this sequence ***(65,71)***. E2F transcription factors are divided into three main groups: E2Fs 1–3, which play a role in progression from G_1 into S-phase of the cell cycle and possess a pRb binding site within their transactivation domain; E2F-4 and -5, which bind to p107 or p130 members of the pocket protein family, and E2F-7these three play a role in differentiation and proliferation; and E2F-6, which is unique since it lacks both the transactivation and the pocket protein binding domains. E2F-6 (also known as EMA) is thought to regulate cell cycle progression via its role as a transcriptional repressor ***(72–74)***. Although overexpression of E2F-6 did not block cycling NIH3T3 fibroblasts from entry into S-phase of the cycle, there was a significant decrease in S-phase entry when G_0-arrested E2F-6 overexpressing cells were stimulated to reenter the cell cycle with serum ***(74)*** More recently, it was suggested that recruitment and interaction of the Polycomb repressor proteins (PcG) are instrumental in mediating the transcriptional repressor function of E2F-6 ***(75)***.

3.2.1. Activation of Transcription by E2F

The precise mechanism by which E2Fs activate transcription is unclear, although studies have shown that the transactivation domain of E2F-1 can interact with cyclic adenosine monophosphate (cAMP) response element binding protein (CBP) ***(76)***. CBP is a transcriptional co-activator and possesses intrinsic histone acetyl transferase (HAT) activity which can modulate chromatin structure and hence gene transcription ***(77)***. Acetylation of histones causes weakening of the interaction between DNA and the nucleosome, thereby making the DNA more accessible for transcription ***(76)***. E2F complexes have also been shown to bend DNA, and this could be important for activation in certain instances ***(78)***.

3.2.2. Subcellular Localization of E2F Transcription Factors

One level of regulation of E2F function occurs through changes in the subcellular localization of individual E2F transcription factors. For example, it has been demonstrated that E2F-4 is expressed in the nucleus and cytoplasm of quiescent cells, but as cells reach S-phase this molecule is found almost exclusively in the cytoplasm ***(79,80)***. This relocation ensures that repressive E2F-4/p107 complexes cannot bind E2F-responsive genes. E2F-4 lacks a nuclear localization sequence (NLS), and therefore it might gain entry to the nucleus by association with DPs or Rb pocket proteins, both of which contain an NLS. Indeed, studies have shown that when it is overexpressed in cells, E2F-4 is only transported to the nucleus when coexpressed with DP-2, p107, or p130 ***(79,81)***. E2F-5 also lacks an NLS, although nuclear localization has been shown to occur in a DP- and Rb-independent manner such that transport of this E2F to the nucleus is mediated via formation of nuclear pore complexes ***(82)***.

3.2.3. Inactivation of E2F Transcription Factors

Inactivation of E2F is as important as E2F activation for continued progression through the cell cycle. It has been shown that expression of a constitutively active mutant of E2F-1 or DP-1 causes accumulation of cells in S-phase, which leads eventually to apoptosis ***(83)***. These results imply that inactivation of E2F is required for exit from S-phase.

Inactivation of E2F may be mediated by phosphorylation of E2F and DP subunits, leading to an inhibition of DNA binding activity ***(83–85)***. E2Fs 1–3 have been shown to contain a conserved region that allows enables interaction with cyclin A/Cdk2 or cyclin E/Cdk2; these interactions lead to inhibitory phosphorylation on these transcription factors ***(84)***. There is also evidence for ubiquitin-directed degradation of E2Fs 1–4 ***(86,87)***, which would lead to regulation of DNA binding activity.

3.2.4. Mechanism of pRb-Dependent Repression of E2F Transcriptional Activity

The exact mechanism of pRb-mediated repression has only recently become understood following the discovery that histone deacetylase-1 (HDAC-1) is involved ***(88–90)***. Recruitment of HDAC-1 to the DNA is thought to repress gene activation by altering chromatin structure. Nucleosomal histones have a high proportion of positively charged amino acids that facilitate interaction with negatively charged DNA. Deacetylation is thought to occur on histone tails protruding from the nucleosome ***(91)***, and this increases their positive charge, causing a tighter interaction with DNA, thereby making the DNA less accessible for transcription. Takahashi et al. ***(92)*** observed that high levels of acetylation correlated with activation of E2F-responsive genes in late G_1 and at the G_1/S border. However, during quiescence, when transcriptional activity is low, histones showed reduced levels of acetylation ***(92)***. They also showed that acetylation of genes occurred in a cell cycle-dependent manner. Thus, during G_0 when transcription levels are low, histones display reduced acetylation levels owing to the recruitment of HDAC-1. However, as cells progress through G_1 into S-phase, this repression is relieved by HAT (**Fig. 3**). As mentioned earlier in **Subheading 3.2.1.**, it has been shown that E2F is able to interact with both CBP and HAT in vitro and also in transiently transfected cells ***(76)***.

The role of HDAC-1 in repressing gene transcription has been demonstrated further such that HDAC-1 physically interacts with the DHFR promoter to affect cell growth. Thus, an association of HDAC-1 with the DHFR promoter was detected in G_0 and early G_1, when the gene was silent and also histone H4 showed low acetylation levels. This association then decreased as cells entered S-phase, consistent with an increase in DHFR mRNA levels ***(93)***.

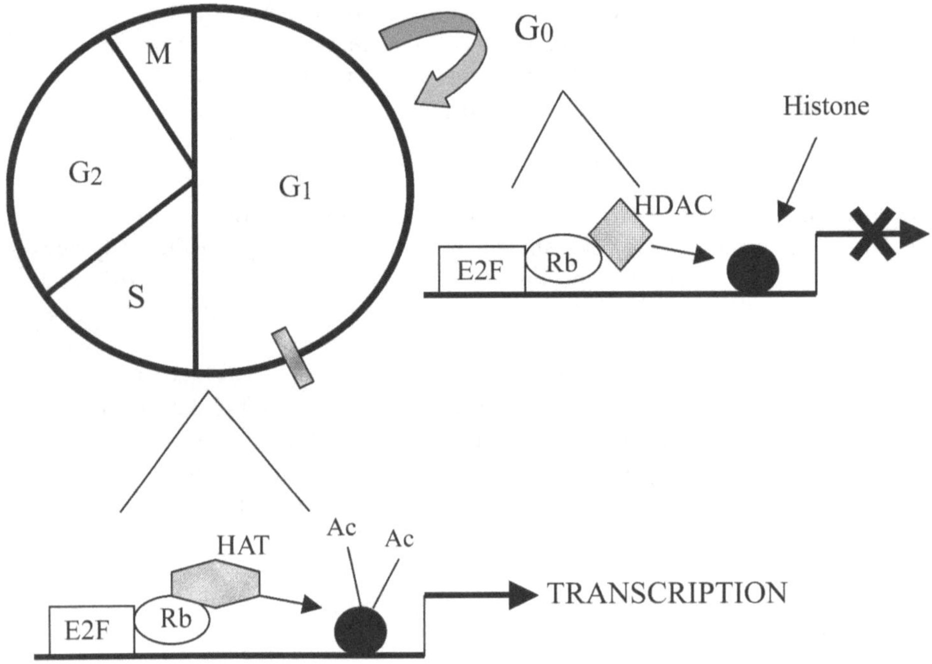

Fig. 3. Histone deacetylase-1 (HDAC-1) is recruited to DNA by retinoblastoma protein (Rb), causing deacetylation and inhibition of transcription during G_0. At the G_1/S transition, this repression is relieved by the action of histone acetyl transferase (HAT), allowing transcription of genes necessary for DNA synthesis.

It has been suggested that chromatin-modifying factors may form multienzyme corepressor complexes at promoter regions. Thus, the modifying factor, mSin3, has been shown to form a corepressor complex that acts as a scaffold for the assembly of HDAC-1 repressor complexes ***(94–96)***. The occupancy of E2F-regulated promoters by HDAC-1 and mSin3B in pocket protein-deficient cells was recently assessed ***(97)***. These studies showed that recruitment of HDAC-1, but not mSin3B, was completely dependent on p107 and p130 but not on pRb. These data suggest that specific E2F/Rb complexes are involved in recruitment of chromatin-modifying factors during G_0/G_1. There is also evidence that the tumor suppressor gene transforming growth factor-β1 (*TGF*-β1), might function by recruitment of HDAC-1 since transgenic mice overexpressing *TGF*-β1 showed enhanced HDAC-1 binding to p130 compared with control animals ***(98)***. Thus, *TGF*-β1 may exert its growth-inhibitory effects by recruitment of HDAC-1.

The model of Rb pocket proteins causing transcriptional repression by association with HDAC-1 is consistent with the model hypothesizing that pRb phosphorylation by cyclin D/Cdk4(Cdk6) complexes relieves E2F transcriptional inhibition. Phosphorylation of pRb by Cdk4(Cdk6) initiates intramolecular interactions between the carboxy-terminus of pRB and the pocket region, which displaces HDAC-1 from the pocket, thereby facilitating subsequent phosphorylation of pRb by Cdk2 complexes followed by disassociation from E2F. These results suggest a sequential phosphorylation of pRb by Cdk4(Cdk6) and Cdk2 ***(99)***.

3.3. Role of the Cyclin E/Cdk2 Complex in the G_1/S Transition

As cells approach the G_1/S border, control of the cell cycle becomes dominated by cyclin E/Cdk2 complexes. It has been demonstrated that overexpression of cyclin E/Cdk2 promotes S-phase entry and that blocking the kinase activity of this complex inhibits progression into S-phase ***(100–102)***. Consistent with its role in S-phase, the cyclin E/Cdk2 complex has been shown to be required for the initiation of DNA replication ***(100–102)***. The importance of phosphorylation of pRb by cyclin E/Cdk2 at the G_1/S border has already been discussed (*see* **Subheading 3.1.**).

A recently discovered substrate for cyclin E/Cdk2 has also been shown to be important for S-phase entry. This substrate is a nuclear protein that maps to the ATM locus (NPAT). NPAT was identified from a phage expression library using cyclin E/Cdk2 as a probe and was shown to associate with cyclin E/Cdk2 in vivo using immunoprecipitation studies. The NPAT protein was shown to be present at all stages of the cell cycle in synchronized cells; however, levels peaked at the G_1/S boundary and decreased as cells progressed through S. Overexpression of NPAT caused an increase in the number of S-phase cells, suggesting that NPAT expression may be a rate-limiting step for S-phase entry ***(103)***.

Histone gene expression is a major event that occurs as cells pass into S-phase. Histones form part of the nucleosomes that are a fundamental subunit of chromatin, and NPAT has been implicated in the regulation of histone gene expression. Both cyclin E and NPAT have been shown to localize to histone gene clusters at the G_1/S border, and phosphorylation of NPAT is required to activate histone gene expression ***(104,105)***. Therefore, evidence exists to show that cyclin E/Cdk2 regulates histone gene expression by phosphorylation of NPAT, a process required for entry into S-phase (*see* **Fig. 4**)

4. S-Phase

S-phase is the point during the cell cycle at which a cell duplicates its chromosomes in readiness for mitosis and cell division ***(1)***. A number of checkpoints exist to ensure that DNA is replicated only once per cycle, that it is fully and correctly replicated, and that replication occurs before cell division. Another important event during S-phase, other than DNA replication, is centrosome duplication. The centrosomes are the primary microtubule organizing center, and failure of cells to coordinate centrosome duplication with DNA replication leads to abnormal segregation of chromosomes, causing genomic instability that can lead to cancer.

4.1. Role of the Cyclin E/Cdk2 Complex in S-Phase Progression

There is much evidence to suggest that DNA synthesis in higher eukaryotes is initiated by activation of Cdk2 ***(23,101,106)***. Cdk2 associates with cyclin E just prior to

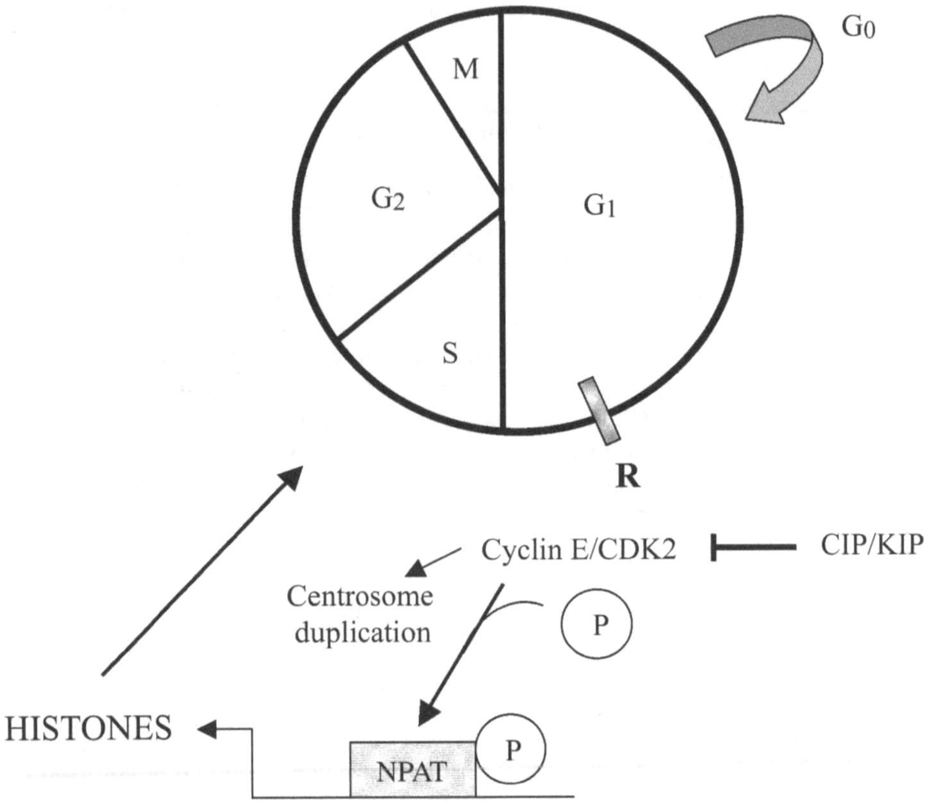

Fig. 4. Activation of NPAT by cyclin E/CDK2 causes histone gene expression necessary for DNA synthesis and S-phase progression. ⊤, inhibition step; ↑, activation step; Cdk, cyclin-dependent kinase; CIP, Cdk-interacting protein; KIP, kinase-inhibitor protein.

the onset of S-phase, and the role of this complex in the activation of NPAT and histone gene expression has already been discussed above (*see* **Subheading 3.3.**). A role for cyclin E/Cdk2 in centrosome duplication has also been suggested ***(107)***. Tarapore and colleagues ***(108)*** developed a cell-free centriole duplication system and demonstrated that centrosome duplication was dependent on the presence of cyclin E/Cdk2 complexes. In addition, cyclin E/Cdk2 was shown to phosphorylate nucleophosmin in this model, causing dissociation from centrosomes and subsequent initiation of centrosome duplication ***(108)***.

4.2. Role of the Cyclin A/Cdk2 in S-Phase Progression

The onset of S-phase correlates with formation of cyclin A/Cdk2 complexes. Microinjection of antibodies against Cdk2 complexed with either cyclin A or cyclin E blocks the initiation of DNA synthesis in mammalian cells ***(109,110)***. Cyclin A might be rate-limiting for DNA replication since it can accelerate entry into S-phase when overexpressed in cells *(111)*. The fact that depletion of cyclin A by injection of anti-cyclin A antibody causes inhibition of DNA synthesis suggests that cyclin A plays a role in this process. It has been shown that CDC6 is an intracellular substrate for cyclin A/CDK2 *(112)*. CDC6 is a protein required for formation of the initiation complex (*see* **Subheading 4.3.2.**), which is necessary for the onset of DNA replication, thereby providing one mechanism by which cyclin A/Cdk2 may regulate DNA replication. CDC6 has been shown to be required for late firing of origins, and this function may be achieved by phosphorylation following cyclin A/Cdk2 activation, suggesting that this complex may be required for continuation of DNA synthesis in addition to the initiation step *(113)*. However, it has been demonstrated that microinjection of cyclin A antibodies into cells already progressing through S-phase causes accumulation of cells in G_2 *(109)*, indicating that, in this instance, cyclin A is not required for cells to complete S-phase, and CDC6 may therefore be regulated by other cyclin/Cdk complexes in the cell.

4.3. Cell Cycle Control of DNA Replication in Mammalian Cells

Eukaryotic genomes are extremely large and can range from 10^7 to greater than 10^9 bp. Because of this large size, duplication of the eukaryotic genome occurs as a multiparallel process with between 10,000 and 100,000 parallel synthesis sites in human somatic cells (reviewed in **refs.** ***114*** and ***115***). Cells need to ensure that DNA replication occurs at the appropriate time in the cell cycle and also that re-replication does not occur before cells undergo mitosis and cytokinesis. Advances in our understanding of the regulation of these sequential processes have come from numerous studies in yeast systems (reviewed in **ref.** ***114***); *see also* Chapter 1, this volume). These simple model systems have provided much information on the protein complexes involved in the activation and inhibition of DNA synthesis, and a number of homologs have since been identified in higher eukaryotes.

Early experiments carried out in mammalian cells by Rao and Johnson *(116)* showed that the initiation of DNA replication is believed to be a two-step process. These investigators showed that fusion of a G_1 cell with an S-phase cell triggered DNA replication but that G_2 cells were unable to undergo DNA initiation *(116)*. This led to the notion that an S-phase–promoting factor was required to push cells from G_1 into S-phase. The two-step process first involves the assembly of initiation factors at origins of replication and second the triggering of these complexes to activate DNA synthesis by the actions of protein kinases. The following sections will give an overview of those molecules involved in driving DNA initiation and replication.

4.3.1. Origins of Replication

DNA synthesis is known to occur at specific sites on the DNA known as origins of replication. The best characterised origins of replication are those found in *Saccharomyces cerevisiae* and are known as autonomous replication sequences (ARS) *(117)*. The ARS contains a highly conserved region of 100–200 bp known as the ARS consensus sequence (ACS), and this is an essential component of the origin of replication to which the origin recognition complex (ORC) binds. The ORC is conserved in all eukaryotes *(118)*.

Three ORC subunits have been identified in humans, HsORC 1, 2, and 4 ***(119)***, all of which are involved in the initiation of DNA replication by recruitment of specific factors to the DNA. Human ORC has been shown to interact with a HAT, and this may be involved in making the initiation site accessible, thereby facilitating replication ***(120)***.

4.3.2. CDC6 and DNA Replication

A key regulator of DNA replication in mammalian cells is CDC6. Immunodepletion of CDC6 in human cells blocks S-phase entry ***(121,122)*** and has been shown to affect the interaction of ORC with minichromosome maintenance (MCM) proteins but not its interactions with DNA ***(123,124)***. These data suggest that CDC6 may act as an adaptor protein for interactions of the ORC with other proteins (e.g., MCM proteins). Levels of CDC6 in cycling human cells remain fairly stable during S-phase, G_2, and mitosis ***(125,126)***, but lower amounts are present in early G_1 when CDC6 is degraded by proteolysis ***(127,128)***. CDC6 does, however, change its subcellular localization during the cell cycle, and it has been shown that nuclear CDC6 is phosphorylated during S-phase and transported to the cytoplasm ***(129)***. Phosphorylation of CDC6 is carried out by cyclin A/Cdk2 and also by Dbf/CDC7. Relocation of CDC6 within the cell might be one way in which cells ensure that re-replication does not occur. However, a substantial amount of CDC6 is found still associated with chromatin during S-phase ***(127)***, suggesting that CDC6 might play roles other than assembly of proteins at the initiation site and may be required for continued synthesis. Because of the relocalization of CDC6 during S-phase, CDC6 must be continually synthesized to account for the fraction associated with chromatin during S-phase ***(130)***.

4.3.3. Minichromosome Maintenance Proteins

The MCM proteins are a complex of six related proteins that form an essential component of the DNA initiation complex. Their requirement for DNA replication has been demonstrated by antibody injection and antisense oligonucleotide experiments ***(131–133)***. The six MCM proteins are not functionally redundant, and deletion of any MCM protein in *S. cerevisiae* or *S. pombe* results in loss of cell viability. In most organisms, the MCM proteins are located in the nucleus throughout the cell cycle ***(134,135)***. In mammalian cells, MCM proteins associate with chromatin in G_1, but as cells progress through S-phase they are phosphorylated, and this reduces their affinity for chromatin ***(131,136)***. This may be one way in which cells ensure that replication occurs only once per cycle. In mammalian cells, some MCM proteins copurify with DNA polymerase α ***(137)***, and evidence exists to suggest that MCMs possess DNA helicase activity ***(138)***. Therefore, it is possible that association of the helicases with a primase forms a mobile primosome that drives discontinuous synthesis.

4.3.4. CDC45

CDC45 is essential for DNA replication in *S. cerevisiae* ***(139–141)*** and this molecule has been shown to interact with MCM family members ***(140,141)***. A human homolog has been identified ***(142)***, and immunoprecipitation experiments indicate that it associates with chromatin periodically throughout the cell cycle. Association of

CDC45 with chromatin may depend on cyclin/CDK complex activity at the G_1/S transition ***(143,144)***.

4.3.5. Regulation of DNA Initiation Complexes

Two classes of protein kinases are essential for the initiation of replication, the Cdks and Dbf4/CDC7 kinase.

4.3.5.1. CDKS

A role for Cdk2 in the initiation of replication in higher eukaryotes has been demonstrated in a number of studies; for example, microinjection of antibodies against certain cyclins and Cdks into mammalian cells inhibits S-phase entry ***(110,145)***.

As mentioned in **Subheading 4.2.**, CDC6 is a substrate for cyclin A/Cdk2, and phosphorylation of CDC6 by this complex possibly contributes to the prevention of DNA reinitiation by causing export of CDC6 from the nucleus ***(126,129)***.

MCM proteins also serve as substrates for certain Cdks, and phosphorylation of MCM proteins causes dissociation from chromatin as cells progress through S-phase. Thus, MCM proteins are substrates for the mitotic complex cyclin B/CDC2 ***(146)***, and this provides a link between mitotic cyclins and the inhibition of reinitiation, ensuring that DNA replication occurs only once before entry into mitosis. It has been shown that MCM2 and -4 are phosphorylated in S-phase and become hyperphosphorylated by G_2/M. Both MCM2 and -4 are good in vitro substrates for phosphorylation by cyclin B/CDC2 ***(146,147)***.

4.3.5.2. DBF4/CDC7 KINASE

CDC7 in *S. cerevisiae* and the *S. pombe* homolog, Hsk 1 have been shown to be essential for viability and are directly involved in DNA replication ***(148,149)***. A human homolog of CDC7 has also been identified ***(150–152)***. The human homolog of Dbf4 is regulated transcriptionally ***(153,154)***, with maximal expression during S-phase, which also corresponds to the kinase activity of the Dbf4/CDC7 complex ***(112,153)***. Studies have shown that inactivation of CDC7 in early S-phase prevents firing from replication origins, implicating CDC7 in the initiation of DNA replication ***(155,156)***. Human MCM2 and -3 are both substrates for CDC7 in vitro ***(151,153)***.

4.4. DNA Replication Checkpoints

Various checkpoints serve to inhibit DNA replication in response to partially replicated DNA or DNA damage, to allow the cell sufficient time to repair the damage before undergoing mitosis (*see* **Fig. 1**). Replication checkpoints have been extensively studied in yeast systems, and homologs for the proteins involved have also been identified in higher eukaryotes, including mammals.

4.4.1. The p53-Dependent Pathway

Several phosphatidylinositol (PI)-3-like kinase proteins are believed to be involved in the DNA replication checkpoint, including ATM, ataxia-telangiectasia related protein (ATR), and DNA-dependent protein kinase (DNA-PK) ***(157–160)*** and these kinases have been shown to be activated by DNA in vitro ***(161,162)***. The tumor suppressor protein p53 is a downstream target of ATM, and immunoprecipitated ATM can phosphorylate p53 on Ser15, a residue that is phosphorylated in vivo in response to DNA damage ***(163–166)***. DNA damage, occurring, for example, in response to ionizing radiation, leads to stabilization and accumulation of p53, which is involved in activation of a number of cellular responses such as cell cycle checkpoints, genomic stability, gene transactivation, and apoptosis ***(162,167–170)***. p53 is normally associated with the ubiquitin ligase MDM2, such that phosphorylation of p53 on Ser15 leads to its dissociation from MDM2, thereby stabilising the p53 protein ***(163)***. Stabilization of p53 leads to transactivation of the CDKI molecule, p21, which leads to cell cycle arrest ***(171)***.

Other regulators of p53 include ATR and Pin1. Thus, the ATR protein is capable of phosphorylating p53 on Ser15 and may also play a part in activating the p53 checkpoint pathway in response to ultraviolet (UV) and ionizing radiation ***(162,172)***. Pin1 has been shown to regulate the G_1/S, G_2/M, and DNA replication checkpoints ***(173)*** and is overexpressed in many human cancers ***(174–176)***. A recent report has shown that Pin1 binds phosphorylated p53 and is involved in stabilization of the protein, probably by interfering with the MDM2 interaction, and is also involved with transactivation of p21 in response to DNA damage ***(177)***.

4.4.2. The p53-Independent Pathway

The p53-independent mechanism of cell cycle block in response to unreplicated DNA or DNA damage involves the Rad proteins (reviewed in **refs. *114*** and ***178***). These proteins were first identified in yeast, and mammalian homologs also have been identified. The proteins involved in recognition and processing of the replication perturbation response are Rad1, Rad9, Rad17, and Hus1. The effects of these proteins are mediated by the protein kinases, CDS1 and CHK1, which target proteins involved in cell cycle regulation, for example, the CDC25 dual-specificity protein phosphatases (*see* **Subheading 5.1.**).

DNA-PK is the human homolog of the fission yeast PI-3–like kinase, Rad3, and is activated by proteins that detect sites of DNA strand breakage. Loss of function of these kinases results in inhibition of the checkpoint, suggesting that DNA-PK is important for sensing DNA damage and initiating the checkpoint mechanism ***(179,180)***.

Rad1 has been shown to be similar to proliferating cell nuclear antigen (PCNA) and possesses exonuclease activity ***(181,182)***. PCNA encircles the DNA during replication and retains the polymerase complex on the DNA. PCNA requires several factors in order to load onto DNA, one of which is known as replication factor C (RFC). Rad17 has been shown to share homology with RFC and also has been shown to interact with Rad1 ***(181)***. Rad1, Rad9, and Hus1 have all been shown to interact physically in mammalian cells ***(183,184)***, and it is believed that Rad17 may serve as a recruitment complex for Rad1, Rad9, and Hus1 to sites of DNA damage ***(185)***. Indeed, a recent study has demonstrated that upon replication block, Rad17 is recruited to the sites of DNA damage during late S-phase and that it binds to the Rad1/Rad9/Hus1 complex, enabling its interaction with PCNA ***(186)***.

The two downstream targets of the Rad proteins are the serine/threonine kinases CHK1 and CDS1. These kinases are activated differentially such that CDS1 is involved in mediating responses to unreplicated DNA, and CHK1 is involved in the G_2 DNA damage response. CDS1 has been shown to be phosphorylated by ATM ***(187,188)***, and following activation it phosphorylates and inhibits the mitotic activator, CDC25C ***(187–189)***, thereby mediating G_2 arrest. CHK1 also phosphorylates CDC25C in vitro ***(190)***. Phosphorylation of CDC25C by CDS1 and CHK1 creates a binding site for the 14-3-3 family of phosphoserine binding proteins (***190***; *see* **Subheading 6.1.2.**). Binding of 14-3-3 has little effect on CDC25C activity, and it is believed that 14-3-3 regulates CDC25C by sequestering it to the cytoplasm, thereby preventing the interactions with cyclin B/CDC2 that are localized to the nucleus at the G_2/M transition ***(190,191)***.

The mechanisms by which DNA replication and DNA damage checkpoints exert their effects on cell cycle progression are now becoming clearer. Both p53-dependent and -independent mechanisms exert their effects via complex pathways on key cell cycle regulatory molecules such as p21 and the mitotic regulator, CDC25C (**Fig. 5**). These events occur at specific points in the cell cycle, ensuring that a cell does not proceed through mitosis without a full complement of replicated and intact DNA, thereby ensuring that the genome is passed equally to each of the daughter cells.

5. The G_2/M Transition

The G_2-phase is another gap phase in the cell cycle in which the cell assesses the state of chromosome replication and prepares to undergo mitosis and cytokinesis. Cyclin B/CDC2 is the key mitotic regulator of the G_2/M transition and was originally identified as the maturation-promoting factor, a factor capable of inducing M-phase in immature *Xenopus* oocytes ***(192–194)***. As is the case with other cyclin/CDK complexes, activation of the cyclin B/CDC2 complex is tightly regulated by phosphorylation and dephosphorylation events and also changes in subcellular localization (reviewed in **refs.** ***195*** and ***196***). The molecules that regulate cyclin B/CDC2 activity receive signals from the checkpoint machinery, as described in **Subheading 4.4.2.** Cyclin A/Cdk complexes also play a role in regulating the G_2/M transition.

5.1. Role of the Cyclin B/CDC2 Complex in the G_2/M Transition

Cyclin B synthesis begins at the end of S-phase ***(197)***. Two cyclin B isoforms exist in mammalian cells, cyclin B1 and B2. Studies in cyclin B1- and cyclin B2-null mice have confirmed that cyclin B2 is non-essential for normal growth and development ***(198)***. This particular isoform associates with the Golgi and may play a role in Golgi remodelling during mitosis ***(198,199)***. In contrast to cyclin B2, cyclin B1 is thought to be responsible for most of the actions of CDC2 in the cytoplasm and nucleus and it appears to compensate for the loss of cyclin B2 in B2-null mice implying that cyclin B1 is capable of targeting CDC2 kinase to the essential substrates of cyclin B2 ***(198)***.

Cyclin B/CDC2 complexes are regulated both positively and negatively by phosphorylation (**Fig. 5**). Phosphorylation of CDC2 on the conserved T-loop region (Thr160) is required for activation, as is the case with all Cdks, and this phosphoryla-

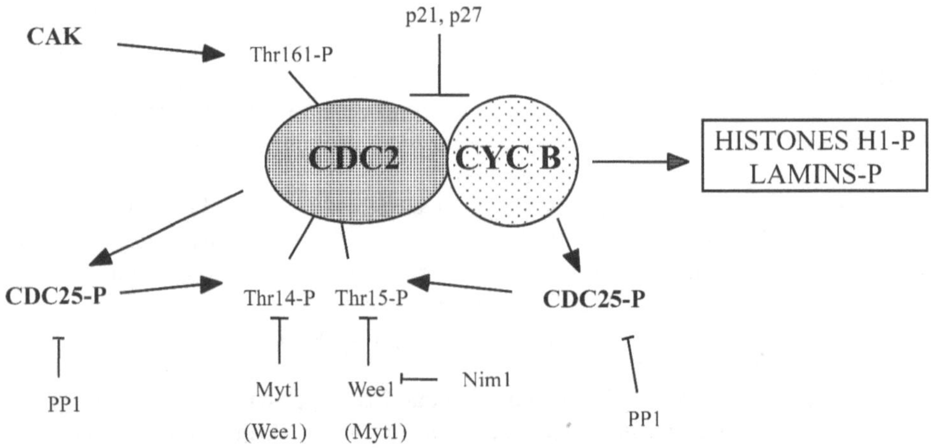

Fig. 5. Regulation of the CDC2/cyclin B complex. The serine/tyrosine kinase, Weel catalyzes phosphorylation of Tyr15 on CDC2. Weel itself is phosphorylated and inactivated by Nim1 and other unidentified kinases to induce mitosis. Thr14 phosphorylation can be mediated by Weel but only once Tyr15 has been phosphorylated. It appears that the Thr/Tyr kinase Mytl is the critical kinase involved here. Inhibition of CDC2 by weel is counteracted by the CDC25 dual-specificity phosphatases. CDC25 is phosphorylated and activated by CDC2/cyclin B (amplification pathway). Protein phosphatase 1 (PP1) inactivates CDC25 by dephosphorylation of the same residue that is phosphorylated by CDC2/cyclin B. Full activation of CDC2 requires Thr161 phosphorylation by Cdk-activating kinase (CAK), which then stabilizes CDC2 association with cyclin A. ⊤, inhibition step; ↑, activation step.

tion event is mediated by CAK. During G_2, cyclin B/CDC2 complexes are held in an inactive state by phosphorylation of CDC2 Thr14 and Tyr15. Phosphorylation on Thr14 prevents ATP binding ***(200)***, whereas that on Tyr15 interferes with phosphate transfer to the substrate owing to its positioning in the ATP binding site on CDC2 ***(201)***. These inhibitory phosphorylation events are carried out by the kinases Weel and Mytl; Weel specifically phosphorylates Tyr15, and Mytl phosphorylates both Tyr15 and Thr14, with a stronger affinity for Thr14 ***(202–204)***. Cyclin B/CDC2 becomes fully activated following dephosphorylation of these sites by the protein phosphatase CDC25C (**Fig. 5**).

6. Mitosis (M-Phase)

Mitosis (also called karyokinesis) and cytokinesis constitute the shortest phase of the eukaryotic cell cycle, typically taking around 1–2 h to complete in a mammalian cell. Mitosis itself comprises five distinct phases as follows:

1. *Prophase*: this stage begins with condensation of the chromosomes in the nucleus and ends with breakdown of the nuclear envelope. (This latter event occurs over a 1–2-min interval.)

2. *Prometaphase*: at this stage the mitotic spindle forms. Three essential events must occur in prometaphase if cell division is to proceed normally: (a) the bipolar spindle axis must be established; (b) the daughter chromatids of each replicated chromosome must become committed to the opposing spindle poles; and (c) the chromosomes must become aligned at, or near to, the spindle equator.

3. *Metaphase*: during this stage all chromosomes are bioriented and positioned near the spindle equator. All chromosomes align themselves along the metaphase plate.

4. *Anaphase*: the sister chromatids that comprise each chromosome separate to form two independent chromosomes. Anaphase is separated into anaphase A and anaphase B.

5. *Telophase*: this is the final stage of mitosis, in which the chromosomes decondense and a nuclear envelope forms around each set of chromatids. The contractile ring begins to form in readiness for the cell to split into two daughter cells, each with one nucleus.

A number of cell cycle regulatory molecules play pivotal roles in promoting progression through mitosis, including the CDC25 protein phosphatases, the polo-like kinases (PLKs), the 14-3-3 proteins, mitotic cyclin/Cdk complexes, and the anaphase-promoting complex (APC). The role that these individual groups of molecules plays in mitosis is now discussed in more detail.

6.1. The CDC25 Protein Phosphatases

The mammalian CDC25 family of dual-specificity phosphatases consists of three members: A, B, and C ***(205,206)***. CDC25B and C are thought to be the main regulators of mitosis, whereas CDC25A plays a role in regulating the G_1/S transition. CDC25B may be involved in the initial dephosphorylation and activation of cyclin B/CDC2, which then initiates the positive feedback loop of CDC25C activation by CDC2 (***207***; *see* **Fig. 5**). CDC25B is also believed to play a role in centrosomal microtubule nucleation during mitosis since overexpression of this molecule causes formation of minispindles in the cytoplasm ***(208)***. CDC25A expression is under transcriptional control of the E2F transcription factors in late G_1 ***(209)*** and is involved in activation of cyclin E/Cdk2 and cyclin A/Cdk2 complexes that regulate entry into S-phase (*see* **Subheadings 4.1.** and **4.2.**).

CDC25C is the protein phosphatase that is mainly responsible for dephosphorylation and activation of the cyclin B/CDC2 complex. Treatment of CDC25C with phosphatases in vitro led to reduced CDC25C phosphatase activity, indicating that hyperphosphorylation of CDC25C is required for phosphatase activity during mitosis ***(207,210,211)***. Cyclin B/CDC2 is able to phosphorylate CDC25C (***207,212***; *see* **Fig. 5**) and this initiates a positive feedback loop that induces rapid activation of cyclin B/CDC2 at the G_2/M transition. However, the initial trigger of CDC25C activation remains unclear, although CDC25C has been shown to be phosphorylated by cyclin E/Cdk2 and cyclin A/Cdk2 in vitro ***(212)***. PLK-1 is another potential upstream regulatory kinase of CDC25C that might function in vivo ***(213)***.

The role of CDC25C as a key mediator of cyclin B/CDC2 activation has recently been questioned owing to studies performed in CDC25C knockout mice. These mice showed no phenotype with respect to regulation of mitosis and showed no differences in CDC2 phosphorylation ***(214)***, suggesting that redundancy exists between CDC25 isoforms. Indeed, a role for CDC25A in CDC2 activation has recently been suggested. Destruction of CDC25A by the ubiquitin-mediated pathway serves to ensure that cells do not undergo premature mitosis, and this is achieved by phosphorylation of CDC25A by CHK1. However, Mailand and co-workers ***(215)*** reported that once cells are committed to mitosis, stability of CDC25A undergoes major changes at the G_2/M transition owing to phosphorylation on Ser17 and Ser115 that uncouples it from the ubiquitin-mediated degradation pathway. Phosphorylation of CDC25A on these specific residues is mediated by cyclin B/CDC2 and therefore forms part of a positive feedback activation loop whereby CDC25A is stabilized by CDC2, and this is followed by dephosphorylation of CDC2 on Thr14 and Tyr15 ***(215)***.

6.1.1. The Polo-Like Kinases

The polo-like kinases are a family of serine/threonine protein kinases, four of which have been described in mammalian cells: PLK-1, PLK-2 (Snk), PLK-3 (Fnk/Prk), and PLK-4 (Sak a/b) ***(216,217)***. The PLKs share a closely related catalytic domain at their N-termini (50–65% identity at the amino acid level) and a homologous C-terminal domain called the polo box domain (PBD)~~of~~ which there are two, PBD1 and PBD2~~that~~ is required for directing subcellular localization of the kinase since mutation of this region has been shown to disrupt localization of PLK-1 ***(218,219)***.

PLK protein levels and phosphorylation status are cell cycle-regulated. Thus, PLK1 is undetectable in cells at the G_1/S-phase transition; however, levels rise during S-phase, and phosphorylation occurs during G_2 ***(220)***. Indeed, PLK1 is important for the G_2/M transition, entry into mitosis and exit from mitosis, and is rapidly degraded as the cell exits mitosis and PLK1 activity levels peak at the metaphase–anaphase transition ***(217,220)***. Activation of PLK1 has been found to occur at a similar time to cyclin B/CDC2 activation, and a recent study by Roshak et al. ***(213)*** demonstrated that human CDC25C is a substrate for PLK1 and that phosphorylation caused activation of the phosphatase and subsequent dephosphorylation of cyclin B/CDC2. Other substrates phosphorylated by PLK1 include cyclin B1, Myt1, and the APC.

PLK2 is the least well characterized of the mammalian PLKs, although it is thought to play a role in cell cycle reentry of G_0-arrested cells ***(221)***. PLK3 has been implicated in DNA damage control at the G_2 checkpoint, where, in mammalian cells, it acts in the p53-mediated stress response pathway ***(222)***. Finally, a cell cycle role for PLK4 has not yet been reported. This particular PLK was isolated from the mouse as two distinct transcripts (a and b) and has since been shown to contain only a single PBD; it also lacks the catalytic motif found in other PLK family members ***(216)***.

6.1.2. The 14-3-3 Proteins

CDC25C phosphatase activity is regulated negatively by phosphorylation on a specific Ser216 residue that creates a binding site for small phosphoserine binding proteins, known as 14-3-3 ***(223)***. A number of 14-3-3 proteins are known to exist, including: 14-3-3 ε, γ, β, σ, ζ, η ***(224)***. CDC25C is localized to the cytoplasm during interphase and is directed to the nucleus just prior to mitosis ***(191,225)***. Binding of 14-

3-3 may prevent nuclear localization of CDC25C by masking the NLS, which is in close proximity to the Ser216 residue. In support of 14-3-3 sequestering CDC25C to the cytoplasm during interphase, Ogg et al. *(226)* demonstrated that Ser216 is the major phosphorylation site of CDC25C during interphase but not during mitosis. Potential candidate kinases for phosphorylation of Ser216 on CDC25C are CHK1, CDS1, and C-TAK1 *(227,228)*. CHK1 and CDS1 are both mediators of G_2 arrest in response to DNA damage or incomplete replication, as discussed above in **Subheading 4.** Therefore, phosphorylation of CDC25C during interphase creates a binding site for 14-3-3, causing cytoplasmic retention. Dephosphorylation of Ser216 (possibly by CDC25B) at the onset of mitosis results in nuclear localization and subsequent activation of cyclin B/CDC2, as described in **Subheading 6.1.1.** above.

6.2. Subcellular Localization of Cyclin B/CDC2 During G_2/M

During interphase, cyclin B/CDC2 complexes are found in the cytoplasm. However, by late prophase, most cyclin B/CDC2 is found in the nucleus following breakdown of the nuclear envelope *(229)*. Cyclin B has a cytoplasmic retention sequence (CRS) in the N-terminal region which, when deleted, causes localization of cyclin B to the nucleus *(230)*. A nuclear export signal (NES) also has been defined within the CRS, and this binds to the export receptor CRM1 *(231)*; it has been that shown that a specific inhibitor of CRM1 causes accumulation of cyclin B in the nucleus *(231)*. During mitosis, cyclin B is hyperphosphorylated within the CRS region, and this phosphorylation is thought to disrupt interactions with CRM1 holding cyclin B in the nucleus *(229,231)*.

The mechanism by which cyclin B enters the nucleus is less well understood. Both cyclin B and CDC2 lack an NLS. Cyclin B1 has, however, been shown to bind importin-b *(232)*, and this could be one mechanism that mediates cyclin B nuclear localization. The CRS of cyclin B1 is also known to interact with cyclin F, which is found predominantly within the nucleus and contains two NLSs *(233)*. Interestingly, overexpression of cyclin F causes relocation of cyclin B to the nucleus, suggesting that it may be involved in the import of cyclin B1.

Although cyclin B contains a CRS and an NES, both of which ensure that cyclin B is localized to the cytoplasm, phosphorylation of cyclin B can lead to association with other molecules, resulting in its relocation to the nucleus and thereby allowing access to nuclear substrates.

6.3. Function of Cyclin B/CDC2 During Mitosis

The cyclin B/CDC2 complex is involved in the initiation of a number of mitotic events in both the cytoplasm and the nucleus. During prophase, cyclin B/CDC2 is associated with duplicated centrosomes, and it promotes centrosome separation by phosphorylation of the centrosome-associated motor protein Eg 5 *(234)*. Cyclin B1/CDC2 and/or cyclin B2/CDC2 complexes are involved in the fragmentation of the Golgi network *(235)*, and cyclin B/CDC2 also is involved in the breakdown of the nuclear lamina and cell rounding *(236)*. Thus, the cyclin B/CDC2 complex is involved in completely reorganizing the cell architecture during mitosis.

6.4. Function of Cyclin A/CDC2 During Mitosis

Cyclin A plays important roles at two distinct phases of the cell cycle, G_1/S (as discussed in **Subheading 4.2.**) and G_2/M. These separate functions coincide with cyclin A binding to two different kinases, CDK2 at the G_1/S border and CDC2 during G_2. Cyclin A levels are undetectable during G_1, and levels begin to rise as cells enter S-phase. By the time a cell enters mitosis, cyclin A levels begin to decline; however, it still is present during prophase, in which it is associated with the centrosomes, although by telophase cyclin A is undetectable ***(109)***. Cyclin A/CDC2 complexes are thought to play a role in activating cyclin B/CDC2 complexes, and recent reports suggest that cyclin A/CDK2 complexes may also act during the G_2 checkpoint ***(237,238)***. Exit from mitosis requires degradation of both cyclin A and cyclin B, and this occurs via a ubiquitin-mediated pathway that itself is regulated by the APC pathway (*see* **Subheading 6.5.** below).

6.5. The Anaphase-Promoting Complex

Exit from mitosis requires ubiquitin-mediated degradation of mitotic cyclins via the *cyclin destruction box* ***(239)***, which is regulated by the APC ubiquitin ligase. APC is a multi-subunit ligase consisting of a number of protein subunits such as APC1, APC2, CDC16, and CDC23 (reviewed in **ref. *240***). APC is inactive in the S- and G_2-phases of the cell cycle but becomes activated in mitosis as a result of phosphorylation that is believed to be carried out by PLK1 ***(241,242)*** and/or cyclin B/CDC2 ***(241)***. APC requires conversion to an active form by CDC20/Fizzy, and this can only occur following phosphorylation of APC ***(243)***. APC is also required for sister chromatid separation during anaphase by causing destruction of securins, the proteins that hold the sister chromatids together. The securins inhibit the highly conserved enzyme separase during the cell cycle until metaphase, in which it is degraded by the APC ***(244)***. Activation of separase is crucial for the onset of anaphase in all eukaryotic cells.

6.6. Other Cell Cycle Regulatory Molecules Involved in Mitosis

A number of other molecules and protein complexes are involved in regulating normal karyokinesis and cytokinesis in mammalian cells. Such molecules include the securins, separase (*see* **Subheading 6.5.**), and rhabdokinesin-6. Expression of RB6K is regulated during the cell cycle at both the mRNA and protein levels and, similar to cyclin B, reaches maximum levels during M-phase ***(18)***. RB6K localizes in the late stages of mitosis to the spindle midzone and appears on the midbodies during cytokinesis. The functional significance of this localization during cell division has been demonstrated by antibody microinjection studies showing exclusive production of binucleated cells that failed to complete cytokinesis ***(245)***.

7. Cytokinesis

At the end of mitosis the cell must ensure that division is taken to completion by a process called cytokinesis. This occurs following assembly of a cleavage furrow at the site of division that contains actin, myosin, and other proteins that eventually form the contractile ring *(246)*. Following chromosome segregation, the microtubules bundle in the midregion of the spindle, forming the spindle midzone. As the contractile ring contracts, it creates a membrane barrier between each cell. The spindle midzone remains connected, forming a cytoplasmic bridge until this is finally cut during abcission. The mid-zone has been shown to contribute to actin ring assembly since placement of an artificial barrier between the spindle midzone and the cell cortex during metaphase caused inhibition of the cleavage furrow, whereas if a barrier was created in early anaphase, cytokinesis proceeded without a problem *(247)*.

Animal cells divide through the formation of an actomyosin contractile ring at the end of anaphase *(248)*. As discussed above, the spindle midzone plays a role in contractile ring assembly. Two major classes of proteins are believed to be important in signaling from the spindle midzone to the contractile ring. The first of these are the chromosomal passenger proteins, e.g., the inner centromere proteins that are initially found localized to chromosomes and centromeres and then translocate to the midzone during anaphase *(249)* and are involved in chromosome alignment, segregation, and cytokinesis. The second class of proteins are the motor-associated proteins, e.g., Eg5, that are required to maintain the midzone. These proteins localize along the spindles during metaphase and concentrate in the spindle midzone during anaphase *(250,251)*.

Specific Cdks also play a role in cytokinesis, and it has been shown that mammalian cells injected with a nondestructible form of cyclin B undergo anaphase and chromosome segregation but do not form a spindle midzone and fail to undergo cytokinesis *(252)*, suggesting a role for cyclin B in inhibiting cytokinesis. Cdks may also inhibit myosin, and thereby contractile ring formation, through inhibitory phosphorylation of the myosin regulatory light chain (RLC) *(253)*. Myosin RLC can be phosphorylated by CDC2, which inhibits its actin-activated ATPase activity in vitro *(254)*. This inhibitory phosphorylation increases in early mitosis and decreases in anaphase simultaneously with a decrease in CDC2 activation *(255)*.

8. Endoreduplication

In most eukaryotic cells, S-phase and mitosis are coupled and occur only once during each cell cycle; however, occasionally the sequence of events is interrupted such that the cell undergoes multiple rounds of DNA synthesis in the absence of mitosis. This process is called endoreduplication. Work in yeast has shown that endoreduplication can occur as a result of multiple initiations within S-phase, reoccurring S-phase, or repeated S- and G-phases *(256)*. In addition, endoreduplication can result in either multiple DNA syntheses within a single nucleus, e.g., megakaryocytes or in multi-nucleated cells, e.g., cardiac myocytes. Little is known about the molecular mechanisms responsible for endoreduplication in multinucleate cells, although a greater understanding of the processes involved might enable cell division to be initiated instead of endoreduplication, which would be useful for replacing damaged cells and would therefore avoid scarring, in terminally differentiated tissues that contain cells such as cardiac myocytes and neurones

9. Destruction of Cell Cycle Regulators During the Cell Cycle: Role Of The SCF Ubiquitin Ligases

The carefully ordered progression of a cell through various stages of the cell cycle is mediated by the timely synthesis and destruction of numerous cell cycle regulatory proteins. Degradation of molecules such as the cyclins occurs via ubiquitin-mediated proteolysis, the specificity of which is controlled by a large number of ubiquitin ligases that themselves are either single subunits or multiprotein complexes that act as recognition factors for substrates to be ubiquitinylated. The SCF complexes represent a large family of ubiquitin ligases that control cell cycle progression. They are so called because they are composed of Skp1, Cul1, and F-box protein as well as Roc1; it is the F-box protein component of the SCF complex that determines substrate recognition. It is well documented that SCF complexes are key in controlling the abundance of cell cycle regulatory proteins, including cyclins, Cdks, and CDKIs ***(257)***. For example, the SCF complex containing the F-box protein Skp2 coordinates the ubiquitinylation of the CDKIs p21 and p27, thereby allowing CDK2 activation at the G_1/S border ***(258,259)***.

10. Summary and Conclusions

The mammalian cell cycle is a highly regulated, conserved, and sequential process that is necessary for normal cell growth and development. Our understanding of the mechanisms involved in cell cycle regulation has increased significantly in recent years, as demonstrated by the award of the Nobel Prize for Physiology and Medicine to Leland Hartwell, Paul Nurse, and Tim Hunt in 2001 for their seminal discoveries relating to the cell cycle machinery. Despite this increased understanding, much remains to be learned about the mechanisms involved in controlling growth and proliferation in specific cell types and organs. Extending our knowledge of cell cycle control in different cell types might help to identify the causes of certain hyperproliferative diseases, including cancer and vascular disease. This then could lead to the development of new therapeutic agents targeting specific cell cycle molecules that become altered in such disorders.

References

1. Alberts, B., Johnson, A., Lewis, J., Raff, M., Roberts, K., and Walter, P. (2002) *Molecular Biology of the Cell*, 4th ed. Garland Science, New York, NY, pp. 983–1026.
2. Bicknell, K. A., Surry, E. L., and Brooks, G. (2003) Targeting the cell cycle machinery for the treatment of cardiovascular disease. *J. Pharm. Pharmacol.* **55,** 571–591.
3. Nurse, P. (2000) A long twentieth century of the cell cycle and beyond. *Cell* **100,** 71–78.
4. Dulic, V., Lees, E., and Reed, S. I. (1992) Association of human cyclin E with a periodic G1-S phase protein kinase. *Science* **257,** 1958–1961.
5. Koff, A., Giordano, A., Desai, D., et al. (1992) Formation and activation of a cyclin E-cdk2 complex during the G1 phase of the human cell cycle. *Science* **257,** 1689–1694.

6. Brooks, G. and La Thangue, N. B. (1999) The cell cycle and drug discovery: the promise and the hope. *Drug Discov. Today* **4,** 455–464.
7. Martinez, A. M., Afshar, M., Martin, F., Cavadore, J. C., Labbe, J. C., and Doree, M. (1997) Dual phosphorylation of the T-loop in cdk7: its role in controlling cyclin H binding and CAK activity. *EMBO J.* **16,** 343–354.
8. Drapkin, R., Le Roy, G., Cho, H., Akoulitchev, S., and Reinberg, D. (1996) Human cyclin-dependent kinase-activating kinase exists in three distinct complexes. *Proc. Natl. Acad. Sci. USA* **93,** 6488–6493.
9. Simone, C. and Giordano, A. (2001) New insight in cdk9 function: from Tat to MyoD. *Front. Biosci.* **6,** D1073–D1082.
10. Napolitano, G., Majello, B., and Lania, L. (2002) Role of cyclinT/Cdk9 complex in basal and regulated transcription (review). *Int. J. Oncol.* **21,** 171–177.
11. Leclerc, V. and Leopold, P. (1996) The cyclin C/Cdk8 kinase. *Prog. Cell Cycle Res.* **2,** 197–204.
12. de Falco, G. and Giordano, A. (1998) CDK9 (PITALRE): a multifunctional cdc2-related kinase. *J. Cell Physiol.* **177,** 501–506.
13. Russo, A. A., Jeffrey, P. D., and Pavletich, N. P. (1996) Structural basis of cyclin-dependent kinase activation by phosphorylation. *Nat. Struct. Biol.* **3,** 696–700.
14. Miller, M. E. and Cross, F. R. (2001) Cyclin specificity: how many wheels do you need on a unicycle? *J. Cell Sci.* **114,** 1811–1120.
15. Peeper, D. S., Parker, L. L., Ewen, M. E., et al. (1993) A- and B-type cyclins differentially modulate substrate specificity of cyclin-cdk complexes. *EMBO J.* **12,** 1947–1954.
16. Dynlacht, B. D., Moberg, K., Lees, J. A., Harlow, E., and Zhu, L. (1997) Specific regulation of E2F family members by cyclin-dependent kinases. *Mol. Cell. Biol.* **17,** 3867–3875.
17. Devault, A., Fesquet, D., Cavadore, J. C., et al. (1992) Cyclin A potentiates maturation-promoting factor activation in the early Xenopus embryo via inhibition of the tyrosine kinase that phosphorylates cdc2. *J. Cell Biol.* **118,** 1109–1120.
18. Reed, S. I., Bailly, E., Dulic, V., Hengst, L., Resnitzky, D., and Slingerland, J. (1994) G1 control in mammalian cells. *J. Cell Sci. Suppl.* **18,** 69–73.
19. Pines, J. (1997) Cyclin-dependent kinase inhibitors: the age of crystals. *Biochim. Biophys. Acta* **1332,** M39–M42.
20. Brooks, G., Poolman, R. A., and Li, J. M. (1998) Arresting developments in the cardiac myocyte cell cycle: role of cyclin-dependent kinase inhibitors. *Cardiovasc. Res.* **39,** 301–311.
21. Pavletich, N. P. (1999) Mechanisms of cyclin-dependent kinase regulation: structures of Cdks, their cyclin activators, and Cip and INK4 inhibitors. *J. Mol. Biol.* **287,** 821–828.
22. Gartel, A. L., Serfas, M. S., and Tyner, A. L. (1996) p21negative regulator of the cell cycle. *Proc. Soc. Exp. Biol. Med.* **213,** 138–149.
23. Sherr, C. J. (1996) Cancer cell cycles. *Science* **274,** 1672–1627.
24. Serrano, M. (1997) The tumor suppressor protein p16INK4a. *Exp. Cell Res.* **237,** 7–13.
25. Zhang, H., Hannon, G., J., and Beach, D. (1994) p21-containing cyclin kinases exist in both active and inactive states. *Genes Dev.* **8,** 1750–1758

26. Prives, C. (1993) Doing the right thing: feedback control and p53. *Curr. Opin. Cell Biol.* **5,** 214–218.
27. Ewen, M. E. (1996) p53-dependent repression of cdk4 synthesis in transforming growth factor-b-induced G1 cell cycle arrest. *J. Lab. Clin. Med.* **128,** 355–360.
28. Weinberg, R. A. (1995) The retinoblastoma protein and cell cycle control. *Cell* **81,** 323–330.
29. Alcorta, D. A., Xiong, Y., Phelps, D., Hannon, G., Beach, D., and Barrett, J. C. (1996) Involvement of the cyclin-dependent kinase inhibitor p16 (INK4a) in replicative senescence of normal human fibroblasts. *Proc. Natl. Acad. Sci. USA* **93,** 13742–13747.
30. Reynisdottir, I., Polyak, K., Iavarone, A., and Massague, J. (1995) Kip/Cip and Ink4 Cdk inhibitors cooperate to induce cell cycle arrest in response to TGF-b. *Genes Dev.* **9,** 1831–1845.
31. el-Deiry, W. S., Tokino, T., Velculescu, V. E., et al. (1993) WAF1, a potential mediator of p53 tumor suppression. *Cell* **75,** 817–825.
32. Kamb, A., Gruis, N. A., Weaver-Feldhaus, J., et al. (1994) A cell cycle regulator potentially involved in genesis of many tumor types. *Science* **264,** 436–440.
33. Nobori, T., Miura, K., Wu, D. J., Lois, A., Takabayashi, K., and Carson, D. A. (1994) Deletions of the cyclin-dependent kinase-4 inhibitor gene in multiple human cancers. *Nature* **368,** 753–756.
34. Levine, A. J. (1997) p53, the cellular gatekeeper for growth and division. *Cell* **88,** 323–331.
35. Porter, P. L., Malone, K. E., Heagerty, P. J., et al. (1997) Expression of cell-cycle regulators p27Kip1 and cyclin E, alone and in combination, correlate with survival in young breast cancer patients. *Nat. Med.* **3,** 222–225.
36. Hunter, T. and Pines, J. (1994) Cyclins and cancer. II: Cyclin D and CDK inhibitors come of age. *Cell* **79,** 573–582.
37. Yu, M., Zhan, Q., and Finn, O. J. (2002) Immune recognition of cyclin B1 as a tumor antigen is a result of its overexpression in human tumors that is caused by non-functional p53. *Mol. Immunol.* **38,** 981–987.
38. Finn, O. J. (2003) Cancer vaccines: between the idea and the reality. *Nat. Rev. Immunol.* **3,** 630–641.
39. Wolfel, T., Hauer, M., Schneider, J., et al. (1995) A p16INK4a-insensitive CDK4 mutant targeted by cytolytic T lymphocytes in a human melanoma. *Science* **269,** 1281–1284.
40. Zuo, L., Weger, J., Yang, Q., et al. (1996) Germline mutations in the p16INK4a binding domain of CDK4 in familial melanoma. *Nat. Genet.* **12,** 97–99.
41. Payton, M., Chung, G., Yakowec, P., et al. (2004) in *The Cell Cycle, Chromosomes and Cancer*, vol. 15. (Deutscher, M. P., Black, S., Boehmer, P. E., et al., eds.), Miami Nature Biotechnology Winter Short Reports, Miami Beach, FLA, pp. 59–60.
42. Marshall, M. S. (1995) Ras target proteins in eukaryotic cells. *FASEB J.* **9,** 1311–1318.
43. Lavoie, J. N., L'Allemain, G., Brunet, A., Muller, R., and Pouyssegur, J. (1996) Cyclin D1 expression is regulated positively by the p42/p44MAPK and negatively by the p38/HOGMAPK pathway. *J. Biol. Chem.* **271,** 20608–20616.

44. Cheng, M., Sexl, V., Sherr, C. J., and Roussel, M. F. (1998) Assembly of cyclin D-dependent kinase and titration of p27Kip1 regulated by mitogen-activated protein kinase kinase (MEK1). *Proc. Natl. Acad. Sci. USA* **95,** 1091–1096.
45. Balmanno, K.. and Cook, S. J. (1999) Sustained MAP kinase activation is required for the expression of cyclin D1, p21Cip1 and a subset of AP-1 proteins in CCL39 cells. *Oncogene* **18,** 3085–3097.
46. Meyer, C. A., Jacobs, H. W., and Lehner, C. F. (2002) Cyclin D-cdk4 is not a master regulator of cell multiplication in *Drosophila* embryos. *Curr. Biol.* **12,** 661–666.
47. Liu, J. J., Chao, J. R., Jiang, M. C., Ng, S. Y., Yen, J. J., and Yang-Yen, H. F. (1995) Ras transformation results in an elevated level of cyclin D1 and acceleration of G1 progression in NIH 3T3 cells. *Mol. Cell. Biol.* **15,** 3654–3663.
48. Serrano, M., Gomez-Lahoz, E., DePinho, R. A., Beach, D., and Bar-Sagi, D. (1995) Inhibition of ras-induced proliferation and cellular transformation by p16INK4. *Science* **267,** 249–252.
49. Sherr, C., J. and Roberts, J., M. (1999) CDK inhibitors: positive and negative regulators of G1-phase progression. *Genes Dev.* **13,** 1501–1512.
50. Cheng, M., Olivier, P., Diehl, J. A., et al. (1999) The p21(Cip1) and p27(Kip1) CDK inhibitors'are essential activators of cyclin D-dependent kinases in murine fibroblasts. *EMBO J.* **18,** 1571–1583.
51. Bottazzi, M. E., Zhu, X., Bohmer, R. M., and Assoian, R. K. (1999) Regulation of p21(cip1) expression by growth factors and the extracellular matrix reveals a role for transient ERK activity in G1 phase. *J. Cell Biol.* **146,** 1255–1264.
52. Sheaff, R. J., Groudine, M., Gordon, M., Roberts, J. M., and Clurman, B. E. (1997) Cyclin E-CDK2 is a regulator of p27Kip1. *Genes Dev.* **11,** 1464–1478.
53. Rivard, N., Boucher, M. J., Asselin, C., and L'Allemain, G. (1999) MAP kinase cascade is required for p27 downregulation and S phase entry in fibroblasts and epithelial cells. *Am. J. Physiol.* **277,** C652C664.
54. Sklar, M. D. (1988) The ras oncogenes increase the intrinsic resistance of NIH 3T3 cells to ionizing radiation. *Science* **239,** 645–647.
55. Kasid, U., Suy, S., Dent, P., Ray, S., Whiteside, T. L., and Sturgill, T. W. (1996) Activation of Raf by ionizing radiation. *Nature* **382,** 813–816.
56. Abbott, D. W. and Holt, J. T. (1999) Mitogen-activated protein kinase kinase 2 activation is essential for progression through the G2/M checkpoint arrest in cells exposed to ionizing radiation. *J. Biol. Chem.* **274,** 2732–2742.
57. Wright, J. H., Munar, E., Jameson, D. R., et al. (1999) Mitogen-activated protein kinase kinase activity is required for the G(2)/M transition of the cell cycle in mammalian fibroblasts. *Proc. Natl. Acad. Sci. USA* **96,** 11335–11340.
58. Lin, A. W., Barradas, M., Stone, J. C., van Aelst, L., Serrano, M., and Lowe, S. W. (1998) Premature senescence involving p53 and p16 is activated in response to constitutive MEK/MAPK mitogenic signaling. *Genes Dev.* **12,** 3008–3019.
59. Serrano, M., Lin, A. W., McCurrach, M. E., Beach, D., and Lowe, S. W. (1997) Oncogenic ras provokes premature cell senescence associated with accumulation of p53 and p16INK4a. *Cell* **88,** 593–602.
60. Vousden, K. H. (1995) Regulation of the cell cycle by viral oncoproteins. *Semin. Cancer Biol.* **6,** 109–116.

61. Grana, X., Garriga, J., and Mayol, X. (1998) Role of the retinoblastoma protein family, pRB, p107 and p130 in the negative control of cell growth. *Oncogene* **17,** 3365–3383.
62. Harlow, E. (1996) A research shortcut from a common cold virus to human cancer. *Cancer* **78,** 558–565.
63. Fearon, E. R. (1997) Human cancer syndromes: clues to the origin and nature of cancer. *Science* **278,** 1043–1050.
64. Hurford, R. K., Jr., Cobrinik, D., Lee, M. H., and Dyson, N. (1997) pRB and p107/p130 are required for the regulated expression of different sets of E2F responsive genes. *Genes Dev.* **11,** 1447–1463.
65. Dyson, N. (1998) The regulation of E2F by pRB-family proteins. *Genes Dev.* **12,** 2245–2262.
66. Nevins, J. R. (1992) E2F: a link between the Rb tumor suppressor protein and viral oncoproteins. *Science* **258,** 424–429.
67. Helin, K., Wu, C. L., Fattaey, A. R., et al. (1993) Heterodimerization of the transcription factors E2F-1 and DP-1 leads to cooperative trans-activation. *Genes Dev.* **7,** 1850–1861.
68. Waga, S., Hannon, G. J., Beach, D., and Stillman, B. (1994) The p21 inhibitor of cyclin-dependent kinases controls DNA replication by interaction with PCNA. *Nature* **369,** 574–578.
69. Luo, Y., Hurwitz, J., and Massague, J. (1995) Cell-cycle inhibition by independent CDK and PCNA binding domains in p21Cip1. *Nature* **375,** 159–161.
70. Knibiehler, M., Goubin, F., Escalas, N., et al. (1996) Interaction studies between the p21Cip1/Waf1 cyclin-dependent kinase inhibitor and proliferating cell nuclear antigen (PCNA) by surface plasmon resonance. *FEBS Lett.* **391**, 66–70.
71. de Bruin, A., Maiti, B., Jakoi, L., Timmers, C., Buerki, R., and Leone, G. (2003) Identification and characterization of E2F7, a novel mammalian E2F family member capable of blocking cellular proliferation. *J. Biol. Chem.* **278,** 42041–42049.
72. Morkel, M., Wenkel, J., Bannister, A. J., Kouzarides, T., and Hagemeier, C. (1997) An E2F-like repressor of transcription. *Nature* **390,** 567–568.
73. Trimarchi, J. M., Fairchild, B., Verona, R., Moberg, K., Andon, N., and Lees, J. A. (1998) E2F-6, a member of the E2F family that can behave as a transcriptional repressor. *Proc. Natl. Acad. Sci. USA* **95,** 2850–2855.
74. Gaubatz, S., Wood, J. G., and Livingston, D. M. (1998) Unusual proliferation arrest and transcriptional control properties of a newly discovered E2F family member, E2F-6. *Proc. Natl. Acad. Sci. USA* **95,** 9190–9195.
75. Trimarchi, J., M.,, Fairchild, B., Wen, J., and Lees, J., A. (2001) The E2F6 transcription factor is a component of the mammalian Bmi1-containing polycomb complex. *Proc. Natl. Acad. Sci. USA* **98,** 1519–1524.
76. Trouche, D., Cook, A., and Kouzarides, T. (1996) The CBP co-activator stimulates E2F1/DP1 activity. *Nucleic Acids Res.* **24,** 4139–4145.
77. Wang, L., Liu, L., and Berger, S. L. (1998) Critical residues for histone acetylation by Gcn5, functioning in Ada and SAGA complexes, are also required for transcriptional function in vivo. *Genes Dev.* **12,** 640–653.

78. Cress, W. D. and Nevins, J. R. (1996) A role for a bent DNA structure in E2F-mediated transcription activation. *Mol. Cell. Biol.* **16,** 2119–2127.
79. Lindeman, G. J., Gaubatz, S., Livingston, D. M., and Ginsberg, D. (1997) The subcellular localization of E2F-4 is cell-cycle dependent. *Proc. Natl. Acad. Sci. USA* **94,** 5095–5100.
80. Verona, R., Moberg, K., Estes, S., Starz, M., Vernon, J. P., and Lees, J. A. (1997) E2F activity is regulated by cell cycle-dependent changes in subcellular localization. *Mol. Cell. Biol.* **17,** 7268–7282.
81. Magae, J., Wu, C. L., Illenye, S., Harlow, E., and Heintz, N. H. (1996) Nuclear localization of DP and E2F transcription factors by heterodimeric partners and retinoblastoma protein family members. *J. Cell Sci.* **109,** 1717–1726.
82. Apostolova, M. D., Ivanova, I. A., Dagnino, C., DSouza, S. J., and Dagnino, L. (2002) Active nuclear import and export pathways regulate E2F-5 subcellular localization. *J. Biol. Chem.* **277,** 34471–3479.
83. Krek, W., Xu, G., and Livingston, D. M. (1995) Cyclin A-kinase regulation of E2F-1 DNA binding function underlies suppression of an S phase checkpoint. *Cell* **83,** 1149–1158.
84. Krek, W., Ewen, M. E., Shirodkar, S., Arany, Z., Kaelin, W. G. Jr., and Livingston, D. M. (1994) Negative regulation of the growth-promoting transcription factor E2F-1 by a stably bound cyclin A-dependent protein kinase. *Cell* **78,** 161–172.
85. Dynlacht, B. D., Flores, O., Lees, J. A., and Harlow, E. (1994) Differential regulation of E2F transactivation by cyclin/cdk2 complexes. *Genes Dev.* **8,** 1772–1786.
86. Hateboer, G., Kerkhoven, R. M., Shvarts, A., Bernards, R., and Beijersbergen, R. L. (1996) Degradation of E2F by the ubiquitin-proteasome pathway: regulation by retinoblastoma family proteins and adenovirus transforming proteins. *Genes Dev.* **10,** 2960–2970.
87. Hofmann, F., Martelli, F., Livingston, D. M., and Wang, Z. (1996) The retinoblastoma gene product protects E2F-1 from degradation by the ubiquitin-proteasome pathway. *Genes Dev.* **10,** 2949–2959.
88. Brehm, A., Miska, E. A., McCance, D. J., Reid, J. L., Bannister, A. J., and Kouzarides, T. (1998) Retinoblastoma protein recruits histone deacetylase to repress transcription. *Nature* **391,** 597–601.
89. Luo, R. X., Postigo, A. A., and Dean, D. C. (1998) Rb interacts with histone deacetylase to repress transcription. *Cell* **92,** 463–473.
90. Magnaghi-Jaulin, L., Groisman, R., Naguibneva, I., et al. (1998) Retinoblastoma protein represses transcription by recruiting a histone deacetylase. *Nature* **391,** 601–605.
91. Loyola, A., LeRoy, G., Wang, Y.-H., and Reinberg, D. (2001) Reconstitution of recombinant chromatin establishes a requirement for histone-tail modifications during chromatin assembly and transcription. *Genes Dev.* **15,** 2837–2851.
92. Takahashi, Y., Rayman, J. B., and Dynlacht, B. D. (2000) Analysis of promoter binding by the E2F and pRB families in vivo: distinct E2F proteins mediate activation and repression. *Genes Dev.* **14,** 804–816.
93. Ferreira, R., Naguibneva, I., Mathieu, M., et al. (2001) Cell cycle-dependent recruitment of HDAC-1 correlates with deacetylation of histone H4 on an Rb-E2F target promoter. *EMBO Rep.* **2,** 794–799.

94. Hassig, C. A., Fleischer, T. C., Billin, A. N., Schreiber, S. L., and Ayer, D. E. (1997) Histone deacetylase activity is required for full transcriptional repression by mSin3A. *Cell* **89,** 341–347.
95. Laherty, C. D., Yang, W. M., Sun, J. M., Davie, J. R., Seto, E., and Eisenman, R. N. (1997) Histone deacetylases associated with the mSin3 corepressor mediate mad transcriptional repression. *Cell* **89,** 349–356.
96. Nagy, L., Kao, H. Y., Chakravarti, D., et al. (1997) Nuclear receptor repression mediated by a complex containing SMRT, mSin3A, and histone deacetylase. *Cell* **89,** 373–380.
97. Rayman, J. B., Takahashi, Y., Indjeian, V. B., et al. (2002) E2F mediates cell cycle-dependent transcriptional repression in vivo by recruitment of an HDAC1/mSin3B corepressor complex. *Genes Dev.* **16,** 933–947.
98. Bouzahzah, B., Fu, M., Iavarone, A., Factor, V. M., Thorgeirsson, S. S., and Pestell, R. G. (2000) Transforming growth factor-b1 recruits histone deacetylase 1 to a p130 repressor complex in transgenic mice in vivo. *Cancer Res.* **60,** 4531–4537.
99. Harbour, J. W., Luo, R. X., Dei Santi, A., Postigo, A. A., and Dean, D. C. (1999) Cdk phosphorylation triggers sequential intramolecular interactions that progressively block Rb functions as cells move through G1. *Cell* **98,** 859–869.
100. Ohtsubo, M. and Roberts, J. M. (1993) Cyclin-dependent regulation of G1 in mammalian fibroblasts. *Science* **259,** 1908–1912.
101. Tsai, L. H., Lees, E., Faha, B., Harlow, E., and Riabowol, K. (1993) The cdk2 kinase is required for the G1-to-S transition in mammalian cells. *Oncogene* **8,** 1593–1602.
102. Resnitzky, D., Gossen, M., Bujard, H., and Reed, S. I. (1994) Acceleration of the G1/S phase transition by expression of cyclins D1 and E with an inducible system. *Mol. Cell. Biol.* **14,** 1669–1679.
103. Zhao, J., Dynlacht, B., Imai, T., Hori, T., and Harlow, E. (1998) Expression of NPAT, a novel substrate of cyclin E-CDK2, promotes S-phase entry. *Genes Dev.* **12,** 456–461.
104. Zhao, J., Kennedy, B. K., Lawrence, B. D., Barbie, D. A., Matera, A. G., Fletcher, J. A., and Harlow, E. (2000) NPAT links cyclin E-Cdk2 to the regulation of replication-dependent histone gene transcription. *Genes Dev.* **14,** 2283–2297.
105. Ma, T., Van Tine, B. A., Wei, Y., et al. (2000) Cell cycle-regulated phosphorylation of p220(NPAT) by cyclin E/Cdk2 in Cajal bodies promotes histone gene transcription. *Genes Dev.* **14,** 2298–2313.
106. Woo, R. A. and Poon, R. Y. (2003) Cyclin-dependent kinases and S phase control in mammalian cells. *Cell Cycle* **2,** 316–324.
107. Lacey, K. R., Jackson, P. K., and Stearns, T. (1999) Cyclin-dependent kinase control of centrosome duplication. *Proc. Natl. Acad. Sci. USA* **96,** 2817–2822.
108. Tarapore, P., Okuda, M., and Fukasawa, K. (2002) A mammalian in vitro centriole duplication system: evidence for involvement of CDK2/cyclin E and nucleophosmin/B23 in centrosome duplication. *Cell Cycle* **1,** 75–81.
109. Pagano, M., Pepperkok, R., Verde, F., Ansorge, W., and Draetta, G. (1992) Cyclin A is required at two points in the human cell cycle. *EMBO J.* **11,** 961–971.
110. Ohtsubo, M., Theodoras, A. M., Schumacher, J., Roberts, J. M., and Pagano, M. (1995) Human cyclin E, a nuclear protein essential for the G1-to-S phase transition. *Mol. Cell. Biol.* **15,** 2612–2624.

111. Resnitzky, D., Hengst, L., and Reed, S. I. (1995) Cyclin A-associated kinase activity is rate limiting for entrance into S phase and is negatively regulated. *Mol. Cell. Biol.* **15,** 4347–4352.
112. Jiang, W., McDonald, D., Hope, T. J., and Hunter, T. (1999) Mammalian Cdc7-Dbf4 protein kinase complex is essential for initiation of DNA replication. *EMBO J.* **18,** 5703–5713.
113. Herbig, U., Griffith, J. W., and Fanning, E. (2000) Mutation of cyclin/cdk phosphorylation sites in HsCdc6 disrupts a late step in initiation of DNA replication in human cells. *Mol. Biol. Cell.* **11,** 4117–4130.
114. Kelly, T. J. and Brown, G. W. (2000) Regulation of chromosome replication. *Annu. Rev. Biochem.* **69,** 829–880.
115. Nasheuer, H. P., Smith, R., Bauerschmidt, C., Grosse, F., and Weisshart, K. (2002) Initiation of eukaryotic DNA replication: regulation and mechanisms. *Prog. Nucleic Acids Res. Mol. Biol.* **72,** 41–94.
116. Rao, P. N. and Johnson, R. T. (1970) Mammalian cell fusion: studies on the regulation of DNA synthesis and mitosis. *Nature* **225,** 159–164.
117. Marahrens, Y. and Stillman, B. (1992) A yeast chromosomal origin of DNA replication defined by multiple functional elements. *Science* **255,** 817–823.
118. Bell, S. P. and Stillman, B. (1992) ATP-dependent recognition of eukaryotic origins of DNA replication by a multiprotein complex. *Nature* **357,** 128–134
119. Gavin, K. A., Hidaka, M., and Stillman, B. (1995) Conserved initiator proteins in eukaryotes. *Science* **270,** 1667–1671.
120. Iizuka, M. and Stillman, B. (1999) Histone acetyltransferase HBO1 interacts with the ORC1 subunit of the human initiator protein. *J. Biol. Chem.* **274,** 23027–23034.
121. Hateboer, G., Wobst, A., Petersen, B. O., et al. (1998) Cell cycle-regulated expression of mammalian CDC6 is dependent on E2F. *Mol. Cell. Biol.* **18,** 6679–6697.
122. Yan, Z., DeGregori, J., Shohet, R., et al. (1998) Cdc6 is regulated by E2F and is essential for DNA replication in mammalian cells. *Proc. Natl. Acad. Sci. USA* **95,** 3603–3608.
123. Coleman, T. R., Carpenter, P. B., and Dunphy, W. G. (1996) The *Xenopus* Cdc6 protein is essential for the initiation of a single round of DNA replication in cell-free extracts. *Cell* **87,** 53–63.
124. Cook, J. G., Park, C. H., Burke, T. W., et al. (2002) Analysis of Cdc6 function in the assembly of mammalian prereplication complexes. *Proc. Natl. Acad. Sci. USA* **99,** 1347–1352.
125. Saha, P., Chen, J., Thome, K. C., et al. (1998) Human CDC6/Cdc18 associates with Orc1 and cyclin-cdk and is selectively eliminated from the nucleus at the onset of S phase. *Mol. Cell. Biol.* **18,** 2758–2767.
126. Jiang, W., Wells, N. J., and Hunter, T. (1999) Multistep regulation of DNA replication by Cdk phosphorylation of HsCdc6. *Proc. Natl. Acad. Sci. USA* **96,** 6193–6198.
127. Mendez, J. and Stillman, B. (2000) Chromatin association of human origin recognition complex, cdc6, and minichromosome maintenance proteins during the cell cycle: assembly of prerepliaction complexes in late mitosis. *Mol. Cell. Biol.* **20,** 8602–8612.
128. Petersen, B. O., Wagener, C., Marinoni, F., et al. (2000) Cell cycle- and cell growth-regulated proteolysis of mammalian CDC6 is dependent on APC-CDH1. *Genes Dev.* **14,** 2330–2343.

129. Petersen, B. O., Lukas, J., Sorensen, C. S., Bartek, J., and Helin, K. (1999) Phosphorylation of mammalian CDC6 by cyclin A/CDK2 regulates its subcellular localization. *EMBO J.* **18,** 396–410.
130. Biermann, E., Baack, M., Kreitz, S., and Knippers, R. (2002) Synthesis and turnover of the replicative Cdc6 protein during the HeLa cell cycle. *Eur. J. Biochem.* **269,** 1040–1046.
131. Kimura, H., Nozaki, N., and Sugimoto, K. (1994) DNA polymerase alpha associated protein P1, a murine homolog of yeast MCM3, changes its intranuclear distribution during the DNA synthetic period. *EMBO J.* **13,** 4311–4320.
132. Todorov, I. T., Pepperkok, R., Philipova, R. N., Kearsey, S. E., Ansorge, W., and Werner, D. (1994) A human nuclear protein with sequence homology to a family of early S phase proteins is required for entry into S phase and for cell division. *J. Cell Sci.* **107,** 253–265.
133. Fujita, M., Kiyono, T., Hayashi, Y., and Ishibashi, M. (1996) Inhibition of S-phase entry of human fibroblasts by an antisense oligomer against hCDC47. *Biochem. Biophys. Res. Commun.* **219,** 604–607.
134. Kearsey, S. E., and Labib, K. (1998) MCM proteins: evolution, properties, and role in DNA replication. *Biochim. Biophys. Acta* **1398,** 113–136.
135. Tye, B. K. (1999) MCM proteins in DNA replication. *Annu. Rev. Biochem.* **68,** 649–686.
136. Todorov, I. T., Attaran, A., and Kearsey, S. E. (1995) BM28, a human member of the MCM2-3-5 family, is displaced from chromatin during DNA replication. *J. Cell Biol.* **129,** 1433–1445.
137. Thommes, P., Fett, R., Schray, B., et al. (1992) Properties of the nuclear P1 protein, a mammalian homologue of the yeast Mcm3 replication protein. *Nucleic Acids Res.* **20,** 1069–1074.
138. Ishimi, Y. (1997) A DNA helicase activity is associated with an MCM4, -6, and -7 protein complex. *J. Biol. Chem.* **272,** 24508–24513.
139. Hopwood, B. and Dalton, S. (1996) Cdc45p assembles into a complex with Cdc46p/Mcm5p, is required for minichromosome maintenance, and is essential for chromosomal DNA replication. *Proc. Natl. Acad. Sci. USA* **93,** 12309–12314.
140. Hardy, C. F. (1997) Identification of Cdc45p, an essential factor required for DNA replication. *Gene* **187,** 239–246.
141. Zou, L., Mitchell, J., and Stillman, B. (1997) CDC45, a novel yeast gene that functions with the origin recogniton complex Mcm proteins in initiation of DNA replication. *Mol. Cell. Biol.* **17,** 553–563.
142. Saha, P., Thome, K. C., Yamaguchi, R., Hou, Z., Weremowicz, S., and Dutta, A. (1998) The human homolog of *Saccharomyces cerevisiae* CDC45. *J. Biol. Chem.* **273,** 18205–18209.
143. Mimura, S. and Takisawa, H. (1998) *Xenopus* Cdc45-dependent loading of DNA polymerase alpha onto chromatin under the control of S-phase Cdk. *EMBO J.* **17,** 5699–5607.
144. Zou, L., and Stillman, B. (1998) Formation of preinitiation complex by S-phase cyclin CDK-dependent loading of Cdc45p onto chromatin. *Science* **280,** 593–596.
145. Pagano, M., Pepperkok, R., Lukas, J., et al. (1993) Regulation of the cell cycle by the cdk2 protein kinase in cultured human fibroblasts. *J. Cell Biol.* **121,** 101–111.

146. Hendrickson, M., Madine, M., Dalton, S., and Gautier, J. (1996) Phosphorylation of MCM4 by cdc2 protein kinase inhibits the activity of the minichromosome maintenance complex. *Proc. Natl. Acad. Sci. USA* **93,** 12223–12228.
147. Fujita, M., Yamada, C., Tsurumi, T., Hanaoka, F., Matsuzawa, K., and Inagaki, M. (1998) Cell cycle- and chromatin binding state-dependent phosphorylation of human MCM heterohexameric complexes. A role for cdc2 kinase. *J. Biol. Chem.* **273,** 17095–17101.
148. Hartwell, L. H. (1973) Three additional genes required for deoxyribonucleic acid synthesis in *Saccharomyces cerevisiae. J. Bacteriol.* **115,** 966–974.
149. Masai, H., Miyake, T., and Arai, K. (1995) hsk1+, a *Schizosaccharomyces pombe* gene related to *Saccharomyces cerevisiae* CDC7, is required for chromosomal replication. *EMBO J.* **14,** 3094–3104.
150. Jiang, W. and Hunter, T. (1997) Identification and characterization of a human protein kinase related to budding yeast Cdc7p. *Proc. Natl. Acad. Sci. USA* **94,** 14320–14325.
151. Sato, N., Arai, K., and Masai, H. (1997) Human and *Xenopus* cDNAs encoding budding yeast Cdc7-related kinases: in vitro phosphorylation of MCM subunits by a putative human homologue of Cdc7. *EMBO J.* **16,** 4340–4351.
152. Hess, G. F., Drong, R. F., Weiland, K. L., Slightom, J. L., Sclafani, R. A., and Hollingsworth, R. E. (1998) A human homolog of the yeast CDC7 gene is overexpressed in some tumors and transformed cell lines. *Gene* **211,** 133–140.
153. Kumagai, H., Sato, N., Yamada, M., et al. (1999) A novel growth- and cell cycle-regulated protein, ASK, activates human Cdc7-related kinase and is essential for G1/S transition in mammalian cells. *Mol. Cell. Biol.* **19,** 5083–5095.
154. Lepke, M., Putter, V., Staib, C., et al. (1999) Identification, characterization and chromosomal localization of the cognate human and murine DBF4 genes. *Mol. Gen. Genet.* **262,** 220–229.
155. Donaldson, A. D., Fangman, W. L., and Brewer, B. J. (1998) Cdc7 is required throughout the yeast S phase to activate replication origins. *Genes Dev.* **12,** 491–501.
156. Bousset, K.. and Diffley, J. F. (1998) The Cdc7 protein kinase is required for origin firing during S phase. *Genes Dev.* **12,** 480–490.
157. Hartley, K. O., Gell, D., Smith, G. C., et al. (1995) DNA-dependent protein kinase catalytic subunit: a relative of phosphatidylinositol 3-kinase and the ataxia telangiectasia gene product. *Cell* **82,** 849–856.
158. Savitsky, K., Bar-Shira, A., Gilad, S., et al. (1995) A single ataxia telangiectasia gene with a product similar to PI-3 kinase. *Science* **268,** 1749–1753.
159. Bentley, N. J., Holtzman, D. A., Flaggs, G., et al. (1996) The Schizosaccharomyces pombe rad3 checkpoint gene. *EMBO J.* **15,** 6641–6651.
160. Keegan, K. S., Holtzman, D. A., Plug, A. W., et al. (1996) The Atr and Atm protein kinases associate with different sites along meiotically pairing chromosomes. *Genes Dev.* **10,** 2423–2437.
161. Gately, D. P., Hittle, J. C., Chan, G. K., and Yen, T. J. (1998) Characterization of ATM expression, localization, and associated DNA-dependent protein kinase activity. *Mol. Biol. Cell.* **9,** 2361–2374.
162. Lakin, N. D., Hann, B. C., and Jackson, S. P. (1999) The ataxia-telangiectasia related protein ATR mediates DNA-dependent phosphorylation of p53. *Oncogene* **18,** 3989–3995.

163. Shieh, S. Y., Ikeda, M., Taya, Y., and Prives, C. (1997) DNA damage-induced phosphorylation of p53 alleviates inhibition by MDM2. *Cell* **91,** 325–334.
164. Siliciano, J. D., Canman, C. E., Taya, Y., Sakaguchi, K., Appella, E., and Kastan, M. B. (1997) DNA damage induces phosphorylation of the amino terminus of p53. *Genes Dev.* **11,** 3471–3481.
165. Banin, S., Moyal, L., Shieh, S., et al. (1998) Enhanced phosphorylation of p53 by ATM in response to DNA damage. *Science* **281,** 1674–1677.
166. Canman, C. E., Lim, D. S., Cimprich, K. A., et al. (1998) Activation of the ATM kinase by ionizing radiation and phosphorylation of p53. *Science* **281,** 1677–1679.
167. Colman, M. S., Afshari, C. A., and Barrett, J. C. (2000) Regulation of p53 stability and activity in response to genotoxic stress. *Mutat. Res.* **462,** 179–188.
168. Ryan, K. M., Phillips, A. C., and Vousden, K. H. (2001) Regulation and function of the p53 tumor suppressor protein. *Curr. Opin. Cell. Biol.* **13,** 332–337.
169. Taylor, W. R.. and Stark, G. R. (2001) Regulation of the G2/M transition by p53. *Oncogene* **20,** 1803–1815.
170. Wahl, G. M. and Carr, A. M. (2001) The evolution of diverse biological responses to DNA damage: insights from yeast and p53. *Nat. Cell Biol.* **3,** E277–E286.
171. Nayak, B. K. and Das, G. M. (2002) Stabilization of p53 and transactivation of its target genes in response to replication blockade. *Oncogene* **21,** 7226–7229.
172. Tibbetts, R. S., Brumbaugh, K. M., Williams, J. M., et al. (1999) A role for ATR in the DNA damage-induced phosphorylation of p53. *Genes Dev.* **13,** 152–157.
173. Winkler, K. E., Swenson, K. I., Kornbluth, S., and Means, A. R. (2000) Requirement of the prolyl isomerase Pin1 for the replication checkpoint. *Science* **287,** 1644–1647.
174. Ryo, A., Nakamura, M., Wulf, G., Liou, Y. C., and Lu, K. P. (2001) Pin1 regulates turnover and subcellular localization of b-catenin by inhibiting its interaction with APC. *Nat. Cell Biol.* **3,** 793–801.
175. Wulf, G. M., Ryo, A., Wulf, G. G., et al. (2001) Pin1 is overexpressed in breast cancer and cooperates with Ras signaling in increasing the transcriptional activity of c-Jun towards cyclin D1. *EMBO J.* **20,** 3459–3472.
176. Liou, Y. C., Ryo, A., Huang, H. K., et al. (2002) Loss of Pin1 function in the mouse causes phenotypes resembling cyclin D1-null phenotypes. *Proc. Natl. Acad. Sci. USA* **99,** 1335–1340.
177. Wulf, G. M., Liou, Y. C., Ryo, A., Lee, S. W., and Lu, K. P. (2002) Role of Pin1 in the regulation of p53 stability and p21 transactivation, and cell cycle checkpoints in response to DNA damage. *J. Biol. Chem.* **17,** 17.
178. Boddy, M. N. and Russell, P. (1999) DNA replication checkpoint control. *Front. Biosci.* **4,** D841–D848.
179. Chan, D. W., Chen, B. P., Prithivirajsingh, S., et al. (2002) Autophosphorylation of the DNA-dependent protein kinase catalytic subunit is required for rejoining of DNA double-strand breaks. *Genes Dev.* **16,** 2333–2338.
180. Woo, R. A., Jack, M. T., Xu, Y., Burma, S., Chen, D. J., and Lee, P. W. (2002) DNA damage-induced apoptosis requires the DNA-dependent protein kinase, and is mediated by the latent population of p53. *EMBO J.* **21,** 3000–3008.
181. Parker, A. E., Van de Weyer, I., Laus, M. C., et al. (1998) A human homologue of the *Schizosaccharomyces pombe* rad1+ checkpoint gene encodes an exonuclease. *J. Biol. Chem.* **273,** 18332–18339.

182. Thelen, M. P., Venclovas, C., and Fidelis, K. (1999) A sliding clamp model for the Rad1 family of cell cycle checkpoint proteins. *Cell* **96,** 769–770.
183. Volkmer, E. and Karnitz, L. M. (1999) Human homologs of Schizosaccharomyces pombe rad1, hus1, and rad9 form a DNA damage-responsive protein complex. *J. Biol. Chem.* **274,** 567–570.
184. Hang, H. and Lieberman, H. B. (2000) Physical interactions among human checkpoint control proteins HUS1p, RAD1p, and RAD9p, and implications for the regulation of cell cycle progression. *Genomics* **65,** 24–33.
185. Zou, L., Cortez, D., and Elledge, S. J. (2002) Regulation of ATR substrate selection by Rad17-dependent loading of Rad9 complexes onto chromatin. *Genes Dev.* **16,** 198–208.
186. Dahm, K., and Hubscher, U. (2002) Colocalization of human Rad17 and PCNA in late S phase of the cell cycle upon replication block. *Oncogene* **21,** 7710–7719.
187. Brown, A. L., Lee, C. H., Schwarz, J. K., Mitiku, N., Piwnica-Worms, H., and Chung, J. H. (1999) A human Cds1-related kinase that functions downstream of ATM protein in the cellular response to DNA damage. *Proc. Natl. Acad. Sci. USA* **96,** 3745–3750.
188. Chaturvedi, P., Eng, W. K., Zhu, Y., et al. (1999) Mammalian Chk2 is a downstream effector of the ATM-dependent DNA damage checkpoint pathway. *Oncogene* **18,** 4047–4054.
189. Matsuoka, S., Huang, M., and Elledge, S. J. (1998) Linkage of ATM to cell cycle regulation by the Chk2 protein kinase. *Science* **282,** 1893–1897.
190. Peng, C. Y., Graves, P. R., Thoma, R. S., Wu, Z., Shaw, A. S., and Piwnica-Worms, H. (1997) Mitotic and G2 checkpoint control: regulation of 14-3-3 protein binding by phosphorylation of Cdc25C on serine-216. *Science* **277,** 1501–1505.
191. Dalal, S. N., Schweitzer, C. M., Gan, J., and DeCaprio, J. A. (1999) Cytoplasmic localization of human cdc25C during interphase requires an intact 14-3-3 binding site. *Mol. Cell. Biol.* **19,** 4465–4479.
192. Masui, Y. and Markert, C. L. (1971) Cytoplasmic control of nuclear behavior during meiotic maturation of frog oocytes. *J. Exp. Zool.* **177,** 129–145.
193. Dunphy, W. G., Brizuela, L., Beach, D., and Newport, J. (1988) The *Xenopus* cdc2 protein is a component of MPF, a cytoplasmic regulator of mitosis. *Cell* **54,** 423–431.
194. Gautier, J., Norbury, C., Lohka, M., Nurse, P., and Maller, J. (1988) Purified maturation-promoting factor contains the product of a *Xenopus* homolog of the fission yeast cell cycle control gene cdc2+. *Cell* **54,** 433–439.
195. Smits, V. A. and Medema, R. H. (2001) Checking out the G(2)/M transition. *Biochim. Biophys. Acta* **1519,** 1–12.
196. Takizawa, C. G. and Morgan, D. O. (2000) Control of mitosis by changes in the subcellular location of cyclin-B1-Cdk1 and Cdc25C. *Curr. Opin. Cell Biol.* **12,** 658–665.
197. Pines, J. and Hunter, T. (1989) Isolation of a human cyclin cDNA: evidence for cyclin mRNA and protein regulation in the cell cycle and for interaction with p34cdc2. *Cell* **58,** 833–846.
198. Brandeis, M., Rosewell, I., Carrington, M., et al. (1998) Cyclin B2-null mice develop normally and are fertile whereas cyclin B1-null mice die in utero. *Proc. Natl. Acad. Sci. USA* **95,** 4344–4349.

199. Jackman, M., Firth, M., and Pines, J. (1995) Human cyclins B1 and B2 are localized to strikingly different structures: B1 to microtubules, B2 primarily to the Golgi apparatus. *EMBO J.* **14,** 1646–1654.
200. Endicott, J. A., Nurse, P., and Johnson, L. N. (1994) Mutational analysis supports a structural model for the cell cycle protein kinase p34. *Protein Eng.* **7,** 243–253.
201. Atherton-Fessler, S., Parker, L. L., Geahlen, R. L., and Piwnica-Worms, H. (1993) Mechanisms of p34cdc2 regulation. *Mol. Cell. Biol.* **13,** 1675–1685.
202. Lundgren, K., Walworth, N., Booher, R., Dembski, M., Kirschner, M., and Beach, D. (1991) mik1 and wee1 cooperate in the inhibitory tyrosine phosphorylation of cdc2. *Cell* **64,** 1111–1122.
203. Parker, L. L., and Piwnica-Worms, H. (1992) Inactivation of the p34cdc2-cyclin B complex by the human WEE1 tyrosine kinase. *Science* **257,** 1955–1957.
204. Li, J., Meyer, A. N., and Donoghue, D. J. (1995) Requirement for phosphorylation of cyclin B1 for *Xenopus* oocyte maturation. *Mol. Biol. Cell.* **6,** 1111–1124.
205. Sadhu, K., Reed, S. I., Richardson, H., and Russell, P. (1990) Human homolog of fission yeast cdc25 mitotic inducer is predominantly expressed in G2. *Proc. Natl. Acad. Sci. USA* **87,** 5139–5143.
206. Galaktionov, K. and Beach, D. (1991) Specific activation of cdc25 tyrosine phosphatases by B-type cyclins: evidence for multiple roles of mitotic cyclins. *Cell* **67,** 1181–1194.
207. Hoffmann, I., Clarke, P. R., Marcote, M. J., Karsenti, E., and Draetta, G. (1993) Phosphorylation and activation of human cdc25-C by cdc2cyclin B and its involvement in the self-amplification of MPF at mitosis. *EMBO J.* **12,** 53–63.
208. Gabrielli, B. G., De Souza, C. P., Tonks, I. D., Clark, J. M., Hayward, N. K., and Ellem, K. A. (1996) Cytoplasmic accumulation of cdc25B phosphatase in mitosis triggers centrosomal microtubule nucleation in HeLa cells. *J. Cell Sci.* **109,** 1081–1093.
209. Vigo, E., Muller, H., Prosperini, E., Hateboer, G., Cartwright, P., Moroni, M. C., and Helin, K. (1999) CDC25A phosphatase is a target of E2F and is required for efficient E2F-induced S phase. *Mol. Cell. Biol.* **19,** 6379–6395.
210. Izumi, T., Walker, D. H., and Maller, J. L. (1992) Periodic changes in phosphorylation of the *Xenopus* cdc25 phosphatase regulate its activity. *Mol. Biol. Cell.* **3,** 927–939.
211. Kumagai, A. and Dunphy, W. G. (1992) Regulation of the cdc25 protein during the cell cycle in *Xenopus* extracts. *Cell* **70,** 139–151.
212. Izumi, T. and Maller, J. L. (1993) Elimination of cdc2 phosphorylation sites in the cdc25 phosphatase blocks initiation of M-phase. *Mol. Biol. Cell.* **4,** 1337–1350.
213. Roshak, A. K., Capper, E. A., Imburgia, C., Fornwald, J., Scott, G., and Marshall, L. A. (2000) The human polo-like kinase, PLK, regulates cdc2/cyclin B through phosphorylation and activation of the cdc25C phosphatase. *Cell Signal* **12,** 405–411.
214. Chen, M. S., Hurov, J., White, L. S., Woodford-Thomas, T., and Piwnica-Worms, H. (2001) Absence of apparent phenotype in mice lacking Cdc25C protein phosphatase. *Mol. Cell. Biol.* **21,** 3853–3861.
215. Mailand, N., Podtelejnikov, A. V., Groth, A., Mann, M., Bartek, J., and Lukas, J. (2002) Regulation of G(2)/M events by Cdc25A through phosphorylation-dependent modulation of its stability. *EMBO J.* **21,** 5911–5920.

216. Fode, C., Motro, B., Yousefi, S., Heffernan, M., and Dennis, J. W. (1994) Sak, a murine protein-serine/threonine kinase that is related to the *Drosophila* polo kinase and involved in cell proliferation. *Proc. Natl. Acad. Sci. USA* **91,** 6388–6392.
217. Glover, D. M., Hagan, I. M., and Tavares, A. A. (1998) Polo-like kinases: a team that plays throughout mitosis. *Genes Dev.* **12,** 3777–377.
218. Lee, K. S., Grenfell, T. Z., Yarm, F. R., and Erikson, R. L. (1998) Mutation of the polo-box disrupts localization and mitotic functions of the mammalian polo kinase Plk. *Proc. Natl. Acad. Sci. USA* **95,** 9301–9306.
219. Nigg, E. A. (1998) Polo-like kinases: positive regulators of cell division from start to finish. *Curr. Opin. Cell. Biol.* **10,** 776–783.
220. Hamanaka, R., Smith, M. R., OConnor, P. M., et al. (1995) Polo-like kinase is a cell cycle-regulated kinase activated during mitosis. *J. Biol. Chem.* **270,** 21086–21091.
221. Simmons, D. L., Neel, B. G., Stevens, R., Evett, G., and Erikson, R. L. (1992) Identification of an early-growth-response gene encoding a novel putative protein kinase. *Mol. Cell. Biol.* **12,** 4164–4169.
222. Xie, S., Wu, H., Wang, Q., et al. (2002) Genotoxic stress-induced activation of Plk3 is partly mediated by Chk2. *Cell Cycle* **1,** 424–429.
223. Peng, C. Y., Graves, P. R., Ogg, S., et al. (1998) C-TAK1 protein kinase phosphorylates human Cdc25C on serine 216 and promotes 14-3-3 protein binding. *Cell Growth Differ.* **9,** 197–208.
224. Aitken, A., Collinge, D. B., van Heusden, B. P., et al. (1992) 14-3-3 proteins: a highly conserved, widespread family of eukaryotic proteins. *Trends Biochem. Sci.* **17,** 498–501.
225. Heald, R., McLoughlin, M., and McKeon, F. (1993) Human wee1 maintains mitotic timing by protecting the nucleus from cytoplasmically activated Cdc2 kinase. *Cell* **74,** 463–474.
226. Ogg, S., Gabrielli, B., and Piwnica-Worms, H. (1994) Purification of a serine kinase that associates with and phosphorylates human Cdc25C on serine 216. *J. Biol. Chem.* **269,** 30461–30469.
227. Zeng, Y., Forbes, K. C., Wu, Z., Moreno, S., Piwnica-Worms, H., and Enoch, T. (1998) Replication checkpoint requires phosphorylation of the phosphatase Cdc25 by Cds1 or Chk1. *Nature* **395,** 507–510.
228. Furnari, B., Blasina, A., Boddy, M. N., McGowan, C. H., and Russell, P. (1999) Cdc25 inhibited in vivo and in vitro by checkpoint kinases Cds1 and Chk1. *Mol. Biol. Cell.* **10,** 833–845.
229. Hagting, A., Jackman, M., Simpson, K., and Pines, J. (1999) Translocation of cyclin B1 to the nucleus at prophase requires a phosphorylation-dependent nuclear import signal. *Curr. Biol.* **9,** 680–689.
230. Pines, J. and Hunter, T. (1994) The differential localization of human cyclins A and B is due to a cytoplasmic retention signal in cyclin B. *EMBO J.* **13,** 3772–3781.
231. Yang, J., Bardes, E. S., Moore, J. D., Brennan, J., Powers, M. A., and Kornbluth, S. (1998) Control of cyclin B1 localization through regulated binding of the nuclear export factor CRM1. *Genes Dev.* **12,** 2131–2143.
232. Moore, J. D., Yang, J., Truant, R., and Kornbluth, S. (1999) Nuclear import of Cdk/cyclin complexes: identification of distinct mechanisms for import of Cdk2/cyclin E and Cdc2/cyclin B1. *J. Cell. Biol.* **144,** 213–224.

233. Kong, M., Barnes, E. A., Ollendorff, V., and Donoghue, D. J. (2000) Cyclin F regulates the nuclear localization of cyclin B1 through a cyclin-cyclin interaction. *EMBO J.* **19,** 1378–1388.
234. Blangy, A., Lane, H. A., dHerin, P., Harper, M., Kress, M., and Nigg, E. A. (1995) Phosphorylation by p34cdc2 regulates spindle association of human Eg5, a kinesin-related motor essential for bipolar spindle formation in vivo. *Cell* **83,** 1159–1169.
235. Lowe, M., Rabouille, C., Nakamura, N., et al. (1998) Cdc2 kinase directly phosphorylates the cis-Golgi matrix protein GM130 and is required for Golgi fragmentation in mitosis. *Cell* **94,** 783–793.
236. Yamashiro, S., Yamakita, Y., Ishikawa, R., and Matsumura, F. (1990) Mitosis-specific phosphorylation causes 83K non-muscle caldesmon to dissociate from microfilaments. *Nature* **344,** 675–678.
237. Furuno, N., den Elzen, N., and Pines, J. (1999) Human cyclin A is required for mitosis until mid prophase. *J. Cell Biol.* **147,** 295–306.
238. Goldstone, S., Pavey, S., Forrest, A., Sinnamon, J., and Gabrielli, B. (2001) Cdc25-dependent activation of cyclin A/cdk2 is blocked in G2 phase arrested cells independently of ATM/ATR. *Oncogene* **20,** 921–932.
239. Glotzer, M., Murray, A. W., and Kirschner, M. W. (1991) Cyclin is degraded by the ubiquitin pathway. *Nature* **349,** 132–138.
240. Harper, J. W., Burton, J. L., and Solomon, M. J. (2002) The anaphase-promoting complex: itš not just for mitosis any more. *Genes Dev.* **16,** 2179–2206.
241. Golan, A., Yudkovsky, Y., and Hershko, A. (2002) The cyclin-ubiquitin ligase activity of cyclosome/APC is jointly activated by protein kinases Cdk1-cyclin B and Plk. *J. Biol. Chem.* **277,** 15552–15557.
242. Kotani, S., Tugendreich, S., Fujii, M., et al. (1998) PKA and MPF-activated polo-like kinase regulate anaphase-promoting complex activity and mitosis progression. *Mol. Cell* **1,** 371–380.
243. Shteinberg, M., Protopopov, Y., Listovsky, T., Brandeis, M., and Hershko, A. (1999) Phosphorylation of the cyclosome is required for its stimulation by Fizzy/cdc20. *Biochem. Biophys. Res. Commun.* **260,** 193–198.
244. Chestukhin, A., Pfeffer, C., Milligan, S., DeCaprio, J. A., and Pellman, D. (2003) Processing, localization, and requirement of human separase for normal anaphase progression. *Proc. Natl. Acad. Sci. USA* **100,** 4574–4579.
245. Fontijn, R. D., Goud, B., Echard, A., et al. (2001) The human kinesin-like protein RB6K is under tight cell cycle control and is essential for cytokinesis. *Mol. Cell. Biol.* **21,** 2944–2955.
246. Noguchi, T., Arai, R., Motegi, F., Nakano, K., and Mabuchi, I. (2001) Contractile ring formation in *Xenopus* egg and fission yeast. *Cell Struct. Funct.* **26,** 545–554.
247. Cao, L. G. and Wang, Y. L. (1996) Signals from the spindle midzone are required for the stimulation of cytokinesis in cultured epithelial cells. *Mol. Biol. Cell.* **7,** 225–232.
248. Gatti, M., Giansanti, M. G., and Bonaccorsi, S. (2000) Relationships between the central spindle and the contractile ring during cytokinesis in animal cells. *Microsc. Res. Tech.* **49,** 202–208.
249. Adams, R. R., Maiato, H., Earnshaw, W. C., and Carmena, M. (2001) Essential roles of *Drosophila* inner centromere protein (INCENP) and aurora B in histone H3 phos-

phorylation, metaphase chromosome alignment, kinetochore disjunction, and chromosome segregation. *J. Cell Biol.* **153,** 865–880.
250. Nislow, C., Sellitto, C., Kuriyama, R., and McIntosh, J. R. (1990) A monoclonal antibody to a mitotic microtubule-associated protein blocks mitotic progression. *J. Cell Biol.* **111,** 511–522.
251. Nislow, C., Lombillo, V. A., Kuriyama, R., and McIntosh, J. R. (1992) A plus-end-directed motor enzyme that moves antiparallel microtubules in vitro localizes to the interzone of mitotic spindles. *Nature* **359,** 543–547.
252. Wheatley, S. P., Hinchcliffe, E. H., Glotzer, M., Hyman, A. A., Sluder, G., and Wang, Y. (1997) CDK1 inactivation regulates anaphase spindle dynamics and cytokinesis in vivo. *J. Cell Biol.* **138,** 385–393.
253. Mishima, M. and Mabuchi, I. (1996) Cell cycle-dependent phosphorylation of smooth muscle myosin light chain in sea urchin egg extracts. *J. Biochem. (Tokyo)* **119,** 906–913.
254. Sellers, J. R. (1991) Regulation of cytoplasmic and smooth muscle myosin. *Curr. Opin. Cell Biol.* **3,** 98–104.
255. Yamakita, Y., Yamashiro, S., and Matsumura, F. (1994) In vivo phosphorylation of regulatory light chain of myosin II during mitosis of cultured cells. *J. Cell Biol.* **124,** 129–137.
256. Grafi, G. (1998) Cell cycle regulation of DNA replication: the endoreduplication perspective. *Exp. Cell Res.* **244,** 372–378.
257. Bloom, J. and Pagano, M. (2004) in *The Cell Cycle, Chromosomes and Cancer* (Deutscher, M. P., Black, S., Boehmer, P. E., et al., eds.), Miami Nature Biotechnology Winter Short Reports, Miami Beach, FLA, pp. 63–64.
258. Carrano, A. C., Eytan, E., Hershko, A., and Pagano, M. (1999) SKP2 is required for ubiquitin-mediated degradation of the CDK inhibitor p27. *Nat. Cell Biol.* **1,** 193–199.
259. Bornstein, G., Bloom, J., Sitry-Shevah, D., Nakayama, K., Pagano, M., and Hershko, A. (2003) Role of the SCFSkp2 ubiquitin ligase in the degradation ofp21Cip1 in S phase. *J. Biol. Chem.* **278,** 25752–25757.

II

Cell Cycle Study Methods and Techniques

7

Synchronization of Cell Populations in G_1/S and G_2/M Phases of the Cell Cycle

Jane V. Harper

Summary

The method described in the following chapter utilizes a double thymidine block (an inhibitor of DNA synthesis) followed by treatment of cells with nocodazole (a mitotic inhibitor) to obtain large cell populations at distinct phases of the cell cycle. Treatment with double thymidine results in a G_1/S-phase arrested cell population, and the use of flow cytometry allows progression of the cells through the cell cycle to be monitored. Flow cytometry enables the calculation of timings for collection of cells at distinct cell cycle phases from G_1/S (following treatment with thymidine) through to G_2/M (owing to the presence of nocodazole).

Key Words

Bromodeoxyuridine (BrdU); cell cycle; cell synchronization; double thymidine block; flow cytometry; nocodazole; propidium iodide (PI).

1. Introduction

This chapter describes methods for the synchronization of cells at the G_1/S and G_2/M phases of the cell cycle using a double thymidine block to synchronize cells at the G_1/S border, followed by addition of nocodazole to block cells in G_2/M. Flow cytometric analysis is used to monitor the progression of cells from G_1/S to G_2/M, allowing calculation of the timing for collection of cells at specific cell cycle phases (**Fig. 1**); these cell populations subsequently can be used in the investigation of various cell cycle-regulated molecules, for example. Treatment of cells with excess thymidine (2 m*M*) causes the arrest of cells at the G_1/S border owing to an inhibition of DNA synthesis that is attributable to feedback inhibition of nucleotide synthesis caused by an imbalance of the nucleotide pool *(1)*. The requirement for two consecutive exposures to thymidine to achieve a G_1 arrest is described in **Fig. 2**; following the first treatment with thymidine for a finite period (e.g., 12 h), those cells arrested within S

From: *Methods in Molecular Biology, vol. 296, Cell Cycle Control: Mechanisms and Protocols*
Edited by: T. Humphrey and G. Brooks © Humana Press Inc., Totowa, NJ

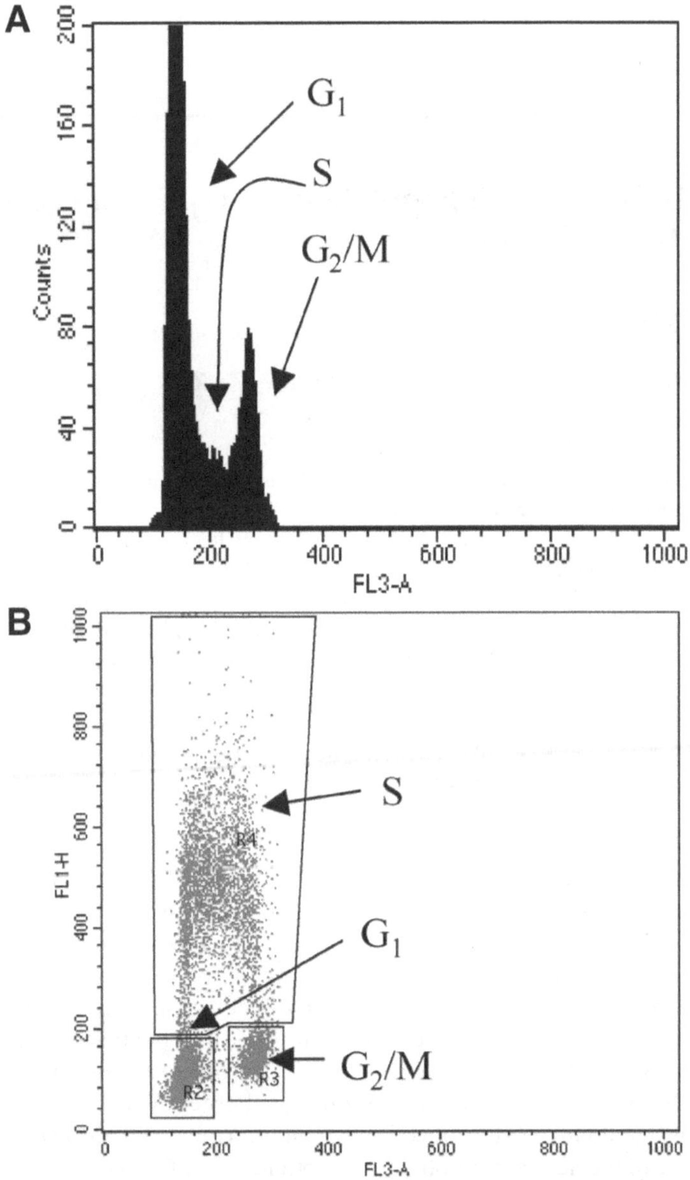

Fig. 1. Determination of cell cycle distribution by flow cytometric analysis. The figure demonstrates the profiles obtained from dual staining with PI and BrdU when samples are analyzed by flow cytometry. **(A)** Histogram representing PI staining (FL3-A). **(B)** Dot plot demonstrating BrdU labeling (FL1-H) vs PI stain (FL3-A); this allows gating and accurate calculation of the percentage of cells in phase each of the cell cycle.

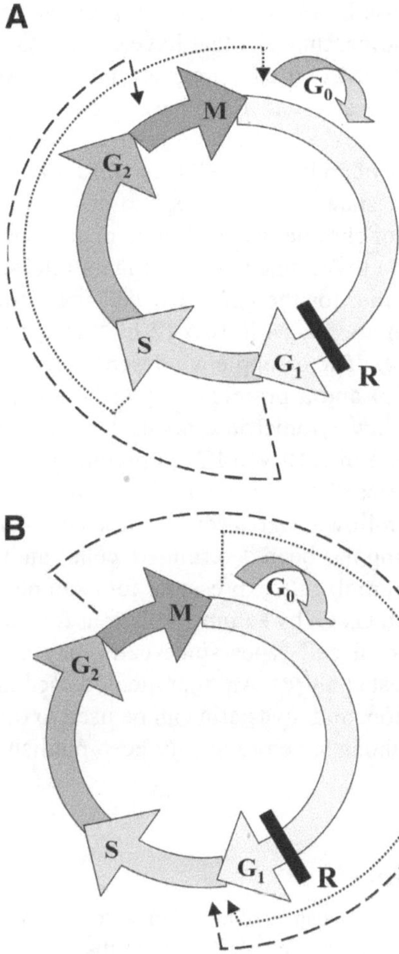

Fig. 2. The mechanism of the double thymidine block. The figure demonstrates the five distinct phases of the cell cycle: G_0, in which cells are resting or quiescent; G_1, in which cells undergo RNA and protein synthesis necessary for entry in to S-phase; S-phase, in which cells undergo DNA replication; G_2, in which cells undergo RNA and protein synthesis in preparation for mitosis; and M-phase, in which cells undergo mitosis and cytokinesis. **(A)** At the end of the first exposure of cells to thymidine, cells become arrested at the G_1/S transition and throughout S-phase owing to the inhibition of DNA synthesis. Following release from the first exposure for 12–16 h, cells arrested in G_1/S and early S-phase will progress through to G_2/M phases and those arrested in late S phase will progress to G_1. **(B)**. Upon the second exposure to thymidine, cells that had progressed to G_2/M will progress into G_1 and eventually become blocked at the G_1/S transition owing to the presence of thymidine; those cells that had entered G_1 (i.e., the late S-phase cells) will also become arrested at the G_1/S transition (*see* **Table 1**). The dashed arrow represents progression of G_1/S and early S-phase cells following release from the first exposure to thymidine. The dotted arrow represents progression of late S-phase cells following the first release from exposure to thymidine. **R**, restriction point.

phase will reenter G_1, and those blocked at G_1/S will progress into G_2/M. Following the second exposure to thymidine, the cells that have entered G_1 will collect at G_1/S, and those that have reached G_2/M will also progress into G_1 and become blocked at G_1/S such that no cells will be arrested in S-phase. Upon the second release from thymidine, nocodazole is added to the medium for up to 24 h (**Table 1**). Following this 24-h incubation period with nocodazole, cells become arrested in M-phase in a prophase to pseudometaphase state owing to the disruption of microtubules that are required for the condensation of chromatin and alignment on the metaphase plate *(2,3)*.

The timings for exposure to thymidine and subsequent release from this agent are dependent on the cell cycle time for the particular cell population under study. For cells with a cell cycle time of approx 24 h, two 12-h blocks separated by a 16–18 h release should be sufficient *(4)*. The example we describe here utilizes A10 vascular smooth muscle cells (VSMCs) and a protocol of two 12-h exposures to thymidine separated by a 12-h release. Flow cytometric analysis demonstrated a block at the G_1/S transition following exposure of A10 VSMCs to thymidine; the example shown in **Fig. 3** demonstrates that approx 90% of cells are arrested at G_1/S. Flow cytometric analysis can then be used to follow progression of these cells through the cell cycle, and once the appropriate timing has been determined, cells can be collected and used to assess cell cycle-dependent molecule expression, for example (*see* **Figs. 3** and **4**).

Cells can also be arrested at G_0/G_1 by serum starvation, for example; however, this method is not successful for all cell types since cells may permanently enter G_0, undergo apoptosis, or not arrest at all *(4)*. Another method used to obtain G_0/G_1 populations is isoleucine deprivation, and lovastatin can be used to obtain cells arrested in early G_1; details of these methods have previously been published by OConnor and Jackman *(4)*.

2. Materials

2.1. Synchronization Protocols

1. Phosphate-buffered saline (PBS) can be obtained in tablet form from Sigma and made up per the manufacturer̀s instructions. This should be sterilized by autoclaving prior to use in cell culture experiments.
2. 100 m*M* Thymidine (Sigma) stock solution in PBS; store at 4C for the duration of a single synchronization experiment. Care should be taken to ensure complete dissolution of thymidine in PBS; this can be achieved by placing the solution into a 37C water bath. Sterilize the solution by filter sterilization.
3. 1 mg/mL Nocodazole (Sigma) stock solution in dimethylsulfoxide (DMSO; Sigma); this can be stored at 4C for 1 mo or at –20C for up to 6 mo.

2.2. Flow Cytometric Analysis

1. 1 m*M* Bromodeoxyuridine (BrdU; Sigma) stock solution in PBS; store in 105- L aliquots at –20C.
2. 70% Ethanol in H_2O; store at –20C.
3. 0.1 *M* HCl solution in PBS; store at room temperature.
4. Anti-BrdU antibody (25 μg/mL) can be obtained from Becton Dickinson and anti-mouse fluorescein isothiocyanate (FITC)-conjugated antibody from Dako.

Table 1
Thymidine Synchronization of A10 VSMCs

	Control	Thymidine 0 h	3 h Release nocodazole	5 h Release nocodazole	7 h Release nocodazole	24 h Release nocodazole
G_0/G_1	51+/–2.5	89+/–1.0*	28+/–8**	26+/–2.3**	27+/–6.4**	27+/–2.6**
S	37+/–2.9	2+/–1.0*	67+/–8**	63+/–2.9**	27+/–18.5**	9+/–1.5**
G_2/M	13+/–1.0	9+/–0.6*	5+/–0**	10+/–0**	46+/–11.8**	64+/–3.1**

The table shows the percentage of cells in each cell cycle phase following double thymidine block in A10 VSMCs. Values were calculated from the PI vs BrdU dot blot (as shown in **Fig. 1B**) in order to give a more accurate estimation of S phase cells. These values clearly show blockade of cells in the G_1 phase following treatment with thymidine. Cells can then be followed through the different phases of the cell cycle at various time points with a large proportion of cells finally collecting in G_2/M because of the presence of nocodazole. *, significantly different from control; **, significantly different from double thymidine-treated cells ($p < 0.005$; n = 3).

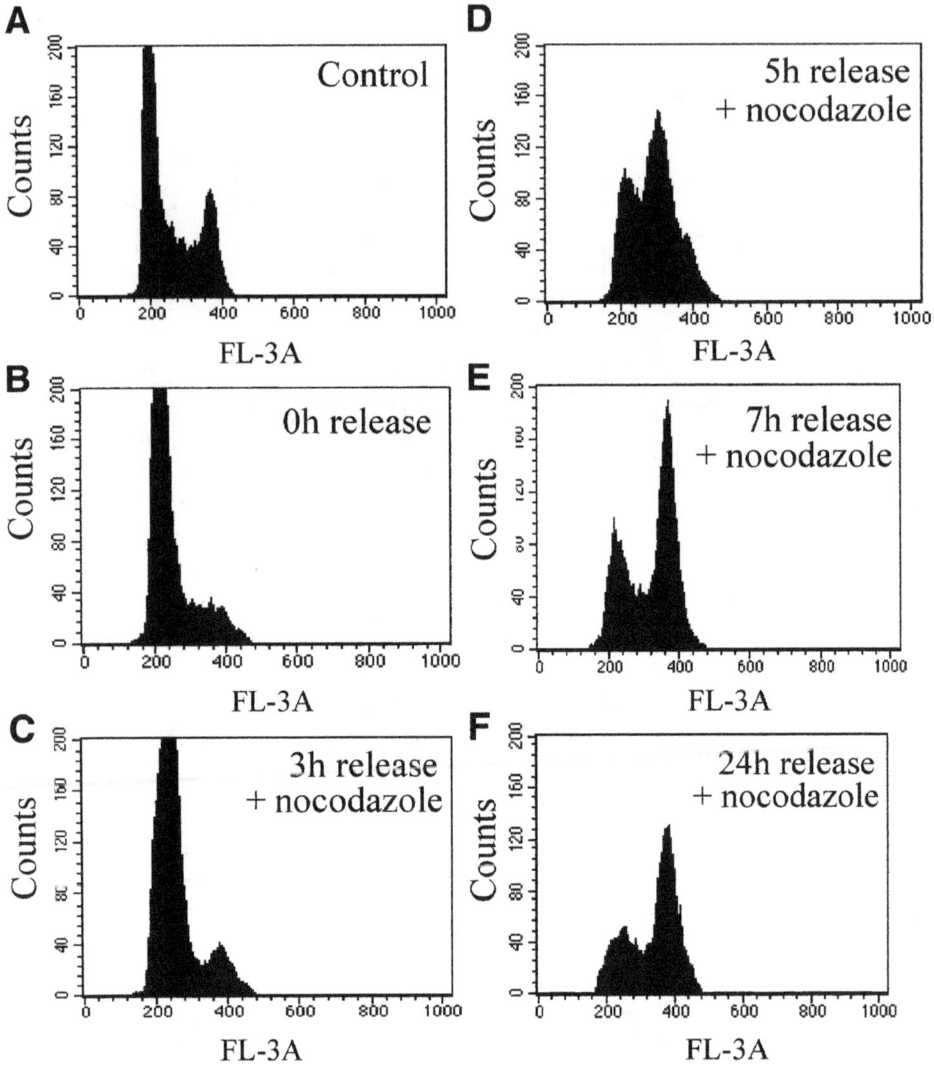

Fig. 3. Thymidine synchronization of A10 VSMCs. The figure shows PI staining of A10 VSMCs following treatment with thymidine. The histograms demonstrate the blockade of cells at the G_1/S border in the presence of thymidine **(B)** compared with untreated cycling cells **(A)**. Upon release from thymidine into medium that contains nocodazole, cells reenter the cell cycle **(C)** and can be collected at S-phase **(D,E)** and G_2/M-phase owing to the presence of nocodazole **(F)**.

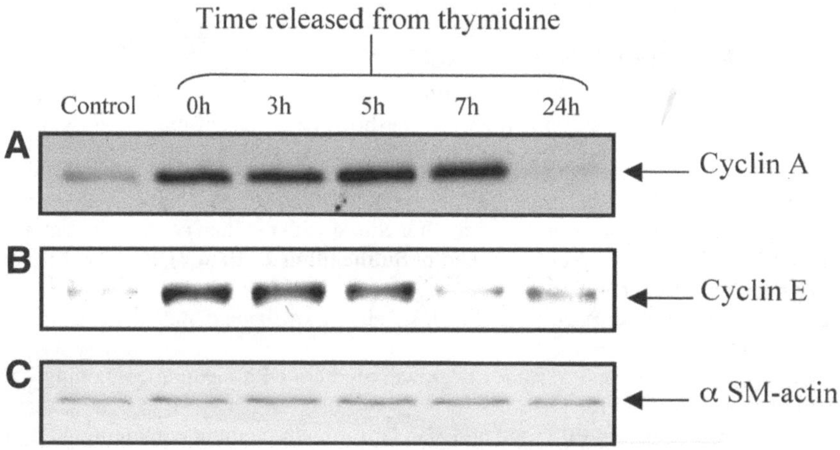

Fig. 4. Cell cycle molecule expression in thymidine-treated cells. The figure demonstrates how the thymidine synchronization protocol can be used to assess the expression of cell-cycle regulated molecules at specific phases of the cell cycle. The example shows the expression patterns of cyclins A and E (A and B, respectively) in thymidine-synchronized A10 VSMCs. Nocodazole was added to the cells at the time of release, causing cells to accumulate at the G_2/M phase by 24 h. Cyclin A levels increase as cells move from G_1 and through the S-phase but drastically decrease as cells reach the G_2/M phases. The expression patterns are consistent with reported functions of cyclin A suggesting that cyclin A/CDK2 complexes are important during S-phase and are thought to play a role in DNA synthesis and that cyclin A/CDC2 complexes are involved in the G_2/M transition (7). Degradation of cyclin A is essential for cells to pass through, and exit, M-phase (8). Cyclin E levels are high when cells are blocked at the G_1/S border and steadily decrease as cells progress through the cell cycle. The expression patterns seen for cyclin E are consistent with the fact that cyclin E levels are known to peak at the restriction point, which is located at the G_1/S transition (***9,10***; *see* **Fig. 2**). Cyclin E levels remained high when cells progressed into S-phase, which is consistent with reports that cyclin E/CDK2 is involved in the processes of initiation of DNA replication (***11–13***) and centrosome duplication ***(14)***, both of which occur during S-phase. Cyclin E levels then diminished as cells passed through S and into G_2/M, which is consistent with the fact that cyclin E is degraded at mid to late S-phase (***9***). SM-actin was used to demonstrate equal loading of A10 VSMC protein (C).

5. Anti-BrdU antibody solution: dilute anti-BrdU antibody 1:5 into PBS containing 0.5% Tween-20 (Sigma) and 1% fetal calf serum (FCS; Invitrogen). Make this solution fresh on the day of use.
6. Anti-mouse FITC antibody solution: dilute anti-mouse FITC antibody 1:10 into PBS containing 0.5% Tween-20 (Sigma) and 1% FCS (Invitrogen). Make this solution fresh on the day of use.
7. 10 mg/mL RNase A (Sigma) solution in water; store in 100-μL aliquots at –20°C.
8. 20 mg/mL propidium iodide (PI; Sigma) stock solution in water; store at 4°C.
9. PI staining solution: 50 μg/mL PI, 200 μg/mL RNase A in PBS. Make this solution fresh on the day of use.

3. Methods

3.1. Synchronization Protocol

The following procedure is a modification of the double thymidine block previously described *(4)* and is a synchronization method for adherent cells (*see* **Note 1**).

1. Plate cells in standard growth medium to achieve approx 40% confluency the following day, for example, approx 2.7×10^3 cells/cm^2 for A10 VSMCs (*see* **Note 2**).
2. The following day, replace standard growth medium with medium containing 2 m*M* thymidine (from the 10 m*M* stock described in **Subheading 2., item 2**), and incubate cells for 12 h under normal conditions.
3. Wash cells three times in PBS, re-feed with standard growth medium, and incubate under normal conditions for 12 h.
4. Following incubation, replace standard growth medium with medium containing 2 m*M* thymidine, and incubate for 12 h.
5. Wash cells three times in PBS, and add standard growth medium containing 40 ng/mL nocodazole (*see* **Note 3**). Cells can be collected at various time points following the second exposure to thymidine allowing collection of cells at specific cell cycle phases (*see* **Note 4** and **Fig. 3**).

3.2. Flow Cytometric Analysis

The following method describes staining for flow cytometric analysis that can be used to determine the cell cycle profiles of synchronized cells from the method described above.

1. Incubate cells in standard growth medium containing 10 μ*M* BrdU for the final 30 min of culture (*see* **Notes 5** and **6**).
2. Collect cells in the usual way, and wash the cell pellet with PBS.
3. Fix the cell pellet in 1 mL 70% ice-cold ethanol, and incubate at 4€ for 30 min (*see* **Note 7**).
4. Centrifuge the cells at 720*g* for 5 min to remove the ethanol.
5. Resuspend the cell pellet in 0.5 mL 0.1 *M* HCl (in PBS), and incubate at 37°C for 10 min (*see* **Note 8**).
6. Stop the reaction by adding 2.5 mL PBS, and centrifuge at 2000 rpm for 5 min.
7. Add 100 μL of anti-BrdU antibody solution, vortex briefly, and incubate for 1 h at room temperature.
8. Add 1 mL of PBS, and centrifuge the cells at 720*g* for 5 min.
9. Wash the cell pellet with PBS, add 100 μL of anti-mouse FITC solution, vortex, and incubate the cells for 30 min at room temperature in the dark.
10. Add 1 mL PBS, and centrifuge the cells at 720*g* for 5 min.
11. Wash the cell pellet with PBS, add 1 mL of PI staining solution, and incubate for 30 min at room temperature in the dark (*see* **Note 9**).
12. Pass samples through the flow cytometer to determine the cell cycle phase.

4. Notes

1. The synchronization protocols can be adapted for use with nonadherent cells.
2. It is important to determine the appropriate cell density for the cell line under investigation, and this will be dependent on the cell cycle time. Ensure that cells do not reach confluence, as this will affect results from subsequent flow cytometric analysis.

3. The optimal concentration of nocodazole and time of exposure need to be determined for each cell line. Concentrations of up to 400 ng/mL have been reported in some cases ***(5)***. The optimal time of exposure should be between 12 and 24 h since exposure of cells to nocodazole can be toxic.
4. The time points for the optimal number of cells in each cell cycle phase are dependent on the cell cycle time and will need to be determined for each cell line; 3, 5, 7, and 24 h should be used as a guideline.
5. It is important to add BrdU directly to the existing cell medium and not to add fresh medium, as this may affect the cell cycle profiles determined by flow cytometry.
6. BrdU is incorporated into DNA during synthesis in S-phase, and incorporation is measured subsequently using an anti-BrdU antibody.
7. The samples can be stored in 70% ethanol at 4€ for up to 1 mo.
8. Antibodies are unable to interact with BrdU incorporated into DNA unless the structure of chromatin is disrupted. Denaturation of DNA is achieved by exposure of cells to acid; exposure to base or heat are other methods suitable for chromatin disruption.
9. PI is a membrane-impermeant dye that will intercalate between the bases of both DNA and RNA ***(6)***; it is for this reason that the PI staining solution also contains RNase A to ensure that only DNA staining is measured.

References

1. Hyland, P. L., Keegan, A. L., Curran, M. D., Middleton, D., McKenna, P. G., and Barnett, Y. A. (2000) Effect of a dCTP:dTTP pool imbalance on DNA replication fidelity in Friend murine erythroleukemia cells. *Environ. Mol. Mutagen.* **36,** 87–96.
2. Jordan, M. A., Thrower, D., and Wilson, L. (1992) Effects of vinblastine, podophyllotoxin and nocodazole on mitotic spindles. Implications for the role of microtubule dynamics in mitosis. *J. Cell Sci.* **102,** 401–416.
3. Vasquez, R. J., Howell, B., Yvon, A. M., Wadsworth, P., and Cassimeris, L. (1997) Nanomolar concentrations of nocodazole alter microtubule dynamic instability in vivo and in vitro. *Mol. Biol. Cell.* **8,** 973–985.
4. OConnor, P. M. and Jackman, J. (1995) Synchronisation of mammalian cells, in *Cell Cycle—Materials and Methods* (Pagano, M. ed.), Springer-Verlag, New York, pp. 61–74.
5. Lee, K. S., Yuan, Y-L. O., Kuriyama, R., and Erikson, R. L. (1995) Plk is an M-phase-specific protein kinase and interacts with a kinesin-like protein, CHO1/MKLP-1. *Mol. Cell. Biol.* **15,** 7143–7151.
6. Tas, J., and Westerneng, G. J. (1981) Fundamental aspects of the interaction of propidium diiodide with nuclei acids studied in a model system of polyacrylamide films. *Histochem. Cytochem.* **29,** 929–936.
7. Pagano, M., Pepperkok, R., Verde, F., Ansorge, W. and Draetta, G. (1992) Cyclin A is required at two points in the human cell cycle. *EMBO J.* 11, 961–971.
8. Murray, A. W. and Kirschner, M. W. (1989) Cyclin synthesis drives the early embryonic cell cycle. *Nature* **339,** 275–280.
9. Dulic, V., Lees, E., and Reed, S. I. (1992) Association of human cyclin E with a periodic G1-S phase protein kinase. *Science* **257,** 1958–1961.
10. Koff, A., Giordano, A., Desai, D., et al. (1992) Formation and activation of a cyclin E-cdk2 complex during the G1 phase of the human cell cycle. *Science* **257,** 1689–1694.
11. Ohtsubo, M. and Roberts, J. M. (1993) Cyclin-dependent regulation of G1 in mammalian fibroblasts. *Science* **259,** 1908–1912.

12. Tsai, L. H., Lees, E., Faha, B., Harlow, E., and Riabowol, K. (1993) The cdk2 kinase is required for the G1-to-S transition in mammalian cells. *Oncogene* **8,** 1593–1602.
13. Resnitzky, D., Gossen, M., Bujard, H., and Reed, S. I. (1994) Acceleration of the G1/S phase transition by expression of cyclins D1 and E with an inducible system. *Mol. Cell. Biol.* **14,** 1669–1679.
14. Lacey, K. R., Jackson, P. K., and Stearns, T. (1999) Cyclin-dependent kinase control of centrosome duplication. *Proc. Natl. Acad. Sci. USA* **96,** 2817–2822.

8

Mapping Origins of DNA Replication in Eukaryotes

Susan A. Gerbi

Summary

Methods are described here to map an origin of replication in eukaryotes. Replicating DNA is enriched by BND cellulose column chromatography and by λ-exonuclease digestion; this approach has largely superceded enrichment by BrdU incorporation. The general area in which replication begins can be deciphered by neutral/neutral 2D gel electrophoresis: a restriction fragment containing the replication bubble will form a bubble arc on these gels. A more sensitive method employs PCR analysis of nascent strands that are size-fractionated. Once the general area containing the origin of bidirectional replication has been mapped, a finer level of resolution can be obtained by replication initiation point (RIP) mapping, in which start sites of DNA synthesis are identified at the nucleotide level.

Key Words

Origin of replication; nascent DNA enrichment; λ-exonuclease digestion; 2D gels; PCR analysis of nascent strands; replication initiation point mapping.

1. Introduction

DNA synthesis in eukaryotes initiates at origins (ORIs) of bidirectional replication. To understand how initiation of DNA synthesis is regulated, it is useful to identify DNA sequences comprising ORIs so that *cis*-acting elements in the vicinity and the *trans*-acting factors that bind them can be determined. This goal was met in the budding yeast *Saccharomyces cerevisiae*, in which 100–200 bp autonomous replication sequence (ARS) elements *(1)* that confer replication ability to episomes were shown by two-dimensional (2D) gels to coincide with the region where DNA replication begins *(2,3)*. Further studies revealed that ARS1 contains an 11-bp ARS consensus sequence (ACS) (reviewed in **ref.** *4*) in the essential element A, which is bound by the origin recognition complex (ORC) *(5)*. The additional elements B1, B2, and B3 are important but individually are dispensable *(6)*. ORC acts as a landing pad for other components of the replication machinery (reviewed in **ref.** *7*), and ultimately DNA

From: *Methods in Molecular Biology, vol. 296, Cell Cycle Control: Mechanisms and Protocols*
Edited by: T. Humphrey and G. Brooks © Humana Press Inc., Totowa, NJ

synthesis starts at a position between elements B1 and B2 that has been identified by replication initiation point (RIP) mapping ***(8,9)***. All ORIs in budding yeast chromosomes can function as ARS elements, but the converse is not true, as some ARS elements are silent with regard to ORI function in their chromosomal context (reviewed in **ref. *10***).

The identification of ORIs in multicellular eukaryotes (and also in the fission yeast *Schizosaccharomyces pombe*) met a stumbling block, as ARS elements have not been readily identified on episomes in metazoan cells. Moreover, it was not even clear whether DNA synthesis initiates at specific DNA sequences. Notably, semiconservative DNA replication initiates randomly with regard to sequence in early embryos ***(11–14)***, before the transcriptional pattern is established by programming chromatin. However, the initiation sites are nonetheless spaced at regular intervals of 9–12 kb ***(15)***. After the midblastula transition and the onset of zygotic transcription, initiation of replication becomes confined to initiation zones ***(16–17)***, which are in intergenic regions (reviewed in **ref. *18***). The size of the initiation zone can vary in different developmental contexts; for example, the approx 8 kb initiation zone shrinks to approx 1 kb during DNA amplification in DNA puff II/9A in the fly *Sciara* ***(19)***. Initiation zones can be as large as 55 kb in the case of the dihydrofolate reductase (DHFR) locus in cultured mammalian cells when studied by 2D gels ***(20)***, but other methods mapped a few preferred sites in which DNA synthesis starts within this large region ***(21,22)***. In fact, RIP mapping has defined the preferred start site for continuous DNA synthesis in *S. pombe* ***(23)***, the fly *Sciara* ***(24)***, and cultured human cells ***(25)***; it remains to be seen whether an ORI species-specific consensus sequence exists.

The methods described below to map an ORI begin first by determining the general location on a restriction map where replication starts. Often this is accomplished by 2D gels of replicating DNA, but recent advances allow this method to be bypassed for others that have better resolution and greater sensitivity and are technically easier. One of these newer approaches is polymerase chain reaction (PCR) mapping of origins. A potential shortcut to identify the region for PCR analysis is chromatin immunoprecipitation (ChIP) to map where ORC is bound, as we have demonstrated that the ORC binding site is adjacent to the start site of leading strand synthesis, both in yeast ***(8,9)*** and in a multicellular organism ***(24)***. Ultimately, once the initiation zone has been identified, the nucleotide position where replication starts can be located by RIP mapping. Some of these selected methods are described in detail below.

2. Materials

2.1. Enrichment of Replicating DNA

1. Benzoylated naphthoylated DEAE (BND) cellulose (Sigma).
2. NET buffer (NaCl, EDTA, Tris-HCl): 1 *M* NaCl, 1 m*M* EDTA, 10 m*M* Tris-HCl, pH 8.0.
3. TE buffer: 10 m*M* Tris-HCl, pH 7.6, 1 m*M* EDTA.
4. 2.5X λ-Exonuclease buffer: 167.5 m*M* glycine/KOH, pH 8.8, 6.25 m*M* $MgCl_2$, 125 g/mL bovine serum albumin (BSA).

2.2. Neutral/Neutral 2D Gels

1. Gel soak I: 0.15 *N* NaOH, 1.5 *M* NaCl.
2. 1X SSC (standard saline citrate buffer): 0.15 *M* NaCl, 0.015 *M* Na acetate, pH 7.0.
3. Prehybridization solution: 0.5 *M* Na phosphate buffer, pH 7.2, 1% sodium dodecyl sulfate (SDS), 1 m*M* EDTA, 1% BSA.
4. Wash solution: 40 m*M* Na phosphate buffer, pH 7.2, 1% SDS, 1 m*M* EDTA.
5. Stripping solution: 0.1X SSC, 0.1% SDS.
6. 10X TBE: 890 m*M* Tris, 890 m*M* boric acid, 25 m*M* EDTA, pH 8.0.

2.3. Mapping the Region Where DNA Synthesis Initiates by PCR

1. 6X Alkaline loading buffer: 300 m*M* NaOH, 6 m*M* EDTA, 18% Ficoll (type 400), 0.15% bromcresol green, 0.25% xylene cyanol.
2. 1X Alkaline buffer: 50 m*M* NaOH, 1 mM EDTA.
3. 10X PCR buffer: 10 m*M* Tris-HCl, pH 8.3, 50 m*M* KCl, 15 m*M* $MgCl_2$, 0.01% gelatin (w/v).
4. SSPE (1X): 0.18 *M* NaCl, 10 m*M* phosphate buffer, pH 7.7, 1 m*M* EDTA.

2.4. RIP Mapping

1. Formamide loading buffer: 95% formamide, 0.025% bromphenol blue, 0.025% xylene cyanol, 0.5 m*M* EDTA, 0.025% SDS.

3. Methods

3.1. Enrichment of Replicating DNA

Certain methods require cell synchronization to select cells that are in S-phase and are actively replicating their DNA. However, cell synchronization is not always practical, and, moreover, it might perturb the processes being studied. The methods described below are designed to map origins of replication in an asynchronous population of cells. Because replicating molecules of DNA will only be a small percentage of thc total population (e.g., 5–10%), various approaches are used for enrichment, such as BND cellulose chromatography or λ-exonuclease digestion, both of which arc described below. The latter has largely replaced the earlier approach of selection of replicating DNA that has incorporated bromodeoxyuridine (BrdU; which is more time-consuming, and not all cell types take up BrdU). In any case, standard methods are used to first isolate DNA from the cells or tissue.

3.1.1. BND Cellulose Chromatography

Newly synthesized DNA can be enriched by making use of the fact that it contains some single-stranded areas. Total DNA is isolated by standard methods (e.g., *see* **ref. *26***) from replicating cells and passed over a BND cellulose column. Both double-stranded and single-stranded DNA are bound to BND cellulose under low salt (300 m*M* NaCl). Double-stranded DNA can be eluted with high salt (1 *M* NaCl), and subsequently single stranded DNA can be eluted with high salt and caffeine (1 *M* NaCl + 1.8% caffeine). Alternatively, one can also apply the DNA sample to BND cellulose in a high salt butter, such that only replication intermediates (RIs) containing single-stranded portions of DNA will bind (*see* **Note 1**). All steps are done at room temperature. The procedure will take a few hours.

3.1.1.1. Preparation of BND Cellulose

1. Boil 4 g BND cellulose in 20 mL dH_2O for 5 min; let it cool to room temperature, break up the particles with a rubber policeman, and spin at 478*g* (2000 rpm in a Sorvall SS-34 rotor) for 2 min in a centrifuge. Decant the supernatant.
2. Suspend and wash the BND cellulose once with 20 mL dH_2O.
3. Wash twice with 20 mL NET buffer.
4. Store at 4°C in 20 mL NET buffer.

3.1.1.2. Enrichment of Replication Intermediates on BND Cellulose

5. Add the BND cellulose suspension to a Bio-Rad polyprep disposable column (cat. no. 731-1550) or an Isolab QS-Q quick-sep column, making a 1 mL bed volume for isolation of up to 20 g RI DNA; RI DNA represents about 10% of the total DNA from asynchronized yeast cells. (Thus, load about 200 g total nuclear yeast DNA onto the 1-mL column.) Wash the column extensively with NET buffer (10 vol) until the OD_{260} is close to zero.
6. Add 5 *M* NaCl to the DNA solution to a final concentration of 1 *M*. Load the DNA solution (about 1 mL) onto the column, and allow it to enter the resin by gravity; collect the flowthrough.
7. Wash the column with 3–5 vol NET buffer or until the OD_{260} is close to zero. Pool this salt wash fraction with the flowthrough (**step 6**) as they both contain mostly nonreplicating double-stranded DNA.
8. Load 1–2 vol (e.g., 1.5 mL) 1.8% caffeine in NET buffer prewarmed to 50°C onto the column; drip off the liquid. (This is the caffeine wash fraction and contains the RI DNA for further analysis.)
9. Spin the caffeine wash 10 min at maximum speed in an Eppendorf microcentrifuge (~ 13,000 rpm), and save the supernatant. The pellet is BND cellulose particles that should be discarded.
10. Add 1 vol isopropanol to the double-stranded DNA and RI fractions, and invert slowly to mix. Let the solution remain at 4°C for at least 30 min.
11. For larger volumes, spin the isopropanol-precipitated solution in a Beckman SW41 rotor at 38,000 rpm for 30 min at 4°C, and decant the supernatant. For smaller volumes, spin down in an Eppendorf centrifuge at 10,000 rpm for 30 min at 4°C.
12. Wash the DNA pellet with 70% ethanol; spin for 2 min in a microfuge, and decant the supernatant. Briefly air-dry the pellet.
13. Redissolve (on ice or in a cold room for ~30 min to several hours) the DNA pellet (a few g; about 5% of total DNA loaded onto the BND cellulose column) in 30–40 L TE buffer. If running a 2D gel, add 4–5 L of 10X loading dye.

3.1.2. λ-Exonuclease Enrichment of Replicating DNA

DNA is phosphorylated by T4 polynucleotide kinase to ensure that all free DNA ends (lacking an RNA primer) carry a phosphate and thus are recognizable as substrates for λ-exonuclease.

3.1.2.1. Phosphorylation of RI DNA by T4 Polynucleotide Kinase

1. Heat-denature RI DNA that was previously enriched by BND cellulose chromatography (*see* **Subheading 3.1.1.2.**) at 100°C for 2 min, and then immediately chill in ice water.
2. Carry out the T4 polynucleotide kinase reaction in a total volume of 20 L: 10 L heat-denatured RI DNA; 2 L ATP (50 *M* final concentration, diluted from a 50 m*M* stock solution); 1 L (10 U) T4 polynucleotide kinase (New England Biolabs); 2 L 10X T4 polynucleotide kinase buffer (New England Biolabs); and 5 L dH_2O (autoclaved and filter-sterilized).
3. Incubate at 37°C for 30 min.
4. To stop the reaction, add: 1 L 5% Sarkosyl, 2 L 250 m*M* EDTA, and 2 L proteinase K (625 g/mL, diluted from stock solution).
5. Incubate further for 30–60 min. at 37°C.
6. Extract the sample once with 25 L phenol/chloroform/isoamyl alcohol (25:24:1, v/v/v) and once with chloroform/isoamyl alcohol (24:1,v/v).
7. Precipitate the DNA with 0.1 vol of 3 *M* Na acetate, pH 5.2, and 2 vol of ethanol at –20°C overnight.
8. Spin DNA in an Eppendorf microcentrifuge at 10,000 rpm at 4°C for 15 min.
9. Wash the pellet once with 70% ethanol, and let it air-dry.
10. Resuspend the DNA in 20 L 10 m*M* Tris-HCl, pH 8.0, and keep it on ice if proceeding immediately to the next step. Alternatively, DNA can be stored at 4°C in 10 L TE and 10 L H_2O added later just before proceeding to the λ-exonuclease digestion (*see* **Notes 2** and **3**).

3.1.2.2. λ-Exonuclease Digestion of DNA

1. Carry out digestion with λ-exonuclease in a total volume of 20 L: 10 L RI DNA (from the phosphorylation procedure above); 8 L 2.5X λ-exonuclease reaction buffer; and 2 L (3–3.5 U/ L) λ-exonuclease (Gibco-BRL Life Technologies).
2. Incubate at 37°C for 12 h. λ-Exonuclease is slow to digest heat-denatured, single-stranded DNA. The optimal pH for λ-exonuclease is 9.4. However, degradation of RNA primers can occur at pH 9.4, since RNA can be hydrolyzed in weak alkali as low as pH 9. Hence, the pH of the reaction buffer used here is pH 8.8.
3. As a control, check λ-exonuclease activity on restricted, nonreplicating phosphorylated DNA from the flowthrough fraction of the BND cellulose column. If the DNA is restricted to produce 5' phosphorylated ends, it has to be heat-denatured for 2 min at 100°C prior to incubation with λ-exonuclease.
4. Treat the control as described in **steps 1** and **2** above; as an additional control, use another sample from **step 3** to treat in parallel to **steps 1** and **2** but omit λ-exonuclease.
5. Run an aliquot of the two controls on a 0.9% agarose gel to check that the λ-exonuclease digestion was complete.
6. If the digestion was incomplete, incubate at 37°C for an additional 3–4 hours, after adding: 2 L 2.5X λ-exonuclease reaction buffer, 1 L λ-exonuclease, and 2 L dH_2O.
7. Repeat **step 5** (0.9% agarose gel) to check the completeness of digestion. Also check that reaction buffers and enzyme preparations do not have RNase activity by incubating 2 g of tRNA in a total volume of 10 L as described in **steps 1** and **2** and running the sample on a 2.5% agarose gel.

8. Once the λ-exonuclease digestion is complete, proceed by heating the samples to 75°C for 10 min to inactivate the λ-exonuclease, and then immediately cool on ice.
9. Extract once with chloroform/ isoamyl alcohol (24:1,v/v) and store at 4°C until use in primer extension reactions.

3.2. Neutral/Neutral 2D Gels

2D gel electrophoresis methods to map origins of replication exploit the idiosyncratic migration of RI on agarose gels. The neutral/neutral (N/N) 2D gel method of Brewer and Fangman ***(2)*** utilizes Southern blot hybridization to find which restriction fragments contain a replication bubble (and hence the ORI) and which contain replication forks. DNA fragments with a fork migrate somewhat more slowly, and those with a bubble migrate much more slowly than their linear counterparts. In the first dimension, genomic DNA from an asynchronous cell population is digested with restriction enzymes and run on a low-percentage agarose gel at low voltage to resolve DNA molecules primarily by mass. In the second dimension, RI DNA is separated from bulk DNA by running it on a higher percentage agarose gel at high voltage to separate DNA molecules on the basis of shape as well as mass. After running the second dimension and blotting the DNA onto a filter, Southern blot hybridization is performed to detect the restriction fragment of interest and visualize it if it contains a replication bubble (and hence an origin of replication), or instead is replicated by a replication fork passing through it.

Another kind of 2D gel is run under neutral/alkaline (N/A) conditions to determine the direction of replication fork movement through a particular region ***(3)***; a protocol for N/A gels is given elsewhere ***(26)***. We developed a 3D gel procedure ***(27)*** that is a composite of an N/N 2D gel followed by an N/A gel (*see* **ref. 26** for the protocol). The 3D gel method can map the ORI by locating the fragment that generates the highest bubble/fork nascent DNA ratio. In addition, it can be used to address some unresolved issues, such as how many initiation events occur on a single DNA molecule.

3.2.1. Day 1: Run Gel (First Dimension)

1. Prepare a 0.4% agarose gel (350–400 mL in 1X TBE) in a large gel tray (e.g., W × L = 20 × 24 cm) in the cold room (*see* **Notes 4** and **5**).
2. Pour in 1X TBE to just even with the gel surface. Do not flood the gel.
3. Load the DNA samples. Run at 0.6–0.7 V/cm. Flood the gel with 1X TBE after the dye has run into the gel (*see* **Note 6**). Run for 36–48 h (Blue dye should have run approx 10–12 cm.). Do not run the first dimension with ethidium bromide.

3.2.2. Day 3: Run Gel (Second Dimension)

4. Boil 350–400 mL 1% agarose in 1X TBE (*see* **Note 7**).
5. For each second dimension gel, prepare 2 L of 1X TBE containing 0.3 g/mL ethidium bromide (from 10 mg/mL stock solution).
6. After the first dimension, cut out the gel lanes with a razor, with the aid of a ruler. Gel lanes should be slightly wider than the wells so that the razor does not touch the DNA. The top 2 cm from the well can be discarded. Take 10-cm-long slices (from 2 to 12 cm starting at the well) for the second dimension. This will include DNA from approx 2 kb and up. Pick up each gel lane with a flexible, thin plastic ruler, rotate it by 90° and put it into the gel tray for the second dimension.

7. Once the melted agarose gel has cooled to 50–60°C, add ethidium bromide stock solution (10 mg/mL) to a final concentration of 0.3 g/mL.
8. In a cold room, pour the 2D gel containing 1% agarose + 0.3 g/mL ethidium bromide around the 1D gel slices. Let it set for approx 20 min. Create wells for size markers with a hot spatula or a glass pipet.
9. Stain the size markers of the first dimension, destain, and photograph (UV-Polaroid), or keep them at 4°C (either in 1X TBE or wrapped on a glass plate) and photograph later, together with the second dimension.
10. Pour in 2 L 1X TBE containing 0.3 g/mL ethidium bromide for each gel box. Load DNA markers.
11. Run gel at 2–3 V/cm. For 35-cm-long gel boxes, 80–100 V is sufficient. Run for 14–18 h. Monitor the run with long-wave UV. The goal is to have the shortest DNA reach the lower right-hand corner of the gel.

3.2.3. Day 4: Southern Transfer

12. Carefully take the gel out of the gel box (slice the gel into top and bottom halves for easier handling), and place it into a large container. Destain the gel in water for approx 30 min. Photograph the gel under short-wave UV.
13. Denature the gel in Gel Soak I for 30 min to 1 h. Do not depurinate the gel in HCl before DNA denaturation to avoid cutting the small nascent DNA strands into pieces that are too small to bind to the filter. Do not neutralize the gel after denaturation in Gel Soak I.
14. Transfer the gel to a Biotran(+) membrane (ICN Biochemicals) with approx 800 mL of 20X SSC, for 18–24 h. Put a glass plate as an even weight on top of a stack of paper towels over the membrane and gel, but do not put extra weight on top of the glass plate, as it may compress the gel too much.

3.2.4. Day 5: Southern Hybridization

15. Remove the membrane from the gel, and fix the DNA onto the membrane by either UV crosslinking or baking at 80°C for 20 min.
16. Rinse the membrane with dH_2O and then 2X SSC; air-dry until damp or dry.
17. Make the random-primed probe: for every two 12 × 20-cm membranes containing two 2D gels, use approx 25 ng template and 5 L [α^{32}P]dATP (3000 mCi/mmol) for a 20–25 L reaction volume, and incubate in the random primer reaction mixture (e.g., Boehringer Mannheim) at 37°C for 30 min to 3 h. Stop the reaction with 2 L 0.5 *M* EDTA, and add TE to 100 L. Pass through a Sephadex G-50 column to remove unincorporated nucleotides.
18. Incubate the membrane in a sealed plastic bag containing prehybridization solution (~10 mL for every 100-cm^2 membrane) at 65°C for 5 min to a few hours.
19. Remove the prehybridization solution, add hybridization solution (~5 mL/100 cm^2 membrane), which is prehybridization solution containing probe and 100 g/mL single-stranded salmon sperm DNA or calf thymus DNA (boil this together with the probe for 10 min), and incubate with shaking at 65°C overnight.

3.2.5. Day 6: Wash Southern Blots

20. Pour hybridization solution into a tube. (Keep it for later reuse if needed.) Rinse the hybridized membrane twice with room temperature wash solution. Add wash solution that was preheated to 65°C, and shake membrane in it at 65°C for 20 min. Repeat the wash two more times with a brief rinse in between. Longer washes (30–60 min) may be needed for blots of DNA from higher eukaryotes.

21. Air-dry the hybridized membrane briefly until damp but not dry. Wrap the membrane in Saran Wrap. Expose it to PhosphorImager plates for 18–24 h, or longer if needed. X-ray films (with intensifying screens) require 3–14 d of exposure.

3.2.6. Stripping and Rehybridization

22. Strip off the previous probe from the hybridized membrane by incubation in a boiling hot solution of 0.1X SSC, 0.1% SDS with shaking for approx 5 min. Repeat two more times. Monitor that the probe has been removed completely by exposing the filter to a PhosphorImager screen.
23. Hybridize the stripped membrane with another probe (*see* **Note 8**).

3.3. Mapping the Region Where DNA Synthesis Initiates by PCR

This very useful method for mapping an ORI by using PCR to analyze nascent DNA strands isolated from an asynchronous population of cells was developed by Vassilev and Johnson *(28)*. The basis of this method is that newly replicated DNA extends bidirectionally from the ORI and progressively increases in length, spanning adjacent sequences on the DNA. Total genomic DNA is isolated from cells, and the nascent DNA is purified away from unreplicated DNA; the nascent DNA is fractionated according to its size. To map an ORI using this method, a minimum of three unique DNA segments are selected that are distributed across the putative ORI region. The size-fractionated nascent DNA is then used as a template in PCR reactions with pairs of oligonucleotide primers specific for each of the segments.

The goal of this analysis is to identify which stretch of DNA has the shortest nascent strands, indicating that the ORI resides in, or very close to, the fragment of DNA produced by that primer pair. The smallest size fraction analyzed should be approx 400–500 nt long, as fragments approx 200 nt will contain Okazaki fragments from all replicating regions. An important control in this PCR mapping procedure is that the smallest nascent strand sizes detected in the other segments should increase in proportion to their distance from the deduced ORI location. The advantage of using PCR analysis of nascent DNA to map an ORI is that the method is more sensitive than 2D gels, thus requiring less starting material; moreover, it does not require cell synchronization. Nascent DNA that was isolated by BrdU incorporation is light-sensitive and subject to breakage; therefore, it is preferable to acquire nascent DNA instead by BND cellulose column chromatography and λ-exonuclease digestion (*see* **Subheading 3.1.**) to remove contaminating parental DNA.

Usually, multicellular eukaryotes contain an initiation zone, throughout which replication can initiate. This zone will be equivalent to the stretch of DNA with different primer pairs that gave PCR products when using small nascent DNA as the template. To determine which site(s) within the initiation zone is preferred for initiation of DNA synthesis, quantitative PCR must be employed. Initially, quantification relied on competitive PCR (*29*; protocol given in **ref.** ***26***). Although this worked, it was time-intensive and required much material. Recent advances permit quantification through the use of real-time PCR, which is far simpler.

*3.3.1. Alkaline Agarose Gel Electrophoresis (see **Note 9**)*

1. Melt agarose in dH_2O; use 1.2–1.5% low melting point agarose.
2. Cool the agarose solution to 60°C. Add NaOH to 50 m *M* and EDTA to 1 m*M* final concentration. Pour the gel (about 25 cm long gives good resolution), and let it harden. Add freshly made 1X alkaline buffer, and let the gel equilibrate with it overnight.
3. Load the aliquots of single-stranded DNA in alkaline loading buffer diluted to 1X from a 6X stock; do not overload lanes (load < 15 g DNA/lane). Also load marker DNA (e.g., 100-bp or 1-kb DNA ladder) in the same loading buffer.
4. Carry out preparative alkaline agarose electrophoresis at 50 V for 16–24 h at 4°C. The gel should be run in a cold room to prevent overheating of the low melting point agarose.
5. Neutralize the gel with three changes of 0.5X TBE buffer, each for 20 min.
6. Cut out the DNA marker lanes, and stain them with ethidium bromide. The unstained sample DNA lanes should be cut into various size fractions, spaced close to one another for best resolution.
7. DNA is recovered after digestion of the gel with GELase (as per the protocol from Epicentre Technologies). This fast procedure recovers DNA of any size in high yields.
8. Resuspend the pellet in dH_2O. Each size-fractionated sample is sufficient for about 10 PCR reactions.

3.3.2. PCR Reactions

1. Set up the PCR reaction mixture as follows: 40 L each dNTPs (10 m*M*), 200 L 10X PCR buffer, 160 L dH_2O, and 10 L Amplitaq DNA polymerase from stock solution (PerkinElmer). Mix well. Aliquot 26.5 L/0.5 mL PCR tube. Store at –20°C until ready to use. This aliquot is enough to do two 50- L PCR reactions.
2. Primers: dilute the required primer sets to 0.5 *M* of each primer per reaction, i.e., 25 pmol primer/50 L reaction. Use at least three primer sets encompassing the suspected ORI.
3. Denature the template before adding it to the PCR reaction by heating at 94°C for 5 min and transferring it immediately to ice. The final reaction mixture contains (*see* **Note 10**): 13.25 L reaction mix (**step 1**), 2.00 L primers (from primer stock, solution of 25 pmol/ L), 1.00 L template (from stock of 100 ng/ L chromosomal DNA), and dH_2O to a final reaction volume of 50 L.
4. Example of PCR cycling conditions: 94°C for 20 s for denaturation; 50–65°C for 90 s for annealing; and 72°C for 30 s for extension. Run the reaction for 25–30 cycles, but do not exceed 30 cycles.
5. Load one-third of the PCR mixture after amplification on a 4% agarose gel, and, after running the gel, stain it with ethidium bromide to visualize the PCR products.
6. Purify the PCR products from the remaining two-thirds of the PCR mixture using the QIAquick PCR Purification kit (Qiagen) (**steps 6** and **7** are optional.)
7. Slot-blot the DNA from **step 6** onto a Nytran Plus nylon filter (Schleicher & Schuell), and hybridize it with oligonucleotides specific for each PCR product. Hybridize overnight with 1–5 × 10^6 cpm/mL probe (5' end-labeled with [γ^{32}P]ATP) at 5°C below the T_m in 6X SSPE, no SDS. The final wash can be done in 1X SSPE, 0.5% SDS at the T_m of the hybrid. Visualize the results with X-ray film and/or a PhosphorImager.

3.4. Replication Initiation Point Mapping

RIP mapping allows the start site of DNA synthesis to be mapped at the nucleotide level. The transition point (TP) between continuous and discontinuous synthesis is the nucleotide position where bidrectional replication starts (origin of bidirectional replication). The principle of RIP mapping is to detect the DNA 5' ends of nascent DNA strands by primer extension using radiolabeled oligonucleotides. Clearly, the ORI should be mapped to less than a 1-kb region (*see* **Subheadings 3.2.** and **3.3.**) before RIP mapping is applied. The key step that allows success in RIP mapping is the treatment of isolated nascent DNA with λ-exonuclease to eliminate nicked DNA that would lead to a high background in the subsequent primer extension reaction. Nascent DNA is protected by its RNA primer from the 5' to 3' exonuclease activity of λ-exonuclease, unlike nicked DNA ends that are digested.

As in the original Hay and DePamphilis approach ***(30)***, RIP mapping identifies the junction of nascent DNA with the RNA primer (because no available DNA polymerase can switch from a DNA to an RNA template). The Hay and DePamphilis method can be used to study animal virus origins, but it lacks the sensitivity needed for mapping origins of replicating DNA from eukaryotic cells. RIP mapping affords the necessary sensitivity for lower eukaryotes ***(8,9,23,24)***; it can be coupled with linker-mediated PCR for use in mammalian genomes ***(25)***.

In general, since the heart of the RIP mapping method is the presence of an RNA primer at the 5'-end of "real" nascent DNA (as opposed to nicked DNA), special care must be taken to avoid anything that might degrade the RNA primers of nascent DNA. After DNA isolation (e.g., *see* **ref. *26***), RIs are enriched by BND cellulose column chromatography followed by λ-exonuclease digestion to remove contaminating parental DNA. This replicative DNA is then used as the template for primer extension, as described below in **Subheading 3.4.2.** with subsequent analysis of the sizes of the extension products by gel electrophoresis. Extension will continue until the junction with an RNA primer on the RI DNA, where it will stop. Because the RI DNA is from asynchronous cells, there will be many positions where such a junction is found, reflecting population polymorphism of the DNA molecules that came from replication bubbles of various sizes. The bottom of the gel will be blank if the nascent DNA used as template for the extension was made by continuous strand synthesis; the first band above the blank space will mark the TP (origin) between continuous and discontinuous strand synthesis of the nascent DNA template. The ladder of bands above the TP represent the start sites for Okazaki fragment synthesis on the nascent DNA template ***(9)***.

Preparation of the end-labeled primers and the primer extension reaction are described in **Subheading 3.4.1.** The length of the primers may vary between 23 and 27 nucleotides, and their GC content should be at least 40% and preferably 50% or more. The T_m should be 68–70°C. Sometimes, primers that fulfill these requirements might not work on RI DNA, although they work in the control reaction. The reason for that is unclear.

3.4.1. Primer Phosphorylation With [γ^{32}P]ATP

1. Place a sterile Eppendorf tube on ice. The labeling reaction is carried out in a total volume of 10 L containing: 2 L primer (200 ng); 1 L 10X T4 polynucleotide reaction buffer

(New England Biolabs); 1 L T4 polynucleotide kinase (10 U/ L; New England Biolabs); 5 L dH_2O; and 1 L [$\gamma^{32}P$]ATP (5000 Ci/mmol, 150 mCi/mL; use only fresh label).
2. Incubate on ice for 1 h.
3. Fill a QS-Q Quick-Sep spin column (Isolab) with preswollen Sephadex G-25. Spin for 3 min at 2000 rpm (478*g*) in a Sorvall GLC-2B centrifuge. Add 1 mL TE buffer, and spin again; do this step twice.
4. Add 40 L TE buffer to the labeling reaction, and load it onto the column.
5. Spin column with reaction mixture for 3 min at 478*g* (2000 rpm in a Sorvall GLC-2B centrifuge).
6. Add 50 L TE buffer to the column, and spin again. This eluate will be pooled with the eluate of **step 5** that remains at the bottom of the tube.
7. Count 1 L of the eluate in a scintillation counter. (Specific radioactivity should be 10^8 cpm/ g, assuming 100% yield.)
8. Store the labeled primer at –20°C until needed; it is best to use it within 1 wk.

3.4.2. Primer Extension Reaction

Use double-stranded, nonreplicating DNA restricted at a unique site as a positive control. This will give you a single extension product of defined size, having extended up to the restriction site. This control will indicate whether the primer anneals at any other places besides the intended place, in which case you will get more than one band. The amount of RI DNA used as template is calculated from the amount that was determined *before* λ-exonuclease treatment.

1. Mix on ice with a sterile pipetman tip containing an aerosol-resistant filter, in this order: 1.0 L dNTP (from each dNTP stock solution); 3.5 L 100 m*M* $MgSO_4$; 3.0 L 10X Vent (exo-) DNA polymerase buffer (New England Biolabs); 1.0 L Vent (exo-) DNA polymerase (2 U/ L) (New England Biolabs); 17.0 L dH_2O (autoclaved and filter-sterilized); 2.0 L template DNA (2–5 ng for highly purified DNA such as viral or plasmid DNA, or 200–500 ng for yeast chromosomal DNA); 2.5 L radiolabeled primer.
2. Overlay with oil (a good precaution, even for PCR machines with heated lids).
3. Run 30 cycles with an initial denaturation step at 95°C for 4 min: 1 min at 94°C, 1 min at 70°C, and 1.5 min at 72°C.
4. Prepare as many Eppendorf tubes as you have primer extension samples and add to each: 1.0 L tRNA (0.5 g/ L), 2.5 L 3 *M* Na acetate, pH 5.2, 16.5 L TE buffer.
5. Add the 30 L primer extension mixture from **steps 1–3**.
6. Extract with 50 L chloroform/isoamyl alcohol (24:1,v/v).
7. Add 100 L 95% ethanol, and precipitate on dry ice for 20 min, or at least 12 h at –20°C.
8. Spin down in an Eppendorf centrifuge at 10,000 rpm for 15 min at 4°C.
9. Decant, and let the pellet air-dry.
10. Dissolve the pellet in formamide loading buffer.
11. Heat samples to 80°C for 5 min. Cool on ice.
12. Load samples onto a 6–8% acrylamide/urea sequencing gel in 1X TBE, and prerun until the gel has reached 40°C.
13. Load sequencing reactions performed with the same primer (T7 Sequenase v 2.0 sequencing kit, Amersham) in adjacent gel lanes.
14. Run gel until bromphenol blue marker has just run out of the gel.
15. Dry the gel under vacuum, and expose it to a PhosphorImager plate overnight or to X-ray film.

4. Notes

1. In general, the less concentrated the sample and the bigger the BND cellulose column, the better the efficiency of purifying RI DNA.
2. The dilution is important, as too much EDTA will inhibit λ-exonuclease.
3. Since the DNA sample still contains free ATP, a calculation of the recovered amount of DNA by spectrophotometry gives unreliable results. However, a 1 L aliquot can be run on a small 0.9% agarose gel to estimate the amount of DNA in the sample.
4. The percentage of the first-dimension gel should vary according to the size of restriction fragments of interest: 0.6% for restriction fragments 1–2 kb, 0.5% for fragments 2–3 kb; 0.4% for fragments 3–7 kb; and 0.3% for fragments 7–10 kb. Ordinary high melting point agarose (e.g., from Gibco-BRL Life Technologies) should suffice if the percentage of the first dimension is not lower than 0.4%. Pulsed-field gel grade agarose (Boehringer Mannheim, cat. no. 1240 609) has higher gel strength and is therefore easier to handle for first-dimension gels.
5. The wells should be more square than rectangular and big enough for a 50- L sample. A large gap between the wells is desirable (alternatively, load every other well to avoid cross-contamination of samples when cutting out the first-dimension gel lanes).
6. If you are short of time, you can run the gel at 2–3 V/cm until the dye enters the gel, flood with 1X TBE, and then decrease the voltage to 0.6–0.7 V/cm. Do not leave the gel for longer than a few hours without flooding it with 1X TBE, or else the gel will dry out.
7. The percentage of the second-dimension gels should also vary according to the size range of interest: 1.5–1.8% for 1–2 kb, 1.2–1.5% for 2–3 kb, 1% for 3–7 kb, and 0.7% for 7–10 kb.
8. The Southern blot membrane can be rehybridized at least seven times. On the last hybridization, use the same probe as for the first hybridization to ensure that the pattern of hybridization is the same (i.e., that no significant amount of DNA has been removed from the membrane by the repeated stripping).
9. The mapping resolution reflects the resolution of the sizing method; alkaline agarose gels give better resolution than alkaline sucrose gradient ultracentrifugation. The resolution also reflects the distance between the primer pairs used for PCR analysis of sized nascent DNA.
10. The concentration of primers, polymerase, and nucleotides must be sufficiently large that they are in excess and essentially constant during the initial PCR cycles and decrease exponentially as the number of cycles increases. It is safest to use the same number of PCR cycles for all fractions and segments analyzed. To ensure that the chosen number of PCR cycles is appropriate for the amount of template DNA, pilot PCR reactions should be performed on a dilution series of plasmid DNA with the ORI region. This will allow one to find the specific conditions that maintain a linear relationship between the various concentrations of template DNA and the intensity of the resulting PCR bands. It would be prudent not to exceed 30 cycles of amplification in order to maintain the reaction in the linear range.

Acknowledgments

I acknowledge support for our replication studies from NIH grant GM35929. Many thanks to former lab members Anja K. Bielinsky, Chun Liang, and Victoria V. Lunyak for their work in developing in my lab the methods described here. I am grateful to many other scientists, too numerous to mention, who shared ideas and protocols with us and provided stimulating discussions.

References

1. Stinchcomb, D. T., Struhl, K., and Davis, R. W. (1979) Isolation and characterisation of a yeast chromosomal replicator. *Nature* **282,** 39–43.
2. Brewer, B. J. and Fangman, W. L. (1987) The localization of replication origins on ARS plasmids in *S. cerevisiae*. *Cell* **51,** 463–471.
3. Huberman, J. A., Spotila, L. D., Nawotka, K. A., El-Assouli, S. M., and Davis, L. R. (1987) The in vivo replication origin of the yeast 2 m plasmid. *Cell* **51,** 463–471.
4. Newlon, C. S. and Theis, J. F. (1993) The structure and function of yeast ARS elements. *Curr. Opin. Genet. Dev.* **3,** 752–758.
5. Bell, S. P. and Stillman, B. (1992) ATP-dependent recognition of eukaryotic origins of replication by a multiprotein complex. *Nature* **357,** 128–134.
6. Marahrens, Y. and Stillman, B. (1992) A yeast chromosomal origin of DNA replication defined by multiple functional elements. *Science* **255,** 817–823.
7. Bell, S. P. and Dutta, A. (2002) DNA replication in eukaryotic cells. *Annu. Rev. Biochem.* **71,** 333–374.
8. Bielinsky, A.-K. and Gerbi, S. A. (1998) Discrete start sites for DNA synthesis in the yeast ARS1 origin. *Science* **279,** 95–98.
9. Bielinsky, A.-K. and Gerbi, S. A. (1999) Chromosomal ARS1 has a single leading strand start site. *Mol. Cell* **3,** 477–486.
10. Newlon, C. S., Collins I., Dershowitz, A. M., Greenfeder, S. A., Ong, L. Y., and Theis, J. F. (1993) Analysis of replication origin function on chromosome III of *Saccharomyces cerevisiae*. *Cold Spring Harbor Symp. Quant. Biol.* **58,** 415–423.
11. Harland, R. M. and Laskey, R. A. (1980) Regulated replication of DNA microinjected into eggs of *Xenopus laevis*. *Cell* **21,** 761–771.
12. Méhali, M. and Kearsey, S. (1984) Lack of specific sequence requirement for DNA replication in *Xenopus* eggs compared with high sequence specificity in yeast. *Cell* **38,** 55–64.
13. Hyrien, O. and Méhali, M. (1993) Chromosomal replication initiates and terminates at random sequences but at regular intervals in the ribosomal DNA of *Xenopus* early embryos. *EMBO J.* **12,** 4511–4520.
14. Shinomiya, T. and Ina, S. (1991) Analysis of chromosomal replicons in early embryos of *Drosophila melanogaster* by two-dimensional gel electrophoresis. *Nucleic Acids Res.* **19,** 3935–3941.
15. Blow, J. J., Gillespie, P. J., Francis, D., and Jackson, D.A. (2001) Replication origins in *Xenopus* egg extract are 5–15 kilobases apart and are activated in clusters that fire at different times. *J. Cell Biol.* **152,** 15–25.
16. Hyrien, O., Maric, C., and Méhali, M. (1995) Transition in specification of embryonic metazoan DNA replication origins. *Science* **270,** 994–997.
17. Sasaki, T., Sawado, T., Yamaguchi, M., and Shinomiya, T. (1999) Specification of regions of DNA replication initiation during embryogenesis in the 65-kilobase DNA pol α-dE2F locus of *Drosophila melanogaster*. *Mol. Cell. Biol.* **19,** 547–555.
18. Gilbert, D. M. (2001) Making sense of eukaryotic DNA replication origins. *Science* **294,** 96–100.
19. Lunyak, V. V., Ezrokhi, M., Smith, H. S., and Gerbi, S. A. (2002) Developmental changes in the *Sciara* II/9A initiation zone for DNA replication. *Mol. Cell. Biol.* **22,** 8426–8437.
20. Dijkwel, P. A. and Hamlin, J. L. (1995) The Chinese hamster dihydrofolate reductase origin consists of multiple potential nascent-strand start sites. *Mol. Cell. Biol.* **15,** 3023–3031.
21. Pelizon, C., Divacco, S., Falaschi, A., and Giacca, M. (1996) High-resolution mapping of the origin of DNA replication in the hamster dihydrofolate reductase gene domain by competitive PCR. *Mol. Cell. Biol.* **16,** 5358–5364.

22. Kobayashi, T., Rein, T., and DePamphilis, M. L. (1998) Identification of primary initiation sites for DNA replication in the hamster dihydrofolate reductase gene initiation zone. *Mol. Cell. Biol.* **18,** 3266–3277.
23. Gomez, M. and Antequera, F. (1999) Organization of DNA replication origins in the fission yeast genome. *EMBO J.* **18,** 5683–5690.
24. Bielinsky, A.-K., Blitzblau, H., Beall, E.L., et al. (2001) Origin recognition complex binding to a metazoan origin. *Curr. Biol.* **11,** 1–20.
25. Abdurashidova, G., Deganuto, M., Klima, R., et al. (2000) Start site of bidirectional DNA synthesis at the human lamin B2 origin. *Science* **287,** 2023–2026.
26. Gerbi, S. A., Bielinsky, A. K., Liang, C., Lunyak, V. V., and Urnov, F.D. (1999) Methods to map origins of replication in eukaryotes, in *Eukaryotic DNA Replication* (Cottterill, S., ed.), Oxford University Press, Oxford, pp. 1–42.
27. Liang, C. and Gerbi, S. A. (1994) Analysis of an origin of DNA amplification in *Sciara coprophila* by a novel three-dimensional gel method. *Mol. Cell. Biol.* **14,** 1520–1559.
28. Vassilev, L. and Johnson, E. M. (1989). Maping intiation sites of DNA replication in vivo using polymerase chain reaction amplification of nascent strand segments. *Nucleic Acids Res.* **19,** 7693–7705.
29. Diviacco, S., Norio, P., Zentilin, L., et al. (1992) A novel procedure for quantitative polymerase chain reaction by coamplification of competitive templates. *Gene* **122,** 313–320.
30. Hay, R. and DePamphilis, M. L. (1982) Initiation of SV40 DNA replication *in vivo*: location and structure of 5' ends of DNA synthesized in the ORI region. *Cell* **28,** 767–779.

9

In Situ Assay for Analyzing the Chromatin Binding of Proteins in Fission Yeast

Stephen E. Kearsey, Lydia Brimage, Mandana Namdar, Emma Ralph, and Xiaowen Yang

Summary

An *in situ* technique for studying the chromatin binding of proteins in single fission yeast cells (*Schizosaccharomyces pombe*) is described. Cells are permeabilized by enzymatic digestion and extracted with a detergent-containing buffer. This procedure removes soluble proteins, but proteins that are bound to insoluble cell structures such as chromatin are retained, and overall cell morphology is maintained. Extraction of proteins is monitored by fluorescence microscopy, either using fluorescently tagged proteins or by indirect immunofluorescence. This method allows the chromatin association of proteins to be correlated with other cell cycle events without the need for cell synchronization.

Key Words

DNA replication; chromatin; cell cycle; *Schizosaccharomyces pombe*; fission yeast; GFP.

1. Introduction

The process of eukaryotic DNA replication involves stepwise assembly of replication proteins at origins in the process leading to initiation. An important protein complex involved in initiation is the origin recognition complex (ORC), which in budding and fission yeasts is bound to chromatin throughout the cell cycle and acts as a site where additional proteins needed for initiation bind. First, Cdt1 and Cdc6/Cdc18 bind to chromatin via ORC, and subsequently the minichromosome maintenance (Mcm2–7) complex assembles onto origins, to form pre-replicative complexes (pre-RCs). The pre-RC is competent for initiation in S-phase upon activation by Cdc7 and Cdc2 protein kinases, and this process involves the assembly of additional proteins onto chromatin such as Cdc45 and the GINS (Sld5-Psf1-Psf2-Psf-3) complex. Thus, monitoring the chromatin binding of Mcm2–7 allows acquisition of replication competence

From: *Methods in Molecular Biology, vol. 296, Cell Cycle Control: Mechanisms and Protocols*
Edited by: T. Humphrey and G. Brooks © Humana Press Inc., Totowa, NJ

to be followed, and Cdc45 can be used to indicate the onset of DNA synthesis. Analyzing the chromatin binding of replication factors in mutants can be used to establish functional dependencies and to help establish when a replication factor functions. For instance, inactivation of Cdc23/Mcm10 has no effect on Mcm2–7 chromatin binding, but Cdc45 chromatin association is blocked, implying that Cdc23/Mcm10 functions after pre-RC formation at around the time of initiation *(1)*.

A number of methods for detecting chromatin association of specific proteins have been developed *(2)*. Chromatin immunoprecipitation uses formaldehyde to crosslink proteins to DNA, and, following shearing and immunoprecipitation of a specific protein, the association of specific DNA sequences with that protein can be established by polymerase chain reaction (PCR; e.g., *see* **ref. *3***). An alternative approach is to use Western blotting to examine whether proteins fractionate with crude chromatin preparations *(2)*. Proteins that are in the chromatin fraction before but not after nuclease treatment can be considered to be chromatin-associated. Finally, cytological methods using fluorescence microscopy allow the chromatin association of proteins in single cells to be examined. Mammalian and other cultured cells can be permeabilized with nonionic detergents prior to fixation, which extracts soluble nuclear proteins. This allowed the demonstration that Mcm2–7 proteins, which remain in the nucleus throughout interphase in mammalian cells, are only bound to chromatin during the telophase/S-phase interval (e.g., *see* **ref. *4***). This approach is more difficult in yeasts, in which the cell wall prevents simple detergent extraction. In the chromosome-spreading technique, yeast cell walls are removed by enzymatic digestion, and spheroplasts are simultaneously lysed and fixed *(5)*. This treatment destroys cell structure, but chromatin association of specific proteins can be detected by subsequent staining with antibodies.

We have developed an *in situ* chromatin binding assay for use with fission yeast that is much less destructive to cell structure *(6)*. The method involves partial removal of the cell wall using a β-glucanase (zymolyase), which leaves α-glucan polymers intact. This enables the cells to withstand detergent extraction, which removes soluble nucleoplasmic proteins, but proteins retained on chromatin can be subsequently detected using a green fluorescent protein (GFP) tag, or by indirect immunofluorescence (**Fig. 1**). Since cell structure is maintained, the chromatin binding of a specific protein can be correlated with other cell cycle events such as anaphase spindle elongation or septation, and in many experiments the need for cell synchronization can be avoided. In addition to analysis of replication proteins *(7,8)* this method has also been used with proteins involved in checkpoint responses *(9,10)* and sister chromatid cohesion *(11)*.

This procedure is simplest when the protein of interest is tagged with a fluorescent protein, since no further processing of the samples is necessary after extraction and fixation (**Subheading 3.1.**). However, indirect immunofluorescence can be used to detect proteins after detergent extraction and fixation (**Subheading 3.2.**). Retention of a nuclear protein after detergent extraction could reflect chromatin binding; alternatively, the protein could be associated with the nuclear matrix or be insoluble under the conditions used. Therefore it is important to show that retention of a protein after detergent extraction is abolished upon nuclease digestion of chromatin (**Subheading**

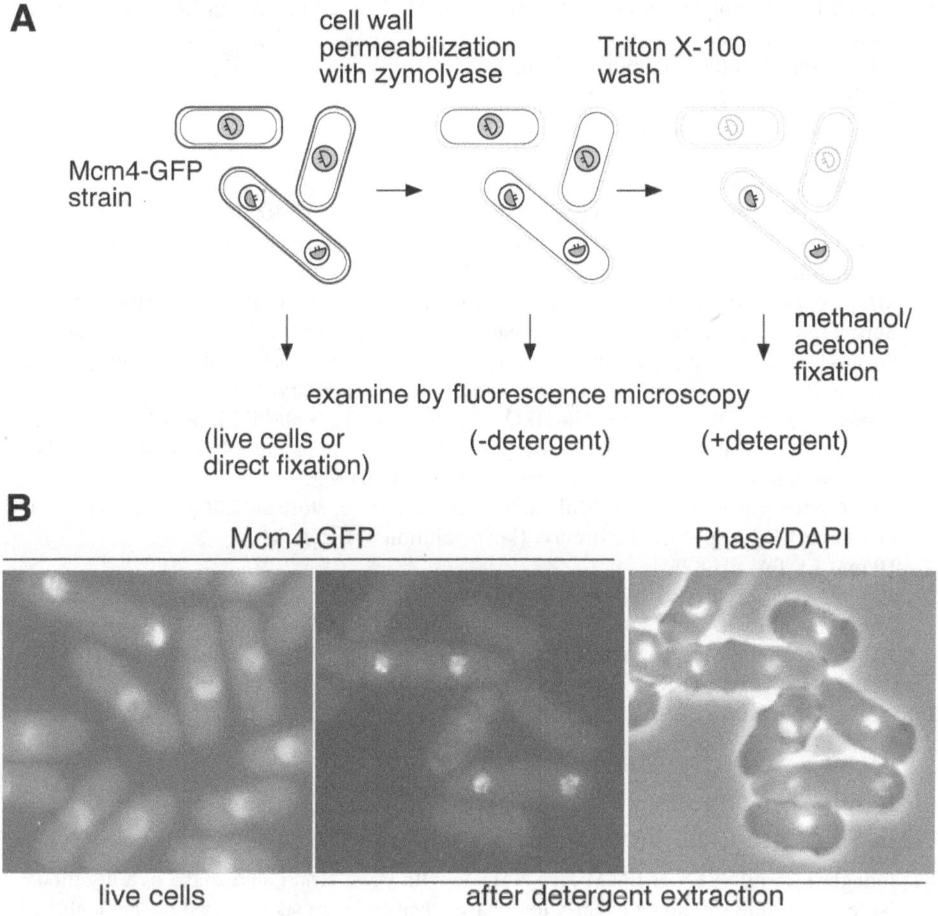

Fig. 1. **(A)** Chromatin-binding assay. **(B)** Example of assay result using an Mcm4/GFP-tagged strain.

3.3.). Cells processed for this assay can also be analyzed by flow cytometry to determine their DNA content (**Subheading 3.4.**).

2. Materials

1. EMMSorb buffer: 15 m*M* KH phallate, 15 m*M* Na_2HPO_4, 90 m*M* NH_4Cl, 1.2 *M* sorbitol, pH 7.0.
2. Extraction buffer 1 (no Mg^{2+}): 20 m*M* PIPES-KOH, pH 6.8, 0.4 *M* sorbitol, 1 m*M* EDTA, 150 m*M* KAc, 0.5 m*M* spermidine HCl, 0.15 m*M* spermine HCl. Store at –20°C.
3. Extraction buffer 2 (containing Mg^{2+}): 20 m*M* PIPES-KOH, pH 6.8, 0.4 *M* sorbitol, 150 m*M* KAc, 2 m*M* MgAc. Store at –20°C.
4. Nuclease extraction buffer: 20 m*M* PIPES-KOH, pH 6.8, 0.4 *M* sorbitol, 150 m*M* KAc, 2 m*M* MgAc, 2 m*M* $CaCl_2$, 250 m*M* NaCl. Store at –20°C.

5. 20,000 U/g Zymolyase 20-T (ICN, cat. no. 320921) at 20 mg/mL in EMMSorb buffer. Store 50- L aliquots at –70°C.
6. 10% Triton X-100 (in extraction buffer).
7. 2% Sodium dodecyl sulfate (SDS).
8. 1 *M* Dithiothreitol (DTT).
9. 1 *M* Sodium azide.
10. 1 *M* deoxy-glucose.
11. Protease inhibitor tablets (Complete mini, cat. no. 1 836 153, Roche).
12. Mounting solution: 50% glycerol in phosphate-buffered saline (PBS), 5–100 ng/mL DAPI (4'6'-diamidino-2-phenylindole).
13. Microscope slides, poly-L-lysine–coated (Sigma, cat. no. P 8920), or alternatively, SuperFrost Plus (Menzel-Glaser, Germany, cat. no. 041300; wash in acetone before use.
14. Poly-L-lysine–coated cover slips. Soak 13-mm cover slips in 0.1% poly-L-lysine (Sigma, cat. no. P 8920) for 10 min. Drain, dry, rinse in water, and dry.
15. PBSBAL buffer: PBS (10 m*M* Na_2HPO_4, 2 m*M* KH_2PO_4, 3 m*M* KCl, 0.14 *M* NaCl) containing 100 m*M* lysine hydrochloride, 10 m*M* NaN_3, 1 % essentially fatty acid-free bovine serum albumin (BSA; Sigma, cat. no. A0281), pH 6.9.
16. 5 m*M* Sytox green in DMSO (Molecular Probes, S-7020; store at –20°C). Dilute to 2 *M* in 10 m*M* EDTA, pH 8.0, before use (keep solution in the dark).
17. 10 m*M* EDTA, pH 8.0.
18. 10 mg/mL RNase A (Roche, cat. no. 109169); boil for 10 min, cool to room temperature, filter, and store at –20°C. Dilute to 0.1 mg/mL in 10 m *M* EDTA, pH 8.0, before use.
19. Micrococcal nuclease (Sigma, cat. no. N3755). Dissolve 50 U in 0.2 mL 20 m*M* HEPES-KOH, pH 7.4, 50 m*M* KCl, 50% glycerol. Store 10- L aliquots at –70°C.
20. 0.2 *M* EGTA, pH 7.5.
21. Yeast EMM or YES media ***(12)***.

3. Methods

3.1. In Situ *Chromatin-Binding Assay*

1. Inoculate 25 mL YES or EMM (*see* **Note 1**) with yeast strain, and shake overnight (*see* **Note 2**). Culture should be in early log phase when cells are taken for assay. Typically 25 mL of culture at OD_{595} = 0.2–0.5 is enough for one to five assays.
2. It is useful to fix an aliquot of cells directly as a control for examining the protein distribution in nonextracted cells and to process the remaining cells for the chromatin binding assay. Fix an aliquot of cells (5 mL) directly by spinning down (3000*g*, 5 min) and resuspending in methanol (0°C). After 10 min, spin down and resuspend in acetone (0°C). Examine cells as in **step 11**.
3. Spin down remaining cells (3000*g*, 5 min), and wash in 2 mL ice-cold EMMSorb containing 10 m*M* DTT (*see* **Note 3**). Transfer to a 2-mL Eppendorf tube.
4. Resuspend in 0.45 mL EMMSorb containing 10 m*M* DTT, and add 50 L of 20 mg/mL zymolyase 20-T. Digest at 32°C for 10–20 min. Test for adequate digestion by mixing a few microliters of cells with an equal volume of 2% SDS on a slide and examining under phase contrast microscopy. Cells should go phase dark. (Aim for >95% cells going phase dark.) If digestion is going slowly, add more zymolyase.
5. When adequate digestion has been achieved, add 1.5 mL EMMsorb buffer (at 4°C), and spin down. (All subsequent spins are 1000*g* for 1 min in an Eppendorf centrifuge at 4°C). Resuspend cells in EMMsorb (4°C), and spin down (*see* **Note 4**).

6. Resuspend in 2 mL extraction buffer, and spin down (*see* **Note 5**).
7. Resuspend in 0.9 mL extraction buffer containing protease inhibitors. Transfer 0.45 mL of this cell suspension to a 2-mL eppendorf tube containing 50 L 10% Triton X-100. Mix and transfer to 20°C water bath for 5 min. Remaining cells can be used as minus detergent control (*see* **Note 6**).
8. Spin cells down (both plus and minus detergent samples), and resuspend in 2 mL methanol (0°C). Keep on ice for 10 min (*see* **Note 7**).
9. Spin down cells and resuspend in 1 mL acetone (0°C). Cells can be stored at this stage at –20°C for a few days.
10. To examine the cells by fluorescence microscopy, vortex cells in acetone to resuspend, and spin down 0.1–0.3 mL. Take off acetone leaving 10–20 L. Cells can be gently sonicated at this stage to break up cell clumps.
11. Spread a thin film of cells on poly-L-lysine–coated or Superfrost microscope slide. Apply 10 L mounting solution and a 22-mm cover slip. The cover slip can be sealed with nail varnish, but this is not necessary if the slides are to be viewed immediately. Obtain images of cells, taking first a phase image, then a green fluorescent protein (GFP) image, and finally a DAPI image for each field of cells (*see* **Note 8**).

3.2. Immunostaining Cells After Detergent Extraction and Fixation

1. Process cells to **step 9** in **Subheading 3.1.** Resuspend cells and transfer approx 0.2 mL to 1.5 mL Eppendorf tube. Spin down (1000*g*, 1 min), and take off most acetone; leave about 20–50 L.
2. Resuspend cells by vortexing, and mildly sonicate to break up cell clumps. Spread about 10 L of the cell suspension onto a poly-L-lysine–coated 13-mm cover slip.
3. Rinse cover slip in PBS, and then incubate in PBSBAL for 30 min. (It is convenient to keep the cover slips in small Petri dishes or multiwell plates for **steps 3–7**.)
4. Remove PBSBAL, add 20 L primary antibody to the cover slip, and incubate in a humid container for at least 1 h.
5. Wash twice in PBS and once in PBSBAL (three times, 5 min).
6. Add 20 L secondary antibody to the cover slip (conjugated with fluorescent dye). Incubate in dark humid container for at least 1 h.
7. Wash three times in PBS (three times, 5 min).
8. Drain the cover slip well. Mount in DAPI/PBS. Seal the edges of cover slip with nail varnish. Obtain images of cells taking first a phase image, then a GFP image, and finally a DAPI image for each field of cells.

3.3. Micrococcal Nuclease Control to Determine Whether Retention of Proteins After Detergent Extraction is Chromatin-Dependent

1. Process cells to **step 6** in **Subheading 3.1.** Resuspend in 0.9 mL nuclease extraction buffer. (This has a higher salt concentration, which is necessary to solubilize digested chromatin ***[13]***, as well as Ca^{2+} for micrococcal nuclease activity.) Add Triton X-100 to 1%, and divide sample into three tubes. To tube 1 (plus nuclease), add micrococcal nuclease to 2.5 U/mL, and to tube 2 add first EGTA to 10 m*M* and then micrococcal nuclease to 2.5 U/mL (plus nuclease, inactive); tube 3 is minus nuclease control. Incubate at 20°C for 5 min, and then proceed as in **Subheading 3.1., step 8**.

Table 1
pSM Vectors

	Selectable marker	Fluorescent protein	Ref. or accession no.
pSMUG2+	*ura4+*	eGFP	AJ306911
pSMRG2+	*kanMX6*	eGFP	AJ306910
pSMUC2+	*ura4+*	CFP	*1*[a]
pSMUY2+	*ura4+*	YFP	*1*[a]
pSMRC2+	*kanMX6*	CFP	
pSMRY2+	*kanMX6*	YFP	

[a]Vector sequences are available at URL: users.ox.ac.uk/~kearsey/plasmids/.

3.4. Flow Cytometric Analysis of Cells After Detergent Extraction and Fixation

1. Process cells to **step 9** in **Subheading 3.1.** Vortex cells and transfer 0.2 mL to 5 mL 10 m*M* EDTA, pH 8.0 (*see* **Note 9**). Spin down (3000*g*, 5 min) and resuspend in 0.5 mL 0.1 mg/mL RNase A in 10 m*M* EDTA, pH 8.0.
2. Incubate at 37℃ for 2–24 h.
3. Add 0.5 mL 2 *M* Sytox green in 10 m*M* EDTA, pH 8.0 (*see* **Note 10**).
4. Sonicate and analyze in flow cytometer.

4. Notes

1. We have found that GFP-tagged strains grown in minimal medium (EMM) give lower cytoplasmic fluorescence in some experiments, such as those involving temperature shifts, than when rich (YES) medium is used.
2. Strain used should have a GFP (or other fluorescent tag) on the protein to be examined; alternatively, use protocol in **Subheading 3.2.** from **step 9** to detect protein via indirect immunofluorescence. A number of vectors have been developed for tagging fission yeast proteins with GFP (e.g., *see* **ref. *14***). **Table 1** and **Fig. 2** give details of vectors that we have developed for GFP, CFP, or YFP tagging. Genes can be tagged by long oligo PCR, using the PCR product for transformation ***(14)***. Alternatively, clone the 3' region of the gene to be tagged into the vector, linearize the construct in this 3' gene region, and transform fission yeast with this DNA.
3. For time-courses, e.g., after a temperature shift, it may be advisable to arrest cellular metabolism by adding 10 m*M* NaN_3 and 10 m*M* deoxy-glucose to cells just before the initial centrifugation.
4. An extra EMMSorb wash can be included at this stage.
5. The Mg^{2+}-containing extraction buffer is suitable for analysis of Mcm2-7 and Cdc45 proteins in log phase cells, whereas the minus Mg^{2+} buffer is recommended for analysis of these proteins in cells that have been synchronized by nitrogen starvation. Buffer modification may be necessary for analysis of other proteins.
6. Cells fixed directly, prior to zymolyase digestion, or viewed live, give the best indication of the in vivo distribution of the tagged protein. Cells fixed after zymolyase digestion and washing in extraction buffer without detergent may show some loss of soluble nucleoplasmic protein (possibly after cell lysis in the hypotonic extraction buffer).

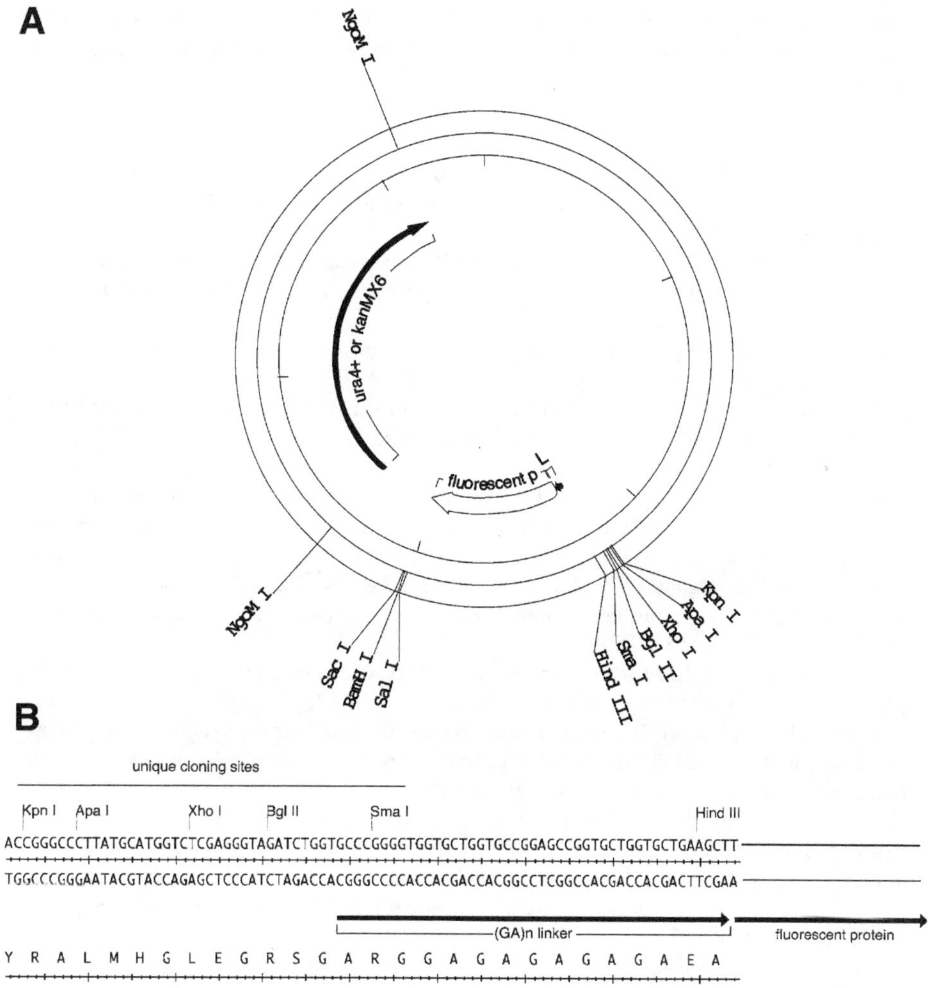

Fig. 2. **(A)** Generic structure of pSM vectors. **(B)** Sequence of pSM vectors in region of cloning sites.

7. Ethanol, 100%, (at 0℃) may be used as an alternative to methanol/acetone fixation, although the fixation procedure may have to be modified depending on the protein being examined.
8. Cells that were not adequately digested during the zymolyase digestion step appear phase bright and should be ignored in subsequent analysis. Suitable filter sets for GFP and its derivatives are: 41017 (eGFP), 31044v2 (CFP), and 41028 (YFP), from Chroma (Brattleboro, VT).
9. In flow cytometric analysis, samples extracted with Triton X-100 will give Sytox green histograms shifted to the left compared with nondetergent-extracted cells, owing to lower cytoplasmic fluorescence. Samples compared should all have had the same extraction and fixation treatment.

10. Propidium iodide (4 *M*) can also be used instead for DNA staining, but Sytox green gives tighter histograms. The GFP fluorescence from proteins expressed at native levels is generally too faint to complicate the analysis of DNA content using Sytox green.

References

1. Gregan, J., Lindner, K., Brimage, L., et al. (2003) Fission yeast Cdc23/Mcm10 functions after pre-replicative complex formation to promote Cdc45 chromatin binding. *Mol. Biol. Cell* **14,** 3876–3887.
2. Donovan, S., Drury, L., and Diffley, J. (1997) Cdc6p-dependent loading of Mcm proteins onto pre-replicative chromatin in budding yeast. *Proc. Natl. Acad. Sci. USA* **94,** 5611–5616.
3. Tanaka, T., Knapp, D., and Nasmyth, K. (1997) Loading of an Mcm protein onto DNA replication origins is regulated by Cdc6p and CDKs. *Cell* **90,** 649–660.
4. Todorov, I. T., Attaran, A., and Kearsey, S. E. (1995) BM28, a human member of the MCM2-3-5 family, is displaced from chromatin during DNA replication. *J. Cell Biol.* **129,** 1433–1446.
5. Klein, F., Laroche, T., Cardenas, M. E., Hofmann, J. F., Schweizer, D., and Gasser, S. M. (1992) Localization of RAP1 and topoisomerase II in nuclei and meiotic chromosomes of yeast. *J. Cell Biol.* **117,** 935–948.
6. Kearsey, S. E., Montgomery, S., Labib, K., and Lindner, K. (2000) Chromatin binding of the fission yeast replication factor mcm4 occurs during anaphase and requires ORC and cdc18. *EMBO J.* **19,** 1681–1690.
7. Lindner, K., Gregan, J., Montgomery, S., and Kearsey, S. (2002) Essential role of MCM proteins in pre-meiotic DNA replication. *Mol. Biol. Cell* **13,** 435–444.
8. Gomez, E. B., Catlett, M. G., and Forsburg, S. L. (2002) Different phenotypes in vivo are associated with ATPase motif mutations in *Schizosaccharomyces pombe* minichromosome maintenance proteins. *Genetics* **160,** 1305–1318.
9. Wolkow, T. and Enoch, T. (2003) Fission yeast Rad26 responds to DNA damage independently of Rad3. *BMC Genet.* **4,** 6.
10. Williams, D. R. and McIntosh, J. R. (2002) mcl1+, the *Schizosaccharomyces pombe* homologue of CTF4, is important for chromosome replication, cohesion, and segregation. *Eukaryot. Cell* **1,** 758–773.
11. Tomonaga, T., Nagao, K., Kawasaki, Y., et al. (2000) Characterization of fission yeast cohesin: essential anaphase proteolysis of Rad21 phosphorylated in the S phase. *Genes Dev.* **14,** 2757–2770.
12. Moreno, S., Klar, A., and Nurse, P. (1991) Molecular genetic analysis of the fission yeast *Schizosaccharomyces pombe. Methods Enzymol.* **194,** 795–823.
13. Wu, L. C., Fisher, P. A., and Broach, J. R. (1987) A yeast plasmid partitioning protein is a karyoskeletal component. *J. Biol. Chem.* **262,** 883–891.
14. Bahler, J., Wu, J. Q., Longtine, M. S., et al. (1998) Heterologous modules for efficient and versatile PCR-based gene targeting in *Schizosaccharomyces pombe. Yeast* **14,** 943–951.

10

Application of Magnetic Beads to Purify Cells Transiently Transfected With Plasmids Encoding Short Hairpin RNAs

Peter D. Adams

Summary

Short interfering (si) RNAs are commonly used to knock down expression of proteins in mammalian cells and thereby investigate protein function. siRNAs were originally introduced into mammalian cells by transient transfection of short, synthetic, double-stranded RNA oligonucleotides. More recently, a convenient, more cost-effective approach has been developed that makes use of plasmids encoding short hairpin (sh) RNAs, which are transiently or stably transfected into cells. After expression in cells, shRNAs are processed by the cell to the corresponding siRNAs. However, most protocols for transient transfection of plasmid DNAs introduce the DNA into a minority of the total cells. Therefore, to investigate the biochemical effects of protein knockdown, it is necessary to purify the transfected cells. This can be done by cotransfection of a plasmid encoding the cell surface marker protein, CD19 or CD20, followed by immunopurification of the CD19- or CD20-expressing cells with magnetic beads. The purified cells can then be used for a wide range of biochemical analyses. In addition, since the CD19/CD20 cell surface marker approach can be readily combined with analysis of cell cycle distribution of propidium iodide-stained cells, it is straightforward to determine simultaneously the biochemical and cell cycle effects of a knocked-down protein.

Key Words

shRNA; transfection; magnetic beads; sorting.

1. Introduction

Recently, with the advent of RNA interference (RNAi) technology, it has become possible to examine the cellular and biochemical effects of RNAi-mediated knockdown of a protein expression, without the need to go through a time-consuming homologous recombination-mediated gene knockout procedure *(1)*. The first demonstration of the feasibility of this approach in mammalian cells used transient transfection of synthetic, double-stranded RNA oligonucleotides *(2)*. However, two major disadvantages of this approach are the cost of oligonucleotide synthesis and the inabil-

From: *Methods in Molecular Biology, vol. 296, Cell Cycle Control: Mechanisms and Protocols*
Edited by: T. Humphrey and G. Brooks © Humana Press Inc., Totowa, NJ

ity to knock down protein expression stably. Subsequently, alternative approaches have been developed, each with their own advantages and disadvantages *(3)*.

We have successfully used the method of plasmid-driven expression of shRNAs developed by Drs. Hannon and Paddison at Cold Spring Harbor Laboratory *(4)*. In this and similar approaches, plasmid vectors are used to direct expression of shRNAs, which are then processed by the cell to form active short interfering RNAs (siRNAs) *(4–8)*. The plasmid DNA can be introduced into the cells by transient transfection, using, for example, calcium phosphate or a liposome-based approach. We favor this plasmid-based approach, at least initially, for the following reasons. First, it is convenient and cost-effective. The only technologies involved are standard molecular cloning techniques and cell transfection. Second, it facilitates a straightforward transition into the retrovirus-, lentivirus, and adenovirus-based delivery approaches, which hold the most promise of RNAi-based methods in general. Once an active short hairpin RNA (shRNA) has been identified through conventional approaches, the same hairpin can easily be transferred into a viral delivery vector. Third, the support provided by Drs. Hannon and Paddison has been excellent, starting with the excellent web site at http://www.cshl.org/public/SCIENCE/hannon.html, which can be used to design the shRNAs.

However, one disadvantage of this approach is that most transient transfection procedures only introduce plasmid DNA into 5–30% of the cell population. Thus, even in a readily transfectable cell line, such as U2OS, any biochemical parameter analyzed in the total cell population reflects the state of primarily the untransfected cells. To overcome this problem, we and others have utilized a cell surface marker to identify the transiently transfected cells (**Fig. 1**). In this method, the plasmid of interest, encoding the shRNA, is cotransfected with another plasmid that encodes the cell surface marker CD19 or CD20. These proteins are normally expressed only on B-cells but are efficiently routed to the cell surface of many other cell types when ectopically expressed. Thus, it serves as a marker of the transfected cells that can be detected in either live or fixed cells. Live, unfixed cells expressing CD19 or CD20 can be physically purified away from the untransfected nonexpressing cells, using anti-CD19- or anti-CD20–coated magnetic beads and/or fluorescence-activated cell sorting (FACS). These purified cells are amenable to analysis by most common biochemical approaches *(9–14)*. Fixed and permeabilized cells can be double stained with fluorescence-conjugated anti-CD19 or anti-CD20 antibodies and propidium iodide to determine the cell cycle distribution of transfected cells *(15,16)*. A single sample of cells can be split and processed by both approaches to obtain biochemical and cell cycle data on the transfected cells.

A major advantage of this approach is its speed and simplicity. The major limitation of this approach is the relatively small number of cells that can easily be obtained. Despite this limitation, using anti-CD19–coated magnetic beads as the sole method of immunopurification, we have obtained sufficient transfected U20S cells to extract RNA for Northern analysis, proteins for immunoprecipitations and Western blots, and chromatin for nuclease analysis *(11–13)*.

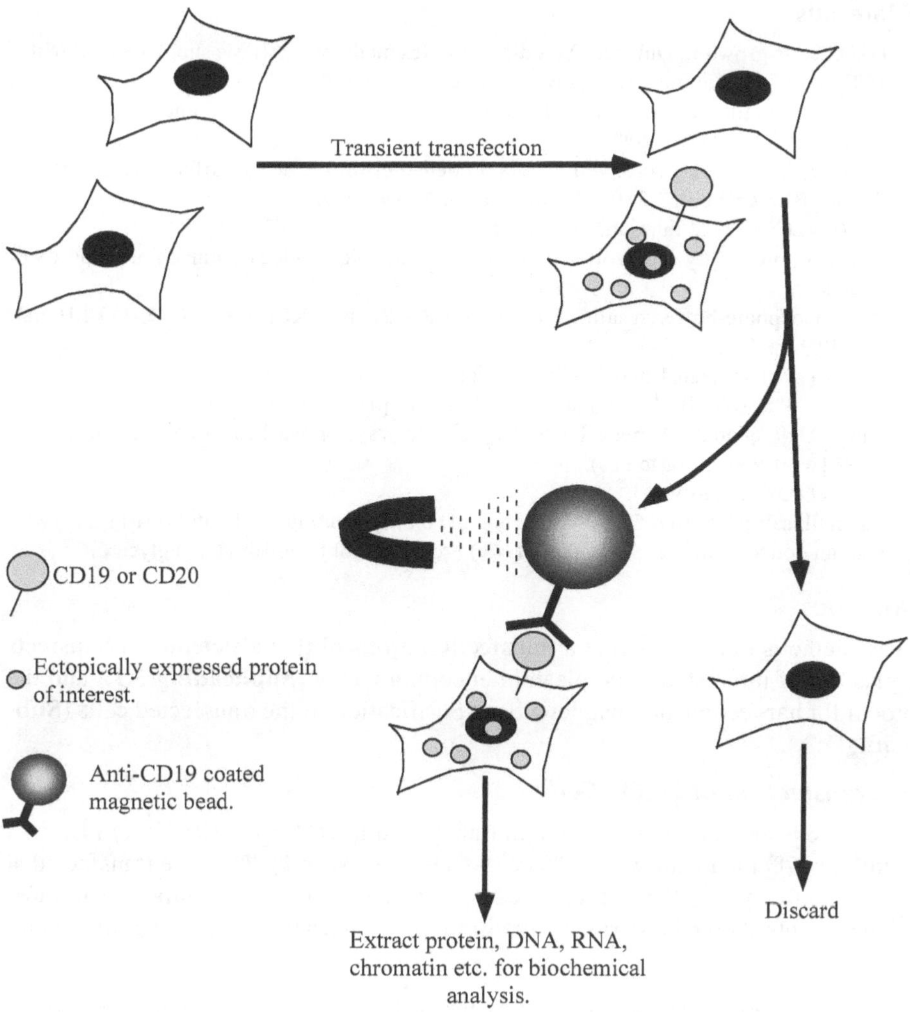

Fig. 1. Magnetic bead sorting of transfected cells. Cells are transiently transfected with plasmids encoding the cell surface marker CD19 or CD20 and the protein of interest. Thirty six hours after transfection, the cells are collected and incubated with magnetic beads coated with an anti-CD19 or anti-CD20 antibody. A magnet is used to immunopurify the CD19-or CD20-expressing cells, which are subsequently processed for biochemical analysis.

2. Materials

1. U2OS cells grown in Dulbecco's modified Eagle's medium (DMEM) supplemented with 10% (v/v) fetal bovine serum (FBS) in a humidified 37°C, 10% (v/v) CO_2 incubator (or most other readily transfectable cell types grown under appropriate conditions).
2. 2.5 *M* $CaCl_2$. Store at –20°C.
3. 2X BES (*N,N-bis*[2-hydroxyethyl]-2-aminoethanesulfonic acid)-buffered saline (BBS): 50 m*M* BES (pH 6.95), 280 m*M* NaCl, 1.5 m*M* Na_2HPO_4.
4. 5% (v/v) CO_2, 37°C, humidified incubator.
5. Transfection quality supercoiled plasmid DNA (double-banded cesium chloride purity or equivalent).
6. PBS (phosphate-buffered saline): 137 m*M* NaCl, 2.7 m*M* KCl, 1.4 m*M* KH_2PO_4, 10 m*M* Na_2HPO_4, pH 7.3.
7. 0.25% (w/v) Trypsin/1 m*M* EDTA solution.
8. PBS + 0.3% (w/v) BSA (bovine serum albumin), pH 7.3.
9. Anti-CD19–coated magnetic beads (Dynal, Norway; prewashed three times in PBS + 0.3% [w/v] BSA prior to use).
10. Magnet (Dynal, Norway).
11. Laemmli sample buffer: 50 m*M* Tris-HCl, pH 6.8, 100 m*M* dithiothreitol (DTT), 2% (w/v) sodium dodecyl sulfate (SDS), 0.1% (w/v) bromophenol blue, 10% (v/v) glycerol.

3. Methods

The methods below describe a transfection protocol that efficiently cotransfects the plasmid of interest and the plasmid encoding CD19 (**Subheading 3.1.**) and the protocol for harvesting and magnetic bead purification of the transfected cells (**Subheading 3.2.**).

3.1. Transfection of U2OS Cells

U2OS cells are routinely grown and transfected in DMEM + 10% (v/v) FBS in a humidified 37°C incubator with 10% (v/v) CO_2 (*see* **Note 1**). They are transfected at 20–60% confluence by the method of Chen and Okayama (*17*; *see* **Note 2**). All transfection reagents should be sterilized by filtration through a 0.22- *M* sterile filter membrane.

1. Four hours prior to transfection, remove the medium from each 10-cm plate of cells and replace with 9 mL of fresh medium (*see* **Note 3**). Return the cells to a humidified 37°C incubator with 10% (v/v) CO_2.
2. Dilute the required amount of 2.5 *M* CaCl2 to 250 m*M*, and aliquot 0.5 mL per transfection to separate 15-mL tubes.
3. Add supercoiled plasmid DNA to a total of 30 g. Typically we add 2 g of pCMV CD19 and 28 g of a plasmid directing expression of the shRNA (based on Invitrogen's pENTR and constructed as detailed at http://www.cshl.org/public/SCIENCE/hannon.html) or control.
4. Add 0.5 mL of 2X BBS dropwise by dripping slowly from a 1-mL pipet vertically down the center of the tube (1–2 drops/s). Do not mix. Wait 15 min. At this time, the precipitate should be barely visible to the naked eye (*see* **Note 4**).

5. Use a 1 mL pipet to blow air bubbles through the solution to mix the precipitate. Distribute the mixture dropwise into the medium evenly over the plate of cells.
6. Rock plates very gently, and place in humidified 37°C incubator with 5% CO_2 (v/v) overnight (*see* **Note 5**).
7. On the following morning, remove the medium, replace with 10 mL of fresh medium, and return to humidified 37°C incubator with 10% CO_2 (v/v).

3.2. Harvesting Cells

In a conventional transient transfection, in which a protein is being ectopically expressed, the cells are usually harvested approx 36 h after transfection. However, when the goal is to use a plasmid-encoded shRNA to knock down expression of an endogenous protein, we have found that more efficient knockdown is generally achieved 60 h after transfection (*see* **Note 6**).

1. Wash the cells with warm PBS and trypsinize by incubation with 1 mL of 0.25% (w/v) trypsin/1 m*M* EDTA solution for 3 min at 37°C (*see* **Note 7**). Add 9 mL of ice-cold (4°C) DME + 10% (v/v) FBS, and repeatedly aspirate with a pipet to disrupt clumps of cells. Visual inspection of the cells through a microscope should confirm that they are predominantly single cells, with few doublets or clumps.
2. Transfer the cells to a 15-mL tube, and pellet them by centrifugation for 3 min at 200*g* at 4°C. Aspirate buffer, leaving approx 50 L, and resuspend cell pellet in this remaining solution. Add 10 mL of PBS + 0.3% (w/v) BSA at 4°C, invert tube to wash cells, pellet again by centrifugation, aspirate the supernatant, resuspend the cells in the remaining 50 L of buffer, and add 950 L of PBS + 0.3% (w/v) BSA at 4°C (*see* **Note 8**). Determine the number of cells/mL by counting with a hemocytometer.
3. Transfer the cells to an Eppendorf tube. Add required volume of anti-CD19–coated magnetic beads. Typically, we add 1.2×10^7 beads per plate of cells. Assuming that at the time of harvesting one 60% confluent, 10-cm plate of U2OS contains 2×10^6 cells and the transfection efficiency is 30%, this would mean 0.7×10^6 target cells per plate and the bead/target cell ratio is approx 17:1. Rotate at 4°C for 15 min (*see* **Notes 9** and **10**).
4. Place in the magnet until all beads are pulled down. Invert tube holder/magnet a few times to get beads and cells out of the liquid meniscus.
5. While tube is still in the magnet, aspirate wash buffer. Remove the tube, and add 1 mL of PBS + 0.3% (w/v) BSA at 4°C; using a blue tip, very gently resuspend cells. Take care not to damage the cells.
6. Repeat **steps 4** and **5** four times.
7. View cells through a microscope. All remaining cells should have a "rosette" of beads around them. Cells without a rosette are probably untransfected. Perform an extra round of sorting if necessary (*see* **Note 11**).
8. Process the immunopurified cells further as desired. Typically, we extract RNA for Northern blots or protein for immunoprecipitations and Western blots or chromatin for nuclease digestion ***(12,13)***, according to standard protocols.

Knockdown of the protein of interest is normally most easily confirmed by Western blotting, using antibodies and conditions previously determined for the protein of interest. **Figure 2** shows a Western blot of cell lysates from immunopurified cells to

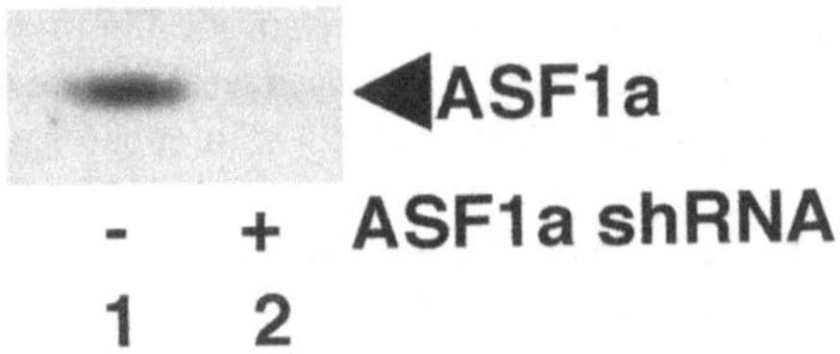

Fig. 2. Analysis of ASF1a expression in transiently transfected, magnetic bead-sorted cells. U2OS cells were transiently transfected with a plasmid encoding CD19 alone or with a plasmid encoding an ASF1a shRNA. CD19+ cells were immunopurified, and extracts were Western-blotted with anti-ASF1a (A88). ASF1a, antisilencing function 1a.

confirm the shRNA-mediated knockdown of the human antisilencing function 1a protein (ASF1a), a regulator of chromatin assembly and histone transcription ***(18)***.

4. Notes

1. The use of a cell surface marker to sort for transfected cells has been successfully applied to a number of different cell types, including SAOS2, MG63 and U2OS osteosarcoma cells, Rat1a fibroblasts, C33A cervical carcinoma cells, T98G glioblastoma cells, J82 bladder carcinoma cells, MCF7 breast carcinoma cells, VA13 cells (human diploid fibroblasts transformed with SV40 virus), NIH3T3 cells, and C2C12 myoblasts ***(10,15,19–22)***. This list is not meant to be exhaustive.
2. This protocol requires that every cell expressing CD19 also expresses the protein of interest. Thus, it depends on efficient cotransfection of the plasmids encoding the CD19 cell surface marker and the protein of interest. Efficient cotransfection is achieved with the calcium phosphate transfection protocol described here. However, it should not automatically be assumed that other transfection protocols reliably cotransfect both plasmids. Most reports using this method have used calcium phosphate as the transfection reagent ***(9–14)***.
3. The number of cells that should be transfected and immunopurified obviously depends on the transfection efficiency of the cells and the number of cells required for the final assay. Typically, the transfection efficiency of U2OS cells is approx 20–30% of the total cells on the plate. Thus, sufficient immunopurified cells are derived from a single transfected plate for Western blotting of moderately abundant proteins, such as ASF1a (**Fig. 2**). However, when working with cells that transfect relatively poorly (transfection efficiency of 5% or less) the number of plates transfected should be increased accordingly.
4. The pH of the BBS is critical to obtain a good transfection efficiency. Normally, we prepare three batches of 500 mL of BBS of pH 6.93, 6.95, and 6.97. A preliminary test is carried out by mixing $CaCl_2$, DNA, and BBS, as described in **Subheading 3.1.** At the end of **step 4**, the precipitate should be barely visible to the naked eye. A batch of BBS that gives such a precipitate in this assay will usually work well. However, this is normally confirmed by transfection of cells with plasmid encoding GFP and visualization by fluorescence microscopy.
5. In the original description of this transfection method by Chen and Okayama ***(17)***, it was reported that transfection efficiency was greatest when the cells were incubated overnight in 2–4% CO_2. We have found that 5% CO_2 also works well. We have not directly compared incubation at 5% and 10% CO_2, although Chen and Okayama ***(17)*** found that this greatly decreased the efficiency of transfection.

6. shRNA-mediated knockdown targets the mRNA rather then the protein. Therefore, the rate at which the protein level is reduced depends on the rate of degradation of the protein. An unstable protein with a short half-life will be knocked down more quickly than a stable protein with a long half-life.
7. Adequate trypsinization of the cells is necessary to generate single cells prior to immunopurification. However, excessive trypsinization could conceivably destroy the cell surface epitope that is recognized by the anti-CD19 antibody. By adhering to the trypsinization conditions described in **Subheading 3.2.**, we have avoided this potential problem.
8. Addition of BSA to the PBS prevents cells being lost during the wash steps through their sticking to the sides of the tube.
9. The number of beads added is in large excess over the number of cells. Consequently, variations in the number of cells, owing to changes in cell density or cell type, do not affect the efficiency of the sorting process.
10. A crude measure of transfection efficiency is obtained at **step 3** of **Subheading 3.2.** When the anti-CD19–coated magnetic beads are added to the suspension of CD19-expressing cells, visible cell clumping occurs within a few minutes. The difference between cells transfected with and without the plasmid encoding CD19 should be readily apparent. If not, it suggests that the transfection efficiency is low, and relatively few immunopurified cells will be obtained at **step 7** of **Subheading 3.2.**
11. It is generally necessary to perform a Western blot of an extract from the immunopurified cells to confirm that protein knockdown has been achieved. An aliquot of the immunopurified cells obtained in **Subheading 3.2.** should be boiled in Laemmli sample buffer and fractionated by SDS-polyacrylamide gel electrophoresis for this purpose.

Acknowledgments

This work was supported by grants RO1 GM62281 and DAMD17-02-1-0726 to P. D. A. P. D. A. is a Scholar of the Leukemia and Lymphoma Society.

References

1. McManus, M. T. and Sharp, P. A. (2002) Gene silencing in mammals by small interfering RNAs. *Nat. Rev. Genet.* **3,** 737–747.
2. Elbashir, S. M., Harborth, J., Lendeckel, W., Yalcin, A., Weber, K., and Tuschl, T. (2001) Duplexes of 21-nucleotide RNAs mediate RNA interference in cultured mammalian cells. *Nature* **411,** 494–498.
3. Dykxhoorn, D. M., Novina, C. D., and Sharp, P. A. (2003) Killing the messenger: short RNAs that silence gene expression. *Nat. Rev. Mol. Cell. Biol.* **4,** 457–467.
4. Paddison, P. J. and Hannon, G. J. (2002) RNA interference: the new somatic cell genetics? *Cancer Cell* **2,** 17–23.
5. Brummelkamp, T. R., Bernards, R., and Agami, R. (2002) A system for stable expression of short interfering RNAs in mammalian cells. *Science* **296,** 550–553.
6. Yu, J. Y., DeRuiter, S. L., and Turner, D. L. (2002) RNA interference by expression of short-interfering RNAs and hairpin RNAs in mammalian cells. *Proc. Natl. Acad. Sci. USA* **99,** 6047–6052.
7. Miyagishi, M. and Taira, K. (2002) U6 promoter driven siRNAs with four uridine 3' overhangs efficiently suppress targeted gene expression in mammalian cells. *Nat. Biotechnol.* **20,** 497–500.

8. Lee, N. S., Dohjima, T., Bauer, G., Li, H., et al. (2002) Expression of small interfering RNAs targeted against HIV-1 rev transcripts in human cells. *Nat. Biotechnol.* **20,** 500–505.
9. Lundberg, A. S. and Weinberg, R. A. (1998) Functional inactivation of the retinoblastoma protein requires sequential modification by at least two distinct cyclin-cdk complexes. *Mol. Cell. Biol.* **18,** 753–761.
10. Peeper, D. S., Upton, T. M., Ladha, M. H., et al. (1997) Ras signalling linked to the cell-cycle machinery by the retinoblastoma protein. *Nature* **386,** 177–181.
11. Hall, C., Nelson, D. M., Ye, X., et al. (2001) HIRA, the human homologue of yeast Hir1p and Hir2p, is a novel cyclin-cdk2 substrate whose expression blocks S-phase progression. *Mol. Cell. Biol.* **21,** 1854–1865.
12. Nelson, D. M., Hall, C., Santos, H., et al. (2002) Coupling of DNA synthesis and histone synthesis in S-phase independent of cyclin/cdk2 activity. *Mol. Cell. Biol.* **22,** 7459–7472.
13. Ye, X., Franco, A. A., Santos, H., Nelson, D. M., Kaufman, P. D., and Adams, P. D. (2003) Defective S-phase chromatin assembly causes DNA damage, activation of the S-phase checkpoint and S-phase arrest. *Mol. Cell* **11,** 341–351.
14. Santoni-Rugiu, E., Falck, J., Mailand, N., Bartek, J., and Lukas, J. (2000) Involvement of Myc activity in a G(1)/S-promoting mechanism parallel to the pRb/E2F pathway. *Mol. Cell. Biol.* **20,** 3497–3509.
15. Zhu, L., Enders, G., Lees, J. A., Beijersbergen, R. L., Bernards, R., and Harlow, E. (1995) The pRB-related protein p107 contains two growth suppression domains: independent interactions with E2F and cyclin/cdk complexes. *EMBO J.* **14,** 1904–1913.
16. Adams, P. D., Sellers, W. R., Sharma, S. K., Wu, A. D., Nalin, C. M., and Kaelin, W. G. Jr. (1996) Identification of a cyclin-cdk2 recognition motif present in substrates and p21-like cyclin-dependent kinase inhibitors. *Mol. Cell. Biol.* **16,** 6623–6633.
17. Chen, C. and Okayama, H. (1987) High-efficiency transformation of mammalian cells by plasmid DNA. *Mol. Cell. Biol.* **7,** 2745–2752.
18. Tyler, J. K. (2002) Chromatin assembly. *Eur. J. Biochem.* **269,** 2268–2274.
19. Sellers, W. R., Rodgers, J. W. & Kaelin, W. G. Jr. (1995) A potent transrepression domain in the retinoblastoma protein induces a cell cycle arrest when bound to E2F sites. *Proc. Natl. Acad. Sci. USA* **92,** 11544–11548.
20. Qin, X. Q., Livingston, D. M., Ewen, M., Sellers, W. R., Arany, Z., and Kaelin, W. G. Jr. (1995) The transcription factor E2F-1 is a downstream target of RB action. *Mol. Cell. Biol.* **15,** 742–755.
21. Adams, P. D. (2003) Unpublished data.
22. Hofmann, F. and Livingston, D. M. (1996) Differential effects of cdk2 and cdk3 on the control of pRb and E2F function during G1 exit. *Genes Dev.* **10,** 851–861.

11

The Cell Cycle and Virus Infection

Stevan R. Emmett, Brian Dove, Laura Mahoney, Torsten Wurm, and Julian A. Hiscox

Summary

A number of different viruses interact with the cell cycle in order to subvert host-cell function and increase the efficiency of virus replication; examples can be found from DNA, retro, and RNA viruses. The majority of studies have been conducted on DNA and retroviruses whose primary site of replication is the nucleus, but increasingly a number of researchers are demonstrating that RNA viruses, whose primary site of replication is normally the cytoplasm, also interfere with the cell cycle. Viral interference with the cell cycle can have a myriad of different effects to improve virus infection, for example to promote replication of viral DNA genomes, or to delay the cell cycle to allow sufficient time for RNA virus assembly. Although cell cycle control is fairly well characterized in terms of checkpoints and control molecules (e.g., cyclins), in recent years several researchers have demonstrated that the nucleolus is also involved in cell cycle control. The nucleolus and associated subnuclear structures can sequester cell cycle regulatory complexes, and nucleolar proteins also have a direct and indirect effect on the cycling cell. Viruses also interact with the nucleolus. In order to study the interactions between a virus and the cell cycle and *vice versa* we have developed and adapted a number of different approaches and strategies. These include determinations of virus yield and measurements of virus replication to flow cytometry and confocal analysis of the host cell. Increasingly we have found that proteomic approaches allow the rapid analysis of a whole plethora of cell cycle proteins that may be affected by virus infection.

Key Words

Cell cycle; virus; infection; confocal microscopy; nucleolus; 2D gel; proteomics.

1. Introduction

A common strategy of viruses is the creation inside the cell of an environment favorable for efficient replication of their genomes; one way of achieving this is by altering the cell cycle. The function of the cell cycle itself comprises a highly con-

From: *Methods in Molecular Biology, vol. 296, Cell Cycle Control: Mechanisms and Protocols*
Edited by: T. Humphrey and G. Brooks © Humana Press Inc., Totowa, NJ

served and sequential series of checkpoints and phases ensuring that conditions are suitable for the proper function of that cell and for DNA replication and cytokinesis. There are three categories of viruses, depending on their genome and replication strategy: DNA, retro, and RNA viruses. DNA viruses employ a number of mechanisms to modify and interfere with the cell cycle regulatory machinery. In some cases viruses are adapted to multiply in resting cells, whereas in others they induce proliferation of arrested cells or just wait until the infected cell replicates.

Two main strategies for DNA viruses interfering with host cell cycle control can be distinguished. One has generally evolved in viruses with large genomes with the potential to encode many proteins, including those required for viral DNA replication. A typical example of this strategy is found in the herpesviruses to block cell cycle progression, preventing entry into S-phase ***(1)***. The other strategy to impinge on cell cycle control is more typical of DNA viruses with small genomes. Here, virus-encoded proteins, which are not homologs of any known cellular protein, interfere directly with the normal function of cell cycle regulatory components to subvert their activity. Typical examples are the viral oncoproteins that sequester the retinoblastoma (RB) tumor suppressor protein as a first step in inducing S-phase entry by activating the expression of E2F-regulated genes; for example, adenoviruses ***(2)***. Disruption of the cell cycle to favor virus replication is not confined to DNA viruses. Retroviruses also disrupt the cell cycle.

Cells infected with human immunodeficiency virus (HIV) do not proliferate but accumulate in the G_2-phase of the cell cycle, resulting in an increase in virus production ***(3)***. The viral protein responsible for this has been identified as Vpr ***(4)***. Interestingly, Henklein et al. ***(5)*** demonstrated that extracellular added Vpr induced G_2-phase arrest, and the authors suggested that free Vpr in the serum of HIV-infected patients may preprogram cells in order to facilitate replication of HIV in infected individuals. Consistent with the perturbation of the cell cycle observed in tissue culture-infected cells, lymphocytes from HIV-infected individuals show high levels of cyclin B and also nucleolar proteins (whose expression is linked to the cell cycle) ***(6)***.

Altering the host cell cycle by RNA viruses has not been described extensively in the literature, and the mechanisms of action are generally not well characterized. For the negative-strand RNA viruses, there are several examples of cell cycle control. For instance, measles virus infection results in a G_0 block ***(7)***, and the paramyxovirus simian virus V protein prolongs the cell cycle by delaying cells in G_1 and G_2 ***(8)***. In the case of positive-strand RNA viruses, the avian coronavirus infectious bronchitis virus (IBV) delays cell growth by inhibiting cytokinesis and also allows cells to accumulate in S/G_2 ***(9,10)***. Several examples of different picornaviruses interacting with the cell cycle have been described. More recently Feuer et al. ***(11)*** have shown that cells arrested in G_1 or G_1/S produced higher levels of infectious coxsackievirus and viral polyproteins compared with cells in the G_0 phase, or cells blocked at the G_2/M boundary. Feuer et al. ***(11)*** suggested that persistence of coxsackie B3 virus (CVB3) in vivo might be dependent on infection of G_0 cells, which do not support replication. If such cells were to reenter the cell cycle, then the virus may reactivate and trigger chronic viral or immune-mediated pathology in the host. Such findings suggest that locally delivered cell cycle inhibitors could form part of an antiviral therapy, similar to the example of interleukin-2 used to correct cell cycle aberrations in HIV-infected individuals ***(12)***.

Viruses also target subnuclear structures involved in cell cycle regulation as part of a strategy to subvert host cell functions such as the cell cycle. For example, many DNA, retro, and RNA viruses target the nucleolus ***(13,14)***. The nucleolus and associated proteins are also implicated in (and regulated by) the cell cycle ***(15)***. Cajal bodies associated with nucleoli can sequester cyclin-dependent kinase 2 (CDK2) and cyclin E in a cell cycle-dependent manner ***(16)***. The concentrations of many nucleolar proteins such as nucleolin ***(17)*** are dependent on the cell cycle ***(18,19)***. Nucleolin itself can also act as a cell cycle regulator ***(20)***.

When studying viruses and the cell cycle, one must consider the interrelationship between the two. The cell cycle will have an effect on virus replication, and virus replication will concomitantly affect the cell cycle. Below we describe experiments that can be used to investigate whether a particular virus interacts with the cell cycle or vice versa. These range from traditional approaches toward measuring the cell cycle and virus production to the use of confocal microscopy to investigate the redistribution of cell cycle factors in virus-infected cells to a proteomic analysis of the nucleolus, which has recently been identified as having roles in cell cycle regulation ***(21,22)***.

2. Materials

2.1. Northern Blot and Detection

1. Ambion BrightStar Psoralen-Biotin kit.
2. Ambion BrightStar BioDetect kit.
3. Ambion NorthernMax Formaldehyde loading dye.
4. Ambion NorthernMax 10X denaturing gel buffer.
5. Ambion NorthernMax 10X MOPS running buffer.
6. Ambion NorthernMax prehybridization/hybridization buffer.
7. 20X Standard saline citrate (SSC) buffer (Invitrogen).
8. 10% Sodium dodecyl sulfate (SDS) solution (Ambion).
9. Ambion BrightStar-Plus positively charged nylon membranes.
10. 3MM blotting paper (Whatman).
11. 1% Neutral red solution in ddH_2O.
12. 2% Low melting point agarose (Sigma) solution in ddH_2O.

2.2. Plaque Assay

1. Vero medium constituents.
    a. 10X Eagle's Minimal Essential Medium (EMEM; Invitrogen).
    b. Foetal bovine serum (Invitrogen).
    c. 200 mM L-Glutamine (Invitrogen).
    d. 10,000 U/mL Penicillin/10,000 mg/mL streptomycin solution (Invitrogen).
    e. 7.5% Sodium bicarbonate solution (Invitrogen).
2. Vero medium is made up to a 2X solution to account for dilution with agarose solution. EMEM with 20% foetal bovine serum, 4 m*M* L-glutamine, 200 U/mL penicillin, 200 g/mL streptomycin, 3 g/L sodium bicarbonate solution.

2.3. Purification of Nucleoli

1. Chemicals to be used in these procedures should be of the best analytical grade commercially available. When applicable, all solutions should be prepared with double-distilled ion-exchange grade water.
2. All solutions in this nucleolar protocol have been supplemented with Complete Protease inhibitor tablets (Roche, cat. no. 1-873-580) at a final concentration of 1 tablet/50 mL).
3. Phosphate-buffered saline (PBS): 1.15 g/L Na_2HPO_4, 8 g/L NaCl, 100 mL/L $MgCl_2$, 200 g/L KH_2PO_4, 200 g/L KCl, 132 mL/L $CaCl_2$.
4. HEPES: 10 m*M* HEPES, pH 7.9, 10 m*M* KCl, 1.5 m*M* MgC12, 0.5 m*M* dithiothreitol (DTT).
5. The sucrose solutions are made up as follows. S1 solution: 0.25 *M* sucrose, 10 m*M* $MgCl_2$; S2 solution: 0.35 *M* sucrose, 0.5 m*M* $MgCl_2$; S3 solution: 0.88 *M* sucrose, 0.5 m*M* $MgCl_2$.

2.4. 2D SDS-PAGE

1. Sample buffer: 8 *M* urea, 2% 3-[(cholamidopropyl)dimethylammonio]-1-propanesulfonate, 50 m*M* dithiothreitol (DTT), 0.2% (w/v) Biolyte 3/10 (Bio-Rad, UK), and a trace of bromophemol blue.
2. Protean IEF cell (Bio-Rad, UK cat. no. 165-4000).
3. 0.1% (w/v) Coomassie brilliant blue solution: 50% methanol, 10% acetic acid, 0.1% (w/v) Coomassie brilliant blue solution (Sigma-Aldrich, UK), 40% water. Store at room temperature. We recommend using the solution only once per gel.
4. Destain buffer: 5% methanol, 7% acetic acid, 88% water. Store at room temperature.
5. IEF gel staining solution (Bio-Rad, UK, cat. no. 161-0434).
6. Equilibration buffer I: 6 *M* urea, 2% SDS, 0.375 *M* Tris-HCl, pH 8.8, 20% glycerol, 37 mg/mL iodoacetamide, and 2% (w/v) DTT. Alternatively, this can be purchased as cat. no. 163-2107 from Bio-Rad, UK.
7. Equilibration buffer II: 6 *M* urea, 2% SDS, 0.375 M Tris-HCl, pH 8.8, 20% glycerol. Alternatively, this can be purchased as cat. no. 163-2108 from Bio-Rad, UK.
8. 1X Tris/glycine/SDS running buffer: prepare a 5X stock solution of SDS-polyacrylamide gel electrophoresis (PAGE) running buffer by mixing 15.1 g Tris base, 72 g glycine, 5 g SDS, and water to 1000 mL. This stock is stable for several weeks; when required dilute it down to 1X for use at the correct pH of 8.3. Alternatively, this stock can be purchased as cat. no. 161-0772 from Bio-Rad, UK.
9. To prepare gels from the second dimension separation, either purchase from Bio-Rad, UK as cat. no. 345-0036 or make them yourself with a suitable gel casting apparatus (based on a 14-cm × 14-cm × 0.75-mm cast):
   a. Briefly, for the stacking gel, 0.65 mL 30% acrylamide: 0.8% *bis*-acrylamide, 1.25 mL 1.5 *M* Tris-HCl, pH 6.8, 50 L 10% SDS, 3.05 mL water, 25 L 10% ammonium persulfate solution, 6 L TEMED (Sigma-Aldrich, UK).
   b. Resolving gel, 6 mL 30% acrylamide: 0.8% *bis*-acrylamide, 3.75 mL 1.5 *M* Tris-HCl, pH 6.8, 150 L 10% SDS, 5 mL water, 150 L 10% ammonium persulfate solution, 6 L TEMED (Sigma-Aldrich, UK).
10. Overlay agarose: 0.5% agarose, 25 m*M* tris-HCl, 192 m*M* glycine, 0.1% SDS, and trace bromophenol blue.

3. Methods

3.1. Measuring the Interrelationship Between Viruses and the Cell Cycle

There are two options for investigating the cell cycle stage and affect on virus replication. The first is to synchronize cells using serum starvation, release the cells, and determine the cell cycle profile by labeling the cells with bromodeoxyuridine (BrdU) and propidium iodide and analyzing on a flow cytometer. Cells can then be infected at different stages of the cell cycle; depending on the length of the virus life cycle virus replication can occur within a particular cell cycle stage. The second alternative is to use cell cycle inhibitors to block a particular stage of the cell cycle. This has the advantage that cells can be inhibited for prolonged lengths of time, perhaps sufficient to cover the infections cycle; it also has several disadvantages, including the fact that not all cells will arrest at the same stage 100% of the time, and cell cycle inhibitors may affect viral processes themselves.

Simple experiments can be performed to investigate whether a virus has an effect on cellular proliferation and/or the cell cycle. For example, a Coulter counter can be used to measure cellular proliferation in infected compared with noninfected cells ***(10)*** and the cell cycle profiles of infected cells compared with noninfected using flow cytometry (**Fig. 1**). Infection protocols vary from virus to virus, and these can be found by consulting specific literature. Below we detail our protocols for determining the cell cycle stage.

3.1.1. Flow Cytometry

3.1.1.1. Harvesting of Cells for Flow Cytometry

1. To harvest cells for flow cytometry, remove the growth medium and wash the monolayer twice with 3 mL PBS.
2. To detach the cells, add 2 mL of PBS/EDTA/trypsin and incubate at 37€ for 5 min or until cells detach (*see* **Note 1**).
3. To remove the remaining cells, scrape using a cell scraper (Sarstcdt). To inactive the trypsin, transfer by pipeting into a canonical tube filled with 8 mL 10% DMEM.
4. The suspension is then centrifuged at 250*g* for 10 min at 4€, the supernatant is removed, and the cell pellet is resuspended in 2 mL ice cold PBS prior spinning at 250*g* for 10 min/4€.
5. The supernatant is removed, and the cells are processed for flow cytometry to detect either incorporated BrdU or cell cycle marker proteins.

3.1.1.2. Analysis of Stained Cells

1. Fixed and stained cells are analyzed by flow cytometry with a fluorescence-activated cell sorter (FACScan; Becton Dickinson; or equivalent).
2. Ten thousand events per sample should be collected, stored, and analyzed using CellQuest software (Becton Dickinson).

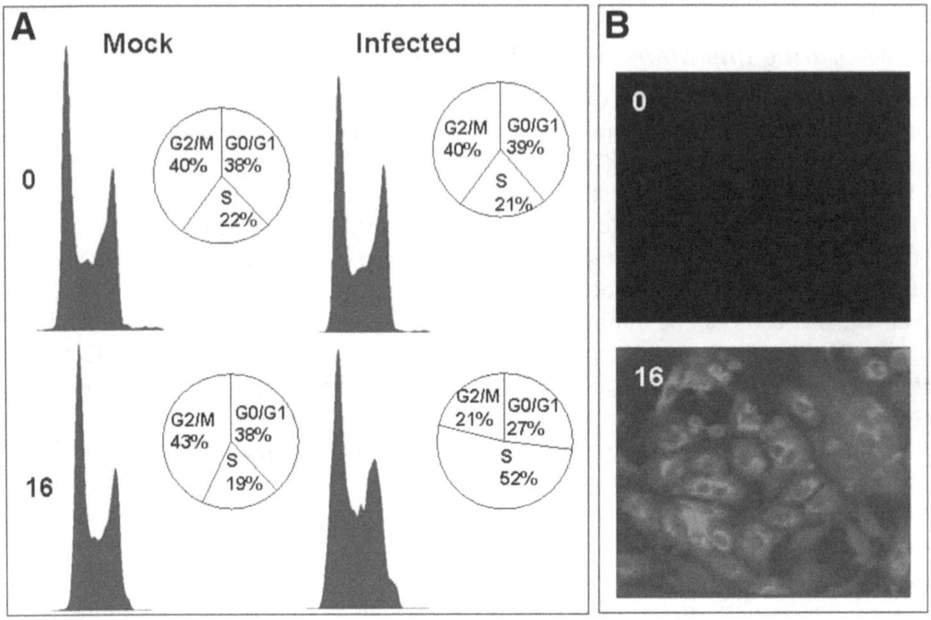

Fig. 1. **(A)** Flow cytometry analysis of mock and avian coronavirus-infected cells at 0 and 16 h post infection. **(B)** Detection of avian coronavirus protein by indirect immunofluorescence in infected cells. Vero cells were infected with coronavirus for 0 h or 16 h, fixed, and analyzed by indirect immunoflourescence using appropriate antibodies. (Original magnification x620.) The data indicate that infected cells accumulate in S-phase compared with mock infected cells.

3.1.1.3. Flow Cytometry Analysis of BrdU Incorporation

1. To determine the proportion of cells undergoing DNA replication, cells are pulsed by addition of 10 m*M* BrdU, an analog of thymidine, for 30 min prior to harvesting.
2. Cells can be harvested by trypsinization and rinsed with PBS as described. After centrifugation at 250*g* for 10 min at 4°C, the pellet is suspended in 1 mL of precooled (–20°C) 70% ethanol (*see* **Note 2**).
3. To remove the fix, cells are centrifuged at 300*g* for 10 min at 4°C and the ethanol removed.
4. To denature the DNA, 500 L of 0.1 *M* HCl is added and incubated at 37°C for 15 min.
5. The reaction is then stopped with the addition of 3 mL PBS and centrifugation at 300*g* for 10 min/4°C.
6. The supernatant can be discarded. Then 100 L of anti-BrdU staining solution (20 L anti-BrdU antibody [Becton Dickinson], 80- L PBS, 0.5% Tween-20, 1% fetal calf serum [FCS]) is added to the nuclei and incubated for 60 min at room temperature.
7. Excess antibody is removed by the addition of 2 mL PBS, and the suspension is centrifuged at 300*g* for 10 min/4°C.
8. Remove the PBS, replace with 100 L of goat antimouse fluorescein isothiocyanate (FITC) antibody staining solution (10 L goat antimouse FITC antibody [Sigma], 90 L PBS, 0.5% Tween-20, 1% FCS), and incubate for 30 min in the dark.

9. Excess antibody should be removed by adding 2 mL PBS and centrifuging at 300*g* for 10 min/4°C.
10. For DNA staining, the cells are incubated in 1 mL staining solution (200 g/mLRNase A [Sigma], 50 g/mL propidium iodide [Sigma] in a FACS flow device [Becton Dickinson]) for 30 min at room temperature in the dark.
11. Fixed and stained cells are then transferred into FACS tubes (Becton Dickinson) and analyzed by flow cytometry.

3.2. Measuring Virus

There are several stages to the virus life cycle, and these include attachment to the host cell, entry, uncoating of the genome, transcription and translation of viral mRNAs, replication of the genome, packaging of new genomes, virus assembly, and release of new virus particles. Several assays can be used to measure these different stages of virus infection; however, a simple analysis measuring replication and total virus production will provide information as to whether a cell cycle stage affects virus infection. One of the simplest ways to measure virus replication is to use Southern (in the case of DNA viruses) and Northern (in the case of RNA viruses) blots. Overall yields of virus can be determined by plaque assay. Using an RNA virus as an example, we detail two protocols that can be used to measure RNA replication (mRNA production) and amount of virus. These can be readily adapted to study other viruses. We routinely work with coronaviruses, which are positive-strand RNA viruses, and principally cause respiratory infection (e.g., severe acute respiratory syndrome.

3.2.1. Methods for Analyzing Virus Production

3.2.1.1. Northern Blot

The following nonisotopic Northern blot protocol has been used in our laboratory for the detection and analysis of coronavirus-derived RNA species. RNA is routinely obtained by extraction of cytoplasmic RNA from coronavirus-infected cells using the Qiagen™Mini-prcp RNA purification kit. For longer RNA species, or for the preparation of RNA that is free from DNA or from tissues and organs, we use the Promega RNAgents® total RNA isolation system coupled with multiple freeze–thaws and homogenization, and routinely include a DNase (RQ1 DNase, Promega) treatment step (*see* **Note 3**). (An example of what purified RNA looks like by nondenaturing agarose electrophoresis is shown in **Fig. 2A**.) In this latter case, all purification steps are conducted at 4°C (i.e., in a cold room). When handling RNA, particular care should be taken to avoid contamination (*see* **Note 4**).

3.2.1.1.1 Sample Preparation and Denaturing Agarose Gel Electrophoresis (see **Note 2***).*

1. Typically 4–6 L (approx 5 g) of each RNA sample are denatured by addition of 3 vol of formaldehyde loading dye, incubation at 65°C for 15 min, and then rapid chilling of samples in a wet-ice bath.
2. Denatured RNA samples are loaded on a 1% formaldehyde denaturing agarose gel, comprised of 90 mL nuclease-free water, 10 mL of 10X formaldehyde denaturing buffer, and 1 g of agarose, within a horizontal submarine gel electrophoresis tank (*see* **Note 5**).

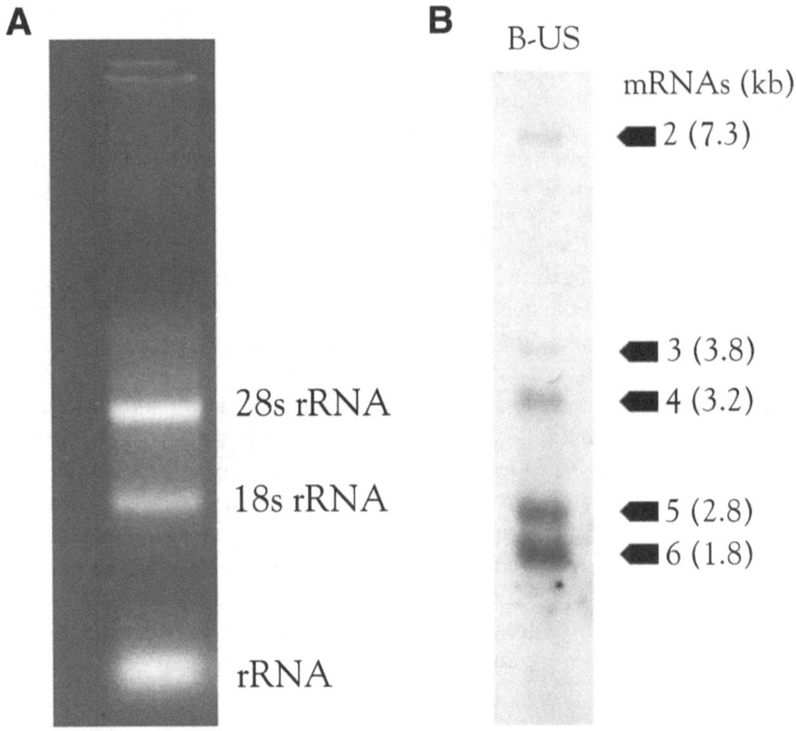

Fig. 2. **(A)** RNA extracted from Vero cells using the Promega RNAgents® total RNA isolation system. This is a nondenaturing 1% agarose gel, and the RNA is visualized by staining with ethidium bromide. In good RNA preparations, the ratio of 28s rRNA to 18s rRNA should be approx 2:1. **(B)** Northern blot analysis of total cytoplasmic RNA extracted from avian coronavirus (IBV B-US)-infected chick kidney (CK) cells. RNA was extracted from IBV B-US-infected CK cells, separated by electrophoresis on a 1% denaturing formaldehyde agarose gel, and transferred to nylon membrane. IBV-derived RNAs were detected by hybridization with an IBV 3' untranslated region genomic probe capable of detecting IBV genomic RNA and IBV mRNAs. The black arrowheads indicate the positions and sizes of the IBV B-US subgenomic mRNAs. The size of the RNA species is indicated in kb.

3. Gels are run in 1X MOPS running buffer overnight (approx 8–16 h) at 30 V (*see* **Note 6**).

3.2.1.1.2. Capillary Transfer of RNA Species to Membrane

1. Excess agarose gel not containing any potential RNA species is trimmed from the gel and nylon membrane, and approx 10 pieces of 3MM blotting paper are cut to the same size as the gel (*see* **Note 7**).
2. The upward capillary transfer method is set up as shown in **Fig. 3**. A 3MM paper wick is cut so that it allows transfer of buffer from the reservoir.
3. The 3MM paper wick is presoaked in 20X SSC buffer, and the nylon membrane and three sheets of 3MM paper are presoaked in 2X SSC prior to capillary transfer assembly.

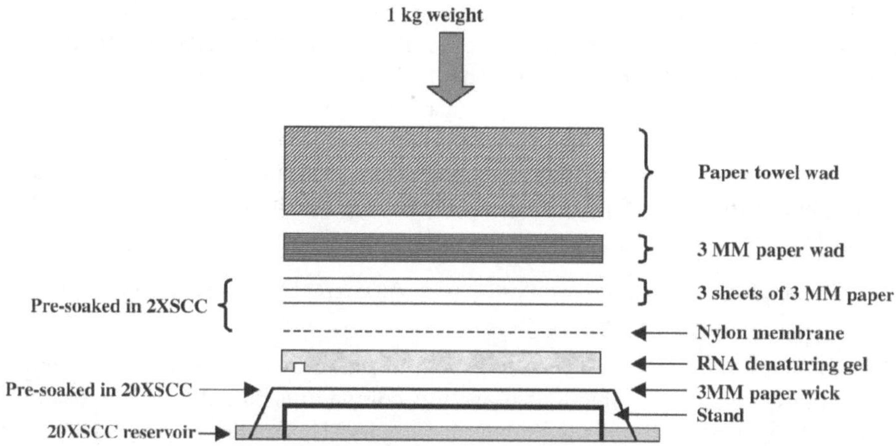

Fig. 3. Diagram showing the arrangement of apparatus for upward capillary transfer for Northern blot. SSC, standard saline citrate.

4. The agarose gel is oriented so that the wells are facing downward to minimize the distance RNA has to migrate through the agarose gel to the nylon membrane.
5. Sufficient dry paper towels must be used draw enough transfer buffer through the gel and membrane to allow efficient transfer to occur. The transfer apparatus is weighed down with a 1-kg weight and left overnight (8–16 h) for transfer to occur.
6. Once transfer is complete, the nylon membrane is washed briefly in 2X SSC buffer, and the RNA is UV-crosslinked (254 nm UV light for 25–50 s) to the nylon membrane (*see* **Note 8**).

3.2.1.1.3. Blot Probing. In recent years nonradioactive detection of nucleic acid species has been as effective as with radioactive methods. The nonradioactive methodologies have a number of advantages, mainly associated with health and safety issues. We routinely use DNA probes nonisotopically labeled using the Ambion BrightStar Psoralen-Biotin kit. The label is composed of a planer, tricyclic psoralen compound covalently attached to biotin. Psoralen intercalates between nucleic acids and covalently binds during irradiation by long-wave UV light to create biotin-labeled DNA.

1. For each labeling reaction 10 L of purified DNA template (0.5 ng/ L–0.5 g/ L) is used, producing five 20 L DNA-labeled probes. Probes are stored at –80°C.
2. Crosslinked membrane is prehybridized prior to probing in 10 mL pre-hybridization/hybridization solution for 30 min at 42°C within a hybridization tube in a roller oven.
3. Following prehybridization, a 20 L prepared probe aliquot is diluted to 100 L with nuclease-free water, incubated at 100°C for 10 min, and then added to the hybridization tube.
4. Membranes are incubated at 37°C overnight (approx 8–16 h). Following hybridization, membranes are washed twice with 20 mL of 2X SSC/1% SDS low-stringency wash solution at 42°C for 5 min and then twice with 20 mL of 0.2X SSC/0.1% SDS high-stringency wash buffer at 42°C for 5 min.

3.2.1.1.4. Probe Detection. Hybridized psoralen-biotin-labeled DNA probe is detected using the Ambion chemiluminescent non-isotopic BrightStar BioDetect kit according to manufacturer's instructions. Hybridized biotinylated DNA probe is bound by a streptavidin/alkaline phosphatase conjugate; the blot is then incubated with detection reagent, resulting in chemiluminescence of labeled probe. Labeled RNA species present on the membrane are then detected following 2–4 h of exposure to photographic film (*see* **Note 9**). An example of a Northern blot of viral RNA is shown in **Fig. 2B**.

3.2.1.2. Quantitative Plaque Assay Infectivity Assay to Determine the Number of Infectious Particles

The following protocol has been successfully used to quantitatively determine the number of infectious avian infectious bronchitis coronavirus particles within a sample by plaque assay using Vero cells within our laboratory. This method can be adapted for calculating the titer of a variety of viruses that induce cytopathic effect in cell culture by adapting the cell type, media, and environmental conditions according to the particular virus. For example, the number of baculovirus infectious particles can be calculated by plaque assay using Grace's insect medium supplemented with FBS on Sf9 cells.

1. An appropriate dilution series of the virus sample is performed, typically in 10-fold dilution steps, for example, addition of 150 L of neat virus supernatant to 1350 L of Vero media to make a 10^{-1} dilution, and so on.
2. Aspirate media from Vero cells grown to confluency in 6-well plate dishes. Duplicate infections per dilution are typically performed with 500 L of innoculum per well of virus.
3. Plates are incubated for 1 h at 37°C within a 5% CO_2 incubator (*see* **Note 10**).
4. During the 1-h incubation, preheat the Vero media to 37°C.
5. Melt the 2% low melting point agarose solution within a microwave or boiling water bath until the agarose is completely in solution, and then equilibrate to 42°C.
6. At 1 h post infection, aspirate media from cells, and wash twice with PBS.
7. Mix an equal volume of equilibrated media and agarose solution and overlay the cells with 2 mL of solution by carefully pipeting the solution down the side of the well (*see* **Note 11**).
8. Allow the agarose overlay to solidify fully, and then incubate the plates for 3 d at 37°C within a 5% CO_2 incubator.
9. At 3 d post infection, prepare media/agarose solution and add 1% neutral red stock solution to a final concentration of 0.01% neutral red.
10. Overlay each well with 2 mL of the media/agarose/neutral red solution, and allow to set fully before incubating the plates for 1 d at 37°C within a 5% CO_2 incubator.
11. Plaques can be visualized, on a light box, as clear zones against a red background. The virus titer can then be calculated at a particular dilution as the number of plaque forming units (PFU)/mL (*see* **Note 12**):

$$\text{Titer (PFU/mL)} = \text{average number of plaques} \times 1/\text{vol of innoculum (mL)} \times 1/\text{dilution}.$$

For example: an average of 25 plaques from 500 l of innoculum per infection at a dilution of 10^{-6}.

$$\text{Titer (PFU/mL)} = 25 \times 1/0.5 \times 1/10^{-6} = \text{Titer of } 5 \times 10^{7} \text{ PFU/mL}$$

3.3. Flow Cytometry and Confocal Analysis of Cell Cycle Factors

Viral effects on the cell cycle can often be attributed to interference with cell cycle factors, such as the cyclins. Here we detail our approaches to investigating the effect of virus on what some might consider nonconventional cell cycle factors, caspase 8 and proliferating cell nuclear antigen (PCNA). Viral effects on these factors (as with any factor) should be studied both at the level of their cellular localization, which in turn can affect function, and also for analysis of expression levels in the cell. To study localization we routinely use confocal microscopy, and to study expression levels, we use flow cytometry. Microscopy has the added advantage that it can be used to check that antibodies recognize proteins of interest in both species-specific and -nonspecific cells. Both monoclonal and polyclonal antibodies to various mammalian proteins can be crossreactive depending on the degree of conservation between the various proteins. For example, a monoclonal antibody to fibrillarin (a nucleolar protein) can be used to detect fibrillarin via immunofluorescence in human, monkey, and avian cells ***(9)***.

3.3.1. Analysis of Caspase 8

Recent work has identified and highlighted the role of caspases in cell proliferation ***(23)***. Caspase 8 has been postulated to activate downstream nuclear caspase 3, which in turn cleave various negative regulators of the cell cycle, such as $p21^{Cip1/Waf1}$ or $p27^{Kip}$ ***(24)***, thereby leading to activation of CDK2. Activation of CDK2 in turn induces its dissociation from cyclin E or cyclin A, thereby inactivating cyclin E function by degrading cyclin E by the proteasome pathway and subsequent G_1 arrest; a number of viruses target this protein to usurp its functions.

3.3.1.1. Indirect Immunofluorescence of Caspase 8

Cells are grown on cover slips, normally in 6-well plates.

1. Vero cells are either treated or not treated with 100 m*M* etoposide (Sigma; as a positive control) and left for 48 h. Cells are analyzed for the distribution of caspase 8, following a protocol published in **ref. *25***.
2. Cells are fixed for 20 min at –20°C with 1 mL methanol/acetone (1:1) following removal of growth medium and washing twice with 2 mL PBS.
3. Cover slips are rinsed twice with PBS and permeabilized with 1 mL 0.1% Triton X-100 (Sigma) in PBS for 30 min at room temperature (RT).
4. The solution is then removed and the cover slips rinsed twice with PBS followed by blocking for 1 h at RT with 4% BSA in PBS.
5. The blocking solution is removed, and rabbit polyclonal anticaspase 8 P-20 (cat. no. H-134, Santa Cruz Biotechnology) diluted 1:100 in blocking solution (4% BSA in PBS) is incubated for 4 h at RT.
6. Excess antibody should be removed and the cover slips rinsed twice with PBS before incubating in PBS for 30 min at RT.
7. Secondary FITC-conjugated antirabbit antibody (Sigma) diluted (1:100) in PBS is then incubated for 1 h at 37°C.
8. Excess antibody is removed by washing the cover slips in 2 mL PBS three times for 10 min each prior to mounting the cover slips on microscope slides using mounting media containing DAPI (Vector).

9. Mounted coverslips can be analyzed using an Axiovision system (Carl Zeiss Jena). Pictures are captured with an Axiocam camera and processed using the Axiovision 3.0 software provided (Carl Zeiss Jena).

Figure 4 provides an example of the distribution of caspase 8 in wild-type cells, cells expressing the avian coronavirus nucleoprotein (which we know has cell cycle effects and redistributes nucleolar proteins), and control cells. As can be seen from the figure, caspase 8 is redistributed to the nucleolus in cells expressing a viral protein (IBVNHis) from an expression plasmid.

3.3.1.2. Flow Cytometry of Caspase 8

1. For flow cytometry of caspase 8 expression, Vero cells either treated with 100 m*M* Etoposide for 48 h for transfected for 24 h (either with pTriEx 1.1 [Novagen] or pTriEx IBVNHis) are harvested by detaching the cells with 2 mL PBS/EDTA/trypsin and transferring them into canonical tubes containing 8 mL 10% DMEM.
2. The supernatant containing detached cells is transferred to canonical centrifuge tubes instead of being discarded.
3. In both cases cells are spun at 250g for 10 min at 4°C, and the cell pellets resuspended in 2 mL ice-cold PBS.
4. Prior centrifugation, both populations are pooled, and the pooled cells are then pelleted by centrifugation at 250*g* for 10 min at 4°C.
5. The supernatant is removed, and the cells are resuspended in 70% precooled (–20°C) methanol and incubated for at least 12 h at –20°C.
6. To detect intracellular caspase 8, cells are centrifuged at 250*g* for 10 min/4°C to remove methanol and washed once with 2 mL ice-cold PBS.
7. Cells are then resuspended in 875 L of ice-cold PBS and fixed by the addition of 175 L of ice-cold PBS with 2% paraformaldehyde, pH 7.4, for 1 h at 4°C.
8. Fixative can be removed by centrifugation for 5 min (250*g*, 4°C) and the supernatant aspirated.
9. Cells are permeabilized by resuspending them in 1 mL of PBS with 0.05% Tween-20 for 15 min at 37°C; then they are washed with 1 mL of PBS and centrifuged for 8 min at 250*g* at 4°C.
10. The presence of caspase 8 can be detected by using rabbit polyclonal antibody against caspase 8 P-20 (1:100; Santa Cruz Biotechnology) and goat antirabbit FITC-conjugated antibody (1:100; Sigma).

3.3.2. Analysis of Proliferating Cell Nuclear Antigen

Proliferating cell nuclear antigen (PCNA) expression is associated with S-phase, and its localization is restricted to sites of DNA replication, as shown by immunofluorescence analysis. During DNA replication, PCNA functions as an auxiliary protein for DNA polymerase α, and its presence is necessary for synthesis of the leading strand, although the precise function has not been clarified.

3.3.2.1. Indirect Immunofluorescence Analysis of PCNA

1. Cells can be grown on glass cover slips and are washed twice with PBS before fixation with 100% precooled (–20°C) methanol at –20°C for 5 min. (In this example we are using Vero and HeLa cells.)

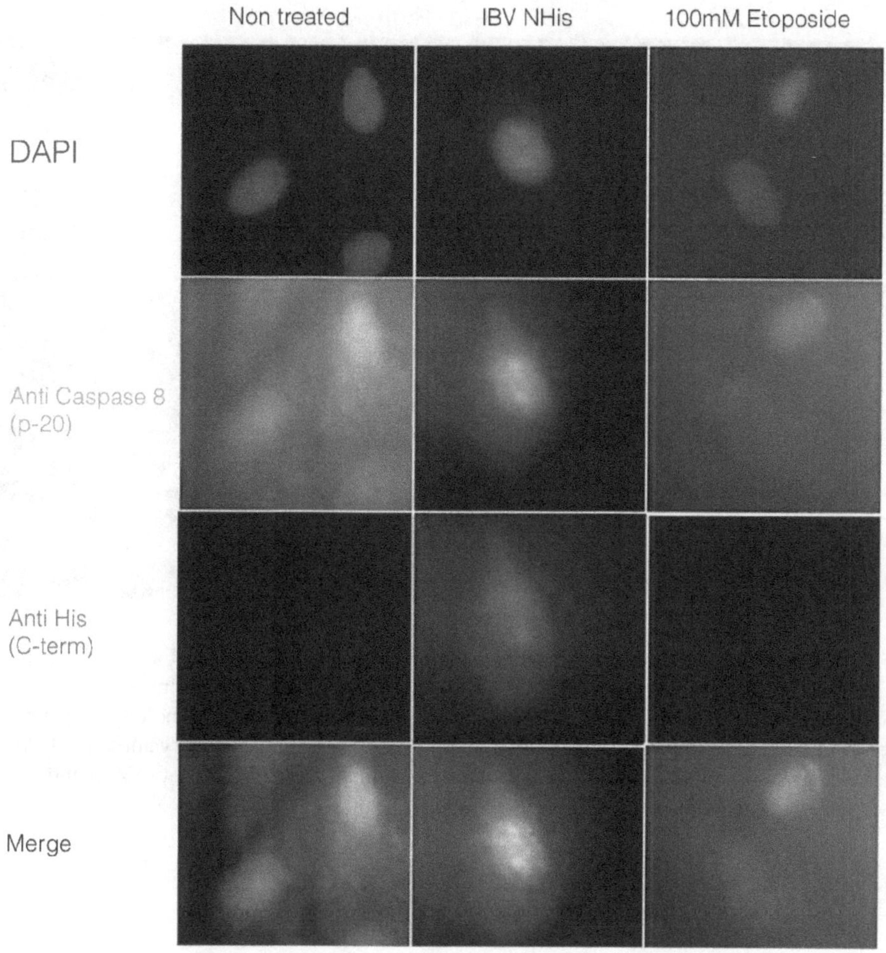

Fig. 4. Localization of caspase 8 in transfected or etoposide-treated Vero cells or cells expressing the avian coronavirus nucleoprotein. Vero cells were transfected with pTriExIBVNHis (a construct that expresses the nucleoprotein under the control of a PolII promoter), treated with 100 m*M* etoposide, or left untreated and analyzed by indirect immunofluorescence for the localization of caspse 8 using a polyclonal anticaspase 8 (P 20) antibody (Santa Cruz Biotechnology). Transfected cells were analyzed 24 h post transfection, whereas etoposide-treated cells were analyzed 48 h post treatment. In nontreated cells (left row) caspase 8 can be detected in both the nucleus and the cytoplasm, in cells transfected with IBVN His (middle row), caspase 8 is localised in the cyto-and nucleoplasm with a prominent signal in the perinuclear region and the nucleolus. In Etoposide treated cells (right row) caspse 8 is almost exclusively localised in the nucleus and the perinuclear region. Primary caspase 8 antibody was detected with FITC-conjugated goat antirabbit antibody (Sigma), and IBV NHis was detected with mouse anti-His (C-term; Sigma) antibody as primary antibodies and Texas red-conjugated goat anti mouse (Harlan Sera Lab) as secondary antibody.

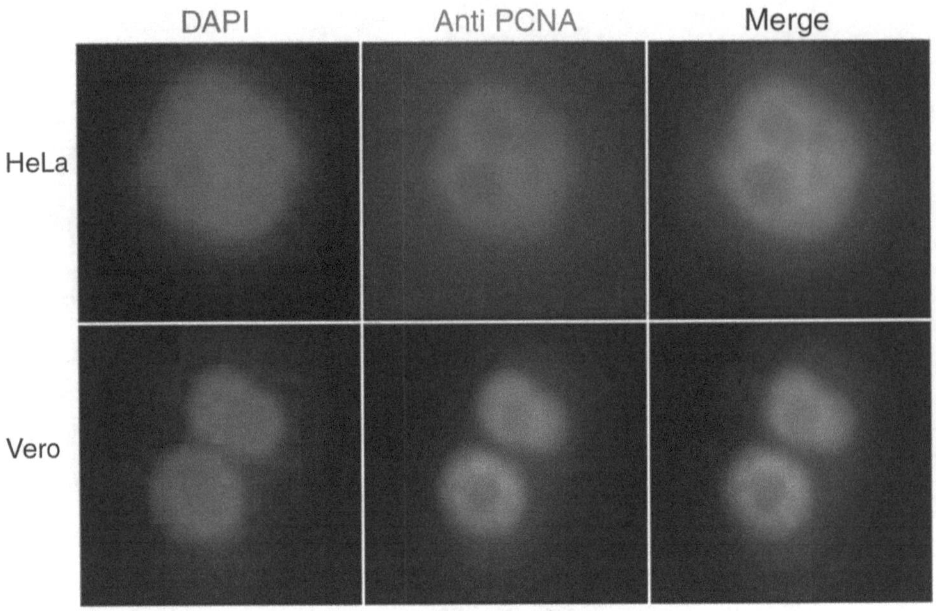

Fig. 5. Localization of proliferating cell nuclear antigen (PCNA) in HeLa and Vero cells. HeLa and Vero cells were grown in 10% DMEM medium on glass cover slips and labeled for PCNA using a mouse monoclonal anti-PCNA antibody (Santa Cruz Biotechnology) as primary antibody and FITC-conjugated goat anti PCNA antibody (Sigma) as secondary antibody. DAPI-containing mounting medium (Vector) was used to counterstain DNA. In both HeLa and Vero cells, PCNA was detected exclusively in the nucleoplasm..

2. Cover slips should be air-dried and washed once with 2 mL PBS prior to addition of 3.5% paraformaldehyde in PBS for 30 min at 4℃.
3. To remove the fix, cells are permeabilized with 0.1% Triton X-100 (Sigma) in PBS for 2 min at RT.
4. After extensively washing four times, each time for 10 min, with PBS, the cover slips should be covered with 500 L of mouse monoclonal anti-PCNA (PC 10) antibody (1:100 in PBS) and incubated for 1 h at 37℃ in a humidified atmosphere.
5. Cover slips are then washed three times in 2 mL PBS and stained with 500 L FITC-conjugated secondary goat antimouse (Sigma) antibody (1:100 in PBS).
6. After 1 h of incubation at 37℃, the cover slips are washed three times with 2 mL PBS and mounted using mounting medium containing DAPI (Vectashield, Vector). (Proteins can be visualized as in **Subheading 3.3.1.1., step 9**).

As can be seen in **Fig. 5** antibody to human PCNA can detect this protein in both human (HeLa) and Vero (a monkey cell line) cells. Thus the antibodies can be used to detect proteins in nonspecies-specific cell lines.

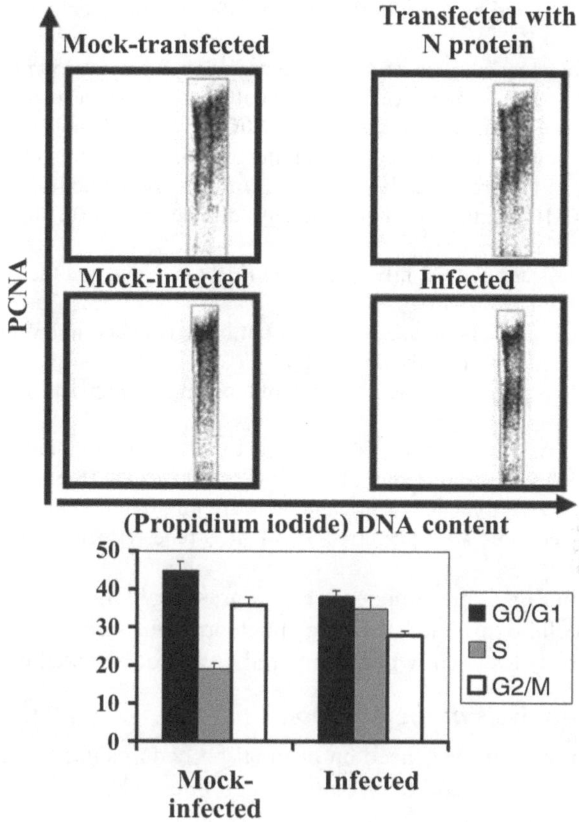

Fig. 6. Analysis of PCNA expression in Vero cells expressing the avian coronavirus nucleocapsid proteins or cells infected with the virus IBV B-US. The cell population was gated in two subpopulations: PCNA-negative (R1) and -positive (R2) cells using CellQuest software. Shown are representative dot plots or mock-transfected cells expressing N protein, control, and IBV Beaudette-infected cell populations. Averages of three experiments for infected cells are shown in the accompanying charts.

3.3.2.2. Flow Cytometry Analysis PCNA

As can be seen in **Fig. 6** the number of cells expressing PCNA in avian coronavirus-infected cells or cells expressing the viral nucleoprotein (when expressed from an expression plasmid) is less than when compared to mock treated cells.

To detect the number of cells expressing PCNA, flow cytometry can be used. Cells were harvested as described above prior to flow cytometry analysis. The protocol used followed essentially a protocol published by **ref.** ***26*** with slight modifications. In our case we have used Vero cells as an example, as these support avian coronavirus infection.

1. Pelleted cells are fixed by the addition of 1 mL 1% paraformaldehyde in PBS and incubated for 2 min at RT.
2. The fix is removed by addition of 2 mL PBS and centrifuging at 250*g* for 10 min at 4℃. To postfix and permeabilize the cells, the cell pellet was then resuspended in 200 L PBS followed by the addition of 2 mL precooled (–20℃) 100% methanol and incubation for 5 min at –20℃. The fix was removed by spinning at 250 *g* for 10 min at 4℃.
3. The cell pellet is then resuspended in 100 L of primary antibody (mouse monoclonal anti-PCNA (PC-10, Santa Cruz Biotechnology) diluted 1:100 PBS and incubated for 1 h at RT.
4. To remove excess antibody, 2 mL of ice-cold PBS are added and the suspension centrifuged at 250*g* for 10 min/4℃.
5. FITC-conjugated goat antimouse secondary antibody is added in a dilution of 1:100 in PBS and incubated for 1 h in the dark at RT.
6. Excess antibody is removed by adding 2 mL of blocking solution and spinning at 250*g* for 10 min/4℃.
7. To stain the DNA, resuspend the cell pellet in 1 mL propidium iodide staining solution (50 g/mL propidium iodide [Sigma] and 200 g/mL RNase [Sigma] in PBS), transfer into FACS tubes (Becton Dickinson), and incubate for 30 min in the dark at RT.
8. Fixed and stained cells are transferred into FACS tubes (Becton Dickinson) and analyzed.

As can be seen in **Fig. 5** the number of cells expressing PCNA in avian coronavirus-infected cells or cells expressing the viral nucleoprotein (when expressed from an expression plasmid) is less than when compared with mock treated cells.

3.4. Proteomic Analysis of the Nucleolus

The following protocols are based on original procedures used by Anderson et al. ***(27)*** for the isolation of nucleoli from cultured cell lines. The methods presented here have been successfully used in our laboratory for isolation of nucleoli for subsequent proteomic analysis by 2D SDS-PAGE. This latter method can be used, in conjunction with mass spectrometry, N-terminal protein sequencing, or Western blotting, to identify proteins isolated from the nucleoli. As discussed in the Introduction, the nucleolus is targeted by many different types of virus ***(13,14)***, and such interactions may cause perturbations to the distribution of cell cycle factors (e.g., **Fig. 4**) and the cell cycle.

3.4.1. Preparation of Tissue Homogenates and Nucleolar Fractions

1. Prior to harvesting cultured cells for the isolation of nucleoli, rinse each flask three times with prewarmed PBS.
2. On the final rinse step, depending on the cell line, trypsinize off the cells by adding 2–5 mL of trypsin/EDTA solution (Invitrogen, UK) per flask, scrape off with a cell disperser, or bang the flask with a sharp vigorous blow to detach cells. In the former case, swirl the flask and return to the incubator for 3–5 min until most cells have detached.
3. Check that cells are detached using phase contrast microscopy. If cells remain attached to the plasticware, either bang the flask again or increase the trypsin/EDTA incubation period.
4. Once most of the cells are detached, add a volume of culture media at incubator temperature and pipet up and down so that a single-cell suspension is generated. This suspension can now be placed into Falcon tubes for further steps.

5. To wash the cell suspension, spin samples using a swingout rotor at 220*g* relative centrifugal force (RCF; e.g., rotor A-4-62 Eppendorf, 220*g* RCF, 1046 rpm), remove supernatant so that a cell pellet remains, and add a volume of 4℃ PBS. Carry out this process a further two times before resuspending cells in HEPES (*see* **Subheading 2.3., item 4** and *see* **Note 13**).
6. Transfer the cell suspension to a precooled tissue homogenizer and homogenize 10 times using a tight-fitting pestle (0.0010–0.0030-inch clearance), while keeping the homogenizer on ice. The number of strokes needed depends on the cell type used, so it is necessary to examine the homogenized cells with a phase contrast microscope after every 10 strokes. Stop when >90% of the cells have burst, leaving intact nuclei.
7. To obtain a pellet containing enriched nuclei, centrifuge the homogenized cells again at 220*g* for 5 min at 4℃.
8. Resuspend the pellet with a volume S1 solution by pipeting up and down. In another tube containing a volume of S2, layer onto the top the resuspended pellet; ensure that the two layers remain cleanly separated.
9. Centrifuge this cushion at 1430*g* for 5 min at 4℃ to obtain a purer nuclear pellet. Following this spin, discard the supernatant, and resuspend the pellet in a volume of S2 solution by pipeting up and down.
10. Appropriate sonication on ice is a crucial stage in the preparation of nucleoli (*see* **Note 14**).
11. To prepare a nucleolar concentrated pellet, layer the sonicated material over a volume of S3 solution and centrifuge at 3000 g for 10 min at 4℃.
12. Further purification of nucleoli can be carried out by re-suspending the pellet with S2 solution, and centrifuging at 1430*g* for 5 min at 4℃. The resulting pellet contains highly purified nucleoli, which can be examined by microscopy and, if required, stored at –80℃.

3.4.2. 2D SDS-PAGE

The use of 2D PAGE has come into widespread usage since the publication of methods combining isoelectric focusing (IEF) in the first dimension and SDS-PAGE in the second dimension. Three separate papers by OFarrell and others have demonstrated that it was possible to combine IEF with gradient SDS-PAGE gels to separate and reveal proteins better in a gel, thus improving the resolution of 1D SDS-PAGE. Two-dimensional SDS-PAGE is particularly useful for the separation of extremely complex protein mixtures.

3.4.2.1. Sample Preparation for 2D SDS-PAGE

The following protocols are based on equipment and materials available from Bio-Rad.

1. Make up prepared nucleoli to 2.5–3 mg total protein in sample buffer.
2. Remove the desired number of pH 4.0–7.0 ReadyStrip IPG strips (Bio-Rad, UK) from the –20℃ freezer, and set them aside to defrost. It is good practice to run two sets of strips per sample, one to be stained following the IEF phase, and the other to be used for the 2D and subsequent analysis.
3. Using a suitable tray, place sufficient sample volume into each well so that each IPG strip is in contact with the solution through its entire length. Lay the IPG strip gel side down into the sample buffer. Ensure that all samples are evenly in contact with the strip since it

will not be absorbed through the plastic backing. It is recommended that for strips of 11 cm, 185 L of sample be used per strip (approx 250 g per strip). When preparing samples, do not place the vials on ice, as the urea will crystallize out of the solution.

4. Leave these strips for 1 hr, then overlay each of the strips with 2–3 mL of mineral oil to prevent evaporation during the strip rehydration process.
5. For rehydration cover the tray with a lid and leave the tray sitting on a level bench overnight (11–16 h) to rehydrate the IPG strips thoroughly with the nucleolar sample; rehydration is crucial to successful 2D.

3.4.2.2. Isoelectric Focusing.

1. In the IEF tray (Bio-Rad, UK; see manufacturer's instructions), place a paper wick at both ends of the channels covering the wire electrodes. Pipet 8 L of pure water onto each wick to wet them.
2. Following strip rehydration, remove strips from the incubation tray using forceps, carefully holding the strips at the end where there is no gel, and hold the strip vertically for 7–8 s until the mineral oil has drained. Each strip can then be transferred to the corresponding channel in the focusing tray, gel side down, with the positive end of the strip adjacent to the positive electrode of the tray, in contact with the wetted electrodes.
3. Each of the strips should again be covered with 2–3 mL of fresh mineral oil.
4. The focusing tray can now be placed into the protean IEF cell (Bio-Rad, UK) with the positive side of the tray corresponding to the positive electrode of the cell.
5. Once the cell cover has been closed, the IEF cell can be programmed (*see* Bio-Rad Protean IEF cell instructions) for a single- or multiple-step focusing protocol. In our laboratory a three-step protocol has proved satisfactory for IEF of nucleolar proteins. Our program follows the basic format of:

   Step 1 linear voltage ramp, 250 V for 20 min.
   Step 2 linear voltage ramp, 8000 V for 2.5 h.
   Step 3 rapid voltage ramp, 20,000 VHrs.

6. When setting up the program, it is satisfactory to use the default IEF cell temperature of 20°C, with a maximum current of 50 A per strip. This three-step program takes approx 6 h. Once the program has finished, it is important either to place the strips under a 500-V holding voltage or remove, cover with tin foil, and freeze at –80°C or stain for protein.

3.4.2.2.1. Staining IPG Strips With IEF Stain or Coomassie Stain

1. For determining whether proteins have isoelectrically focused correctly, it is possible to transfer IPG strips to a clean, dry piece of blotting filter paper with the gel side up, thoroughly wet a second filter paper of the same size with pure water, and carefully lay the wet filter paper onto the IPG strips. Then 'peel' back the top filter paper. This blotting step removes mineral oil on the surface of the IPG.
2. The IPG strips can then be stained for protein using 0.1% (w/v) Coomassie brilliant blue solution (1 h) and then destained with destain buffer until proteins are revealed (1–3 h; *see* **Subheading 2.4.**). Alternatively, strips can be stained with Bio-Rad's IEF stain (Bio-Rad, UK).

3.4.2.3. Separating Samples in the Second Dimension

1. If strips were frozen following the first dimension separation, then remove from the –80°C freezer and allow to defrost thoroughly. It is best to not leave the thawed IPG strips for longer than 15–20 min, as diffusion of the proteins can result in reduced sharpness of the protein spots.

2. Take either the thawed or the freshly prepared strips and place each in an incubation tray with 4 mL of equilibration buffer I for 10 min on a slowly rotating orbital shaker. At the end of the 10-min incubation, discard the used equilibration buffer I in its entirety by carefully decanting the liquid from the tray. To each strip then add 4 mL of equilibration buffer II, and place on an orbital shaker for a further 10 min.
3. During this incubation either prepare SDS-PAGE gels and ensure that the stacking gel is of sufficient size to take the IPG strip, or obtain a suitable number of precast polyacrylamide gels for your samples (*see* **Subheading 2.4.**).
4. Once the combs have been removed from the gels, rinse each well briefly with nanopure water using a water bottle. Working quickly, prepare sufficient 1X Tris-HCl/glycine/SDS running buffer to run the number of gels you have decided upon.
5. Once the strips have finished incubating and the gels are prepared, melt sufficient overlay agarose in a microwave to cover the IPG strips once they are inserted into the wells.

3.4.2.4. Running the IPG Strips in the Second Dimension

1. Ensure that the IPG wells of the gels are free of any liquid by blotting with Whatman 3MM paper.
2. Remove the incubated strip from the incubation tray, and dip each IPG strip into a tube of suitable length to take the entire length containing 1X Tris-HCl/glycine/SDS running buffer.
3. Carefully lay each strip gel side up and onto the back plate of the SDS-PAGE gel above the IPG well, and pipet into the well the liquid overlay agarose. Once the well is full, gently move the IPG strip down until it is in contact with the top of the SDS-PAGE gel (avoid trapping any air bubbles).
4. Allow the agarose to solidify for 5 min before proceeding.
5. Once the agarose has solidified, mount the gel, fill the reservoirs with running buffer, and begin the electrophoresis, run at the appropriate voltage for the gel size used (*see* manufacturer's instructions; or 150 V for a 14-cm 12% Tris-HCl SDS-PAGE gel).

3.4.2.5. Revealing Proteins in 2D Gel

1. Once the sample front has reach the bottom of the gel, you can proceed to reveal the proteins in the nucleolar sample by using commercially available stains such as Coomassie blue (*see* **Subheading 2.**) or silver stain plus (Bio-Rad, UK; carry out staining as recommended by the manufacturer) (**Fig.** 7). If you wish to examine individual, known proteins, the gel can simply be blotted onto nitrocellulose or polyvinyl idene difluoride membranes for probing with specific antibodies. For further information on this latter technique of western blotting, *see* **ref. *33***.

4. Notes

1. This time will vary depending on the cell type and passage number.
2. Fixed samples can be kept at 4℃ until used.
3. DNase should be removed prior to any cloning by reverse transcriptase (RT)-PCR. A phenol/chloroform extraction followed by ethanol precipitation is suitable for this.
4. To avoid RNase contamination, at all stages it is highly advantageous to use either nuclease-free or diethyl pyrocarbonae (DEPC)-treated water to make up all solutions. All work areas and equipment should be cleaned with an RNase inhibitor (such as Ambion RNaseZap), and gloves should be worn at all times. It is also advisable to use preracked

Fig. 7. Silver-stained 2D gel of purified nucleoli focused in the first dimension using a 3-10 strip (Bio-Rad) and then subsequently subjected to electrophoresis on a 10–20% SDS-PAGE gel.

RNase-free tips and prepacked RNase-free tubes. When conducting RNA work, ensure that all liquid dispensers (i.e., Gilson pipets or equivalents) are wiped with 75% ethanol, and wear suitable gloves. Be sure not to touch any part of exposed skin (i.e., your face) with the gloves; this is a bad habit that many laboratory workers have. Many people secrete DNase and RNases.

5. It is advisable to use as thin a gel as possible; the thinner the gel, the faster and more efficient the transfer of RNA to membrane.
6. Running agarose gels for Northern blot analysis at high voltages can result in gel warping owing to heat. It is advisable to run the gel overnight at low voltages to minimize temperatures, and it is advantageous to use a buffer recirculation pump.
7. Touch the nylon membrane as little as possible to prevent nuclease contamination.
8. Once they are crosslinked, membranes can be stored at –20°C for several months.
9. Unlike isotopic ^{32}P-labeled probes, placing the film cassette containing the membrane exposed to film at –80°C will not increase the signal of weak RNA-labeled species.
10. Every 15 min, gently agitate the plates to ensure that the Vero cells are fully overlaid with innoculum to prevent cell desiccation.
11. Ensure that the agarose is fully equilibrated to 42°C before addition to the media. Addition of agarose solution at temperatures higher than 42°C can damage the cells.
12. Virus titer can vary depending on the cell type used.
13. It is good practice to ensure that cells have not lysed but have become swollen in the buffer conditions by examining with an inverted microscope. Extra care should be taken when working with mammalian cells, as these are particularly prone to lysis at 37°C in hypotonic conditions; therefore preparation on ice is imperative.
14. With most cell preparations, sonication of the nuclear suspension for six 10-s bursts (with 10-s intervals between each burst) has proved sufficient in our hands. In our laboratory we use a Misonix XL 2020 sonicator fitted with a microtip probe set at a power setting of 5. The optimal sonication conditions do, however, depend on the cell types used; oversonication leads to destruction of nucleoli, whereas undersonication leaves the sub-

cellular component intact. For best results, examine the suspension by microscopy after each round of sonication; there should be virtually no intact cells, and the nucleoli should be seen as dense, refractive bodies.

References

1. Flemington, E. K. (2001) Herpesvirus lytic replication and the cell cycle: arresting new developments. *J. Virol.* **75,** 4475–4481.
2. Moran, E. (1993) Interactions of adenoviral proteins with pRB and p53. *FASEB J.* **7,** 880–885.
3. Re, F., Braaten, D., Franke, E. K., and Luban, J. (1995) Human immunodeficiency virus type 1 Vpr arrests the cell cycle in G_2 by inhibiting the activation of $p34^{cdc2}$-cyclin B. *J. Virol.* **69,** 6859–6864.
4. Poon, B., Grovit-Ferbans, K., Stewart, S. A., and Chen, I. S. Y. (1998) Cell cycle arrest by Vpr in HIV-1 virions and insensitivity to antiretroviral agents. *Science* **281,** 266–269.
5. Henklein, P., Bruns, K., Sherman, M. P., et al. (2000) Functional and structural characterization of synthetic HIV-1 Vpr that transduces cells, localizes to the nucleus and induces G_2 cell cycle arrest. *J. Biol. Chem.* **275,** 32016–32026.
6. Cannavo, G., M. Paiardini, D., Galati, B., et al. (2001) Abnormal intracellular kinetics of cell-cycle-dependent proteins in lymphocytes from patients infected with human immunodeficiency virus: a novel biologic link between immune activation, accelerated T-cell turnover, and high levels of apoptosis. *Blood* **97,** 1756–1764.
7. Naniche, D., Reed, S. I., and Oldstone, M. B. A. (1999) Cell cycle arrest during measles virus infection: a G_0-like block leads to suppression of retinoblastoma protein expression. *J. Virol.* **73,** 1894–1901.
8. Lin, G. Y. and Lamb, R. A. (2000) The paramyxovirus simian virus 5 V protein slows progression of the cell cycle. *J. Virol.* **74,** 9152–9166.
9. Chen, H., Wurm, T., Britton, P., Brooks, G., and Hiscox, J. A. (2002) Interaction of the coronavirus nucleoprotein with nucleolar antigens and the host cell. *J. Virol.* **76,** 5233–5250.
10. Wurm, T., Chen, H., Britton, P., Brooks, G., and Hiscox, J. A. (2001) Localisation to the nucleolus is a common feature of coronavirus nucleoproteins and the protein may disrupt host cell division. *J. Virol.* **75,** 9345–9356.
11. Feuer, R., Mena, I., Pagarigan, R., Slifka, M. K., and Whitton, J. L. (2002) Cell cycle status affects coxsackievirus replication, persistence, and reactivation in vitro. *J. Virol.* **76,** 4430–4440.
12. Paiardini, M., Galati, D., Cervasi, B., et al. (2001) Exogenous interleukin-2 administration corrects the cell cycle perturbation of lymphocytes from human immunodeficiency virus-infected individuals. *J. Virol.* **75,** 10843–10855.
13. Hiscox, J. A. (2002) Brief review: the nucleolusa gateway to viral infection? *Arch. Virol.* **147,** 1077–1089.
14. Hiscox, J. A. (2003) The interaction of animal cytoplasmic RNA viruses with the nucleus to facilitate replication. *Virus Res.* **95,** 13–22.
15. Carmo-Fonseca, M., Mendes-Soares, L., and Campos, I. (2000) To be or not to be in the nucleolus. *Nat. Cell Biol.* **2,** E107–E112.
16. Liu, J.-L., Hebert, M. B., Ye, Y., Templeton, D. J., King, H.-J., and Matera, A. G. (2000) Cell cycle-dependent localization of the CDK2-cyclin E complex in Cajal (coiled) bodies. *J. Cell Sci.* **113,** 1543–1552.
17. Sirri, V., Roussel, P., Gendron, M. C., and Hernandez-Verdun, D. (1997) Amount of the two major Ag-NOR proteins, nucleolin and protein B23 is cell-cycle dependent. *Cytometry* **28,** 147–156.

18. Azum-Gelade, M.-C., Noaillac-Depeyre, J., Caizergues-Ferrer, M., and Gas, N. (1994) Cell cycle redistribution of U3 snRNA and fibrillarin. *J. Cell Sci.* **107,** 463–475.
19. Fomproix, N., Gebrane-Younes, J., and Hernandez-Verdun, D. (1998) Effects of anti-fibrillarin antibodies on building of functional nucleoli at the end of mitosis. *J. Cell Sci.* **111,** 359–372.
20. Srivastava, M. and Pollard. H. B. (1999) Molecular dissection of nucleolinś role in growth and cell proliferation: new insights. *FASEB J.* **13,** 1911–1922.
21. Olson, M. O., Dundr, M., and Szebeni, A. (2000) The nucleolus: an old factory with unexpected capabilities. *Trends Cell Biol.* **10,** 189–196.
22. Olson, M. O., Hingorani, K., and Szebeni, A. (2002) Conventional and nonconventional roles of the nucleolus. *Int. Rev. Cytol.* **219,** 199–266.
23. Algeciras-Schimnich, A., Barnhart, B. C., and Peter, M. E. (2002) Apoptosis-independent functions of killer caspases. *Curr. Opin. Cell Biol.* **14,** 721–726.
24. Levkau, B., Koyama, H., Raines, E. W., et al. (1998) Cleavage of p21Cip1/Waf1 and p27Kip1 mediates apoptosis in endothelial cells through activation of Cdk2: role of a caspase cascade. *Mol. Cell.* **1,** 553–563.
25. MacFarlane, M., Merrison, W., Dinsdale, D., and Cohen, G. M. (2000) Active caspases and cleaved cytokeratins are sequestered into cytoplasmic inclusions in TRAIL-induced apoptosis. *J. Cell Biol.* **148,** 1239–1254.
26. Kurki, P. Ogata, K., and Tan., E. M. (1988) Monoclonal antibodies to proliferating cell nuclear antigen (PCNA)/cyclin as probes for proliferating cells by immunofluorescence microscopy and flow cytometry. *J. Immunol. Methods* **109,** 49–59.
27. Andersen, J. S., Lyon, C. E., Fox, A. H., et al. (2002) Directed proteomic analysis of the human nucleolus. *Curr. Biol.* **12,** 1–11.
33. Gershoni, J. M. and Palade, G. E. (1983) Protein blotting: principles and applications. *Anal. Biochem.* **131,** 1–15.

12

Methods for Preparation of Proteins and Protein Complexes That Regulate the Eukaryotic Cell Cycle for Structural Studies

Julie Welburn and Jane Endicott

Summary

The determination of structures for proteins that control the eukaryotic cell cycle by nuclear magnetic resonance (NMR) spectroscopy and X-ray crystallography has made a significant contribution to our understanding of the molecular mechanisms that control cell cycle progression. CDK2 has proved particularly tractable to structural analysis, and CDK2 in complex with various regulatory proteins and in different phosphorylation states provides a paradigm for the control of this important kinase family. This chapter describes a number of protocols that can be used to prepare CDKs and selected CDK binding proteins suitable for structural studies by heterologous expression in either *E. coli* or insect cells.

Key Words

CDK; cyclin; cell cycle; coexpression; polo-like kinase-1; phosphoprotein; X-ray crystallography; NMR.

1. Introduction

Structural studies on proteins that control passage through the eukaryotic cell cycle have provided insights into the molecular biochemistry that underlies its fine and complex regulation. The determination by X-ray crystallography of a number of cyclin-dependent kinase (CDK) structures in alternative phosphorylation states and bound to activating and/or inhibitory partners has made a major contribution to our current understanding of how CDKs are regulated (*1–7*; reviewed in **refs. *8*** and *9*). A number of potent and selective CDK inhibitors have been developed using a rational drug design approach starting from structures for CDK2 and CDK2/cyclin A bound to ATP-competitive small molecule inhibitors. These compounds have provided an opportunity to evaluate cell cycle regulators as potential targets for pharmaceutical

From: *Methods in Molecular Biology, vol. 296, Cell Cycle Control: Mechanisms and Protocols*
Edited by: T. Humphrey and G. Brooks © Humana Press Inc., Totowa, NJ

intervention for the treatment of cancer and other diseases ***(10)***. Highlights of recent X-ray crystallographic and electron microscopy (EM) studies within the cell cycle field are the determination of structures for the E3 ubiquitin ligases Skp1/cullin/F-box protein^{Skp2} (SCF^{Skp2}) ***(11)*** and the anaphase-promoting complex/cyclosome (APC/C) ***(12)***. These E3s are large macromolecular complexes that, together with an ubiquitin-conjugating enzyme (E2), polyubiquitinate selected cell cycle regulatory proteinsa—move that ensures irreversible cell cycle progression as a result of the proteinŝ subsequent degradation by the proteasome.

Cryo-electron microscopy (cryo-EM), X-ray crystallography, and nuclear magnetic resonance (NMR) are powerful techniques for protein structure determination. A challenge for structural studies in the cell cycle field is to offer explanations for the multiple alternative complexes that many cell cycle regulators form in response to the cellular environment. A second challenge is to elaborate the biology of macromolecular assemblies of ever increasing size. To overcome both these challenges, a combination of techniques will be required. For example, although NMR methods focus on determining structures for small proteins (less than ~40 kDa), the technique is increasingly being used to study the structures of larger complexes and to probe the dynamics of protein–protein interactions (reviewed in **refs.** ***13*** and ***14***). Docking of the structures of individual complex components determined by X-ray crystallography into low-resolution cryo-EM maps can provide an opportunity to elaborate the architecture of large macromolecular assemblies at high resolution ***(15)***. For example, within the cell cycle field, cryo-EM has provided a snapshot at subatomic resolution of the APC/C ***(12)***, and the first APC/C subunit structure determined by X-ray crystallography has been reported in **refs.** ***16*** and ***17***.

Structural studies require soluble proteins that can be readily purified in milligram amounts. The protein sample also needs to be homogeneous (although the requirements for NMR are not as stringent as for X-ray crystallography and cryo-EM). Macromolecules subjected to an X-ray beam and equally to magnetic fields give a weak signal, and so the signals contributed by many molecules in the crystal or in solution have to be recorded and averaged. Image reconstruction of a sample by cryo-EM requires approx 10,000 particles. Usually, protein concentrated to approx 10 mg/mL or more is a starting point for crystallization trials. Automation of crystallization trials allows many more conditions to be screened with less material, the volume of a typical drop in a manual screen being 500 nL, compared with 50–200 nL for automated screens. For NMR experiments, the protein ideally should be concentrated to 500 *M* (i.e., ~5 mg of a 20-kDa protein for 500- L sample volume). For cryo-EM, the sample should be around approx 1 mg/mL. The major limitation of this technique is the size of the sample (>250 kDa).

These requirements have ensured that the starting material for most structural studies is protein produced by heterologous expression in *E. coli*, yeast, insect, or mammalian cell lines. Expression in *E. coli* cells is the method of choice for its cost and capacity, but for studying proteins that regulate the cell cycle it has several disadvantages. Many cell cycle regulatory proteins are reversibly phosphorylated at multiple sites, and their state can dramatically affect the proteinŝ structure and biology. To be able to elaborate the role of phosphorylation in protein function, a homogenously phosphorylated sample has to be produced. As one solution, coexpression systems for producing the protein and its regulatory kinase can be considered, and as an example the use of such an approach to purify phosphorylated CDK2 from *E. coli* cells will be described.

A second drawback to expressing eukaryotic proteins in prokaryotic cell lines is the propensity for the cellular environment to induce protein aggregation into insoluble inclusion bodies. If a refolding protocol can be determined, inclusion body formation can be advantageous in subsequent purification, as the starting material is usually in high yield and relatively pure. This is a particular problem for cell cycle regulatory proteins, as many function as components of large complexes, and their surface properties reflect this. To circumvent this problem, one approach is to coexpress an insoluble protein with a more soluble binding partner; another is to identify a biologically active but soluble protein fragment. Examples are given below of how both these strategies have been used successfully to produce proteins for structure determination.

At all stages in protein preparation, methods to assess protein integrity and activity should be available. Techniques that are particularly useful for protein characterization include mass spectrometry, surface plasmon resonance, analytical gel filtration, and analytical ultracentrifugation. We will confine the scope of this chapter to the preparation of CDKs and CDK-associated proteins for use in X-ray crystallography and NMR. However, the structures of a number of cell cycle regulatory complexes have been determined, and **Fig. 1** illustrates how various cell cycle regulatory proteins, whose structures have been determined, act within the cell cycle.

2. Materials

Chemicals to be used in these procedures should be of the best grade available commercially.

1. pET21d, pETDuet™ and pACYCDuet™ plasmids (Novagen).
2. pGEX plasmids (Amersham Biosciences).
3. pBacPAK8 baculovirus transfer vector (Clontech).
4. Luria broth base (Miller's LB broth base; Invitrogen).
5. Sf900-II serum-free medium (Gibco, Life Technologies).
6. *E. coli* strain BL21(DE3) (Novagen).
7. Insect cell lines (ATCC).
8. Modified HEPES-buffered saline (HBS): 200 m*M* NaCl, 40 m*M* HEPES, pH 7.0, 0.01% monothiolglycerol (MTG).
9. Buffer A: 300 m*M* NaCl, 50 m*M* Tris-HCl, pH 7.5 at 20°C, 10 m *M* $MgCl_2$, 0.01% MTG.
10. Buffer B: 50 m*M* Tris-HCl, pH 8.5 at 20°C, 100 m *M* $MgCl_2$, 0.01% MTG, 0.01% NaN_3.
11. Buffer C: 25 m*M* NaCl, 10 m*M* Tris-HCl, pH 7.4, 1 m*M* EDTA, 1 m*M* dithiothreitol (DTT).
12. Buffer D: 25 m*M* NaCl, 10 m*M* HEPES, pH 7.4, 1 m*M* EDTA, 10% glycerol, 0.5 m*M* DTT.
13. Buffer E: 10 m*M* HEPES, pH 7.4, 0.5 m*M* DTT.
14. Buffer F: 25 m*M* NaCl, 50 m*M* Tris-HCl, pH 7.4, 1 m*M* DTT.
15. Buffer G: 200 m*M* NaCl, 50 m*M* Tris-HCl, pH 7.4, 1 m*M* DTT.
16. Sodium dodecyl sulfate (SDS) polyacrylamide gels, prepared and run as described in **ref. *18***.

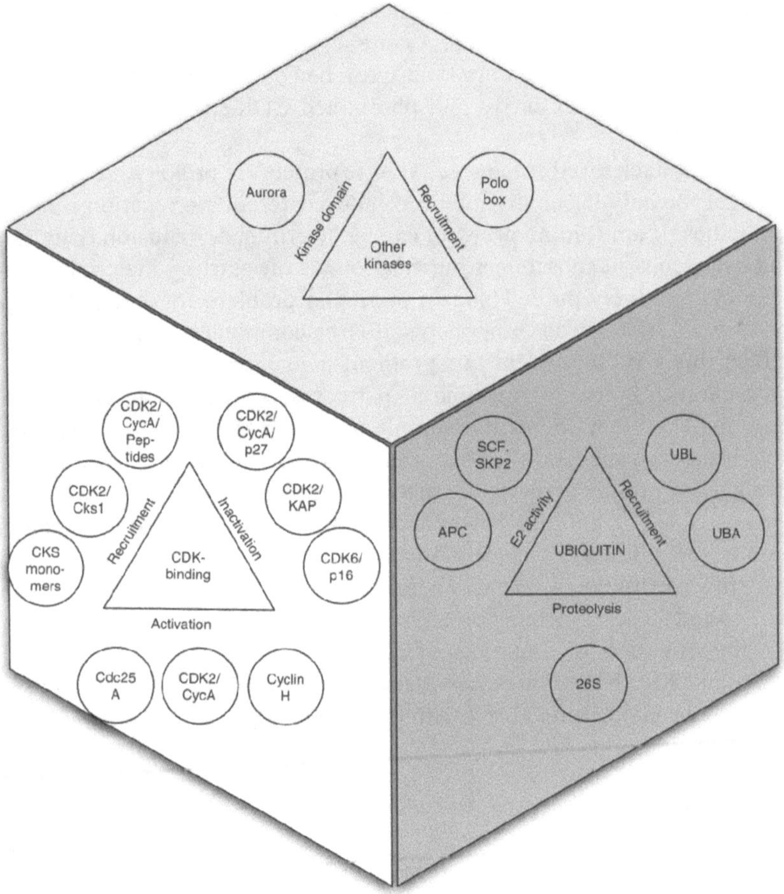

Fig. 1. The roles of selected cell cycle regulatory proteins and protein complexes for which structures have been determined. Aurora, Aurora-2-kinase catalytic domain *(32)*; Polo box, Polo box domains from SAK *(33)* and Plk1 *(31)*; SCF-SKP2, subcomplexes of the SCF in which Skp2 acts as the F-box component; APC, anaphase-promoting complex; UBA, ubiquitin-associated domains; UBL, ubiquitin-like domains; 26S, the 26S proteasome. CDK-containing complexes are described in the text, with the exception of cyclin H (described in **ref. *34***) and CDK2/cyclin A/peptide complexes (described in **ref. *35***).

17. Glutathione sepharose 4B, DEAE Sepharose™ Fast Flow and SP-Sepharose™ Fast Flow matrices (Amersham Biosciences).
18. Ni^{2+}-NTA (Qiagen).
19. Adenosine 5'-triphosphate-agarose (11-atom matrix spacer, coupled through ribose hydroxyls; Sigma).
20. HiTrap chelating Sepharose columns and HiLoad Superdex 75 pg, 26/60 size-exclusion column (Amersham Biosciences).

21. Protease inhibitor solutions prepared as recommended by suppliers (Roche or Sigma).
22. PreScission™ Protease (Amersham Biosciences).
23. A range of reagents for determination of solution protein concentrations (Pierce).
24. Vivaspin Concentrators (various volumes and membrane molecular weight cutoffs; Vivascience).
25. Antibodies: anti-CDK2 (H-298, Santa Cruz Biotechnology) or anti-Cdc2 (CDK1), anti-phosphotyrosine 15 (Cell Signaling Technology).

3. Methods

3.1. Preparation of Monomeric CDK and CDK-Associated Proteins

3.1.1. Expression and Purification of CDK2 and Cyclin A From E. coli *Cells*

A number of human CDKs and CDK-associated proteins can be expressed in soluble form in *E. coli* cells. For example, members of the Cks family are small (~13 kDa) proteins that express in high yield as tagged (usually N-terminal glutathione-S-transferase [GST] or C-terminal hexahistidine [His6]) or untagged species ***(19–21)***. Various cyclin-dependent kinase inhibitor (CKI) structures have also been reported using both full-length and truncated fragments expressed in *E. coli* cells (***5–7***; reviewed in **ref. *8***). Human CDK2 expressed in *E. coli* cells as untagged or C-terminally His6-tagged protein forms inclusion bodies. This result can be avoided by expressing the protein fused to a N-terminal GST tag at lower (20–25°C) temperatures. A number of vectors are available for expressing GST fusion proteins. CDK2 is proteolytically sensitive, and, in our hands, the best results have been obtained using the pGEX6P vectors that introduce a PreScission protease site between the GST and CDK2 domains. CDK2 (residues 1-298) was cloned into pGEX6P-1. Following cleavage with PreScission™ protease CDK2 carries an N-terminal tag of four residues (sequence Gly-Pro-Gly-Ser) before residue Met1 (The cloning strategy used deleted the triplet encoding a leucine residue 5' to the *Bam*HI site.) An inspection of the monomeric CDK2 crystal lattice shows that it cannot accommodate these residues, and to date we have been unable to crystallize this monomeric CDK2 species. However, it does crystallize readily in multiple crystal lattices when bound to human cyclin A.

The structure of monomeric cyclin A was first determined using the bovine ortholog ***(22)***, but subsequent structural studies have used the human protein exclusively. Early biochemical studies had shown that a proteolytically stable C-terminal cyclin A fragment can bind to CDK2 and activate its kinase activity. This fragment (termed cyclin A3), derived from either the human or bovine cyclin A sequence, is prone to aggregation, a phenomenon that was shown with bovine cyclin A3 to be prevented by inclusion of low concentrations of $MgCl_2$ in the buffer ***(22)***. In our hands, for structural studies on monomeric cyclin A3, bovine cyclin A3 is a more stable reagent than human cyclin A3 (unpublished observations). Both cyclins tagged at the C-terminus with His6 express in the soluble fraction and can be subsequently purified on an Ni-NTA affinity column (*see* **Subheading 3.1.1.3.** below). Bovine cyclin A3 (equivalent to human cyclin A3 residues 171–432) is expressed from pET21d.

Untagged human cyclin A3 is recommended for preparation of CDK2/cyclin A3-containing complexes. Cyclin A3 is purified from the *E. coli*-soluble fraction on a CDK2 affinity column (*see* **Subheading 3.2.1.** below). Human cyclin A3 is cloned into pET21d following restriction enzyme digestion of a polymerase chain reaction (PCR) fragment (forward and reverse primers, respectively: 5'-CGCGGATCC ATGGAAGTACCAGACTACCATGAGG-3' and 5'-GCGGGTACCTCGAGTT ACAGATTTAGTGTCTCTGG-3') with the restriction enzymes *Nco*I and *Xho*I. This strategy introduces an *Nco*I site at the 5'-end of the coding sequence, and cyclin A residue Asn173 is changed to the start methionine. The reverse primer sequence can be modified by omitting the stop codon to generate a His6-tagged protein following subcloning into the pET21d vector.

3.1.1.1. CDK2 and Cyclin A3 Expression

1. *E. coli* strain BL21(DE3)/pLYS-S is used for protein expression. Transform cells with plasmid of interest, plate, and leave to grow overnight. For long-term storage, transformed BL21(DE3)/pLYS-S cell cultures are stable kept as 30% glycerol (final) stocks at –80°C.
2. To prepare a starter culture, pick a single colony into 10–25 mL LB broth supplemented with chloramphenicol (34 g/mL) and ampicillin (100 mg/mL) (*see* **Notes 1** and **2**), and leave to grow overnight at 37°C with vigorous shaking.
3. Inoculate expression culture flasks by adding 5–10 mL starter culture to 500 mL LB broth supplemented with ampicillin to 50 g/mL in a 2-L flask.
4. Grow cells at 37°C with vigorous shaking (220 rpm) to ensure aeration of the growing culture. Once the optical density at 600 nm ($OD_{600\ nm}$) is between 0.6 and 0.8, decrease the incubator temperature to 25°C, and continue incubation for a further 30 min. This step increases the yield of soluble protein.
5. Induce cells by adding isopropyl β-D-1-thiogalactopyranoside (IPTG; 400 *M* final; *see* **Note 2**), and continue the incubation at 25°C for at least 3 h. The yield of recombinant protein is decreased if cells are induced at $OD_{600\ nm} > 1.0$, and maximum expression is usually achieved after 5 h.
6. Harvest cells by centrifugation (4000*g*, 4°C, 20 min).
7. Resuspend cell pellet in modified HBS (CDK2) or buffer A (cyclin A3) containing a protease inhibitor cocktail (*see* **Note 3**) using 50 mL buffer/1 L cell culture.
8. Transfer cell suspension to a 50-mL Falcon tube, flash freeze in an ethanol/dry ice bath or in liquid nitrogen, and store at –20°C.

3.1.1.2. CDK2 Purification

1. Prepare a glutathione-sepharose 4B column (8 mL 50% bead slurry), and equilibrate at 4°C in modified HBS at a flow rate of 0.5–1.0 mL/min (*see* **Notes 1** and **4**).
2. Thaw the GST-CDK2 cell suspension (0.5 L culture volume, 25 mL) in an ice-cold water bath, and when just thawed add $MgCl_2$ (to 10 m*M*) and DNaseI (10 g/mL final).
3. Lyse the cells by sonication on ice.
4. Centrifuge the cell lysate (100,000*g*, 4°C, 1 h).
5. Quickly and carefully remove the supernatant (S100 fraction), and dilute twofold with ice-cold modified HBS.
6. Load slowly onto the glutathione-sepharose 4B column at a flow rate of 0.5–1 mL/min.
7. Wash the column with modified HBS until a baseline $OD_{280\ nm}$ value is reached.
8. Prepare 50 mL of a 20 m*M* glutathione solution in ice-cold modified HBS (*see* **Note 5**).
9. Elute GST-CDK2 with modified HBS containing 20 m*M* glutathione at a flow rate of 1 mL/min.

10. Pool the peak fractions, and determine the approximate yield of GST-CDK2 by protein assay (for example, a Coomassie-based method; *see* **Note 6**).
11. Cleave GST tag by addition of PreScission protease, 1/50 w/w at 4°C overnight.
12. Rejuvenate the glutathione-sepharose 4B column by washing with 5 column vol of 6 *M* urea in 10 m*M* HEPES, pH 7.0, and then re-equilibrating in modified HBS.
13. Separate the CDK2 from GST dimers and contaminating proteins by size-exclusion chromatography. (For example, on a HiLoad 26/60 Superdex 75 pg column at a flow rate of 3 mL/min). Pre-equilibrate the column in modified HBS supplemented with 0.01% NaN_3 at 20°C (*see* **Note 7**).
14. CDK2 elutes after the GST dimer. Pool CDK2-containing fractions (assessed by SDS-PAGE; *see* **Note 7**), and reapply to the glutathione-sepharose 4B column equilibrated in modified HBS (flow rate, 1 mL/min). Any contaminating GST binds to the column, and the CDK2 is collected in the flowthrough.
15. CDK2 can be concentrated to approx 5 mg/mL and stored in 50% glycerol at –20°C.

3.1.1.3. Purification of Bovine Cyclin A3

1. Prepare an Ni^{2+}-NTA column (8 mL 50% bead slurry), and equilibrate at 4°C in buffer A at 0.5–1 mL/min (*see* **Notes 3**, **8**, and **9**).
2. Thaw the bovine cyclin $A3_{His6}$ cell suspension (0.5 L culture volume, 25 mL) in an ice-cold water bath, and when just thawed add DNaseI (10 g/mL final).
3. Lyse the cells by sonication on ice.
4. Prepare an S100 cell lysate by centrifugation (100,000*g*, 4°C, 1 h).
5. Adjust imidazole concentration to (~) 50 m*M* by addition of an appropriate volume of 2 *M* imidazole stock.
6. Load the S100 fraction onto the Ni-NTA column at 1 mL/min (*see* **Note 1**).
7. Wash column with buffer A (supplemented to 50 m*M* imidazole) until a baseline $OD_{280\ nm}$ value has been reached.
8. Elute bovine cyclin $A3_{His6}$ with an imidazole gradient (50–500 m*M* imidazole) in buffer A.
9. Pool the cyclin $A3_{His6}$-containing fractions (as judged by SDS-PAGE analysis), and adjust the $MgCl_2$ concentration to 100 m*M* to prevent aggregation.
10. Separate monomeric cyclin $A3_{His6}$ from aggregates and contaminating proteins by size-exclusion chromatography (for example, HiLoad 26/60 Superdex 75 pg run at 3 mL/min). Equilibrate, and run the column in buffer B.
11. Cyclin $A3_{His6}$ is stable stored in buffer B for up to 14 d at 4°C. For longer term storage, add glycerol to 50% final, and store at –20°C.
12. For crystallization trials, concentrate the sample to 5–10 mg/mL (*see* **Note 6**).

3.1.2. Expression and Purification of Human CDK2 From Recombinant Baculovirus-Infected Insect Cells

Protein expression in insect cells has a number of advantages over prokaryotic expression systems. Notably, proteins fold with greater efficiency, they may undergo post-translational modifications, and the cells contain low levels of proteases. However, this protein expression system is expensive, and the yield may not be as high. Monomeric human CDK2 crystals are readily grown from CDK2 produced from recombinant baculovirus-infected insect cells. The CDK2 contains no additional residues at either the N- or C-terminus. Human CDK2 cloned into the baculovirus transfer vector pVL1393 was used to create a recombinant baculovirus (*Ac*CDK2) expressing CDK2 *(23)*.

1. Shake flasks (2 L) containing *Spodoptera frugiperda* Sf900 cells (500 mL) at 28°C in Sf900-II serum-free medium supplemented with 0.1% gentamycin (10 g/mL final; *see* **Note 2**).
2. Infect cells (2×10^9) with *Ac*CDK2 at a multiplicity of infection (moi) of 10.
3. Continue to grow cells at 28°C with gentle rotation for 48 h.
4. Harvest virus-infected cells (400*g* for 10 min at 20°C), and resuspend the pellet in 25 mL buffer C supplemented with protease inhibitors (*see* **Note 1**).
5. Alternatively, fast freeze the pellet in a dry ice/ethanol bath or in liquid nitrogen and store at –80°C. To thaw the cell pellet, add 25 mL buffer C to the frozen pellet, and continue to thaw cells in an ice-cold water bath, gently resuspending with a pipet.
6. Gently homogenize cells using a hand-held Dounce homogenizer on ice.
7. Clarify the cell lysate by ultracentrifugation (100,000g, 1 h, 4°C)
8. Load clarified lysate at 0.5-1 mL/min onto a DEAE Sepharose™ Fast Flow column (1.6 × 13 cm) preequilibrated in buffer C (*see* **Note 4**).
9. Collect the flow-through, and adjust the NaCl concentration to 50 m*M*. Load at 1 mL/min onto an S-Sepharose™ Fast Flow column (1.6 × 30 cm) preequilibrated in buffer C supplemented to 50 m*M* NaCl. Wash the column extensively until the $OD_{280\ nm}$ absorption returns to a baseline value. CDK2 elutes from this column as two discrete (but overlapping) peaks: the first is in the flowthrough volume, and the second is slightly delayed (*see* **Note 10**). The two CDK2 species can only be distinguished by analysis of each fraction by SDS-PAGE (*see* **Note 7**). Fractions containing each species should be pooled separately before each is purified further.
10. Load one of the CDK2 pools onto an ATP-agarose (1.0 × 5.0 cm) column preequilibrated with buffer D at 1 mL/min. The two CDK2 populations that elute from the SP-sepharose™ column behave equivalently on this column.
11. Wash the column until a baseline $OD_{280\ nm}$ value has been reached and then elute CDK2 with a salt gradient (25–500 m*M* NaCl) in buffer D.
12. Pool the CDK2-containing fractions.
13. For storage (longer than 48 h), concentrate the CDK2 to 5–15 mg/mL (*see* **Note 6**), dilute with glycerol to 50% glycerol (final), and transfer to –20°C.
14. For crystallization trials, the NaCl concentration in the CDK2 solution has to be gradually reduced. Concentrate CDK2-containing fractions to 2 mL and dilute twofold with buffer E. Repeat this step four times in total to reduce the NaCl concentration 16-fold to approx 15 m*M* NaCl (presuming that the CDK2 elutes from the ATP-agarose column at ~250 m*M* NaCl). Continue to concentrate the CDK2 to 10 mg/mL, and set up crystallization trays immediately (*see* **Note 11**).
15. Rejuvenate columns according to the Manufacturer's instructions (*see* **Notes 4** and **7**).

3.2. Preparation of Protein Complexes

Protein complexes can be prepared by separately purifying the individual components and then mixing them in stoichiometric ratios prior to a chromatographic step to separate the complex from the individual components. As an example, the structure of CDK2/Cks1 *(24)* was determined from a protein complex prepared in this way.

Alternatively, if using an *E. coli* cell-based system, the proteins can be coexpressed and the complex purified from the cell lysate. Proteins can be coexpressed from one vector either using a single promoter to produce one polycistronic transcript or from independent promoters to produce multiple transcripts. Alternatively, proteins can be cloned into compatible plasmids. To be perpetuated through successive rounds of cell division, each plasmid must contain different replicons and drug resistance genes. For example pGEX and pET vectors that carry a ColE1 compatability group origin of replication sequence are compatible with pACYC and pLysS vectors that carry the origin of replication from plasmid p15A. The pETDuet and pACYCDuet vectors have been designed to allow expression of four proteins in *E. coli* cells and are compatible with other commonly used vector series. Transfer vectors to produce recombinant baculoviruses capable of producing multiple proteins can also be used (reviewed in **ref. *25***). Alternatively, multiple recombinant baculoviruses can be used simultaneously to infect insect cells (*see* **Subheading 3.2.3.**).

3.2.1. Preparation and Purification of a Human CDK2/Cyclin A Complex From E. coli *Cells*

In the following protocol, the high affinity of CDK2 for cyclin A is exploited to purify human cyclin A3. It is important that all the GST-CDK2 immobilized on the glutathione-sepharose 4B column be bound to cyclin A3, as the subsequent chromatographic steps cannot separate excess CDK2 from CDK2/cyclin A.

1. Grow, induce, harvest, and lyse recombinant *E. coli* cells expressing human GST-CDK2 and human cyclin A3, and prepare S100 extracts as described above (*see* **Note 12**).
2. Prepare a glutathione-sepharose 4B column (10 mL 50% bead slurry), and equilibrate at 4°C in modified HBS at a flow rate of 0.5–1.0 mL/min (*see* **Notes 1** and **4**).
3. Dilute the GST-CDK2 clarified lysate with an equal volume of ice-cold modified HBS, load onto the affinity column at 0.5–1 mL/min, and then wash the column with modified HBS until a baseline $OD_{280\ nm}$ value has been reached.
4. Now apply the cyclin A3 S100 fraction to the column, maintaining the flow rate at 0.5–1 mL/min. Again, wash the column extensively with modified HBS until a baseline $OD_{280\ nm}$ value is reached.
5. Prepare 50 mL of a 20 m*M* glutathione solution in ice-cold modified HBS (*see* **Note 5**).
6. Elute GST-CDK2/cyclin A3 with modified HBS containing 20 m*M* glutathione at a flow rate of 1 mL/min.
7. Determine the approximate GST-CDK2/cyclin A3 yield by protein assay (*see* **Note 6**).
8. Cleave GST tag by addition of PreScission protease, 1/50 w/w at 4°C overnight.
9. Rejuvenate the glutathione-sepharose 4B column by washing with 5 column vol of 6 *M* urea in 10 m*M* HEPES, pH 7.0, and then reequilibrating in modified HBS.
10. CDK2/cyclin A3 and GST dimers coelute on a size exclusion column (for example, HiLoad 26/60 Superdex 75 pg). However, this step does remove contaminating proteins and buffer-exchanges the complex to remove glutathione. Preequilibrate, and run the gel filtration column, in modified HBS supplemented with 0.01% NaN_3 at 20°C at 3 mL/min (*see* **Note 7**).
11. Pool CDK2/cyclin A-containing fractions (assessed by SDS-PAGE, *see* **Note 7**), and reapply to the rejuvenated glutathione-sepharose 4B column equilibrated in modified HBS (1 mL/min). GST binds to the column, and CDK2/cyclin A3 is collected in the flowthrough.
11. CDK2/cyclin A3 can be concentrated to approx 12–15 mg/mL. To store at –20°C, dilute appropriately with 100% glycerol and 5X modified HBS stock to 50% glycerol in 1X modified HBS (*see* **Note 6**). Alternatively, for crystallization trials, take 25- L aliquots

of CDK2/cyclin A at 12–15 mg/mL in modified HBS, plunge freeze in liquid nitrogen, and store at –80°C. To set up crystallization trials, thaw quickly between the fingers, and place immediately on ice.

3.2.2. Preparation and Purification of a Human Skp1/Skp2 Complex From E. coli Cells

Full-length Skp2 is insoluble when expressed in *E. coli* cells, but parts of the protein do express in the soluble fraction when coexpressed with its smaller (19-kDa) binding partner, Skp1. The structure of a Skp1/Skp2 complex has recently been reported ***(26)***. In this example, human Skp1 and a Skp2 fragment were cloned into the vector pGEX-2T to create a dicistronic message in which the region coding for the GST/Skp2 fusion protein preceded that of Skp1.

A similar approach can be taken to express an N-terminal fragment of Skp2 that includes the F-box sequence that binds to Skp1. Skp2 (residues 1–152) was cloned into pGEX6P-1 as a *Bam*HI/*Xho*I fragment following PCR amplification with suitably modified primers. Untagged Skp1 was cloned into pET21d. Using this construct as a template, a Skp1 expression cassette encoding a ribosome binding site was amplified by PCR to introduce *Xho*I sites at each end. This DNA sequence was subsequently digested with XhoI and ligated into pGEX6P-1/GST-Skp2, also cut with *Xho*I. Colonies were screened to select for plasmids that had incorporated the insert in the correct orientation. Using this construct, a single RNA transcript is synthesized that codes for both proteins. The complex expresses to high yield in the soluble fraction following transformation of the plasmid into BL21(DE3)/pLYS-S *E. coli* cells and subsequent induction at 30°C for 5 h once the cell density has reached $OD_{600\ nm}$ 0.6–0.8. The complex can be purified using the GST affinity tag and following the protocol described above for the purification GST-CDK2 (*see* **Subheading 3.1.1.2.**).

3.2.3. Preparation of X. laevis CDK7/Cyclin H From Insect Cells

The following example illustrates how two recombinant baculoviruses, each driving expression of one protein, can be used to produce a protein complex (in this case *X. laevis* CDK7/cyclin H) following dual infection of insect cells. The proteins have been engineered to each contain a His_6 tag immediately preceding the protein sequence ***(27)***. Each virus was derived from the baculovirus transfer vector pBacPAK8.

1. Shake 2-L flasks containing *Spodoptera frugiperda* Sf9 cells (500 mL) at 28°C in Sf900-II serum-free medium supplemented with 0.1% gentamycin (*see* **Note 2**).
2. Coinfect cells (2 × 10^9) with *Ac*CDK7 and *Ac*cyclin H each at an moi of 5.
3. Continue to grow cells at 28°C with gentle rotation for 48 h.
4. Prepare a 5-mL HiTrap chelating sepharose column charged with $CoCl_2$, and preequilibrate in buffer F supplemented with 20 m*M* imidazole (*see* **Notes 1**, **8**, and **9**).
5. Harvest virus-infected cells (400*g* for 10 min at 20°C), and resuspend the pellet in 25 mL buffer F supplemented with protease inhibitors (*see* **Note 3**). Alternatively, fast freeze the pellet in a dry ice/ethanol bath or in liquid nitrogen, and store at –80°C. To thaw the cell pellet, add 25 mL ice-cold buffer F, and continue to thaw cells in an ice-cold water bath, gently resuspending with a pipet.

6. Gently homogenise cells using a hand-held homogenizer on ice. Increase the concentration of NaCl to 200 m*M* by addition of an appropriate volume of 5 *M* NaCl stock.
7. Clarify the cell lysate by ultracentrifugation (100,000*g*, 1 h, 4°C).
8. Adjust the imidazole concentration in the cell lysate to 20 m*M*, and load onto the Co^{2+} affinity column at 1 mL/min. Wash column with buffer G supplemented to 20 m*M* imidazole until the $OD_{280\ nm}$ reaches a baseline value.
9. Elute the CDK7/cyclin H complex with an imidazole gradient developed from 20–200 m*M* imidazole in buffer G.
10. Pool the CDK7/cyclin H-containing fractions as judged by SDS-PAGE analysis (*see* **Note 7**).
11. Further purify CDK7/cyclin H by size-exclusion chromatography (for example, using a HiLoad 26/60 Superdex 75 pg at a flow rate of 3 mL/min). Preequilibrate the column in buffer G supplemented with 0.01% NaN_3 (*see* **Note 7**). This chromatographic step removes residual contaminants and excess CDK7 or cyclin H monomers.
12. Pool the CDK7/cyclin H-containing fractions as judged by SDS-PAGE, and concentrate the sample to approx 5 mg/mL and store in buffer G/50% glycerol (final) at –20°C.

3.3. Preparation of Phosphorylated CDK2 in E. coli Cells

CDK2 associated with cyclin A requires Thr160 phosphorylation for full activation. This complex can subsequently be inhibited by phosphorylation of Tyr15 and Thr14 (reviewed in **ref. *28***). This section describes methods to produce alternatively phosphorylated forms of CDK2 in *E. coli* cells by coexpressing GST-CDK2 with various protein kinase domains. As an example, the preparation of CDK2 phosphorylated on Thr160 by CDK-activating kinase (CAK) in vivo 1 (CIV1) ***(29)*** will be described, but an identical approach has also been used to prepare CDK2 phosphorylated on Tyr15. In this case, GST-CDK2 was coexpressed with a GST fusion of the human Wee1 kinase catalytic domain ***(30)***.

The GST-CDK2 and GST-CIV1 sequences are encoded on the same plasmid (pGEX6P-1 backbone), but in this instance the two proteins are expressed from independent expression cassettes (pGEX6P-CDK2/CIV1). These cassettes are in tandem, separated by rrn transcription termination sequences, and both promoters are directly IPTG-inducible with transcription dependent on *E. coli* RNA polymerase. Notably, the GST tag can be cleaved from CDK2 but not from CIV1 by PreScission protease. The CIV1 expression cassette was derived from pGEX-2T and has a thrombin cleavage site.

3.3.1. Expression of CDK2 Phosphorylated on Thr160

1. *E. coli* strain BL21(DE3)/pLYS-S is used for protein expression. Transform cells with (pGEX6P-CDK2/CIV1), plate, and leave to grow overnight (*see* **Note 13**).
2. To prepare a starter culture, pick a single colony into 10–25 mL LB broth supplemented with chloramphenicol (34 g/mL) and ampicillin (100 g/mL) (*see* **Notes 1** and **2**), and leave to grow overnight at 37°C with vigorous shaking.
3. Inoculate expression culture flasks by adding 5–10 mL starter culture to 500 mL LB broth supplemented with ampicillin to 50 g/mL in a 2-L flask.
4. Grow cells at 37°C with vigorous shaking to ensure aeration of the growing culture. Once the $OD_{600\ nm}$ is 0.8, decrease the incubator temperature to 20°C as rapidly as possible, and continue incubation for a further 60 min. This step increases the yield of soluble protein.

5. Induce cells by adding IPTG (80 *M* final; *see* **Note 2**), and continue the incubation at 20°C for at least 24 h (*see* **Note 14**).
6. Harvest cells by centrifugation (4000*g*, 4°C, 20 min).
7. Resuspend cell pellet in modified HBS containing a protease inhibitor cocktail (*see* **Note 3**) using 50 mL buffer/1 L cell culture.
8. Transfer cell suspension to a 50-mL Falcon tube, flash freeze in an ethanol/dry ice bath or in liquid nitrogen, and store at –20°C.

3.3.2. Purification of T160pCDK2 and T160pCDK2/Cyclin A3

Monomeric CDK2 phosphorylated on Thr160 and a T160pCDK2/cyclin A3 complex can be purified for crystallization trials following the protocols described in **Subheadings 3.1.1.2.** and **3.2.1.**, respectively.

3.3.3. Evaluation of the Extent of CDK2 Phosphorylation

CDK2 phosphorylated on Thr160 migrates more quickly than unphosphorylated CDK2 on SDS-PAGE (*see* **Note 15**). The extent of Thr160 phosphorylation can be estimated from the relative intensities of the two protein bands following Coomassie staining of the gel. Phosphorylation of CDK2 on Tyr15 does not produce a gel shift, but it can be detected by Western blot analysis. The best resolution is provided by 15% SDS polyacrylamide denaturating gels. Alternatively, mass spectrometry can be used to estimate the extent of phosphorylation.

3.4. Identification of Protein Fragments Suitable for Protein Expression and Crystallization

For X-ray crystallographic studies of cell cycle proteins, it is essential that the protein sample be pure and homogeneous. In certain cases the full-length protein is not tractable to heterologous expression, and so a more suitable fragment must be identified (for example, cyclin A3). Limited proteolysis using selected proteases (of which subtilisin has proved to be particularly useful) in combination with mass spectrometry and N-terminal sequencing of the resultant products can be an informative method to identify stable proteolytic fragments amenable to structural studies. Typically 10 g of target protein is incubated with subtilisin (0.05–0.2 % w/w) at room temperature or at 37°C for 1 h, the resulting fragments are resolved by SDS-PAGE, and the sample is then submitted for N-terminal sequence analysis.

3.4.1. Preparation of Protein for Structural Studies Using Limited Proteolysis

In certain circumstances, limited proteolysis can be used to improve the quality of a protein sample following preparative scale expression and purification. The proteolytic enzyme is presumed to attack unstructured protein regions preferentially (for example, surface loops), the presence of which might be a hindrance to successful crystallization.

The polo-like kinase-1 polo box domain (Plk1 PBD) formed needle-like crystals that did not diffract (**Fig. 2**). The diffraction quality of these crystals was dramatically improved after the protein was subjected to limited subtilisin digestion. The crystal structure indicated that the subtilisin treatment had removed surface loops that had

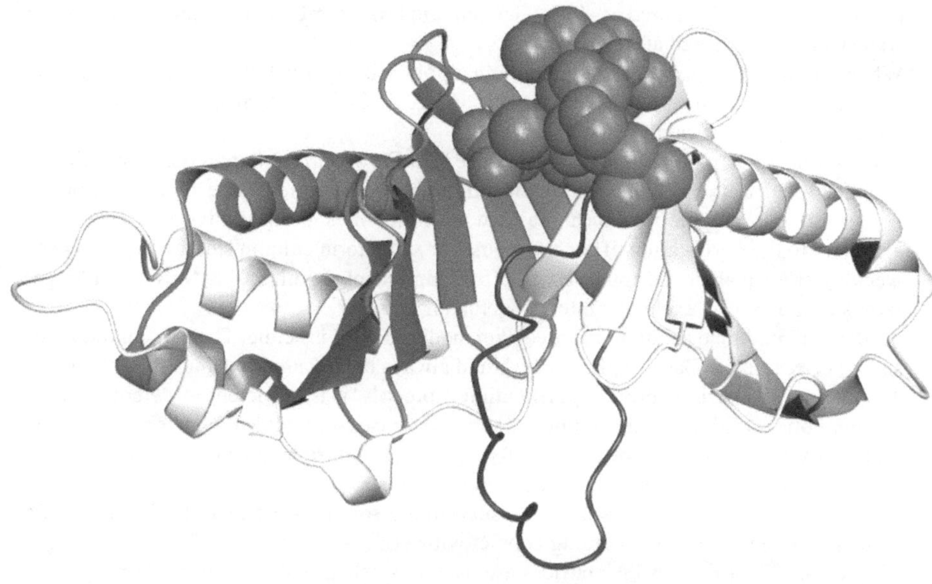

Fig. 2. The structure of the polo-box domain of Plk1 in complex with a peptide (31). The N-terminal polo box (residues 371–490) is in white, a linker that is flexible in the absence of bound peptide is black, and the C-terminal polo box (residues 509–593) is a gray. A phosphothreonine–containing peptide ligand is drawn in CPK representation.

presumably deleteriously affected crystal growth and, importantly, did not affect the tertiary structure of PBD ***(31)***.

1. Incubate pure PBD with subtilisin (1:200 w/w) for 1 h at room temperature.
2. Stop the reaction by 1:1000 addition of phenylmethylsulfonyl fluoride (PMSF) stock solution (100 m*M*), (*see* **Note 16**).
3. Analyze the reaction products by 12% SDS-PAGE (*see* **Note 7**). In this example, two fragments of 10 and 12 kDa should be obtained, starting, respectively, at ^{367}GEVVDCH and ^{495}ANITPREGDE.
4. Load the PBD onto a size-exclusion column, equilibrated in modified HBS (for example, a HiLoad 26/60 Superdex 75 pg) to buffer-exchange the protein and to remove the subtilisin and any remaining contaminants. The two fragments remain noncovalently associated and elute as a single peak.

4. Notes

1. Buffers should be made fresh on the day of use, filtered and thoroughly degassed and equilibrated to temperature before use. Antibiotics are heat-labile, and MTG and DTT oxidize in aqueous solutions. MTG has advantages over DTT in that it is easier to handle, more persistent, and compatible (at 0.01%) with Ni^{2+}-NTA columns. However, it is not as effective.
2. Ammonium persulfate (APS; 10% w/w in water, filter sterile), IPTG (0.1 *M* in water), ampicillin (100 mg/mL in water, filter sterile), and chloramphenicol (34 mg/mL in etha-

nol) stocks should be stored at –20°C. Gentamycin (10 mg/mL stock solution prepared in water) should be stored at 4°C.

3. When one is purifying a protein containing a poly-His tag, the buffer should not contain any EDTA, which interferes with the Ni^{2+}-NTA or Co^{2+}-NTA matrix. An EDTA-free protease inhibitor cocktail should be used.
4. Glutathione Sepharose 4B, DEAE-Sepharose Fast Flow, and SP-Sepharose Fast Flow are supplied in 50% ethanol. They should be thoroughly equilibrated in the appropriate buffer before use. GST-sepharose 4B can be used multiple times, but it is advisable to dedicate a column to the purification of each protein. For short-term column storage (column used weekly), store at 4°C in the appropriate buffer supplemented with 0.01% NaN_3. For longer term storage, buffer-exchange into 20% ethanol in water.
5. A 20 m*M* glutathione solution prepared with modified HBS is acidic. The pH of the solution should be readjusted back to pH 7.0. It should always be prepared fresh just prior to use.
6. Throughout the various protein purification protocols, it is sufficient to determine total protein concentrations using a quick Coomassie-based protocol. For determining final yields in advance of using the material for crystallization trials, the protein concentration should be determined using a more accurate method.
7. Caution: NaN_3 and unpolymerized polyacrylamide solutions are neurotoxic and should be manipulated with care, avoiding contact with skin.
8. The Ni^{2+}-NTA or Co^{2+}-NTA matrices are usually washed with buffers containing a low concentration of imidazole to reduce the background binding of bacterial proteins. Bovine cyclin A_{His6} binds tightly to Ni^{2+}-NTA, and so an unusually high concentration of imidazole can be used in the loading buffer. As a result, the cyclin $A3_{His6}$ is relatively pure following this one purification step. Lower concentrations of imidazole (5–20 m*M*) should be used in most circumstances, the concentration being determined by carrying out batch binding experiments prior to preparative-scale purification.
9. The pH of the imidazole stock (2 M in water) is basic; on addition to a running buffer, the pH of the solution should be checked before use.
10. The SP-sepharose cation exchange column is of large volume to maximize the separation of these two CDK2 species, and the best results (in our hands) have been achieved by preparing a fresh column for every preparation. The separation is critically dependent on pH and can be problematic. If the two species do not separate, the CDK2 fractions can be pooled and reapplied to the rejuvenated column. Each CDK2 species crystallizes under similar conditions, but the mixture produces rapidly nucleating showers of crystals that are too small for analysis by X-ray diffraction.
11. CDK2 is prone to precipitation at low NaCl concentrations and should only be concentrated and buffer-exchanged immediately prior to setting up crystallization trays.
12. To ensure that cyclin A3 is in excess, grow culture volumes in a ratio of 1:3, GST-CDK2/cyclin A3. Pellets for cells expressing GST-CDK2 and cyclin A3 can be processed at the same time, as the cyclin A3 is stable in the *E. coli* cell lysate when kept on ice at 4°C. However, as CDK2 is less stable, sonicate cyclin A3-expressing cell lysates first.
13. BL21(DE3)/pLYS-S cells transformed with pGEX6P-CDK2/CIV1 tend to be unstable. Although stocks can be kept at 30% glycerol (final) at –80°C, it is advisable to transform the plasmid into fresh cells for every protein preparation and use the plates within a week.
14. The induction of phosphorylated CDK2 is carried out at 20°C for 24 h to ensure complete CDK2 phosphorylation. GST-CDK2 expression is readily detectable by SDS-PAGE analysis of whole cell extract, but GST-CIV1 expresses at very low levels. However, there is sufficient GST-CIV1 to phosphorylate the excess CDK2. Shortening the incubation time results in incomplete CDK2 phosphorylation.

15. For optimal separation (using a Bio-Rad Mini-Protean[R] 3 electrophoresis system), prepare a 12% gel.
16. PMSF is extremely unstable in water, and solutions should be prepared immediately before use. It is also toxic and should be handled with care.

Acknowledgments

The authors would like to thank all their collaborators who generously provided reagents and colleagues at the Laboratory of Molecular Biophysics who developed the protein purification strategies described. Nick Brown is thanked for his expert advice and proof reading of the manuscript. The authors are very grateful to Martin Noble for his assistance in the manuscripts̀ preparation. J. Welburn is supported by a studentship from the Wellcome Trust (grant number 065956). Work in the authors'laboratory is supported by the MRC, BBSRC, the Wellcome Trust, and Oxford University.

References

1. De Bondt, H. L., Rosenblatt, J., Jancarik, J., Jones, H. D., Morgan, D. O., and Kim, S. H. (1993) Crystal structure of cyclin-dependent kinase 2. *Nature* **363,** 595–602.
2. Jeffrey, P. D., Russo, A. A., Polvak, K., et al. (1995) Mechanism of CDK activation revealed by the structure of a cyclinA-CDK2 complex. *Nature* **376,** 313–320.
3. Brown, N. R., Noble, M. E., Endicott, J. A., et al. (1999) Effects of phosphorylation of threonine 160 on cyclin-dependent kinase 2 structure and activity. *J. Biol. Chem.* **274,** 8746–8756.
4. Russo, A. A., Jeffrey, P. D., and Pavletich, N. P. (1996) Structural basis of cyclin-dependent kinase activation by phosphorylation. *Nat. Struct. Biol.* **3,** 696–700.
5. Russo, A. A., L. Tong, Lee, J. O., Jeffrey, P. D., and Pavletich, N. P (1998) Structural basis for inhibition of the cyclin-dependent kinase Cdk6 by the tumour suppressor p16INK4a. *Nature* **395,** 237–243.
6. Brotherton, D. H., Dhanaraj, V., Wick, S., et al. (1998). Crystal structure of the complex of the cyclin D-dependent kinase Cdk6 bound to the cell-cycle inhibitor p19INK4d. *Nature* **395,** 244–250.
7. Russo, A. A., Jeffrey, P. D., Patten, A. K., Massague, J., and Pavletich, N. P. (1996). Crystal structure of the p27Kip1 cyclin-dependent-kinase inhibitor bound to the cyclin A-Cdk2 complex. *Nature* **382,** 325–331.
8. Endicott, J. A., Noble, M. E. and Tucker, J. A. (1999) Cyclin-dependent kinases: inhibition and substrate recognition. *Curr. Opin. Struct. Biol.* **9,** 738–744.
9. Pavletich, N. P. (1999) Mechanisms of cyclin-dependent kinase regulation: structures of Cdks, their cyclin activators, and Cip and INK4 inhibitors. *J. Mol. Biol.* **287,** 821–828.
10. Davies, T. G., Pratt, D. J., Endicott, J. A., Johnson, L. N., and Noble, M. E. (2002) Structure-based design of cyclin-dependent kinase inhibitors. *Pharmacol. Ther.* **93,** 125–133.
11. Zheng, N., Schulman, B. A., Song, L., et al. (2002) Structure of the Cul1-Rbx1-Skp1-F boxSkp2 SCF ubiquitin ligase complex. *Nature* **416,** 703–709.
12. Gieffers, C., Dube, P., Harris, J. R., Stark, H., and Peters, J. M. P. (2001) Three-dimensional structure of the anaphase-promoting complex. *Mol. Cell* **7,** 907–913.
13. Fernandez, C. and Wider G. (2003) TROSY in NMR studies of the structure and function of large biological macromolecules. *Curr. Opin. Struct. Biol.* **13,** 570–580.
14. Post, C. B. (2003) Exchange-transferred NOE spectroscopy and bound ligand structure determination. *Curr. Opin. Struct. Biol.* **13,** 581–588.

15. Roseman, A. M. (2000) Docking structures of domains into maps from cryo-electron microscopy using local correlation. *Acta Crystallogr. D. Biol. Crystallogr.* **56,** 1332–1340.
16. Wendt, K. S., Vodermaier, H. C., Jacob, U., et al. (2001) Crystal structure of the APC10/DOC1 subunit of the human anaphase-promoting complex. *Nat. Struct. Biol.* **8,** 784–788.
17. Au, S. W., Leng, X., Harper, J. W., and Barford, D. (2002) Implications for the ubiquitination reaction of the anaphase-promoting complex from the crystal structure of the Doc1/Apc10 subunit. *J. Mol. Biol.* **316,** 955–968.
18. Sambrook, J., Fritsch, E. F., et al. (1989) *Molecular Cloning: A Laboratory Manual.* Cold Spring Harbor, Cold Spring Harbor Laboratory Press.
19. Parge, H. E., A. S. Arvai, Murtari, D. J., Reed, S. I., and Tainer, J. A. (1993) Human CksHs2 atomic structure: a role for its hexameric assembly in cell cycle control. *Science.* **262,** 387–395.
20. Arvai, A. S., Bourne, Y., Hickey, M. J., and Tainer, J. A. (1995) Crystal structure of the human cell cycle protein CksHs1: single domain fold with similarity to kinase N-lobe domain. *J. Mol. Biol.* **249,** 835–842.
21. Endicott, J. A., Noble, M. E., and Garman, E. F. (1995) The crystal structure of p13suc1, a p34cdc2-interacting cell cycle control protein. *EMBO J.* **14,** 1004–1014.
22. Brown, N. R., Noble, M. E., Endicott, J. A., et al. (1995) The crystal structure of cyclin A. *Structure.* **3,** 1235–1247.
23. Rosenblatt, J., De Bondt, H., Jancarik, J., Morgan, D. O., and Kim, S. H. (1993) Purification and crystallization of human cyclin-dependent kinase 2. *J. Mol. Biol.* **230,** 1317–1319.
24. Bourne, Y., Watson, M. H., Hickey, M. J., et al. (1996) Crystal structure and mutational analysis of the human CDK2 kinase complex with cell cycle-regulatory protein CksHs1. *Cell* **84,** 863–74.
25. Roy, P., Mikhailov, M., and Bishop, D. H. (1997) Baculovirus multigene expression vectors and their use for understanding the assembly process of architecturally complex virus particles. *Gene* **190,** 119–129.
26. Schulman, B. A., Carrano, A. C., Jeffrey, P. D., et al. (2000) Insights into SCF ubiquitin ligases from the structure of the Skp1-Skp2 complex. *Nature* **408,** 381–386.
27. Lawrie, A. M., Tito, P., Hernandez, H., et al. (2001) Xenopus phospho-CDK7/cyclin H expressed in baculoviral-infected insect cells. *Protein Expr. Purif.* **23,** 252–260.
28. Morgan, D. O. (1997) Cyclin-dependent kinases: engines, clocks and microprocessors. *Annu. Rev. Cell Dev. Biol.* **13,** 261–291.
29. Thuret, J. Y., Valay, J. G., Faye, G., and Mann, C. (1996) Civ1 (CAK in vivo), a novel Cdk-activating kinase. *Cell* **86,** 565–576.
30. Tucker, J. (2001) Control of the cyclin-dependent family: structural studies on the molecular mechanisms that regulate active site phosphorylation. *Biochemistry* Oxford University Press, Oxford, UK.
31. Cheng, K. Y., Lowe, E. D., Sinclair, J., Nigg, E. A., and Johnson, L. N. (2003) The crystal structure of the human polo-like kinase-1 polo box domain and its phospho-peptide complex. *EMBO J.* **22,** 5757–5768.
32. Nowakowski, J., Cronin, C. N., McRee, D. E., et al. (2002) Structures of the cancer-111related Aurora-A, FAK, and EphA2 protein kinases from nanovolume crystallography. *Structure (Camb.)* **10,** 1659–1667.
33. Leung, G. C., J. W. Hudson, Kozarova, A., Davidson, A., Dennis, J. W., and Sicheri, F. (2002) The Sak polo-box comprises a structural domain sufficient for mitotic subcellular localization. *Nat. Struct. Biol.* **9,** 719–724.

34. Kim, K. K., Chamberlin, H. M., Morgan, D. O., and Kim, S. H. (1996) Three-dimensional structure of human cyclin H, a positive regulator of the CDK-activating kinase. *Nat. Struct. Biol.* **3,** 849–855.
35. Lowe, E. D., Tews, I., Cheng, K. Y., et al. (2002) Specificity determinants of recruitment peptides bound to phospho-CDK2/cyclin A. *Biochemistry* **41,** 15625–15634.

III

Detection and Measurement of Cell Cycle Molecules

13

E2F Transcription Factors and pRb Pocket Proteins in Cell Cycle Progression

Ludger Hauck and Rüdiger von Harsdorf

Summary

The E2F-family of transcripion factors exerts fascinating and contrasting functions in transcriptional repression and activation of genes regulating proliferation, apoptosis, and differentiation. E2F is principally regulated by its temporal association with retinoblastoma pocket protein (pRb) family members. In turn, pRb is regulated through phosphorylation by cyclin-dependent kinase (cdk). The activity of cdk is negatively regulated by cdk-inhibitors, exemplified by $p16^{INK4a}$, $p21^{CIP1}$, and $p27^{KIP1}$. Therefore, positive and negative signaling events converge on E2F activity resulting in distinct growth-controling and apoptotic activities. Here we describe the immunocytochemical detection of E2F, genomic DNA, BrdU-incorporation, and mitosis in cardiomyoctes. A detailed protocol is given to illustrate this technique in primary heart muscle cells.

Key Words

E2F; cell cycle; cardiomyocytes; immunocytochemistry; S-phase; BrdU; phospho-histone H3; mitosis.

1. Introduction

The transcription factor E2F1 plays a pivotal role in the coordinated expression of genes necessary for cell cycle progression and division. Overexpression of E2F1–3 drives quiescent cells into S-phase *(1)*. Ectopic expression of E2F1 also leads to apoptosis, which is specific to E2F1 and not other E2F family members *(2)*. E2F interacts with DP proteins to form a heterodimer and binds to DNA in a sequence-specific manner. E2F-dependent transcriptional activation is principally regulated by interaction with a member of the retinoblastoma gene family (pRb). Transcriptional studies in vitro have identified the existence of three types of E2F/pRb complexes *(3)*. In *activator* E2F complexes, the E2F transactivation domain drives transcription in the absence of pRb. In *inhibited* E2F/pRb complexes, the transactivation domain of E2F1

From: *Methods in Molecular Biology, vol. 296, Cell Cycle Control: Mechanisms and Protocols*
Edited by: T. Humphrey and G. Brooks © Humana Press Inc., Totowa, NJ

is blocked by bound pRb, rendering E2F transcriptionally inactive. In *repressor* E2F complexes, pRb is recruited to E2F binding sites by E2F and inhibits promoter activation by E2F with or without further utilization of histone-modifying enzymes. In a simplified model, pRb is hypophosphorylated and active in G_0 and carries a stably bound E2F ***(4)***. During mid-G_1, pRb is inhibited by subsequent phosphorylation through the cyclin-dependent kinases (cdks) cdk4/6 and cdk2, liberating the inhibition of E2F activity. Released E2F then activates cellular genes involved in DNA replication and cell cycle progression.

Little is known about the function of E2F1 during development and differentiation of cardiomyocytes. Adenovirus mediated gene delivery of E2F1 or E1A induced DNA synthesis in isolated rat ventricular cardiomyocytes ***(5,6)***. Ectopic expression of E2F1 in cardiomyocytes also resulted in an increase in the expression of key cell cycle activators and the induction of apoptosis, which was accompanied by concomitant loss of cdk inhibitors $p21^{CIP1}$ and $p27^{KIP1}$ ***(5)***. During hypoxia-induced cell death, cardiomyocyte apoptosis selectively activated cdk2/3, leading to the inactivation of pRb and induction of E2F-dependent gene transcription ***(7)***. Under these conditions, apoptosis was inhibited by ectopically expressed $p21^{CIP1}$/ $p27^{KIP1}$, but not $p16^{INK4}$, and dn.cdk2/3. Importantly, cardiac myocytes could be rescued from hypoxia-induced cell death by dominant-negative pRb and transcriptionally inactive E2F1/DP1. Moreover, inhibition of E2F abrogated the hypertrophic growth response to serum and phenylephrine in neonatal cardiomyocytes ***(8)***. Collectively, these data imply that targeting the E2F/pRb function might provide a useful strategy for treatment of pathophysiological heart diseases.

2. Materials

This chapter describes a detailed method for the intracellular detection of proteins in vivo. Immunocytochemistry is an easily performed and powerful tool. It allows one to conveniently study the intracellular localization and expression levels of any fluorochrome-labeled molecule, even in living cells. Immunocytochemistry is highly reproducible and delivers important information about the regulation of cell cycle factors in response to extracellular stimuli of various origins. Therefore, it is reasonable to organize the immunocytochemistry around your ordinary work day as you already do with your blotting applications.

1. Chemicals mentioned in this chapter should be of the best grade commercially available. Prepare all solutions preferentially in MilliQ-H_2O or double-distilled water. Phosphate-buffered saline (PBS) contains Ca/Mg unless otherwise stated (*see* **Subheading 3.2., step 1**). All solutions should be at room temperature and all incubation steps are carried out at ambient temperature, unless stated otherwise.
2. Cover slips should not exceed 12 mm in diameter. Glasses should be extensively washed with 1 *N* HCl and rinsed with H_2O prior to use. Sterilize by autoclaving.
3. Collagen R (Serva): dilute fivefold in sterile H_2O. Apply 500 L/well for 60 min, remove the solution, and air-dry under a sterile bench.
4. Fixation solution: 3.7% buffered formalin. Dilute 37% formaldehyde solution with an appropriate volume of PBS (8.0 g/L NaCl, 0.2 g/L KCl, 1.15 g/L Na_2HPO_4, 0.2 g/L KH_2PO_4, 0.1 g/L $MgCl_2 \cdot 6H_2O$, 0.1 g/L $CaCl_2$). Prepare fresh on the day you use it.

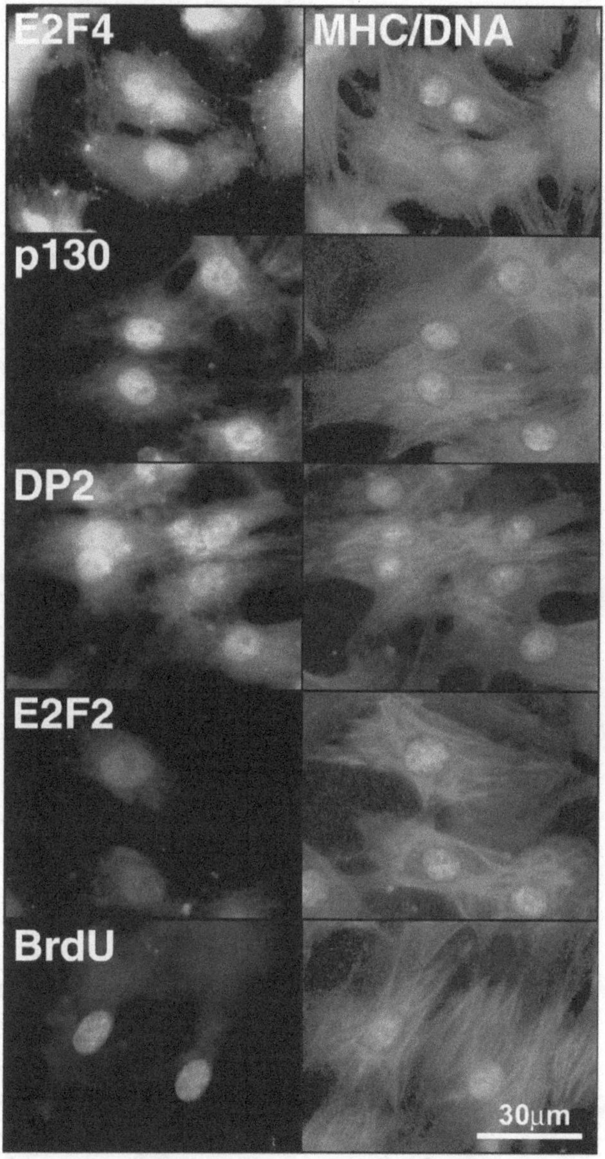

Fig. 1. Expression and nuclear localization of cell cycle factors in rat neonatal ventricular cardiomyocytes. After cultivation for 24 h in the presence of fetal calf serum (10%), cells were fixed, permeabilized, and costained by indirect immunofluorescence with specific antibodies to E2F4, p130, DP2, and E2F2 as indicated and TRITC-conjugated cardiac specific mouse monoclonal antibody to sarcomeric myosin heavy chain (MF20). Genomic DNA was stained with Hoechst 33258. Cardiomyocytes were infected with adenovirus wild-type E2F2 (100 PFU/cell) and cultivated with bromodeoxyuridine (BrdU, 10 mM) for 24 h (bottom section). Cells were prepared for indirect immunofluorescence microscopy using anti-BrdU antibody as described.

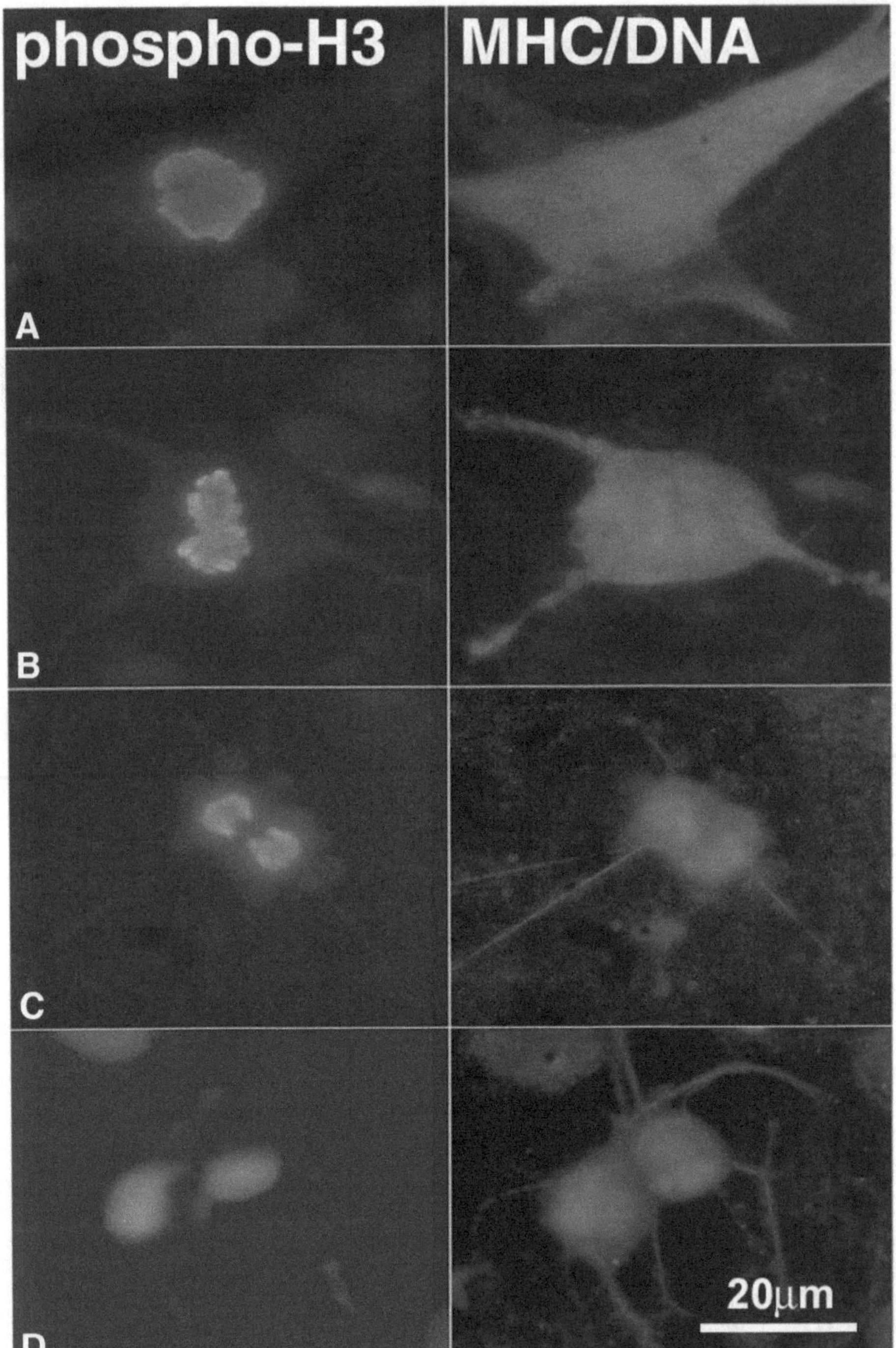
phospho-H3
MHC/DNA
A
B
C
D
20µm

5. Permeabilization buffer: 0.1% Triton X-100 in PBS. Prepare fresh on the day you intend to use it.
6. Blocking buffer: 5% (v/v) goat serum (*see* **Note 1**) or 5% (v/v) bovine serum albumin (BSA) fraction V, 0.2% Tween-20 in PBS. Stir for 60 min, and store aliquots at –20°C.
7. Wash buffer: 0.1% NP-40 (Roche) in PBS. Prepare fresh on the day you use it (*see* **Note 1**).
8. DNA stain: prepare a stock solution of Hoechst 33258 (*bis*-benzimide) at 1.0 mg/mL in H_2O. Store at 4°C protected from light. *Hoechst 33258 precipitates in phosphate buffers.* Working solution is diluted fresh 1000-fold in H_2O.
9. Antifade: DABCO (triethylenediamine; Sigma, cat. no. D 2522). DABCO is an antifade reagent. DABCO acts as a scavenger of free radicals produced by the excitation of fluorochromes and retards photobleaching of fluorescent dyes such as fluorescein isothiocyanate/tetrarhodamine isothiocyanate (FITC/TRITC). Also use glycerol, 99% and Tris-HCl, 1 *M*, pH 8.0. Combine 233 mg DABCO, 200 L Tris-HCl, 800 L H_2O, 9.0 mL glycerol. Dissolve at 70°C while intermittently vortexing. Aliquot and store at –20°C.
10. Antibodies: anti-E2F2 (mouse monoclonal antibody, 66711A, Transduction Laboratories), anti-E2F4 (sc-512), anti-p130 (sc-317), and anti-DP2, (sc-829) rabbit polyclonal antibodies (all from Santa Cruz Biotechnology) anti-phospho histone H3 (Ser10; NEB, cat. no. 9701), antisarcomeric myosin heavy chain (clone MF20; the University of Iowa Hybridoma Bank; *see* **Note 2**), and monoclonal rat anti-bromodeoxyuridine (BrdU; Roche. FITC/TRITC-conjugated secondary antibodies to rabbit, mouse, or rat immunoglobulins (whole molecule) were purchased from Transduction Laboratories.

3. Methods

3.1. Isolation of Neonatal Cardiomyocytes

For the isolation of ventricular cardiomyocytes from neonatal rats, please refer to **ref. 5**. Since cardiomyocytes withdraw from the cell cycle during the early postnatal stage, 4–5-d-old animals should be used. For elimination of noncardiomyoctes, cell suspensions are preplated twice (for 60 and 30 min) and held in serum-free medium in the presence of 10 mol/L cytosine arabinoside (AraC; Sigma) for 48 h. This procedure usually results in cultures containing 3–8% noncardiomyocyte cells and should be routinely monitored in parallel by indirect immunofluorescence staining with monoclonal antibody to sarcomeric myosin heavy chain (MHC; MAb1628, Chemicon; clone MF20, University of Iowa Hybridoma Bank).

Fig. 2. *(continued from opposite page)* Ectopic E2F2 induces cell cycle reentry and mitosis in cardiomyocytes. Cardiomyocytes were infected with adenovirus wild-type E2F2 (100 PFU/cell) and continued to be cultivated for 48 h. For the detection of M-phase–specific phosphorylation of histone H3, fixed cells were stained with rabbit polyclonal antibody to phospho-histone H3, which recognizes serine 10 phosphorylation of histone H3. Cardiomyocytes were identified by costaining for the expression of MHC (MF20) and genomic DNA (Hoechst 33258). Description of mitotic figures identified in cardiomyocytes: A, prophase; B, metaphase; C, early telophase; D, late telophase.

3.2. Indirect Immunofluorescence Staining of Fixed Cardiomyocytes Immobilized on Cover Slips

1. Cardiomyocytes are seeded at a density of 50,000 cells/well in a 24-well plate on collagen R-coated cover slips in a total volume of 600 L/well culture medium. On the day of analysis, wells are rinsed three times with 500 L PBS. *Use PBS lacking Ca/Mg when adult cardiomyocytes are to be analyzed.* For analysis of BrdU incorporation, cells are metabolically labeled with 10 *M* BrdU (Sigma, cat. no. B 9285).
2. Fixate the cells using 500 L/well fixation solution for 8 min with gentle agitation.
3. Discard the formalin fixate and rehydrate with 500 L/well PBS.
4. Cardiomyocytes are permeabilized with 500 mL/well PBS/0.1% Triton X-100 for 8 min with gentle agitation.
5. Remove the permeabilization solution, and rinse with 500 L/well PBS.
6. For detection of incorporated BrdU, denaturate the DNA by incubating the sample in 500 L/well of 2 *N* HCl/PBS for 30 min at 37°C. Remove the HCl/PBS solution, and rinse with 500 L/well PBS.
7. Add 500 L/well blocking solution (containing either goat serum or BSA), and incubate for 60 min while rocking.
8. Discard the blocking solution, and incubate the sample with the first primary antibody (e.g., anti-E2F, anti-BrdU) at a final concentration of 0.5–2.0 g/mL in blocking solution employing 150 L/well. The incubation time is 60 min at room temperature or preferentially overnight at 4°C with shaking. Both methods will give similar results, but the overnight step allows you to arrange the immunocytochemistry conveniently around your usual timetable in the lab. Use a humidified chamber for the overnight procedure to prevent samples from drying out. This is easily done by adding 5.0 mL water to the interspace between the wells and sealing the 24-well plate with Parafilm.
9. Remove the antibody solution, and wash once with 500 mL/well PBS/0.1% NP-40. In our lab, the antibody solution is stored at –20°C and reused two times for screening applications without significant loss of sensitivity.
10. Incubate the cells with 150 L/well of FITC-conjugated secondary antibody to rabbit IgG (diluted 100-fold in PBS/0.1% NP-40) when your primary antibody was raised in rabbit.
11. Remove the antibody solution, and wash once with 500 L/well PBS/0.1% NP-40.
12. Dilute the second primary antibody (usually a mouse monoclonal cardiac-specific antibody, e.g., MF20 directed against MHC) to 0.5–2.0 g (usually 100-fold) in a final volume of 150 L/well in PBS/0.1% NP-40, and incubate for 60 min while rocking.
13. Remove the antibody solution, and wash once with 500 L/well PBS/0.1% NP-40.
14. Incubate the cells with 150 L/well of TRITC-conjugated secondary antibody to mouse IgG (diluted 100-fold in PBS/0.1% NP-40), when your primary antibody was raised in mouse, for 60 min with constant agitation.
15. Remove the antibody solution, and wash once with 500 L/well PBS/0.1% NP-40.
16. Rinse with 500 L H_2O. (*Never do this with your Western blot.*)
17. For staining of genomic DNA, incubate your sample in 500 L/well Hoechst 33258 diluted 1000-fold in H_2O (1.0 mg/ml stock solution in H_2O). Keep in mind that Hoechst precipitates in PBS.
18. Add a little mounting solution (~30 L), sufficient to completely cover your cover slip on a microscope slide.
19. Remove the cover slip from the well, and briefly submerge it in H_2O. For your convenience, use a 500-mL beaker. Carefully remove excessive liquid by draining the edge and the back side of your slide with a paper towel.

20. Carefully place your cover slip (cells downside) directly onto the mounting solution drop applied on the microscope slide. *Avoid getting bubbles under your cover slip.*
21. Carefully seal the edge of your cover slip with nail polish. You must not shift the cover slips with the brush while applying the polish. Your sample is now ready for microscopic inspection. Specimens can be stored for short-term storage (<1 yr) at 4€ or indefinitely at –20€.

4. Notes

1. The choice of blocking reagent used in **step 7** depends on the source of your secondary fluorochrome-labeled antibody. When your antibody was raised in goat you must use goat serum; when your antibyody was raised in donkey you must use donkey serum for the blocking step. Alternatively, BSA fraction V may be used when a secondary antibody from another species is employed. The amount of BSA in the blocking buffer must be empirically determined for reducing background signals. As a starting point 5% BSA is recommended.
2. Cardiomyocyte-specific antibodies are mostly mouse monoclonal antibodies. Therefore, direct fluorochrome-conjugation is preferred when using, e.g., mouse antisarcomeric myosin in combination with other mouse monoclonals. In our lab, MF20-producing hybridomas were obtained from the University of Iowa Hybridoma Bank. Monoclonal antibodies are secreted into the culture medium of MF20 cultures. (Serum-free hybridoma culture medium is available from Gibco.) MF20 antibody is easily purified through Protein G-columns (Pharmacia). The isolated MF20 antibodies are then conveniently coupled to TRITC/rhodamine by a simple one-step reaction (instructions are available from Pierce), and unbound dye is removed by desalting columns (Pharmacia). The entire procedure does not require any special equipment.

References

1. Johnson, D. G., Schwarz, J. K., Cress, W. D., and Nevins, J. R. (1993) Expression of transcription factor E2F1 induces quiescent cells to enter S phase. *Nature* **365,** 349–352.
2. Kowalik, T. F., DeGregori, J., Schwarz, J. K., and Nevins, J. R. (1995) E2F1 overexpression in quiescent fibroblasts leads to induction of cellular DNA synthesis and apoptosis. *J. Virol.* **69,** 2491–2500.
3. Dyson, N. (1998) The regulation of E2F by pRB-family proteins. *Genes Dev.* **12,** 2245–2262.
4. Brown, V. D., Phillips, R. A., and Gallie, B. L. (1999) Cumulative effect of phosphorylation of pRB on regulation of E2F activity. *Mol. Cell. Biol.* **19,** 3246–3256.
5. von Harsdorf, R., Hauck, L., Mehrhof, F., Wegenka, U., Cardoso, M. C., and Dietz, R. (1999) E2F-1 overexpression in cardiomyocytes induces downregulation of p21CIP1 and p27KIP1 and release of active cyclin-dependent kinases in the presence of insulin-like growth factor I. *Circ. Res.* **85,** 128–136.
6. Kirshenbaum, L. A., Abdellatif, M., Chakraborty, S., and Schneider, M. D. (1996) Human E2F-1 reactivates cell cycle progression in ventricular myocytes and represses cardiac gene transcription. *Dev. Biol.* **179,** 402–411.
7. Hauck, L., Hansmann, G., Dietz, R., and von Harsdorf, R. (2002) Inhibition of hypoxia-induced apoptosis by modulation of retinoblastoma protein-dependent signaling in cardiomyocytes. *Circ. Res.* **91,** 782–789.
8. Vara, D., Bicknel, K. A., Coxon, C. H., and Brooks, G. (2003) Inhibition of E2F abrogates the development of cardiac myocyte hypoertrophy. *J. Biol. Chem.* **278,** 21388–21394.

14

Forkhead (FOX) Transcription Factors and the Cell Cycle

Measurement of DNA Binding by FoxO and FoxM Transcription Factors

Katrina A. Bicknell

Summary

Certain forkhead (FOX) transcription factors have been shown to play an intrinsic role in controlling cell cycle progression. In particular, the FoxO subclass has been shown to regulate cell cycle entry and exit, whereas the expression and activity of FoxM1 is important for the correct coupling of DNA synthesis to mitosis. In this chapter, I describe a method for measuring FoxO and FoxM1 transcription factor DNA binding in nuclear extracts from mammalian cells.

Key Words

Cell cycle; DNA binding; electromobility shift assays; forkhead transcription factors; FoxO; FoxM1.

1. Introduction

Members of the winged helix or forkhead box (FOX) transcription factor family have been shown to play essential roles in a diverse range of developmental and cellular processes, including cellular proliferation, differentiation, longevity, and cellular transformation (reviewed in **ref. *1***). Characterized by a highly conserved 110-amino acid 'winged helix' monomeric DNA binding domain, more than 100 FOX transcription factors have been described ***(1)***. In 2000, a unified nomenclature was proposed to simplify the identification and naming of the FOX transcription factors, which divided

From: *Methods in Molecular Biology, vol. 296, Cell Cycle Control: Mechanisms and Protocols*
Edited by: T. Humphrey and G. Brooks © Humana Press Inc., Totowa, NJ

all known chordate Fox transcription factors into 15 subclasses based on phylogenetic analysis *(2)*.

A number of different FOX subclasses have been directly or indirectly linked to cell cycle control, including the FoxO, FoxM, and FoxK subclasses. For example, the FoxO subclass has been implicated in controlling cell cycle entry and exit *(3,4)*, FoxM1 appears to control progression from DNA synthesis to mitosis *(5–7)*, and FoxK transcription factors have been shown to control cell cycle progression in myogenic cells via p21 *(8)*. In this chapter, I have focused on measurement of the DNA binding activity of FoxM1 and the FoxO transcription factors.

Members of the FoxO subclass of forkhead transcription factors have been implicated in the control of both cell cycle entry and cell cycle exit and include the *Caenorhabditis elegans* forkhead transcription factor DAF16 and five mammalian orthologs: FoxO1 (also known as forkhead in human rhabdomyosarcoma [FKHR]), FoxO2 (also called AF6q21), FoxO3 (also called FKHR-like 1 [FKHR-L1]), FoxO4 (also called acute lymphocytic leukemia-1 Fused gene from chromosome X [AFX]), and a recently described member, FoxO6 *(9–15)*. The dual role of FoxO transcription factors in controlling cell cycle entry and exit is reflected in the regulation of their subcellular localization and hence, activity by phosphorylation events (*see* **Fig. 1A** and **refs.** ***3***, ***4***, ***16***, and ***17***). Maintaining a cell cycle arrest, FoxO transcription factors have been reported to bind to and/or transactivate a number of promoters, including the p27 promoter *(16)*, insulin-like growth factor binding protein 1 (IGFBP1) *(18)*, and retinoblastoma pocket protein, p130 *(3)*. Moreover, FoxO transcription factors have been shown to control mitotic exit by inducing the expressions of cyclin B1 and polo-like kinase 1 mRNAs *(4)*.

The DNA binding and transcriptional activity of FoxO transcription factors is mediated, at least in part, by the activation of the phosphatidylinositol 3-kinase (PI3K)/protein kinase B (PKB; also known as Akt), which results in the inactivation of FoxO transcription factors (**Fig. 1**; reviewed in **ref.** ***19***, and also *see* **refs.** ***18*** and ***20–23***). PKB-mediated phosphorylation of FoxO transcription factors occurs in the nucleus and induces the relocation of FoxO from the nucleus to the cytoplasm *(20,21,24)*. The intracellular trafficking of FoxO is believed to involve 14-3-3 proteins, which bind to nuclear FoxO in a phosphorylation-dependent manner, inhibiting DNA binding and masking FoxO transcription factors nuclear localization signal (NLS) *(21,23,24)*. Moreover, the cytoplasmic retention of phosphorylated FoxO transcription factors enhances their ubiquitination-mediated degradation *(25,26)*. However, prolonged activation of the PI3K/PKB pathway, and hence inactivation of FoxO transcription factors, results in an accumulation of cells in G_2/M, defective cytokinesis, and a delayed M- to G_1-phase transition *(4)*. Hence, it appears that attenuation of the PI3K/PKB signaling pathway and subsequent reactivation of FoxO transcription factors is required for G_2- to M-phase progression and cytokinesis *(4)*.

Another forkhead transcription factor that plays a role in regulating cell progression is FoxM1 (also known as Trident, WIN, HFH-11, and MPP2; **27–30**). FoxM1 is ubiquitously expressed in all proliferating mammalian cells, with its expression levels rising at the start of DNA synthesis and persisting until the end of mitosis *(5)*. In M-

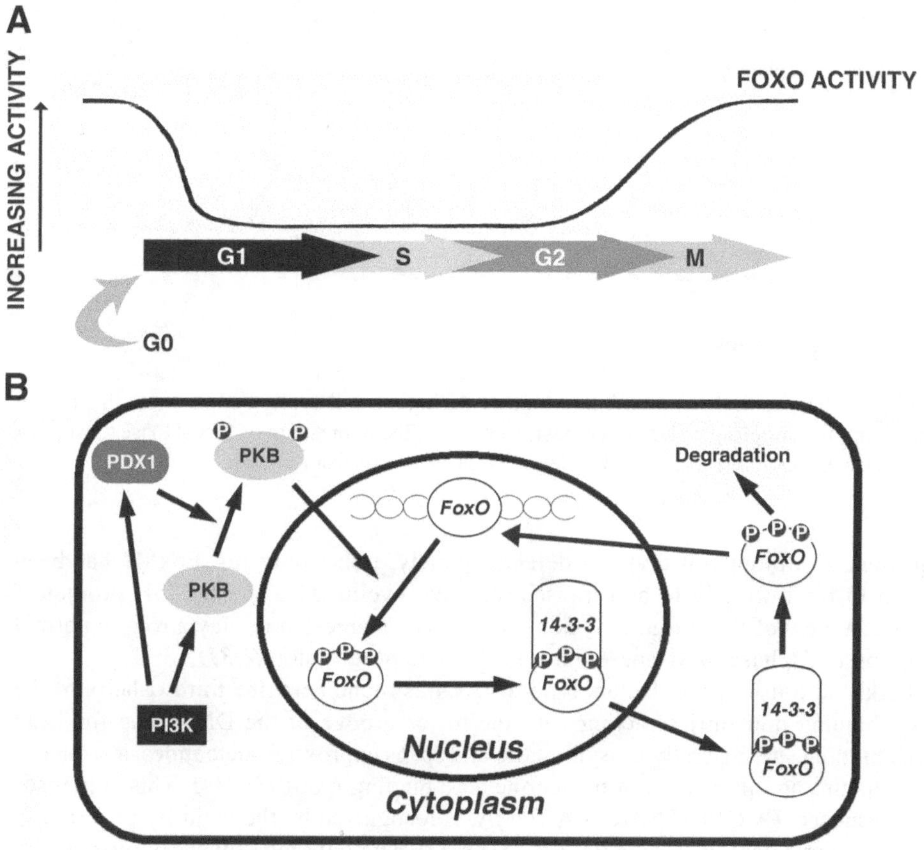

Fig. 1. Regulation of the activity of FoxO transcription factors during the cell cycle. **(A)** The activity of FoxO transcription factors is regulated throughout the cell cycle. In G_0, the activities of FoxO transcription factors maintain cells in quiescence. Upon mitogenic stimulation, FoxO transcription factors are phosphorylated and translocate from the nucleus into the cytoplasm, and cells progress into G_1. FoxO transcription factors also play a role in exit from mitosis, with activity levels rising in the latter stages of the cell cycle. **(B)** The activity of FoxO transcription factors is regulated by the phosphatidylinositol-3 kinase/protein kinase B (PI3K/PKB) signaling pathway. Phosphorylation of FoxO by activated PKB in the nucleus results in the association of FoxO with 14-3-3 proteins and translocation of this resulting complex to the cytoplasm. In the cytoplasm, the FoxO-14-3-3 complex disassociates, and FoxO is either degraded or imported back into the nucleus.

phase, FoxM1 is phosphorylated and degraded before M-phase exit (**Fig. 2**) ***(5)***. Overexpression of FoxM1 accelerated cell cycle time and, consistent with a role in the latter phases of the cell cycle, elevated FoxM1 levels and increased cyclin B1, but not cyclin D1 expression ***(28,30,31)***. Generation of FoxM1 null mice further implicated a role for FoxM1 in cell cycle control ***(6)***. Dying shortly after birth, homozygous FoxM1

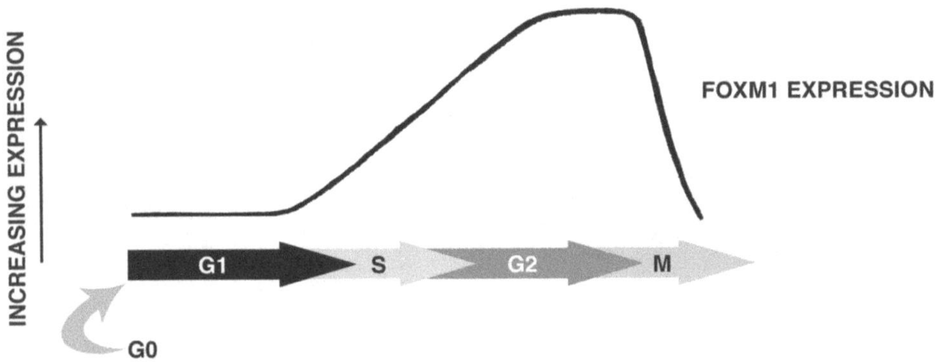

Fig. 2. Changes in the expression of FoxM1 during the cell cycle. Consistent with a role in controlling the coupling of DNA synthesis to mitosis, the expression of FoxM1 rises just prior to S-phase, peaks in G_2-phase, and is degraded prior to exit from mitosis.

null mice exhibited cell division defects, namely, polyploidy *(6)*. FoxM1 has been shown to bind directly to and transactivate the cyclin B1 and CDC25B promoters *(7,31)*. In view of these findings, FoxM1 has been suggested to play a role in normal coupling of S-phase to M-phase during cell cycle progression *(6,31)*.

Forkhead transcription factors bind to DNA as monomers, the third α-helix of the DNA binding domain interacting with the major groove of the DNA. The forkhead DNA binding site generally is asymmetric and spans approx 15 nucleotides, a sequence that contains an optimal seven-nucleotide core binding motif *(32–34)*. This consensus core sequence, (A/G)(C/T)(A/C)AA(C/T)A, is recognized by the majority of forkhead transcription factors, including FoxM1 *(5,32–34)*. The flanking sequences surrounding the core consensus sequence also play a role in the binding affinity of the forkhead transcription factor family *(32)*. Some subclasses of the Forkhead transcription factor family, including the FoxO subclass, bind sequences that partially diverge from this consensus motif, such as the insulin response elements, CAAAACAA or TTATTTG *(18,21)*. Indeed, the binding efficiencies of different FoxO family members to these sequences also have been reported to vary. For example, FoxO1 will bind to and activate the DAF16 family protein binding element 5'-AAGTAACAACTATGTAAACAA-3', whereas FoxO4 is only minimally responsive to this element, and FoxO3a will not bind *(35)*. The consensus DNA binding motif of FoxO members has been described as TTGTTTAC *(36)*.

In this chapter, an electromobility shift assay (EMSA) protocol for measuring DNA binding of FoxM1 or FoxO transcription factors in mammalian cell nuclear extracts is described. The method has also been used to measure the DNA binding of a variety of transcription factors including E2F transcription factors *(37)*. The electrophoretic mobility shift assay, also called gel retardation or gel shift assay, is one of the most common techniques used to study specific interactions between transcription factors and DNA. This method measures the binding of a specific transcription factor to a

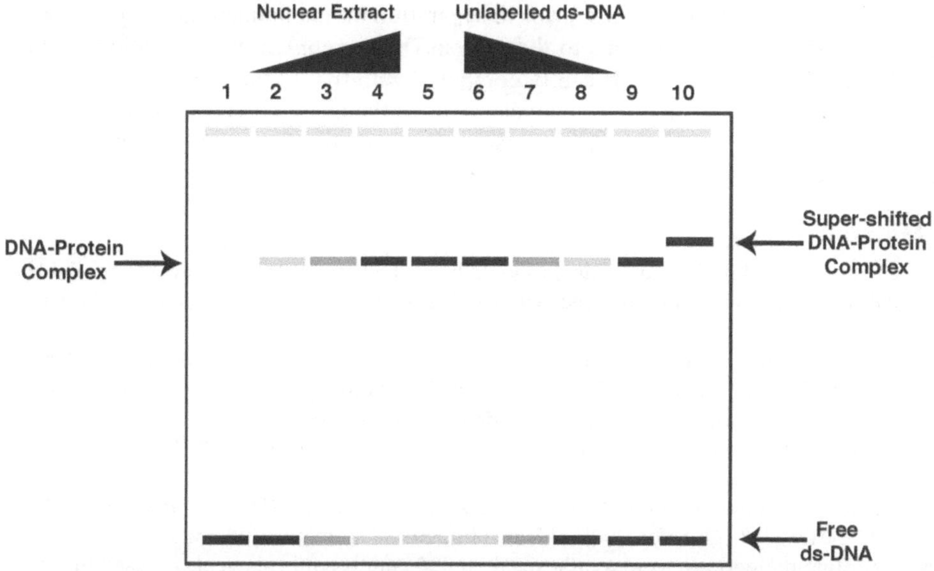

Fig. 3. Diagrammatic representation of an electromobility shift assay. DNA/protein binding reactions are separated by nondenaturing electrophoresis. Unbound double-stranded (ds) DNA probe migrates unhindered and is indicated at the bottom of the gel. When the nuclear extract is omitted from the binding reaction, only the free probe is observed (lane 1). DNA/protein complexes migrate more slowly, and the intensity of the shifted band increases with increasing amounts of protein bound (lanes 2–4). Nonspecific binding of proteins to the DNA probe can be visualized by the use of several important controls. Addition of an unlabeled noncompetitive double-stranded DNA should not affect the shift or intensity of the observed shifted DNA-protein complex (lane 5). However, addition of increasing amounts of unlabeled competitive probe should result in a progressive decrease in the intensity of the shifted band (lanes 6–8). If an appropriate antibody is available, supershift assays can also be performed to verify the identity of the transcription factor(s) present in the shifted DNA/protein complex. Addition of antibody results in an additional shift in mobility (lane 10) compared with DNA/protein complexes alone (lane 9).

radioactively labeled double-stranded oligonucleotide that contains the transcription factor's consensus binding motif. Stable transcription factor–DNA interactions are visualized following nondenaturing polyacrylamide gel electrophoresis, since binding of the transcription factor results in a shift in the mobility of the labeled DNA (**Fig. 3**). Specificity of binding is determined using competition studies, i.e., unlabeled double-stranded DNA oligonucleotides containing the binding consensus sequence, and/or unrelated or mutated DNA sequences. The addition of unlabeled double-stranded DNA containing the transcription factor binding motif should result in a reduction in the intensity of the shifted band. The identity of bound transcription factors can also be determined using super-shift EMSA assays. In super-shift assays, a specific antibody that recognizes the transcription factor of interest is incubated with the protein/DNA complexes, prior to separation on the nondenaturing polyacrylamide gel. If the antibody binds to the protein/DNA complex, the migration of the complexes will be shifted significantly compared with that of the protein/DNA complexes alone (**Fig. 3**). Super-shift analysis requires the availability of a suitable antibody, an antibody that recognizes the native form of the transcription factor when bound to DNA.

2. Materials

1. Chemicals used in these procedures should be of the best grade commercially available. All solutions should be prepared with double-distilled sterile water unless otherwise stated.
2. Nuclear extraction buffer 1: 10 m*M* HEPES, pH 7.9, 10 m*M* KCl, 1.5 m*M* $MgCl_2$, 0.3 m*M* Na_3VO_4, 200 *M* leupeptin, 10 *M* aprotinin, 5 m*M* dithiothreitol (DTT), 300 *M* 4-(2-aminoethyl)benzenesulfonyl fluoride hydrochloride (AEBSF).
3. Nuclear extraction buffer 2: nuclear extraction buffer 1 containing 0.1% IGEPAL (an NP-40 substitute).
4. Nuclear extraction buffer 3: 20 m*M* HEPES, pH 7.9, 420 m*M* NaCl, 1.5 m*M* $MgCl_2$, 0.2 m*M* EDTA, 25% glycerol, 200 *M* leupeptin, 10 *M* aprotinin, 300 *M* AEBSF.
5. All oligonucleotides were synthesized (0.025 mol synthesis scale), desalted, and deprotected by Sigma-Genosys (Cambridgeshire, UK). The sequences of the oligonucleotides used are listed in **Table 1**. Lyophilized oligonucleotides must be resuspended prior to use. Oligonucleotides were dissolved in sterile double-distilled water at a concentration of 100 *M* and stored at –20°C until use.
6. Redivue [γ-^{32}P]ATP was purchased from Amersham Biosciences (Buckinghamshire, UK). Appropriate care must be taken when using [^{32}P]ATP, and local rules governing the use of radiochemicals must be followed. For example, appropriate Perspex shielding should be employed at all times when [^{32}P]ATP is present, such as during the labeling of double-stranded oligonucleotides or when using labeled double-stranded DNA probes in binding reactions. Following use of [^{32}P]ATP, work areas should be monitored for ^{32}P contamination and, if detected, decontaminated appropriately.
7. T4 polynucleotide kinase (T4-PNK) and 10X T4-PNK reaction buffer (Promega, Southampton, UK).
8. Labeled double-stranded oligonucleotides were purified using Microspin G25 columns (Amersham Biosciences). These prepacked Sephadex G-25 columns are used for speed and convenience. Other similar methods for probe purification can be utilized and have been described ***(38)***.
9. TE buffer: 10 m*M* Tris-HCl, pH 8, 1 m*M* EDTA, pH 8.
10. 4X EMSA binding buffer: 400 m*M* KCl, 80 m*M* HEPES, pH 7.9, 2 m*M* DTT, 0.8 m*M* EDTA, pH 8, 80% glycerol. A stock solution of EMSA binding buffer can be made, omitting the DTT. Immediately prior to use, add an appropriate amount of DTT to a small volume of the stock buffer.
11. Double-stranded alternating copolymer, poly-deoxy-inosinic-deoxy-cytidylic acid (poly[dI-dC]) was purchased from Roche Applied Science (East Sussex, UK). Poly(dI-dC) is often sold in units, where 1 U is equivalent to 50 g poly(dI-dC) and will yield an absorbance of 1 at OD_{260}.
12. 5X Tris-borate-EDTA (TBE) buffer: Add 54 g tris base, 27.5 g boric acid, and 20 mL 0.5 *M* EDTA, pH 8.0, to 800 mL double-distilled water. Mix until all components dissolve and adjust to 1L using double distilled water. Store the 5X stock solution in a glass bottle

Table 1
Oligonucleotides for Measuring FoxO and FoxM1 Binding by EMSA

Transcription factor	Oligonucleotide name	Sense oligonucleotide (5'–3')	Complementary oligonucleotide (5'–3')	Ref.
FoxO1 (FoxO4)	DAF16	AAGTAAACAACTATGTAAACAA	TTGTTTACATAGTTGTTTACTT	***35,36***
FoxO	IGFBP1	ATTAGATCTTAAATAAATAGATCTTTA	TAAAGATCTATTTATTTAAGATCTATT	***18***
FoxO	p130	CATAAATAAATAAGTAAACAAATAAA	TTTATTTGTTTACTTATTTATTTATG	***21***
FoxM1	2X	AGCTTGATTGTTTATAAACAGCCCGGG	CCCGGGCTGTTTATAAACAATCAAGCT	***5***
FoxM1	0X	AGCTTGATTGCCCATCCCCACCGGG	CCCGGTGGGGATGGGCAATCAAGCT	***5***

at room temperature. Upon storage of concentrated TBE solutions, a precipitate is sometimes observed. If a precipitate forms, discard the batch and replace.

13. 10% Ammonium persulfate (APS): dissolve 1 g APS in sterile double-distilled water to a final volume of 10 mL. Aliquots can be stored at –20°C until use.
14. 6% Polyacrylamide gel solution: 6% (v/v) Accugel acrylamide/*bis*-acrylamide (29:1) solution (National Diagnostics, Yorkshire, UK), 0.5X TBE, 0.1% (w/v) APS, 0.01% (v/v) N,N,N′,N′-tetramethylethylenediamine (TEMED).
15. 10X Gel loading dye: 250 m*M* Tris-HCl, pH 7.5, 0.2% bromophenol blue, 40% glycerol.

3. Methods

3.1. Preparation of Nuclear Extracts

3.1.1. Extraction of Nuclear Proteins

Nuclear extracts were prepared using a method based on a protocol first described by Dignam et al. ***(39)***. This modified method allows the rapid preparation of nuclear extracts without the need for dounce homogenization or dialysis. This method has been successfully employed by both our laboratory and others to study a range of different transcription factors, including E2Fs and ATF2 and c-Jun ***(37,40)***.

1. Harvest individual cells using standard techniques (e.g., trypsinization of adherent cell monolayers). Collect individual cells by gentle centrifugation (500*g* for 5 min; *see* **Note 1**).
2. Wash cells in 1 mL 1X PBS, and transfer to a 1.5-mL tube. Centrifuge at low speed (500*g*) for 5 min. Discard supernatant, and estimate the packed volume of cells. A maximum of 250 L packed cell volume can be prepared in a 1.5-mL tube.
3. For every 50 L packed cell volume, resuspend the cell pellet in 150 L ice-cold nuclear extraction buffer 1. Adjust volumes used accordingly.
4. Vortex cell pellet for 15 s at maximum speed. Incubate on ice for 10 min, vortex again for 15 s, and then centrifuge at 10,000*g* for 5 min at 4°C.
5. Remove the supernatant, the cytoplasmic fraction, and store, if required, in aliquots at –80°C. It might be necessary to wash the pellet to remove additional cytoplasmic proteins. If this is required, wash the pellet once with 500 L ice-cold 1X PBS
6. Resuspend the pellet in 150 L nuclear extraction buffer 2. Vortex to mix, incubate on ice for 10 min, and then vortex again. Centrifuge at 10,000*g* for 5 min at 4°C. Discard the supernatant fraction.
7. Resuspend the nuclear pellet in 50 L nuclear extraction buffer 3 and incubate samples on ice for 1 h, with regular vortexing (every 5–10 min).
8. Centrifuge samples at 10,000*g* for 5 min at 4°C. Transfer nuclear extracts (supernatant) to fresh tube, aliquot into 5 L aliquots, and snap freeze in liquid nitrogen (*see* **Note 2**). Store aliquots of nuclear extracts at –80°C. Retain one aliquot for protein quantification (*see* **Note 3**).

3.2. Estimation of Protein Concentrations of Nuclear Extracts

Protein concentrations can be determined by the method described by Bradford ***(41)***. This is a simple colorimetric assay that measures the color change occurring when proteins are complexed with Brilliant Blue G dye.

1. Prepare a protein solution of known concentration, e.g., a solution of bovine serum albumin (BSA), type V (Sigma-Aldrich) at 1 mg/mL in sterile double-distilled water. This protein standard should be stored frozen at –20°C in 80 L aliquots for future use.

2. In duplicate, prepare a series of dilutions of the protein standard (e.g., 1, 2, 5, 10, and 20 L) in 800 L double-distilled water. Similarly, prepare suitable dilutions of the protein sample of unknown protein concentration in 800 L double-distilled water. The dilution required will depend on the concentration of unknown protein; a suitable starting dilution for nuclear extracts is 1 L protein in 800 L double-distilled water.
3. Add 200 L Bradford reagent (Sigma), vortex to mix, and incubate for 5–30 min.
4. Using disposable plastic spectrophotometer cuvets, measure the absorbance at 595 nm using a spectophotometer.
5. Protein concentrations of unknown samples then can be determined using the BSA standard curve.

3.3. Electrophoretic Mobility Shift Assays

EMSAs measure the effect that binding of protein(s) has on the electrophoretic mobility of the double-stranded DNA probe. This section describes how to prepare and radioactively label double-stranded oligonucleotide probes, DNA–protein binding reactions, and nondenaturing polyacrylamide gel electrophoresis (PAGE).

3.3.1. Annealing of Single-Stranded Complementary Oligonucleotides

1. Resuspend lyophilized oligonucleotides in sterile double-distilled water to generate a stock solution of 100 *M*. Transfer equal volumes of sense and antisense complementary oligonucleotides to a sterile 1.5-mL plastic tube.
2. Incubate oligonucleotides at 95°C for 5 min in a heating block or water bath. Turn off heating block or water bath, and allow complementary oligonucleotides to anneal by cooling slowly to room temperature (*see* **Note 4**).
3. Store double-stranded oligonucleotides at –20°C until required.

3.3.2. Radioactive Labeling of Double-Stranded Oligonucleotides

Double-stranded oligonucleotides are labeled using T4 PNK. T4 PNK catalyzes transfer of the γ-phosphate group from [γ-^{32}P]ATP to the 5'-hydroxyl terminus of the oligonucleotide.

1. In a sterile microcentrifuge tube, add in the following order: 10 pmol double-stranded oligonucleotide, 2.5 L 10X T4 PNK kinase buffer, 10 U T4 PNK enzyme, 25 Ci [γ-^{32}P]ATP. Adjust the reaction volume to 25 L using sterile double distilled water, and mix gently.
2. Incubate labeling reaction at 37°C for 30 min.
3. Stop labelling reaction by adding 2 L of 0.5 *M* EDTA, pH 8.0. Mix well (*see* **Note 5**).
4. Briefly spin labeling reaction in a microcentrifuge to collect any condensation that has formed. Store reaction on ice prior to purification (*see* **Subheading 3.3.3.**).

3.3.3. Purification of Labeled Probe

Although often considered an optional step, the removal of unincorporated [γ-^{32}P]ATP from the labeled double-stranded oligonucleotide probe will significantly reduce the background observed in EMSAs. There are many methods for purifying radioactive probes, and the choice of which will depend on the type of probe used, cost, and time constraints. For oligonucleotide probes greater than 10 nucleotides in length, G25 Sephadex columns are routinely used. Ethanol precipitation can also be used to purify larger double-stranded probes. In our laboratory, Microspin G-25 columns (Amersham Biosciences, Buckinghamshire, UK) are used for convenience.

1. Prepare Microspin G-25 column by resuspending the resin by vortexing gently. Loosen the cap, and break off the bottom closure.
2. Place column in a sterile 1.5-mL microcentrifuge tube, and spin column at 700*g* for 1 min (*see* **Note 6**).
3. Wash column once with 200 L TE buffer.
4. Place column in a fresh sterile 1.5-mL microcentrifuge tube, and carefully apply the probe to the center of the resin bed. Do not disturb the resin (*see* **Note 7**).
5. Spin the column for 2 min at 700*g*. Unincorporated nucleotides will be retained in the column, and purified probe will be collected in the fresh microcentrifuge tube.
6. Discard the column containing unincorporated radioactive nucleotides.
7. Purified labeled double-stranded oligonucleotide probes should be stored at –20°C (*see* **Note 8**).

3.3.4. Preparation of a Nondenaturing Polyacrylamide Gel

A 6% nondenaturing polyacrylamide gel should be prepared before setting up binding reactions, as it takes at least 2 h to prepare and prerun the gel prior to loading the DNA/protein complexes. Routinely, our laboratory separates DNA/protein complexes on 20 × 20-cm, 1.5-mm-thick nondenaturing polyacrylamide gels. Other gel sizes also can be used.

1. Clean glass plates, spacers, and comb well, and assemble the plates and spacers ready to pour the gel (*see* **Note 9**).
2. For a 20 × 20-cm gel, combine in the following order: 8 mL 30% acrylamide/*bis*-acrylamide (29:1) solution, 4 mL 10X TBE, 800 L 10% (w/v) APS, and 80 L TEMED. Briefly mix the gel solution prior to pouring the gel (*see* **Note 10**). Insert the comb, ensuring that bubbles are excluded. Allow gel to polymerize for 1 h at room temperature (*see* **Note 11**).
3. Once it is polymerized, carefully remove the comb from the gel. Carefully wash any unpolymerized acrylamide from the wells using 0.5X TBE buffer.
4. Prerun the gel in 0.5X TBE for 1 h at 100 V in a cold room (4°C; *see* **Note 12**).

3.3.5. Protein/DNA Binding Reactions

The DNA binding reactions should be performed while the polyacrylamide gel is prerun (*see* **Subheading 3.3.4.**). Nuclear extracts are mixed with the radioactively labeled DNA probe, and the mixture is incubated. During this incubation, specific proteins bind to the DNA consensus binding sequences in the probe. Nonspecific DNA binding is reduced by the addition of carrier DNA, poly(dI-dC) (*see* **Note 13**). Specificity of the DNA-protein binding is verified using a series of control reactions. First, in one binding reaction, the nuclear extract is omitted, to demonstrate the migration of the free DNA probe. The addition of unlabeled DNA probe to another binding reaction verifies the specificity of the DNA/protein complex because the "cold" probe should compete for transcription factor binding and thus reduce the intensity of the shifted DNA/protein complex. Finally, an unlabeled DNA probe, known not to contain the binding motif for that transcription factor, is added to a binding reaction. In this noncompetitive control binding reaction, the shifted DNA/protein complexes should be unaffected, since the transcription factor(s) measured should only bind to the labeled DNA probe.

1. Thaw Nuclear Extracts on ice (*see* **Note 14**).
2. To prepare a working stock of EMSA binding buffer, add 1 L 1 *M* DTT to 500 L EMSA binding buffer, and mix well.
3. For each binding reaction, to a fresh 1.5-mL microcentifuge tube, add 5 L EMSA binding buffer, 1 L poly(dI-dC) (*see* **Note 13**), and 10–20 g nuclear extract. Adjust reaction volume to 20 L using sterile double-distilled water.
4. For noncompetitive and competitive control binding reactions, to fresh 1.5-mL microcentrifuge tubes, add 1 L cold competitive or noncompetitive double-stranded oligonucleotides (*see* **Subheading 3.3.1.**) to 5 L EMSA binding buffer, 1 L poly(dI-dC), and 10–20 g nuclear extract. Adjust reaction volume to 20 L using sterile double-distilled water. Mix well, and incubate on ice for 20 min prior to addition of the radioactively labeled probe.
5. For a probe-only control reaction, to a fresh 1.5-mL microcentrifuge tube add 5 L EMSA binding buffer and 1 L poly(dI-dC). Adjust reaction volume to 20 L using sterile double-distilled water.
6. Add 1 L ^{32}P-labeled double-stranded oligonucleotide to each binding or control reaction. Mix gently, and incubate for 10 min at room temperature (*see* **Note 15**).
7. To each binding reaction, add 5 L 5X EMSA loading dye, and mix gently. Store reactions on ice until polyacrylamide gel is ready to load.

3.3.6. Polyacrylamide Gel Electrophoresis

1. Load 25 L samples on the prerun gel, and run the gel at 4°C in 0.5X TBE buffer at 200 V for 2.5 h. The bromophenol blue dye front should have migrated approximately three-fourths of the way down the gel (*see* **Note 16**).
2. Once the gel has run, carefully transfer the gel onto 3MM paper, and cover with plastic wrap. Dry the gel in a gel dryer under vacuum for 2 h at 80°C.
3. Expose the dried gel to a storage phosphor screen overnight, and scan the screen using a PhosphorImager system. Alternatively, the dried gel can be exposed to X-ray film with intensifying screens overnight at –70°C (*see* **Note 17**). Examples of representative FoxM1 and FoxO electromobility shift assays are shown in **Figs. 4** and **5**, respectively.

4. Notes

1. Alternatively, cell monolayers can be washed with ice-cold PBS and cells scraped directly into ice-cold nuclear extraction buffer 1. Use 150 L nuclear extraction buffer 1 for cells cultured in a single 60-mm dish. Adjust volumes accordingly.
2. Nuclear extracts must be snap-frozen in liquid nitrogen and stored at –80°C to retain their integrity. Frozen aliquots are stable for up to 6 mo at –80°C. Once thawed, nuclear extracts should not be refrozen and should be discarded.
3. Other methods of protein quantification can be used. If using an alternative method, ensure that the method is compatible with the detergents used.
4. Annealing of oligonucleotides can also be performed in a water bath or thermocycler. If you are using a water bath, heat oligonucleotides and then switch off the water bath. If you are using a thermocycler, establish a program that holds the oligonucleotide mixture at 95°C for 5 min and then gradually drops the temperature 1°C every minute until room temperature is reached.
5. Some end-labeling protocols will recommend that the labeling reaction be heated at 65°C for 10 min to inactivate the T4 PNK enzyme. This will result in denaturation of the double-stranded oligonucleotide probe and should be avoided.

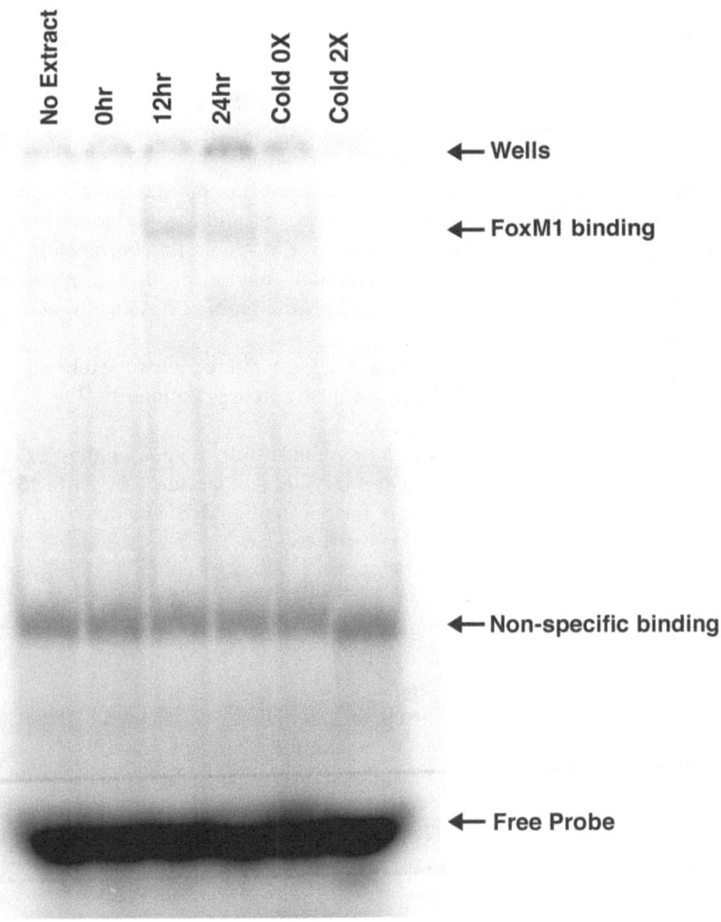

Fig. 4. A representative electromobility shift assay demonstrates FoxM1 binding in the Rat1 fibroblast cell line. Rat1 cultures were serum-starved for 48 h prior to stimulation with 10% FCS for 0, 12, or 24 h. Treated cells were harvested and FoxM1 binding measured by EMSA performed as described in **Subheading 3.3.** DNA binding to labeled double-stranded “2X” oligonucleotide (*see* **Table 1**) is shown. Migrating free probe (Free Probe) and the shifted band indicating specific FoxM1 binding (FoxM1) are shown by arrows. The probe-only (No Extract) and cold “2X” (Cold 2X) and “0X” (Cold OX) competition assay controls are also indicated. The FoxM1 DNA binding observed is consistent with the reported expression of FoxM1 during cell cycle progression, with FoxM1 levels rising in late G_1, through S-phase into G_2/M, where it is degraded prior to exit from mitosis.

6. In a standard benchtop microcentrifuge, 700*g* corresponds to approx 2000 rpm.
7. Application of the labeling reaction to the center of the resin bed without disturbing the resin is important to achieve maximum separation of labeled probe from unincorporated nucleotides.

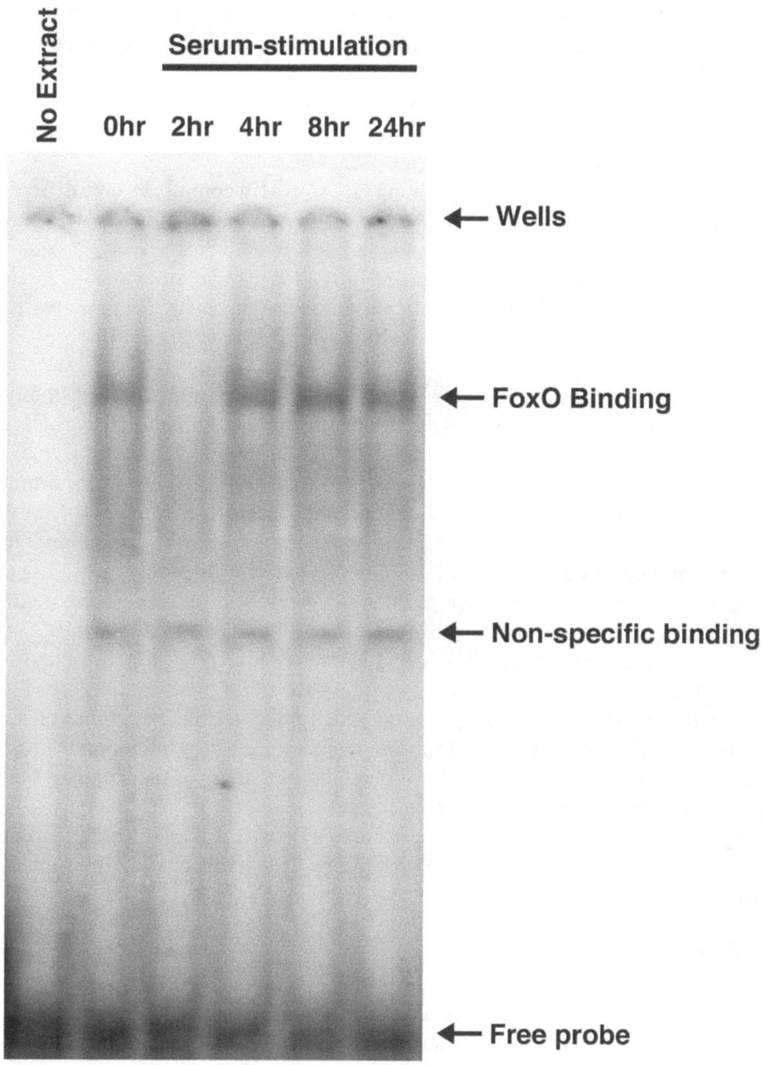

Fig. 5. A representative electromobility shift assay demonstrates changes in FoxO binding in nuclear extracts from serum-stimulated rat cardiomyocytes. Primary rat neonatal cardiomyocytes were serum-starved from 48 h prior to stimulation with 20% FCS for 0, 2, 4, 8 or 24 h. Treated cells were harvested and FoxO binding measured by EMSA performed as described in **Subheading 3.3.** DNA binding to labeled double-stranded DAF16 oligonucleotide (*see* **Table 1**) is shown. Migrating free probe (Free probe), a nonspecific shifted band, and the shifted band indicating specific FoxO binding (FoxO) is shown by arrows. Probe-only control reaction (No Extract) is also indicated. FoxO DNA binding in cardiomyocytes decreases transiently with serum stimulation but levels return to that of serum-starved cultures within 4 h.

8. The labeled double-stranded oligonucleotide probes can be stored at –20°C for up to 1 wk. After this time, the intensity and quality of the shifted DNA/protein complexes on resultant EMSA will deteriorate.
9. The binding interaction between DNA and certain transcription binding factors is very sensitive to detergents. It is necessary to ensure that all polyacrylamide electrophoresis equipment and glassware used is detergent-free.
10. Other sizes of gels can be used. If separating DNA/protein complexes on a different sized gel, please adjust volumes of gel mixture. For a minigel, 10 mL of 6% nondenaturing gel solution will be sufficient.
11. It is important to allow the gel to polymerize fully. It is also possible to pour a gel the day before it is needed and allow it to polymerize overnight at 4°C. If you are storing a prepoured gel overnight, wrap the gel well in plastic film to ensure that gel shrinkage does not occur.
12. The stability of DNA/protein complexes can be affected by temperature. Although some EMSAs can be performed successfully at room temperature, our laboratory routinely runs all EMSAs at 4°C.
13. Sheared salmon sperm DNA is sometimes used as a carrier DNA. For EMSAs this should be avoided, as the sheared DNA might contain the consensus binding motif for the transcription factor being studied and interfere with the transcription factor binding, much like a cold competitor control. Use of poly(dI-dC) avoids this problem.
14. It is important to thaw nuclear extracts on ice, since it will reduce protein degradation. Protein degradation might result in the observation of multiple shifted bands.
15. The binding interaction between DNA and certain transcription binding factors can be very sensitive to temperature; binding reactions are often incubated on ice. FoxO and FoxM1 transcription factors appear to bind equally well to the DNA binding sequences used in this protocol when incubated at room temperature for 10 min or on ice for 30 min.
16. If the bromophenol blue dye front has migrated three-quarters of the length of the gel, the free probe front should still be visible on the gel.
17. Exposure times will vary. Overnight exposure to a Phosphorimager storage screen is sufficient to visualize FoxMI and FoxO DNA binding. If you are using X-ray film, extended exposure time might be necessary.

References

1. Carlsson, P. and Mahlapuu M. (2002) Forkhead transcription factors: key players in development and metabolism. *Dev. Biol.* **250,** 1–23.
2. Kaestner, K. H., Knochel, W., and Martinez, D. E. (2000) Unified nomenclature for the winged helix/forkhead transcription factors. *Genes Dev.* **14,** 142–146.
3. Kops, G. J. P. L., Medema, R. H., Glassford, J., et al. (2002) Control of cell cycle exit and entry by protein kinase B-regulated forkhead transcription factors. *Mol. Cell. Biol.* **22,** 2025–2036.
4. Alverez, B., Martinez-A., C., Burgering, B. M. T., and Carrera, A. C. (2001) Forkhead transcription factors contribute to execution of the mitotic programme in mammals. *Nature* **413,** 744–747.
5. Korver, W., Roose, J., and Clevers, H. (1997) The winged-helix transcription factor Trident is expressed in cyclin cells. *Nucleic Acids Res.* **25,** 1715–1719.
6. Korver, W., Schilham, M. W., Moerer, P., et al. (1998) Uncoupling of S phase and mitosis in cardiomyocytes and hepatocytes lacking the winged-helix transcription factor Trident. *Curr. Biol.* **8,** 1327–1320.

7. Wang, X., Kiyokawa, H., Dennewitz, M. B., and Costa, R. H. (2002) The forkhead box M1b transcription factor is essential for hepatocyte DNA replication and mitosis during mouse liver regeneration. *Proc. Natl. Acad. Sci. USA* **99,** 16881–16886.
8. Hawke T. J. Jiang, N., and Garry, D. J. (2003) Absence of p21^{CIP} rescues myogenic progenitor cell proliferative and regenerative capacity in Foxk1 null mice. *J. Biol. Chem.* **278,** 4015–4020.
9. Ogg, S., Paradis, S., Gottlieb, S., et al. (1997) The Fork head transcription factor DAF-16 transduces insulin-like metabolic and longevity signals in *C. elegans*. *Nature* **389,** 994–999.
10. Lin K, Dorman JB, Rodan A, Kenyon C. (1997) Daf-16: an HNF-3/forkhead family member that can function to double the life-span of *Caenorhabditis elegans*. *Science* **278,** 1319–1322.
11. Galini, N., Davis, R. J., Fredericks, W. J., et al. (1993) Fusion of a fork head domain gene to PAX3 in the solid tumour alveolar rhabdomyosarcoma. *Nat. Genet.* **5,** 230–235.
12. Hillion, J., Le Coniat, M., Jonveaux, P., Berger, R., and Bernard, O.A. (1997) AF6q21, a novel partner of the MLL gene in t(6;11)(q21;q23), defines a forkhead transcriptional factor subfamily. *Blood* **90,** 3714–3719.
13. Anderson, M. J., Viars, C. S., Czekay, S., Cavenee, W. K., and Arden, K. C. (1998) Cloning and characterization of three human forkhead genes that comprise an FKHR-like gene subfamily. *Genomics* **47,** 187–199.
14. Borkhardt, A., Repp, R., Haas, O.A., Leis, T., Harbott, J., Kreuder, J., Hammermann, J., Henn, T. and Lampert, F. (1997) Cloning and characterization of AFX, the gene that fuses to MLL in acute leukemias with a t(X;11)(q13;q23). Oncogene. 14, 195202.
15. Jacobs, F. M. J., van der Heide, L. P., Wijchers, P. J. E. C., Burbach, P. H., Hoekman, M. F. M., and Smidt, M. P. (2003) FoxO6, a novel member of the FoxO class of transcription factors with distinct shuttling dynamics. *J. Biol. Chem.* **278,** 35959–35967.
16. Dijkers, P. F., Medema, R. H., Pals, C., et al. (2000) Forkhead transcription factor FKHR-L1 modulates cytokine-dependent transcriptional regulation of p27^{KIP1}. *Mol. Cell. Biol.* **20,** 9138–9148.
17. Schmidt, M., Fernandezde Mattos, S., van der Horst, A., et al. (2002) Cell cycle inhibition by FoxO forkhead transcription factors involves downregulation of cyclin D. *Mol. Cell. Biol.* **22,** 7842–7852.
18. Brunet, A., Bonni, A., Zigmond, M. J., et al. (1999) Akt promotes cell survival by phosphorylating and inhibiting a forkhead transcription factor. *Cell* **96,** 857–868.
19. Burgering, B. M. T. and Kops, G. J. P. L. (2002) Cell cycle and death control: long live forkheads. *Trends Biochem. Sci.* **27,** 352–360.
20. Biggs, W.H., Meisenhelder, J., Hunter, T., Cavenee, W.K. and Arden, K.C. (1999) Protein kinase B/Akt-mediated phosphorylation promotes nuclear exclusion of the winged helix transcription factor FKHR1. *Proc. Natl. Acad. Sci. USA* **96,** 742–7426.
21. Kops, G. J., de Ruiter, N. D., De Vries-Smits, A. M., Powell, D. R., Bos, J. L., and Burgering, B. M. (1999) Direct control of the forkhead-like transcription factor AFX by protein kinase B. *Nature* **398,** 630–634.
22. Rena, G., Guo, S., Cichy, S. C., Unterman, T. G., and Cohen, P. (1999) Phosphorylation of the transcription factor forkhead family member FKHR by protein kinase B. *J. Biol. Chem.* **274,** 17179–17183.
23. Brownawell, A. M., Kops, G. J., Macara, I. G., and Burgering, B. M. (2001) Inhibition of nuclear import by protein kinase B (Akt) regulates the subcellular distribution and activity of the forkhead transcription factor AFX. *Mol. Cell. Biol.* **21,** 3534–3546.

24. Rena, G., Prescott, A. R., Guo, S., Cohen, P., and Unterman, T. G. (2001) Roles of forkhead in rhabdomyosarcoma (FKHR) phosphorylation sites in regulating 14-3-3 binding, transactivation and nuclear targetting. *Biochem. J.* **354,** 605–612.
25. Matsuzaki, H., Daitoku, H., Hatta, M., Tanaka, K., and Fukamizu, A. (2003) Insulin-induced phosphorylation of FKHR (Foxo1) targets to proteasomal degradation. *Proc. Natl. Acad. Sci. USA* **100,** 11285–11290.
26. Plas, D. R. and Thompson, C. B. (2003) Akt activation promotes degradation of tuberin and FOXO3a via the proteasome. *J. Biol. Chem.* **278,** 12361–12366.
27. Yao, K-M., Sha, M., Lu, Z., and Wong, G. G. (1997) Molecular analysis of a novel winged helix protein, WIN: expression pattern, DNA binding property, and alternative splicing within the DNA binding domain. *J. Biol. Chem.* **272,** 19827–19836.
28. Ye H., Kelly T. F., Samadani U., et al. (1997) Hepatocyte nuclear factor 3/fork head homolog 11 is expressed in proliferating epithelial and mesenchymal cells of embryonic and adult tissues. *Mol. Cell. Biol.* **17,** 1626–1641.
29. Westendorf, J. M., Rao, P. N., and Gerace, L. (1994) Cloning of cDNAs for M-phase phosphoproteins recognized by the MPM2 monoclonal antibody and determination of the phosphorylated epitope. *Proc. Natl. Acad. Sci. USA* **91,** 714–718.
30. Ye, H., Holterman, A. X., Yoo, K. W., Franks, R. R., and Costa, R. H. (1999) Premature expression of the winged helix transcription factor HFH-11B in regenerating mouse liver accelerates hepatocyte entry into S phase. *Mol. Cell. Biol.* **19,** 8570–8580.
31. Leung T. W. C., Lin, S. S. W., Tsang, A. C. C., et al. (2001) Over-expression of FoxM1 stimulates cyclin B1 expression. *FEBS Lett.* **507,** 59–66.
32. Overdier, D. G., Porcella, A., and Costa, R. H. (1994) The DNA-binding specificity of the hepatocyte nuclear factor 3/forkhead domain is influenced by amino-acid residues adjacent to the recognition helix. *Mol. Cell. Biol.* **14,** 2755–2766.
33. Pierrou, S., Hellqvist, M., Samuelsson, L., Enerback, S., and Carlsson, P. (1994) Cloning and characterization of seven human forkhead proteins: binding site specificity and DNA bending. *EMBO J.* **13,** 5002–5012.
34. Kaufmann, E., Muller, D., and Knochel, W. (1995) DNA recognition site analysis of *Xenopus* winged helix proteins. *J. Mol. Biol.* **248,** 239–254.
35. Bois, P. R. and Grosveld, G. C. (2003) FKHR (FOXO1a) is required for myotube fusion of primary mouse myoblasts. *EMBO J.* **22,** 1147–1157.
36. Furuyama, T., Nakazawa, T., Nakano, I., and Mori, N. (2000) Identification of the differential distribution patterns of mRNAs and consensus binding sequences for mouse DAF-16 homologues. *Biochem. J.* **349,** 629–634.
37. Vara, D., Bicknell, K.A., Coxon, C.H. and Brooks, G. (2003) Inhibition of E2F abrogates the development of cardiac myocyte hypertrophy. J. Biol. Chem. 278, 2138821394.
38. Ausubel, F. M., Brent, R., Kinston, R. E., et al. (eds.) (1987) *Current Protocols in Molecular Biology*. Wiley Interscience, New York.
39. Dignam, J. D., Lebowvitz, R. M., and Roeder, R. G. (1983) Accurate transcription initiation by RNA polymerase II in a soluble extract from isolated mammalian nuclei. *Nucleic Acids Res.* **11,** 1475–1489.
40. Clerk, A. and Sugden, P. H. (1997) Cell stress-induced phosphorylation of ATF2 and c-Jun transcription factors in rat ventricular myocytes. *Biochem. J.* **325,** 801–810.
41. Bradford, M. M. (1976) A rapid and sensitive method for the quantitation of microgram quantities of protein utilizing the principle of protein-dye binding. *Anal. Biochem.* **72,** 248–254.

15

Measurement of Geminin Activity in *Xenopus* Egg Extracts

Thomas J. McGarry

Summary

Geminin is an unstable protein that inhibits DNA replication by interfering with the assembly of prereplication complex at replication origins. Geminin is thought to prevent a second round of replication during late S- or G_2-phase. The protein is destroyed by ubiquitin-dependent proteolysis during mitosis, allowing a new round of replication in the next cell cycle. This chapter describes protocols for measuring the stability of geminin and two of its activities, the inhibition of replication and the inhibition of pre-RC loading.

Key Words

Geminin; DNA Replication; ubiquitin-dependent proteolysis; prereplication complex; re-replication; cell cycle; S-phase; mitosis; anaphase-promoting complex.

1. Introduction

To maintain a stable genome, it is vitally important that every origin of replication initiates DNA synthesis only once during each S-phase. The protein geminin provides one mechanism that prevents a second round of origin firing *(1)*. Geminin is an unstable protein that inhibits DNA replication by interfering with the assembly of prereplication complex (pre-RC). Pre-RC is a collection of essential replication factors that assembles on origins of replication before DNA synthesis begins. The components of pre-RC include the origin recognition complex (ORC), Cdc6, and the minichromosome maintenance (MCM) complex (**Fig. 1**). Geminin directly binds to a component of pre-RC called Cdt1 *(2,3)*. The exact function of Cdt1 is not clear, but the protein is required for the addition of the MCM complex, the putative replication helicase.

From: *Methods in Molecular Biology, vol. 296, Cell Cycle Control: Mechanisms and Protocols*
Edited by: T. Humphrey and G. Brooks © Humana Press Inc., Totowa, NJ

A single round of origin firing is permitted because the concentration of geminin is regulated by selective protein degradation as cells pass through the different stages of the cell cycle *(1)*. Geminin is absent during G_1-phase, the time when pre-RC assembles. At the G_1/S transition, geminin begins to accumulate, and further origin loading is inhibited. Geminin persists throughout S-, G_2-, and early M-phase, and at the metaphase/anaphase transition geminin is abruptly destroyed by ubiquitin-dependent proteolysis. The destruction of geminin allows the origins to load replication factors in the next cell cycle, and the process begins anew.

Geminin is ubiquitylated by a multisubunit E3 ubiquitin–protein ligase called the anaphase-promoting complex (APC) or the cyclosome *(1)*. APC ubiquitylates a small group of regulatory proteins that includes geminin and the B-type cyclins. It recognizes a nine-amino acid motif on its substrates called the destruction box. The activity of APC is dependent on the stage of the cell cycle. The complex is switched on at the metaphase/anaphase transition and remains active until the G_1/S transition, when it is switched off. The mechanisms that regulate APC activity are the subject of current research. The pattern of APC activation dictates that geminin will accumulate only during S-, G2-, and early M-phase, the times in the cell cycle when origin refiring must be prevented.

This review describes several protocols that demonstrate various aspects of geminin$ biology. The first procedure demonstrates the degradation of geminin in mitotic cell extracts. The second procedure demonstrates the inhibition of DNA replication by geminin, and the third procedure demonstrates how geminin interferes with pre-RC assembly.

The procedures are performed using extracts that are prepared by crushing *Xenopus* eggs in a centrifuge (**Fig. 2**). These extracts reproduce cell cycle reactions in vitro, and they can be manipulated to carry out the reactions characteristic of different cell cycle stages *(4)*. When the eggs are laid, they are arrested in metaphase of the second meiotic division by cytostatic factor (CSF). CSF-arrested extracts can condense chromosomes and assemble a mitotic spindle in vitro but will not replicate DNA or degrade geminin. When an egg is fertilized, an influx of extracellular calcium breaks the CSF arrest and triggers progression from metaphase to anaphase and subsequently into S-phase of the next cell cycle. In the same way, geminin degradation and DNA replication can be triggered in vitro by adding calcium to a CSF-arrested extract (**Fig. 2**, right panel; protocol in **Subheading 3.2.**). Eggs can also be activated before crushing by placing them in a calcium-rich solution and adding a calcium ionophore. Extracts made from activated eggs can be arrested in either interphase or mitosis by controlling the concentration of B-type cyclins, which are absolutely required for the activity of the mitotic protein kinase Cdc2 (**Fig. 2**, left panel). The extract is trapped in interphase by adding cycloheximide, which prevents the synthesis of B-type cyclins. Alternatively, the extract is trapped in mitosis by adding cyclin B Δ90, a nondegradable mutant that lacks a destruction box *(3)*. Because of the differential activity of APC, geminin is degraded by mitotic extract but not by interphase extract (protocol in **Subheading 3.1.**).

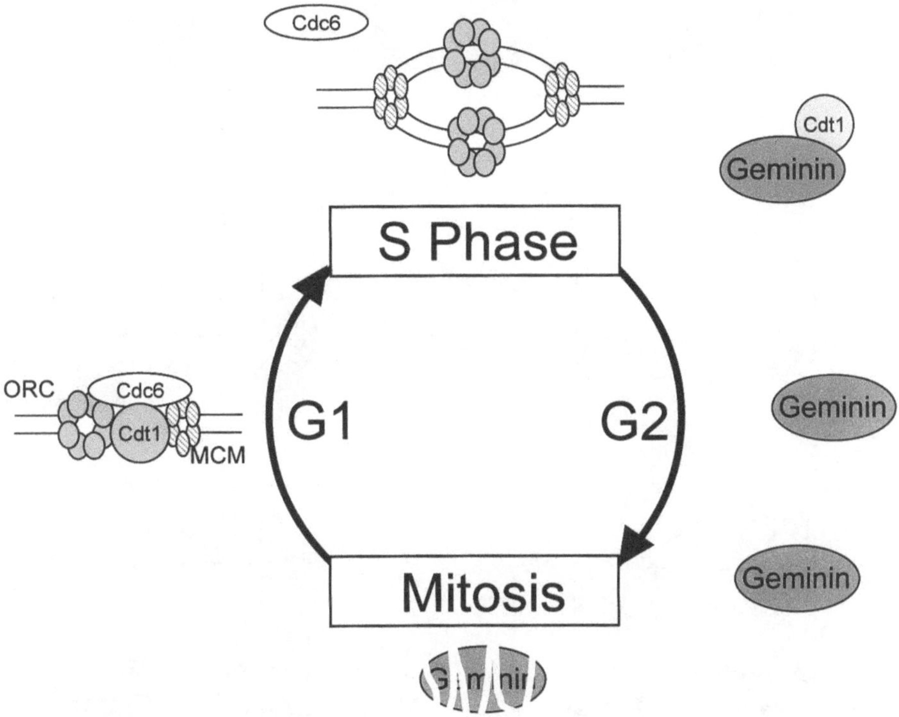

Fig. 1. Geminin inhibits rereplication. During G_1 phase, prereplication complex (pre-RC) loads onto origins of replication. When DNA synthesis begins during S-phase, Cdt1 is bound by geminin. This prevents a second round of pre-RC loading. The degradation of geminin during mitosis allows pre-RC to load during the following G_1 phase. MCM, minichromosome maintenance complex; ORC, origin recognition complex. (Adapted from **ref. *10*.**)

2. Materials

1. The water for making solutions should be double distilled or from a Milli-Q still (Millipore). Chlorinated tap water is quite toxic to frogs and should never be used. Frogs and their eggs maintain their health better at temperatures slightly cooler than room temperature (18–22°C). Water taken directly from the still is often slightly warmer than room temperature and should be allowed to cool before use.
2. Details on the care and feeding of *Xenopus* adults are given by Wu and Gerhart ***(6)***.

2.1. Determination of Mitotic Stability

1. 1X MMR (Marc's Modified Ringer's): 100 m*M* NaCl, 2 m*M* KCl, 2 m*M* $CaCl_2$, 1 m*M* $MgCl_2$, 5 m*M* HEPES-NaOH pH 7.4. Make 5 L of a 10X stock, and autoclave. Stable for >1 yr at room temperature.

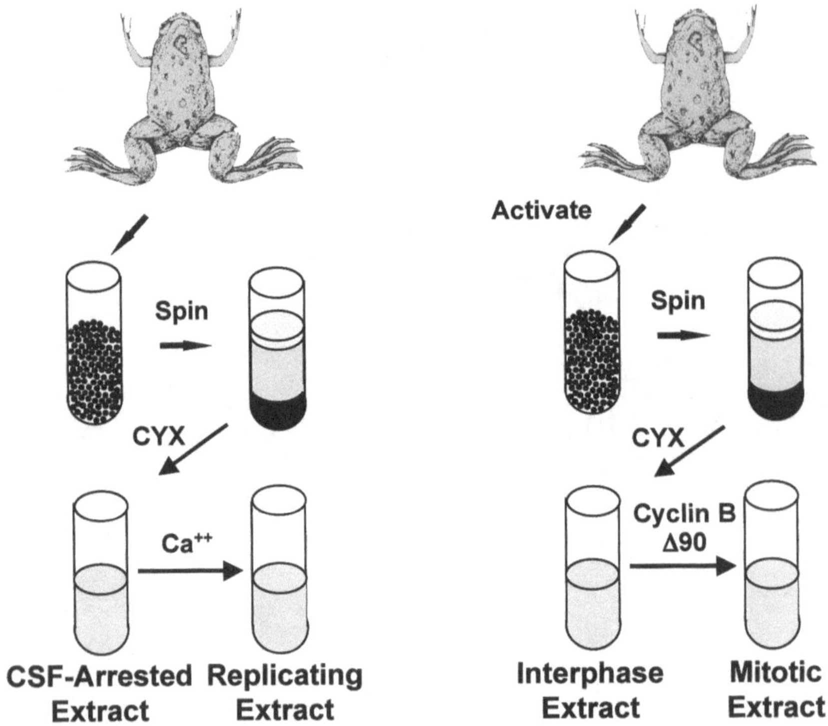

Fig. 2. Preparation of *Xenopus* egg extracts. (Right) Interphase and mitotic extracts for protein stability studies. Unfertilized eggs are activated with calcium ionophore, packed in a centrifuge tube, and crushed by spinning. To make interphase extract, the cytoplasmic (middle) layer is collected and cycloheximide (CYX) is added to inhibit the synthesis of B-type cyclins. To make mitotic extract, an nondegradable mutant of B-type cyclin is added to interphase extract. (Left) Cytostatic factor (CSF)-arrested extract for DNA replication studies. Metaphase-arrested eggs are crushed without activation. Cycloheximide is added to prevent the extract from resynthesizing B-type cyclins after activation. The replication reaction is started by adding calcium chloride to the extract.

2. Human chorionic gonadotropin (HCG; Sigma, cat. no. CG-10). Add 10 mL of water to a vial containing 10,000 U lyophilized solid. Stable at 4°C for several weeks.
3. Pregnant mare serum gonadotropin (PMSG; Calbiochem, cat. no. 367222). Add water to solid to make 1000 U/mL. Freeze in small aliquots, and store at –20°C. Stable for at least several months.
4. 5% Gelatin (Fisher Scientific, cat. no. G8-500 or equivalent). Make a slurry of 2.5 g gelatin in 50 mL Milli-Q water in a sterile 250-mL Erlenmeyer flask. Dissolve the gelatin by placing the flask in boiling water. Store the molten solution at 37°C so that the gelatin does not set. The solution may be reused for 1–2 wk provided that it remains sterile.
5. Dejellying solution: 2% L-cysteine (Sigma, cat. no. C7352), pH 7.80. Oxidized by air. Prepare fresh on the day of the experiment.

6. Ionophore A23187 (Sigma, cat. no. C7522): 10 mg/mL in dimethylsulfoxide (DMSO). Divide into 10- L aliquots and freeze. Stable for >5 yr at –80°C.
7. Extract buffer (XB): 100 m*M* KCl, 0.1 m*M* $CaCl_2$, 1 m*M* $MgCl_2$, 10 m*M* HEPES-KOH, pH 7.7, 50 m*M* sucrose. Prepare fresh from stocks of 1 *M* HEPES-KOH, pH 7.7 (filter-sterilize and store aliquots at –20°C), 1.5 *M* sucrose (filter-sterilize and store aliquots at –20°C), and 20X XB salts (2 *M* KCl, 2 m*M* $CaCl_2$, 20 m*M* $MgCl_2$; filter-sterilize and store at 4°C). Each component solution is stable for months.
8. 1000X Protease inhibitors: 10 mg/mL each of leupeptin (Chemicon, cat. no. EI8), pepstatin (Chemicon, cat. no. EI9), and chymostatin (Chemicon, cat. no. EI6) in DMSO. Stable for years at –80°C.
9. 1000X Cytochalasin B (Sigma, cat. no. C-6762). Make a 10 mg/mL stock in DMSO. Stable for years at –80°C.
10. Oil: we use Niosil M25, distributed by TAI Lubricants (PO Box 1579, Hockessin DE 19707; 302-326-0200).
11. 100X Cycloheximide (Sigma, cat. no. C-7698): 10 mg/mL in water. Make fresh every time.
12. 20X Energy mix: 150 m*M* creatine phosphate (= phosphocreatine, Sigma, cat. no. P 7936), 20 m*M* ATP (diluted from a 100 mM ATP solution; Amersham cat. no. 27-2056-01), 2 m*M* EGTA, 20 m*M* $MgCl_2$. Store in 50- L aliquots at –80°C. Stable for years.
13. 1X Polyacrylamide gel sample buffer: 50 m*M* Tris-HCl, pH 6.8, 100 m*M* dithiothreitol (DTT), 2% (w/v) sodium dodecyl sulfate (SDS), 0.1% or less bromphenol blue, 10% (v/v) glycerol. Store in 1-mL aliquots at –20°C. Stable for at least a year.

2.2. Preparation of Demembranated Sperm DNA

1. Tricaine anesthetic solution: dissolve 1.0 g sodium bicarbonate in 1 L of water. Add 1.0 g of tricaine (ethyl-3-aminobenzoate, methane sulfonate salt; Sigma, cat. no. A 5040). The bicarbonate must be in solution before the tricaine is added or an insoluble oil forms. Wear gloves while handling the anesthetic solution. Store at room temperature. The solution slowly becomes discolored but is usable for about 4 mo.
2. XN buffer: 50 m*M* HEPES-KOH, pH 7.0, 250 m*M* sucrose, 75 m*M* NaCl, 0.5 m*M* spermidine hydrochloride (Sigma, cat. no. S2501), 0.15 m*M* spermine hydrochloride (Sigma S1141).
3. XN with 3% BSA: Float 3g of powdered bovine serum albumin (BSA; Roche, cat. no. 100 350) on top of 100 mL of XN in a beaker with a stir bar. Stir gently until the BSA is in solution. Filter-sterilize, and store at 4°C. Stable for at least a year as long as it remains uncontaminated.
4. XN with 50% glycerol: mix 120 g of glycerol with 120 mL of a 2X stock of XN. Filter-sterilize, and store at 4°C. Stable for at least a year.
5. Lysolecithin (lyso-phosphatidylcholine; Sigma, cat. no. L4129). Make a fresh 2 mg/mL solution in XN each time.
6. Hoechst 33342 (Sigma, cat. no. B2261): 10 mg/mL stock in water.

2.3. Preparation of CSF-Arrested Extracts for DNA Replication

1. CSF extract buffer (CSF-XB): 100 m*M* KCl, 0.1 m*M* $CaCl_2$, 2 m*M* $MgCl_2$, 10 m*M* HEPES-KOH, pH 7.7, 50 m*M* sucrose, 5 m*M* EGTA, pH 7.7. Prepare fresh from stock solutions as for extract buffer (XB, **Subheading 2.1., item 7**). Note that the composition of the salts is slightly different than for XB and that EGTA must be added.

2. 20X CSF Energy mix: 150 m*M* creatine phosphate (= phosphocreatine, Sigma, cat. no. P7936), 20 m*M* ATP (diluted from a 100 m*M* ATP solution; Amersham, cat. no. 27-2056-01), 20 m*M* $MgCl_2$. Note that the composition is slightly different from that of energy mix (**Subheading 2.1., item 12**). Stable for years at –80°C.
3. Sperm dilution buffer (SDB): 5 m*M* HEPES-KOH, pH 7.7, 150 m*M* sucrose, 100 m*M* KCl, 1 m*M* $MgCl_2$. Aliquot and store at –20°C.
4. Nuclear fixation buffer: Mix 10 L 1X MMR, 30 L 37% formaldehyde (handle in a fume hood), 60 L of 80% glycerol (w/w), and 1 L of 100 ng/mL Hoechst 33342. Prepare fresh every time. Dispose of formaldehyde as organic waste.

2.4. Inhibition of DNA Replication by Geminin

1. Geminin dilution buffer: 10 m*M* HEPES-NaOH, pH 7.7, 300 m*M* NaCl.
2. Replication start mix: for each reaction, mix 0.1 L [α-^{32}P]dATP or dCTP (10 mCi/mL, 3000 Ci/mmol), 1.3 mL of 10X calcium solution in SDB (concentration determined in **Subheading 3.2.2., step 4**), and 21,000 sperm DNA templates in a volume of 0.4 L (Dilute sperm stock with SDB to get appropriate concentration.)
3. Replication stop mix: 20 m*M* Tris-HCl, pH 8.0, 20 m*M* EDTA, 0.5% SDS. Stable at room temperature for years.
4. Protease K (Roche, cat. no. 745 723): make 10 mg/mL in 10 m*M* HEPES-KOH, pH 7.4, 50% glycerol (w/v).
5. 100% Trichloroacetic acid (TCA): add 227 mL of water directly to a stock bottle containing 500 g of TCA (Fisher, cat. no. A322-500). Stable at 4°C for at least a year. **Caution:** Handle all TCA-containing solutions with gloves and eye protection.
6. 10% TCA/sodium pyrophosphate: dilute 100% TCA 1:10, and add 2 g sodium pyrophosphate decahydrate (Sigma, cat. no. S 6642) per 100 mL.

2.5. Analysis of Pre-RC Loading

1. Chromatin binding buffer: 100 m*M* KCl, 2.5 mM $MgCl_2$, 50 mM HEPES-KOH, pH 7.5, 0.25% Triton X-100. Store cold; stable for at least a few weeks.
2. Cushion solution: 100 m*M* KCl, 2.5 m*M* $MgCl_2$, 50 m*M* HEPES-KOH pH 7.5, 30% sucrose. Filter-sterilize and store at 4°C. Stable for at least a few weeks.

3. Methods

3.1. Determination of Mitotic Stability

This assay is used to measure the half-life of geminin (or any protein) during interphase or mitosis *(7)*. Radioactively labeled geminin is prepared by transcribing and translating a geminin-encoding plasmid in reticulocyte lysate in the presence of [^{35}S]methionine. The crude reticulocyte lysate containing the radioactive protein is mixed with interphase or mitotic cell extracts and incubated at room temperature. Aliquots are removed at timed intervals, and the amount of protein remaining is quantitated by electrophoresis and autoradiography (**Fig. 3**).

3.1.1. Preparation of Interphase and Mitotic Egg Extracts

This procedure is slightly modified from the method described by Murray *(4)*.

1. The day before the experiment, inject two to four adult female frogs with 800–1000 U of HCG to induce ovulation. Place each frog in approx 2 L of MMR at 19–20°C overnight.

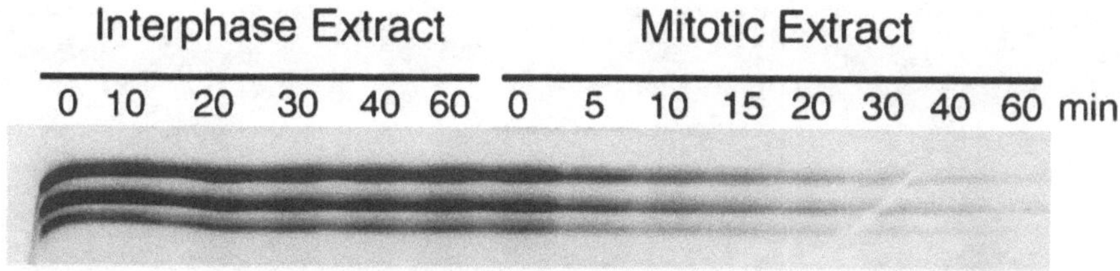

Fig. 3. Degradation of geminin C180 by mitotic extract. A plasmid encoding a geminin mutant in which amino acids 180–219 are deleted was translated in vitro and added to interphase or mitotic extract as described in the text. The protein is stable in interphase extract but is rapidly degraded by mitotic extract. Three radioactive bands are seen because of false initiation at two internal methionines.

2. The next morning inspect the eggs laid by each frog under a dissecting microscope to determine their quality (**Fig. 4**). The success of the experiment depends in large part on the selection of the highest quality eggs. High-quality eggs are uniformly pigmented throughout the animal pole, do not burst open or activate spontaneously (evidenced by contraction of the cortical pigment), and are firm to pressure (do not collapse flat when placed in a dry Petri dish). For the best results, use only the eggs from a single frog (a single malt'extract in the vernacular). For optimal results, use only eggs that are freshly expressed from the frogs cloaca by squeezing her flanks.
3. Collect the eggs in a 600-mL beaker that has been coated with a 5% gelatin solution and allowed to set (*see* **Note 1**). If necessary, rinse the eggs in 1X MMR to remove dead skin, feces, and other debris (~200 mL/rinse, pour on–pour off).
4. To remove the jelly coat surrounding the eggs, pour off the MMR, and replace it with approx 200 mL of freshly prepared dejellying solution. Gently agitate the eggs every minute or so. Once the jelly has been removed, the eggs will sink to the bottom of the beaker more quickly and will pack together more closely (**Fig. 3**, compare **A** and **B**). When the jelly appears to be completely removed, replace about half the supernatant with fresh dejellying solution, and continue the incubation for another 1–2 min with intermittent swirling. Once the jelly coat has been removed, it is necessary to keep the eggs submerged at all times.
5. Carefully pour off the cysteine dejellying solution, and rinse the eggs three times with 1X MMR (pour on–pour off).
6. To activate the eggs, pour off the MMR, and add a solution of 0.4 g/mL calcium ionophore A23187 in 1X MMR. Incubate the eggs in this solution for exactly 3 min, and then rinse three times with fresh 1X MMR. Successful activation is indicated by the simultaneous contraction of the animal pigment in all eggs (**Fig. 3C**). The eggs usually activate during the MMR rinses. Let the activated eggs sit in MMR for 15 min. During this time remove any eggs that do not show cortical contraction. Act quickly because the contraction is only sustained for a few minutes. Rinse three times with XB buffer (pour on–pour off). During the rinses, remove any dead or damaged eggs (bloated white eggs that work their way to the top of the pile).
7. Place 1 mL of XB containing 1X protease inhibitors and 10X cytochalasin B (10X = 100 g/mL) in a Beckman Sw55 centrifuge tube (Beckman, cat. no. 344057) or equivalent (*see* **Note 2**). Transfer the eggs to these tubes using a Pasteur pipet that has been broken

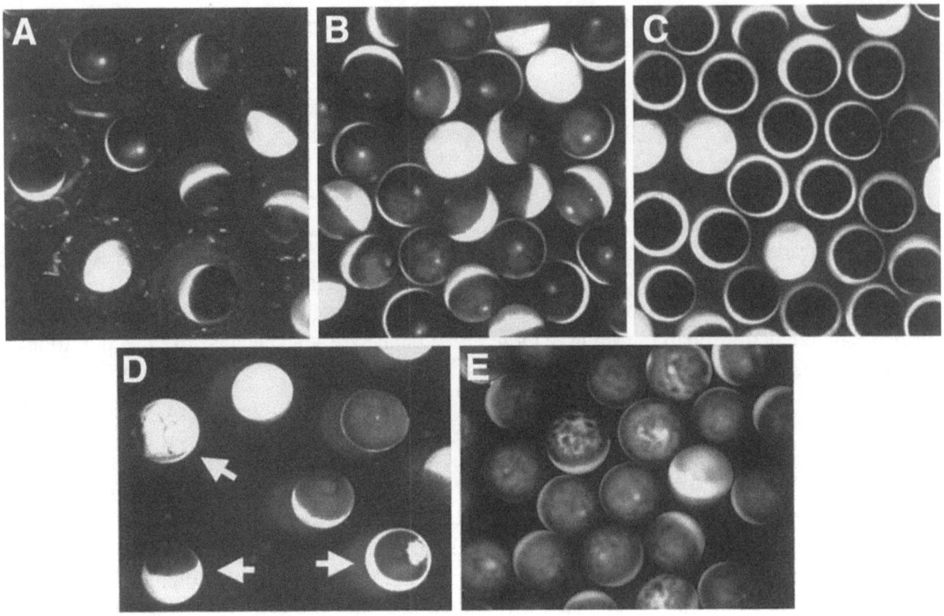

Fig. 4. A gallery of *Xenopus* eggs. **(A)** *Xenopus* eggs before removal of the jelly coat, which is rendered visible by a slight coating of debris. Note that the jelly prevents the eggs from touching each other. **(B)** *Xenopus* eggs after removal of the jelly coat. The eggs pack together more tightly. **(C)** *Xenopus* eggs immediately after activation with ionophore. The surface pigment has contracted so that the animal pole appears darker and smaller. This cortical contraction relaxes after a few minutes. **(D)** Damaged *Xenopus* eggs. Three of the eggs in this field were damaged when the female was squeezed (arrows). One has completely lysed (top), and the other two have spontaneously activated, as evidenced by contraction of the surface pigment and a splotch of white cytoplasm on the egg at the right. Small numbers of damaged eggs can be removed during extract preparation, but if too many are damaged this becomes impractical. **(E)** Mottled *Xenopus* eggs. The mottled pigmentation pattern indicates that these eggs have poor quality. Healthy eggs have an even coating of surface pigment with a single white spot near the apex **(B)**.

off and flame-polished at the tip to give an opening of 3–4 mm. Be careful to transfer as little liquid as possible.

8. After transfer, withdraw as much liquid as possible and spin the eggs in a clinical centrifuge at 1000*g* for 15 s. Remove as much liquid as possible and layer 0.5–1.0 mL of oil over the eggs. Spin at 1000*g* for 2 min, and again remove as much liquid as possible (*see* **Note 3**).
9. Chill the packed eggs in ice water for 10 min.
10. Spin the packed eggs at 9500*g* for 10 min at 4°C in a Beckman SW55 rotor or equivalent. The centrifugation crushes the eggs and separates their components into three phases that layer on top of each other like a parfait (**Fig. 2**). The top (yellow) layer is lipid, the middle (tan or gray) layer is egg cytoplasm, and the bottom (green or black) layer is yolk platelets

and pigment granules. There is usually a thin layer of light-colored material at the interface between the bottom and middle layers.

11. To collect the extract puncture the tube with a 20-gage needle just above the interface between the middle and bottom layer, and allow the cytoplasmic fraction to drip by gravity into a collection tube on ice. The drip rate can be increased by draping a small piece of Parafilm over the hole to break surface tension. Avoid contaminating the cytoplasmic fraction with the lipid layer. From this point onward the extract should be kept on ice.
12. Measure the volume of extract and add protease inhibitors, cytochalasin B, and energy mix to a final concentration of 1X. Mix by inversion, and collect the extract in the bottom of the tube by spinning for 1–2 s in a nanofuge.
13. The extract can be used directly if it is straw-colored, uncontaminated by lipid, and slightly turbid. If the extract is gray-black or contaminated by significant amounts of lipid or oil, it can be clarified by a second spin in the SW55 at a slightly lower speed (7500g) for 10 min. For this step we routinely use long, thin centrifuge tubes (Beckman, cat. no. 344090) that fit into adapters for the SW55. Collect the clarified extract by puncturing the tube as before.

3.1.2. Degradation Reactions

1. Add cycloheximide to the extract to make the final concentration 100 g/mL. Remove one aliquot (the interphase extract).
2. To another aliquot (the mitotic extract), add recombinant cyclin B Δ90 to a final concentration of 50 g/mL. Cyclin B Δ90 is a mutant of sea urchin cyclin B drives the extract into a stable mitotic state *(3)*. The Δ90 protein cannot be destroyed because it lacks a destruction box. Incubate both the interphase and mitotic extract at room temperature for 45 min.
3. Prepare radioactive geminin (or any other test protein) by transcribing and translating a plasmid cDNA clone in reticulocyte lysate in the presence of [^{35}S]methionine (Promega TnT Kit; *see* **Note 4**). At the end of the reaction, add cycloheximide to 100 g/mL to inhibit further translation.
4. Combine the reticulocyte lysate containing radioactive geminin with either the mitotic extract or the interphase extract in a ratio of 1:3 by volume. Typically we use 8 L of translation lysate and 24 L of egg extract. Mix the two viscous solutions thoroughly by stirring and pipeting up and down, taking care not to introduce air bubbles.
5. Incubate the mixture at room temperature. At designated times (t = 0, 5, 10, 20, and 40 min after mixing) remove a 6- L aliquot from the mixture and add it to 30 L of 1X polyacrylamide gel sample buffer. Vortex and store on ice.
6. After all the samples have been collected, run about one-third of each time-point on a polyacrylamide gel, and visualize the radioactive proteins by autoradiography
   Typical results are shown in **Fig. 3**. Geminin is stable in interphase extract but is rapidly degraded by mitotic extract, with a half-life of about 15 min. Half-lives are measured by quantitating the band intensity at each time-point with a PhosphorImager and fitting the results to an exponential decay curve. The half-life is calculated from the initial portion of the curve if the decay if the plot of log(intensity) vs time is not linear.

3.2. Inhibition of DNA Replication

Xenopus egg extracts contain adequate concentrations of deoxyribonucleotide triphosphates and replication proteins (e.g., pre-RC components, polymerases, and so on) to replicate many different types of DNA *(3)*. Typically, demembranated *Xenopus* sperm heads are chosen as the template. Upon being placed in the extract, the highly compacted sperm DNA chromatin decondenses and becomes surrounded by a normal double-membrane nucleus. The extract replicates the sperm DNA completely and only once, mirroring replication in vivo.

3.2.1. Preparation of Demembranated Sperm DNA

1. Inject an adult male frog (*see* **Note 5**) with 25 U PMSG 3–5 d before isolating sperm. Inject the same frog with 200 U HCG the day before the procedure.
2. Anesthetize the frog by immersion in 1 L of tricaine anesthetic solution for 10–15 min. When adequately anesthetized, the animal will be completely limp and unresponsive. If the animal remains active, prepare fresh anesthetic solution and repeat.
3. Pith the frog by dislocating the cervical spine with a bone shears (Fisher Scientific, cat. no. S17342) and macerating the brain and spinal cord with a dissecting needle. Puncture the skin with a pointed scalpel in the lower abdomen at the level of the hip about 1 cm from the midline. Grasp the skin with a toothed forceps inserted into the puncture hole. Tent up the skin, and extend the incision vertically to a length of about 2 cm. Repeat the process to open the underlying abdominal muscle layer. Reach into the abdominal cavity with the forceps, and pull out the fat body, which consists of many lobules of yellow gelatinous adipose tissue. The testis is a bean-shaped white organ covered by a delicate lacy network of red blood vessels that is attached to the base of the fat body near the midline. The testis is attached by a transparent membrane (the mesotestis) along one side. Grasp the mesotestis with a forceps and gently snip it away from the fat body with a sharp pair of dissecting scissors. Avoid leaving any fat tissue attached to the testis. Place the testis in cold 1X MMR, and repeat the dissection on the other side. After removing the testes, freeze the carcass at –80°and dispose of it properly.
4. Rinse the testes in cold MMR, and then place them in a clean Petri dish sitting open in an ice bucket. Macerate the testes with two sharp forceps (e.g., Dumont #5) until a thick paste forms. Resuspend the paste in 2 mL cold XN solution, and homogenize in a Dounce homogenizer for a few strokes using a loose-fitting pestle (*see* **Note 6**). Filter through a 40- m cell strainer (Falcon, cat. no. 352340) to remove large chunks of tissue.
5. Pellet the sperm by spinning in a swinging bucket rotor at 1000*g* for 10 min. Aspirate off the supernate, and wash the pellet twice in 10 mL XN buffer. Resuspend the final pellet in 1 mL XN buffer, and warm to room temperature. Spot a 1- L aliquot onto a microscope slide, add 4 L of XN containing1 g/mL Hoechst 33342, and put a cover slip on. Observe the sample under a fluorescence microscope. A few damaged sperm will stain with Hoechst dye, but most sperm will not stain until demembranated. Sperm nuclei have a curlicue or “integral-sign”shaped nucleus with tapered ends (**Fig. 5A**).
6. Add 200 L of lysolecithin solution, mix by inversion, and start a timer. Every 2–3 min withdraw a 1- L aliquot and stain with Hoechst dye as above. When all the sperm appear to stain, stop the reaction by adding 10 mL cold XN with 3% BSA. Work quickly; complete demembranation typically takes 3–5 min. The demembranated sperm should be kept on ice from this point onward.
7. Pellet the sperm (1000*g*, 10 min) and rinse with XN buffer three times. Resuspend the final pellet in 1 mL XN containing 50% glycerol (w/w). Dilute an aliquot approx 100-fold with XN buffer containing 1 g/mL Hoechst 33342, and count the sperm under a fluorescent microscope using a hemocytometer. Count only the corkscrew-shaped sperm nuclei, ignoring round nuclei that come from blood cell contamination (**Fig. 5A**). Typical yields

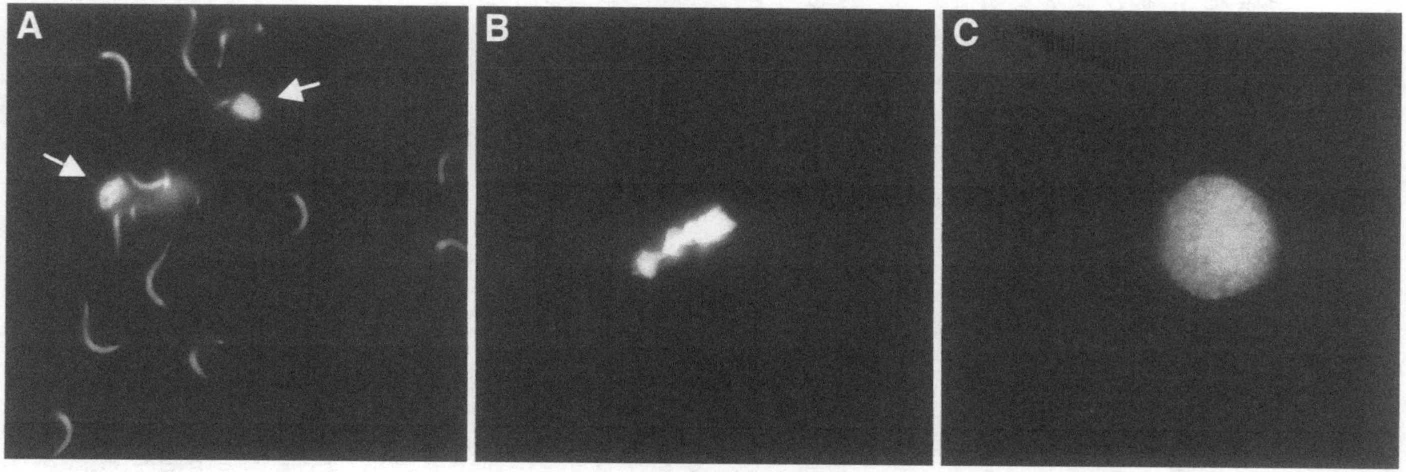

Fig. 5. The appearance of nuclei in *Xenopus* egg extract. **(A)** *Xenopus* sperm nuclei stained with DAPI. Notice the elongate curlicue shape. The two round nuclei are from contaminating blood cells. **(B)** Nucleus in mitotic or CSF-arrested extract. The chromatin is condensed and stains brightly with Hoechst dye. **(D)** Nucleus in interphase extract or CSF-arrested extract activated with calcium. The chromatin is diffuse and surrounded by an intact nuclear membrane.

are 2–10 × 10^7 sperm/mL. Divide the preparation into small (~10 L) aliquots, and store at –80°C. Do not snap-freeze with liquid nitrogen. Sperm are stable at –80°C for years.

3.2.2. Preparation of CSF-Arrested Extracts for DNA Replication

Interphase extracts made from activated eggs (**Subheading 3.1.1.**) can replicate DNA, but we prefer to use CSF-arrested extracts since in this case the replication reaction can be started at will by adding calcium. The protocol for preparing CSF extracts is the same as in **Subheading 3.1.1.** except that great care is exercised to prevent the premature activation. EGTA is added to the extract buffer to chelate calcium, and the eggs are not incubated in ionophore. The concentration of calcium that activates the extract must be determined empirically for each experiment. This protocol is slightly modified from that described by Murray *(4)*.

1. Collect and dejelly *Xenopus* eggs as in **Subheading 3.1.1., steps 1–5**. Rinse the eggs three times with MMR, three times with XB, and three times with CSF-XB. The final rinse should contain 1X protease inhibitors. Remove any activated or dead eggs during the rinses.
2. Collect the eggs in a SW55 centrifuge tube (Beckman 344057) containing 1 mL CSF-XB containing 1X protease inhibitors and 10X cytochalasin B (100 g/mL).
3. Pack and crush the eggs, and collect the cytoplasmic extract as in **Subheading 3.1.1., steps 7–10**. Add protease inhibitors, cytochalasin B, and CSF energy mix to 1X. Perform a clarifying spin if necessary. Add cycloheximide to 100 g/mL (*see* **Note 7**).
4. To determine how much calcium is needed to activate the extract, prepare 1 mL of four different 10X calcium solutions containing 0, 2, 4, or 10 m*M* $CaCl_2$ in SDB. Aliquot 1.0 L of each solution to one of four 0.5-mL microfuge tubes. Take 50 L of extract, and add about 10,000 demembranated sperm to give a concentration of 200 sperm/ L. Warm the extract to room temperature, then add 9.0 L to each of the four small microfuge tubes, and mix gently by stirring and pipeting up and down. Take care not to introduce air bubbles. Incubate at room temperature for 60 min. At the end of the incubation, spot 1 L of each reaction onto a microscope slide, add 4 L of fixation buffer, and cover with an 18-mm^2 cover slip. Inspect each slide under a fluorescence microscope. The extract with 0 m*M* $CaCl_2$ should have condensed nuclear DNA, whereas at least one of the extracts that had calcium added should have formed a round interphase nucleus (**Fig. 5B** and **C**). To initiate the replication reactions, use the lowest amount of calcium that activates the extract (*see* **Note 8**). If the control reaction does not show condensed DNA, or if an interphase nucleus did not form at any calcium concentration, then the extract cannot be used.

3.2.3. Inhibition of DNA Replication by Geminin

1. Dilute recombinant geminin to 52 g/mL in geminin dilution buffer. Serially dilute 1:2 to make stocks at 26 g/mL, 13 g/mL, and so on.
2. For each replication reaction, place 10.2 L of CSF-arrested extract in a 1.5-mL microfuge tube. Add 1.0 L of diluted geminin stock to each, and mix well (*see* **Note 9**).
3. Prepare a replication start mixture containing for each replication assay (*see* **Note 10**): 0.1 L [α^{32}P]dATP or dCTP (10 mCi/mL, 3000 Ci/mmol); 1.3 L of 10X calcium solution in SDB (*see* **Subheading 3.2.2., step 4** for concentration); and 0.4 L sperm template DNA (16,300 sperm).
4. Warm the replication reactions to 20–23°C for 5 min. Start the reaction by adding 1.8 L

of replication start mix and mixing well. Incubate the reactions at 20–23°C for 2–3 h.

5. At the conclusion of the incubation, add 185 L replication stop mix, and mix by pipeting up and down (*see* **Note 11**). The reactions may be frozen at –20°C at this point.
6. Add 10 L of 10 mg/mL protease K, and incubate at 37°C for 60 min. Extract the reaction once with 1:1 chloroform/water-saturated phenol and once with chloroform (*see* **Note 12**).
7. Label with India ink two glass fiber filter discs (Whatman GF-C or equivalent) for each reaction (*see* **Note 13**). Spot 10 L of the extracted reaction onto one. Dry under a heat lamp, and count in a scintillant (e.g., Research Products International 3a70B) to obtain total counts.
8. Spot 50 L of the extracted reaction onto the other filter, and dry under a heat lamp. After drying, precipitate the DNA by immersing the discs in 10% TCA containing 2% sodium pyrophosphate (~10 mL/disc) for 30 min at 4°C with occasional swirling.
9. Pour off the supernatant, and wash the discs three times with cold 5% TCA and then twice with cold 95% ethanol. Each wash is for 5 min. Dispose of the washes as radioactive waste.
10. Dry the discs under a heat lamp, and then count in a scintillant to obtain incorporated counts.
11. Calculate the percentage of counts incorporated for each reaction. The degree of inhibition of replication may be expressed as a percent of the positive control or as an absolute percentage of input sperm DNA replicated. The latter number is obtained by this formula (*see* **Note 14**):

$$\% \text{ replication} = [(\text{incorporated counts})/(\text{total counts} \times 5)] \times 1310$$

Typical results are shown in the top panel of **Fig. 6**. There should be very few counts in the negative controls (no calcium and no sperm, lanes 2 and 3) and around 100% incorporation in the positive control (lane 1). The calculated absolute percentage of DNA replicated may be above or below 100% if there are inaccuracies in counting the number of sperm. Replication is completely inhibited by geminin concentrations from 0.5–2.0 g/mL (20 to 80 n*M*) depending on the extract. A control protein (glutathione-*S*-transferase) has no effect (lane 10).

3.3. Analysis of Pre-RC Loading

Several protocols have been developed to monitor the assembly of pre-RC in *Xenopus* egg extracts. These protocols share the same basic outline: sperm template DNA is incubated in the extract for a short period to allow pre-RCs to form, and then the DNA and its attached proteins are pelleted through a sucrose cushion to separate them from the other proteins in the extract. The DNA-bound proteins in the pellet are detected by immunoblotting. The difficulty of this assay lies in reducing the background to an undetectable level. Every experiment should include a negative control that contains extract but no added sperm DNA. This protocol is modeled after that described by Kubota et al. *(9)*.

1. For each binding assay, add 2.25 L of diluted geminin protein (**Subheading 3.2.3., step 1**) to 25 L CSF extract (**Subheading 3.2.1.**) in a 1.5-mL microfuge tube. Incubate at room temperature for 5 min.
2. Add approx 60,000 demembranated sperm (**Subheading 3.2.1.**) in a volume of 1.25 L (final concentration ~2000/ L; *see* **Note 15**).
3. Start the binding reaction by adding 1.5 L of 20X $CaCl_2$ solution in SDB (**Subheading**

3.2.2.; *see* **Note 16**).

4. After 20 min, stop the binding reaction by adding 1 mL of cold (4℃) chromatin binding buffer. Mix by pipeting up and down. Keep the solution cold from this time onward.
5. Carefully underlayer the binding mixture with 100 L of cushion solution containing 30% sucrose. Spin in a swinging bucket rotor at 10,000*g* for 15 min.
6. Aspirate the supernatant down to the interface. Wash the sides of the tube with 800 L fresh stop solution, and aspirate again down to the bottom of the tube, leaving 2–3 L and avoiding any pelleted material.
7. Add SDS protein sample buffer. Load about 1/5 of the sample onto a lane of a polyacrylamide gel.
8. Detect the bound proteins by immunblotting using ORC2, MCM3, or Cdc6 antibodies *(1)*.

A typical result is shown in the bottom panel of **Fig. 6**. The MCM complex associates with the pelleted chromatin (lane 1), but the complex is not found if sperm are omitted or if calcium is not added to the extract (lanes 2 and 3). The latter control shows that the pre-RC assembly is cell cycle-regulated in these extracts. Geminin inhibits MCM association with chromatin at the same concentration that inhibits replication.

4. Notes

1. Pour 10–25 mL of a warm (37℃) 5% gelatin solution into the beaker, and quickly tilt it to coat the sides, Pipet off as much gelatin solution as possible, and let sit at room temperature for about 10 min. Return the gelatin solution to 37℃ promptly so that it does not solidify.
2. If there are insufficient eggs to fill a full-size SW55 tube, the eggs can be spun in a smaller tube (Beckman 347356) that can be placed inside the SW55 tube.
3. Excess water dilutes the extract and makes it less likely to replicate DNA or degrade proteins.
4. The coding sequence must be preceded by a bacteriophage SP6 or T7 promoter in order to be transcribed using the kit.
5. Males are much smaller than females and have no cloacal opening between their legs. Occasionally the supplier will misclassify a juvenile female as a male.
6. The homogenization step may be omitted if a suitable Dounce is not available.
7. The cycloheximide is to prevent the extract from entering mitosis.
8. If the control reaction does not show condensed DNA, or if an interphase nucleus did not form at any calcium concentration, then the extract cannot be used. If the extract forms a nucleus even if calcium was not added, make sure you remembered to add EGTA to the CSF-XB.
9. We routinely perform two positive control reactions: (1) no additions, and (2) geminin dilution buffer without geminin.
10. We routinely perform two negative control reactions: (1) SDB added instead of calcium solution, and (2) no sperm DNA. Make up any difference in volume with SDB.
11. Typically the reaction mixture has congealed at this point.
12. Disposal of organic radioactive waste is problematic. One approach is to store the waste for 6 mo until the ^{32}P has decayed and then dispose of it as regular organic waste. Check with your institution for proper disposal technique.
13. Use India ink for labeling. Pencil and other inks are removed during washing.

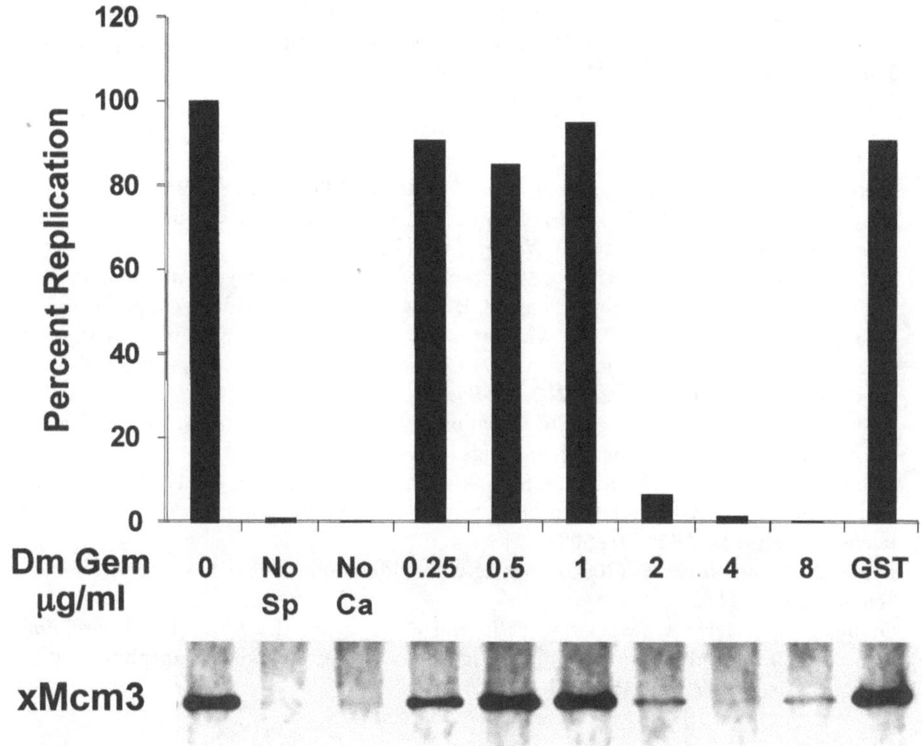

Fig. 6. Inhibition of DNA replication and MCM loading by geminin. *Drosophila* geminin (Dm Gem) was added to replication extract at various concentrations. (Top) Extent of DNA replication at various geminin concentrations. Replication is completely inhibited at concentrations above 2 g/mL. Addition of glutathione-S-transferase (GST) protein has no effect. (Bottom) Minichromosome maintenance complex (MCM) loading onto chromatin is inhibited by the same concentration of geminin that inhibits replication. (Reprinted with permission from **ref.** ***11***.)

14. The number 1310 takes into account the concentration of template DNA (5 ng/ L) and the concentration of cold dNTPs present naturally in *Xenopus* egg extract (about 50 *M* each dNTP) ***(3)***. If more or less template is added, adjust the number accordingly.
15. The sperm must be added after the geminin because geminin has no effect on pre-RCs that are already assembled ***(1)***.
16. This solution is made 20X instead of 10X to avoid diluting the extract. If the calcium concentration required to activate the extract is 0.2 m*M*, the 20X stock should be 4 m*M* in SDB.

References

1. McGarry, T. J. and Kirschner, M. W. (1998) Geminin, an inhibitor of DNA replication, is degraded during mitosis. *Cell* **93,** 1043–1053.

2. Wohlschlegel, J. A., Dwyer, B. T., Dhar, S. K., Cvetic, C., Walter, J. C., and Dutta, A. (2000) Inhibition of eukaryotic DNA replication by geminin binding to cdt1. *Science* **290,** 2309–2312.
3. Tada, S., Li, A., Maiorano, D., Mechali, M., and Blow, J. J. (2001) Repression of origin assembly in metaphase depends on inhibition of RLF-B/Cdt1 by geminin. *Nat. Cell Biol.* **3,** 107–113.
4. Murray, A. W. (1991) Cell cycle extracts. *Methods Cell Biol.* **36,** 581–605.
5. Glotzer, M., Murray, A. W., and Kirschner, M. W. (1991) Cyclin is degraded by the ubiquitin pathway [see comments]. *Nature* **349,** 132–8.
6. Wu, M. and Gerhart, J. (1991). Raising *Xenopus* in the laboratory, in *Methods in Cell Biology*, Vol. 36 (Kay, B. K. and Peng, H. B., eds.), Academic, San Diego, pp. 3–18.
7. Lustig, K. D., Stukenberg, P. T., McGarry, T. J., et al. (1997) Small pool expression screening: identification of genes involved in cell cycle control, apoptosis, and early development. *Methods Enzymol.* **283,** 83–99.
8. Blow, J. J. and Laskey, R. A. (1986) Initiation of DNA replication in nuclei and purified DNA by a cell-free extract of *Xenopus* eggs. *Cell* **47,** 577–587.
9. Kubota, Y., Mimura, S., Nishimoto, S., Takisawa, H., and Nojima, H. (1995) Identification of the yeast MCM3-related protein as a component of *Xenopus* DNA replication licensing factor. *Cell* **81,** 601–609.
10. Lygerou, Z. and Nurse, P. (2000) License withheldgeminin blocks DNA replication. *Science* **290,** 2271–2273.
11. Quinn, L. M., Herr, A., McGarry, T. J., and Richardson, H. (2001) The *Drosophila* geminin homolog: roles for geminin in limiting DNA replication, in anaphase and in neurogenesis. *Genes Dev.* **15,** 2741–2754.

16

CDK-Activating Kinases

Detection and Activity Measurements

Stéphane Larochelle and Robert P. Fisher

Summary

All cyclin-dependent kinases (CDKs) involved in eukaryotic cell cycle control require phosphorylation at a conserved threonine (or serine) residue within the activation- or T-loop to attain full enzymatic activity. The enzyme responsible for this activating phosphorylation, the CDK-activating kinase (CAK), is therefore essential for proliferation of all eukaryotic cells. We describe methods to assess the T-loop phosphorylation state of the major mammalian CDKs in vivo; to measure the levels of CAK activity in cell-free extracts; and to analyze the abundance, subunit composition, and phosphorylation state of CAK complexes in metazoan cells. When derangement of normal CDK regulation is suspected as a cause of disturbed cell cycle progression, the combination of these methodologies can ascertain whether a primary CAK defect is the explanation.

Key Words

Cell cycle; cyclin; cyclin-dependent kinase; CDK-activating kinase; protein phosphorylation.

1. Introduction

A required step in the activation pathway for all cyclin-dependent kinases (CDKs) involved in cell cycle control is the phosphorylation of a conserved threonine residue within the activation segment, or T-loop, by a CDK-activating kinase (CAK) *(1)*. Preventing T-loop modificationeither by mutation of the phosphoacceptor amino acid residue within the target CDK, or by inactivation of CAK(s)blocks CDK activation and cell cycle progression *(2–8)*. Aside from targeted experimental perturbations of the CAK–CDK pathway, however, instances in which a block to cell cycle progression is

From: *Methods in Molecular Biology, vol. 296, Cell Cycle Control: Mechanisms and Protocols*
Edited by: T. Humphrey and G. Brooks © Humana Press Inc., Totowa, NJ

the direct result of CAK failure are surprisingly rare, and a definitive demonstration that CAK activity is ever rate-limiting for cell cycle progression in vivo remains elusive. From the accumulation of negative data on this question, a picture emerged of CAK as a constitutive activator, regulated, if at all, only by the availability and accessibility of its targets ***(9)***. Recent advances in model genetic organismsboth fission yeast and *Drosophila melanogaster*suggest a more satisfying (and interesting) explanation: CAKs and CDKs in eukaryotic cells constitute a network, which links cell cycle progression to gene expression programs ***(10)*** and, as such, must be regulated in more subtle, complex ways than are the dedicated components of either the cell cycle or transcriptional machineries. There has been, for example, no obvious periodicity observed for CAK expression or activity levels in any system. On the other hand, a recent report on the *Drosophila* CAK CDK7 showed that its subcellular localization and ability to promote mitotic progression were dramatically influenced by a regulated association with the XPD helicase ***(11)***.

What the preceding overview of the CAK–CDK pathway means, in practical terms, is that any analysis of CAK function in a biological setting only *begins* with the measurement of T-loop phosphorylation on endogenous CDKs. Whereas normal levels of T-loop phosphorylation on the major CDKs can probably be taken as evidence that the CAK pathway is not affected by a given perturbation (barring effects restricted to specific CDK/cyclin complexes, or subpopulations thereof, which have not yet been described), finding that T-loop phosphorylation *does* change could indicate a primary CAK deficiency but is equally (or more) likely to be a secondary effect. Decreased T-loop phosphorylation in metazoans is a predictable consequence of increased binding of CDK/cyclin complexes by CDK inhibitors (CKIs) of the Cip/Kip class ***(12)*** and of decreased CDK/cyclin complex assembly ***(6,13)***. Before concluding that diminished T-loop phosphorylation in vivo is owing to reduced CAK activity, it is therefore necessary to exclude these other possible scenarios. Fortunately, the past several years have seen the development of powerful and readily available analytic tools and assays for this purpose, many of which we will describe in **Subheading 3.1.**

Once other possible mechanisms have been ruled out, a change in cellular CAK activity can be measured by several well-established biochemical assays with recombinant CDK/cyclin complexes as targets for CAK in cell-free extracts. As is the case in vivo, failure to activate a CDK in vitro is not necessarily due to insufficient CAK; it is therefore always advisable to assay CAK activity with multiple methods. **Subheading 3.2.** describes several direct and indirect CAK assays that differ with respect to sensitivity, selectivity, and susceptibility to factors other than CAK that may reside in crude extracts. Finally, antibodies to the major mammalian CAK catalytic subunit, CDK7, are commercially available and allow easy detection and measurement of CDK7 in both human and mouse cell extracts. The same antibodies will immunoprecipitate CDK7 complexes from mammalian cell extracts in active form. In **Subheading 3.3.**, we will describe methods to ascertain whether a change in cellular CAK activity measured in a whole-cell extract is correlated with changes in the abundance or activity of CDK7, the major (and so far, the only) CAK in mammalian cells.

2. Materials

1. Sodium dodecyl sulfate polyacrylamide gel electrophoresis (SDS-PAGE) resolving gel buffer: to resolve CDK1(CDC2) and CDK7 isoforms, use 1.5 *M* Tris-HCl, pH 9.2, instead of pH 8.8. Stacking gel buffer: 0.5 *M* Tris-HCl, pH 6.8.
2. Piperazine diacrylamide (PDA) crosslinker (Bio-Rad, cat. no. 161-0202).
3. Electrophoretic-grade solid acrylamide from various suppliers.
4. Liquid acrylamide (30%) stock solution: 29.5 g acrylamide, 0.5 g PDA in 100 mL H_2O.
5. Homogenization buffer (HoB): 25 m*M* HEPES, pH 7.4, 150 m*M* NaCl, 20 m*M* Na β-glycerophosphate, 50 m*M* NaF, 1 m*M* EDTA, 0.5 m*M* dithiothreitol (DTT), 0.1% Triton X-100, 1 m*M* phenylmethylsulfonyl fluoride (PMSF), 2 g/mL aprotinin, 1 g/mL leupeptin.
6. Anti-phospho-CDC2 (T161) and anti-phospho-CDK2 (T160) polyclonal antibodies (Cell Signaling Technologies, cat. nos. 9114 and 2561, respectively).
7. Anti-CDC2 and -CDK2 polyclonal antibodies (Santa Cruz Biotechnology, cat. nos. sc-954 and sc-163, respectively).
8. Anti-CDK7 monoclonal antibody from Sigma (cat. no. C7089) for use with human protein only or from Santa Cruz Biotechnology (cat. no. sc-7344), for human or mouse proteins. Both antibodies can immunoprecipitate active CDK7 protein complexes.
9. Anti-cyclin A, B, or E (Santa-Cruz Biotechnology, cat. nos. sc-751, sc-752, or sc-248, respectively) can immunoprecipitate active complexes.
10. HBS: 10 m*M* HEPES, pH 7.4, 150 m*M* NaCl. HBST: HBS + 0.1% Triton X-100.
11. HD: 10 m*M* HEPES-NaOH, pH 7.4, 1 m*M* DTT.
12. RIPA buffer: 50 m*M* HEPES-NaOH, pH 7.4, 150 m*M* NaCl, 1% (v/v) Triton X-100, 1% (w/v) sodium deoxycholate, 0.1% (w/v) SDS.
13. Protein A-Sepharose (Sigma) or Protein G-Sepharose (Pharmacia), swollen in HBS.
14. Monoclonal anti-HA antibody, 16B12, supplied as ascites fluid at approx 7–20 mg/mL total protein concentration (Covance).
15. Histone H1 (Roche, cat. no. 1004875) dissolved in H_2O or HBS at 5 mg/mL and stored in aliquots at –80€.

3. Methods

3.1. Detecting T-Loop Phosphorylation of CDKs In Vivo

A distinguishing feature of several CDKs is that phosphorylation at the activating site within the T-loop *increases* their electrophoretic mobility in standard SDS-PAGE gels. Although the basis for this behavior is unknown, it provides a very simple and convenient measurement for the activation state of metazoan CDK2 and budding yeast CDK1 in vivo. Unfortunately, many CDKs do not shift appreciably upon T-loop phosphorylation when they are analyzed in SDS-PAGE gels containing the standard crosslinking agent *bis*-acrylamide. In several of these cases, shifts to increased mobility can be revealed in a modified SDS-PAGE system with the crosslinking agent, PDA substituted for *bis*-acrylamide ***(14)*** and with the separating gel pH raised from 8.8 to 9.2 ***(15)***. We and others have successfully resolved phosphoisoforms of metazoan CDK1 ***(6,14)*** and CDK7 ***(16,17)*** with this gel system, which is described in detail in **Subheading 3.1.1.**

Other CDKs, including fission yeast CDK1, have resisted our attempts to resolve their phosphorylated and unphosphorylated isoforms by 1D gel electrophoresis. Fortunately, the development in recent years of phosphoisoform-specific antibodies to several CDKs has provided an excellent alternative for detecting the activated forms of CDK1 and CDK2 from various sources. The first such antibody was raised by Solomon and co-workers against budding yeast CDK1 but crossreacted with metazoan CDKs ***(18)*** and with fission yeast CDK1 ***(10)***. More recently, antibodies against the activated phosphosioforms of metazoan CDKs 1 and 2 have become commercially available, and we provide a protocol for their use, as well as an analysis of their specificities, in **Subheading 3.1.2.**

3.1.1. Detecting T-Loop Phosphorylation by Electrophoretic Mobility Shift in 1D SDS Polyacrylamide Gels Containing the Crosslinker PDA

Although the increased mobility of CDK2 caused by T-loop phosphorylation is readily observed under most electrophoretic conditions, such is not the case with CDK1 and CDK7. Although CDK7 can assemble into an active complex with an unmodified T-loop ***(19–21)***, T-loop phosphorylation is required for full function in vivo and was also shown to regulate the catalytic activity of CDK7 in a substrate-dependent manner ***(17)***. We have therefore devoted some effort to developing conditions that will allow the separation of phosphoisoforms of CDK7 (and of its target, CDK1) and thus provide a convenient assay of activity state in vivo.

In addition to increasing the length of the gel used for separation, we find that two modifications to the standard SDS-PAGE conditions favor resolution of different phosphoisoforms of CDK1 and CDK7. The first is the use of PDA instead of *bis*-acrylamide as the crosslinker in the acrylamide gel solution. This modification was first described by Kumagai and Dunphy ***(14)*** and by itself affords more reliable separation of CDK1 isoforms.

A second modification to standard conditions is a pH increase of the resolving gel from 8.8 to 9.2, which results in a dramatic increase in the resolution of CDK7 isoforms ***(16,17)***. Elevated pH alters the sieving properties of the gel so that a 10% gel with a pH of 9.2 will approximate a 12% gel with a pH of 8.8 in its resolving capacity. Interestingly, CDK2 isoforms are not resolved under such conditions. We therefore stress that the optimal conditions to resolve phosphoisoforms of different CDKs (or any other phosphoprotein) must be empirically determined by the investigator.

3.1.1.1. Gel Electrophoresis

1. Prepare the acrylamide stock solution by dissolving 29.5 g of acrylamide and 0.5 g of PDA in 60 mL of ultrapure water. Once the solids have dissolved, adjust the volume to 100 mL with water (*see* **Note 1**).
2. Prepare a 9% gel according to standard protocols using a Tris-HCl buffer at pH 9.2 for the separation of CDK1 and CDK7 isoforms, or at pH 8.8 for CDK2 isoforms (*see* **Note 2**).
3. The gel is poured in any commercially available apparatus with a length of at least 10 cm.
4. Total cell lysates should be prepared in the presence of protein phosphatase inhibitors to prevent the dephosphorylation of the T-loop during manipulations.

5. Load the sample along with a prestained standard marker, and run the gel at 200 V until the protein of interest is close to the bottom of the gel or has migrated at least 6 cm through the resolving gel.
6. Transfer and detect by immunoblotting according to standard methods with the appropriate antibody.

3.1.2. Antiphospho-T-Loop Antibodies

The availability of phosphoisoform-specific antibodies may eventually render the eletrophoretic mobility shift assay unnecessary except to measure the relative abundance of different isoforms within the same sample. (There are also CDKs for which phosphoisoform-specific antibodies may remain unavailable.) To our knowledge, the only T-loop specific antibodies available commercially are from Cell Signaling Technologies. Although the CDK2 antibody (cat. no. 2564) does not appear to recognize T-loop phosphorylated CDK1, the same, unfortunately, cannot be said for the antibody raised against T-loop phosphorylated CDK1 (cat. no. 9114), which crossreacts with phosphorylated CDK2. In any case, it may be preferable to immunoprecipitate the kinase of interest prior to detection with phosphoisoform-specific antibodies in order to reduce the background that can be present in whole-cell lysates. We find that the antibodies detect CDKs phosphorylated on their T-loops semiquantitatively, with an approx 20-fold dynamic range (**Fig. 1**).

Below is an abbreviated protocol to immunoprecipitate cyclin B-bound CDK1 in order to assess its state of T-loop phosphorylation. Refer to **Subheading 3.1.1.** for general procedural details of immunoprecipitation. If CDK2 is the protein of interest, then cyclin E and/or A should be immunoprecipitated with appropriate antibodies prior to immunoblotting (*see* **Note 3**).

3.1.2.1. Immunoprecipitation

1. Prepare in advance Protein A beads coupled noncovalently to the antibody of choice (e.g., anti-cyclin B1, sc-752) by incubating together at a ratio of 200 ng of antibody per 20 L of beads. The beads can be kept at 4C for several months.
2. Lyse cells or tissue in HoB and measure protein concentration. As little as 100 g of total extracted protein can be used per assay (*see* **Note 4**).
3. Add each sample to 10–20 L Protein A–antibody beads, and incubate with gentle rocking at 4C for 1–3 h (*see* **Note 5**).
4. Wash the beads three times with 0.4–1.0 mL of ice-cold HoB.

3.1.2.2. Immunobloting

5. Add 1 vol of 2X SDS-PAGE sample buffer, and run identical aliquots (1/4 to 1/2 of total) of the boiled samples on duplicate 10% SDS-PAGE gels.
6. Transfer to nitrocellulose.
7. Following transfer, cut the membrane at approx 50 kDa using the prestained marker as guide. This will allow for the simultaneous probing of cyclin B and CDK1, and helps ensure that the immunoprecipitations were equally efficient between different samples.
8. Proceed to detection with phosphoisoform-specific and -nonspecific CDK1 antibodies. The antibodies described in **Subheading 2.** are diluted 1:1000 for this purpose.

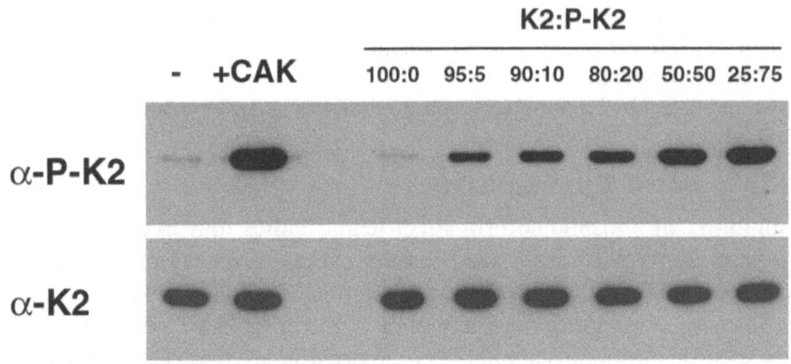

Fig. 1. Detection of T-loop phosphorylation with phosphoisoform-specific antibodies. CDK2 was phosphorylated by incubating pure CDK2/cyclin A complexes with CAK (CDK7/cyclin H). The untreated (–) and phosphorylated (+CAK) CDK2 complexes were mixed in the proportions indicated above each lane. Each lane contains a total of 4 ng CDK2. The top panel was probed with the phospho-CDK2(T160)-specific antibody at 1:1000. The bottom panel was probed with an antibody to the carboxyl-terminus of CDK2 to ensure equal loading. Note: the electrophoresis conditions do not allow separation of CDK2 isoforms, which therefore run as a single band.

3.1.2.3. Histone H1 Kinase Assay

9. It may be important to test that the activity of the isolated CDK1/cyclin B complex correlates with the level of T-loop phosphorylation. For this, follow the protocol in **Subheading 3.2.1.** (below) from **step 16** onwards.

3.2. Activating Phosphorylation of CDKs In Vitro

Various assays have been developed to measure the T-loop phosphorylation of CDKs in vitro. All depend on recombinant sources of the substrates, because endogenous CDKs are often extensively phosphorylated on their T-loops in vivo. Likewise, CDKs coexpressed with their partner cyclins in mammalian or insect cells are often too heavily phosphorylated on their T-loops to be satisfactory substrates for the CAK reaction in vitro. We therefore express CDKs and cyclins in separate infections with recombinant baculoviruses and either: (1) purify them separately for sequential or simultaneous addition to the activation reactions, or (2) mix crude preparations of CDK and cyclin subunits and then purify the assembled complexes chromatographically. There are now commercial sources for recombinant baculoviruses expressing CDKs and cyclins and several published protocols for their purification ***(13,22,23)***.

For many purposes, the simplest, most convenient assay of CAK activity is an indirect one, in which a substrate CDK/cyclin is first activated, in Mg/ATP-dependent fashion, by preincubation with CAK, and then tested for its ability to phosphorylate a model substrate, such as histone H1, in a subsequent reaction containing radiolabeled ATP. Among the advantages of the indirect CAK assay, described in **Subheading 3.2.1.**, are sensitivity, because the signal is amplified by the two rounds of catalysis; reasonable specificity without an absolute need for an activation-site mutant, because no other phosphorylation event on CDK is known to stimulate its enzymatic activity; and simplicity.

Despite the advantages of the indirect assay, it may still be important to confirm that the T-loop threonine is indeed being phosphorylated. A direct CDK phosphorylation assay may become essential when the source of CAK is a crude extract or fraction, which also contains inhibitory activities that may limit or even abolish activation. T-loop phosphorylation may still be detectable under these conditions but must be distinguished from phosphorylation on other sites, including inhibitory tyrosine and threonine residues targeted by the negative regulators Weel and Mytl ***(1)***. In the case of metazoan CDK2 (and budding yeast CDK1), that distinction can be based on the shift to increased electrophoretic mobility caused by activating but not by inhibitory phosphorylations. For other CDKs, specialized gel systems (*see* **Subheading 3.1.1.** above) may facilitate this determination, as may immunoblotting with the anti-phospho-T-loop antibodies (*see* **Subheading 3.1.2.**). A well-established method of distinguishing T-loop phosphorylation from other modifications is to perform parallel phosphorylation reactions on activation-site mutant CDKs (e.g., the Thr160-to-Ala mutant of human CDK2). In **Subheading 3.2.2.**, we describe both radioactive and nonradioactive methods for detecting direct phosphorylation of the CDK T-loop in vitro.

3.2.1. Indirect CAK Assays: Activating Recombinant CDK/Cyclin Complexes With Crude Subcellular Fractions

3.2.1.1. Immobilization of CDK–Cyclin Complexes

1. Aliquot Protein G-Sepharose into 2-mL microcentrifuge tubes. Make sure the slurry is well mixed (vortex), and take out 50 L of 1:1 slurry per reaction plus approx 200 L extra.
2. Spin for 1 min in a microcentrifuge (1000–10,000*g*), and aspirate the supernatant. Wash beads twice with 1 mL HBST (mix, spin, and aspirate the supernatant) and resuspend to the original volume (i.e., reconstitute the 1:1 slurry) with fresh HBST.
3. Add antibody (0.25–0.5 L per reaction) to beads and mix well.
4. Add CDK2HA/cyclin A complex to the antibody–bead mixture. Typically, we use 0.1–0.2 g of CDK/cyclin per reaction. (If the CDK/cyclin complex is to be formed just prior to the reaction, the purified subunits should be mixed and incubated for ~20 min at room temperature to allow assembly prior to adding to antibody–bead mixture.) Mix well by inversion or gentle vortexing.
5. Incubate for 1–2 h with gentle rocking at 4°C.
6. Recover beads by centrifugation and aspirate supernatant.
7. Wash beads 2X with 1 mL HBST and 2X with 1 mL HD.
8. After the last wash, resuspend beads to original volume in HD.
9. Aliquot 50 L of bead slurry into 1.5-mL microcentrifuge tubes; be sure to mix stock tube between each withdrawal of beads to ensure uniform bead delivery to reaction tubes.

3.2.1.2. Activation

10. Make 10X Mg/ATP mix: 100 m*M* $MgCl_2$, 10 m*M* cold ATP in HD.
11. Add to reaction tubes in order: HD (calculate volume to make total reaction volume 100 L); cell-free extracts or fractions to be assayed for CAK activity; 10 L of 10X Mg/ATP mix.

12. Incubate for 30 min at room temperature, with occasional mixing (flick tubes every 5 min or so).
13. Recover beads by centrifugation, and aspirate supernatant.
14. Wash beads twice with 1 mL HD. (Optionally, preequilibrate complexes in kinase reaction conditions by substituting for last HD wash a single wash with mock kinase buffer: 10 m*M* HEPES, pH 7.4, 1 m*M* DTT, 10 m*M* $MgCl_2$, 50 *M* cold ATP.)
15. After the last wash, aspirate all excluded liquid from surface of beads; be careful not to aspirate beads.

3.2.1.3. Histone H1 Kinase Assay

16. Make 1X kinase mix: 10 m*M* HEPES, pH 7.4, 1 m*M* DTT, 10 m*M* $MgCl_2$, 50 *M* cold ATP, 1 Ci/reaction [γ-^{32}P]ATP; 5 g/reaction histone H1. Note: total reaction volume is 30 L; make enough mix for all your reactions *plus one*.
17. Add 30 L kinase mix to beads; mix gently.
18. Incubate for 10 min at room temperature.
19. Add 10 L 4X SDS-PAGE sample buffer, mix gently, and quick-spin.
20. Pierce caps with a 16-gage needle.
21. Heat at >95C for 1–5 min.
22. Load on 10% SDS-PAGE gel.
23. Electrophorese at 200 V, keeping unincorporated ATP on the gel; separate plates, trim the bottom of the gel at the dye front, and discard the bottom strip containing unincorporated label; notch the lower left corner, and remove and discard stacking gel. Stain gel with Coomassie brilliant blue; destain briefly; dry, and autoradiograph.
24. Quantify incorporation either by a PhosphorImager or scintillation counting of the excised histone H1 band.

3.2.2. Measuring T-Loop Phosphorylation Directly by Radioactive and Nonradioactive Methods

3.2.2.1. Direct Radiolabeling of CDK by CAK in Extracts or Partially Purified Fractions

1. Prepare protein G/Sepharose/antibody mixture by following **steps 1–3** of protocol in **Subheading 3.2.1.**; keep on ice while performing kinase reactions.
2. Assemble CDK/cyclin complexes by incubating CDK (~1.0 g per reaction) with appropriate cyclin (~1.0 g per reaction) in HD and incubating for 20 min at room temperature. Alternatively, preformed, unphosphorylated CDK/cyclin complexes may be added directly to kinase reactions.
3. Make 10X Mg/ATP mix containing radioactive ATP: 100 m*M* $MgCl_2$, 500 *M* cold ATP, 5 Ci/ reaction [γ-^{32}P]ATP. You will need 5 L of this mix per reaction, so make enough for one or two more than the planned number of assays.
4. Assemble the kinase reactions in 1.5-mL microcentrifuge tubes by addition, in order, of: HD buffer, calculated to bring total reaction volume to 50 L; the CDK/cyclin assembly mix, or the preformed, purified complex; the cell-free extract or protein fraction to be assayed for CAK activity; and the radioactive, 10X Mg/ATP mix (*see* **Note 6**).
5. Incubate for 30 min at room temperature.
6. Place reactions on ice and add, to each tube, 50 L of bead–antibody mixture. (Keep beads well suspended in stock tube while aliquoting.)
7. Immunoprecipitate by gently rocking for 1–2 h at 4C.
8. Spin briefly in microcentrifuge; add to each tube 1 mL ice-cold RIPA buffer, and mix by inversion.

9. Recover beads by centrifugation; remove supernatant, and discard as radioactive waste.
10. Wash beads 2X with 1 mL HBST and 2X with 1 mL HBS. Aspirate dry.
11. Add 30 L 1X SDS-PAGE sample buffer, mix gently, and quick-spin.
12. Pierce caps with a 16-gage needle.
13. Heat at >95°C for 1–5 min.
14. Load on 10% SDS-PAGE gel.
15. Follow **steps 23–24** of protocol in **Subheading 3.2.1.**

3.3. Measuring Protein and Activity Levels of Metazoan CAK

It has been over a decade since the identification of a metazoan CAK containing CDK7 ***(24–26)***. In that time, no additional CAK activities have been positively identified in any metazoan species, although trace amounts of such enzymes have been reported in mammalian cell extracts ***(27,28)***. Conversely, there have been sporadic reports of cell cycle-arresting agents that seemingly downregulate the steady-state level of T-loop phosphorylation in vivo but leave CDK7 and its associated subunits (and activity) undiminished ***(27,29)***. That the CDK7 complex can be targeted or sequestered within the cell by proteins such as Xeroderma Pigmentosum complementation group D (XPD) ***(11)*** could explain these discrepancies and so, until a clear demonstration of another metazoan CAK, the presumption that cellular CAK activity equates with CDK7 activity remains a valid one. In this section, we will describe methods to detect and quantify the level of CDK7 protein in cell-free extracts with commercially available antibodies to CDK7 and its regulatory subunits, cyclin H and ménage-à-trois 1 (MAT1). By combining anti-CDK7 immunoblotting with the specialized SDS-PAGE system described above in **Subheading 3.1.1.**, moreover, we can precisely measure the distribution of CDK7 among its different T-loop phosphoisoforms. Finally, we will present a method for measuring the CAK activity associated with CDK7 in immunoprecipitates from cell extracts.

3.3.1. Immunoblotting With Antibodies to Components of the Mammalian CDK7 Complex

See **Subheading 3.1.1.** above for SDS-PAGE and immunoblotting; by separating cellular proteins in conventional SDS-PAGE gels (separating gel pH 8.8; crosslinker *bis*-acrylamide), one can prevent separation of the CDK7 band into multiple bands representing distinct isoforms, and thus better quantify total CDK7 in the extract.

3.3.2. Kinase Assays on Anti-CDK7 Immunoprecipitates

CDK7 is a fairly abundant protein in most cells and tissues and can readily be immunoprecipitated as an active complex with antibodies to cyclin H, MAT1, or CDK7 itself. We routinely use the anti-CDK7 monoclonal MO1-1 (Sigma, cat. no. C7089) to immunoprecipitate active CDK7.

3.3.2.1. Immunoprecipitation

1. Prepare in advance Protein G–antibody beads by incubating 0.25–0.5 L of antibody (supplied as crude ascites fluid) for each 20 L of packed beads. The beads can be kept at 4°C for several months.
2. Lyse cells or tissue in HoB, and measure extracted protein concentration. As little as 100 g of total protein can be used per assay.

3. Add each sample to 20 L protein G–antibody beads, and rock gently at 4°C for 1 h.
4. Keeping tubes on ice whenever possible, wash the beads three times with 0.4–1.0 mL of HoB.
5. Wash the beads twice with 0.4–1.0 mL of HBS.
6. Wash the beads once with 0.4 mL of HBS + 10 m*M* $MgCl_2$.

3.3.2.2. Activity Assay

7. While washing, prepare the kinase assay mix: 20 L per assay (25 m*M* HEPES pH 7.4, 150 m*M* NaCl, 1 m*M* DTT, 10 m*M* $MgCl_2$, 50–200 *M* ATP) with 5 Ci [γ^{32}P[ATP and 1–2 g CDK2/cyclin A complex (or 5 g GST-CTD substrate).
8. Remove as much of the supernatant from the immunoprecipitate as possible and transfer the tubes to room temperature.
9. Start a timer, and add 20 L of the kinase assay mix to each sample. Mix by pipeting gently up and down a few times (*see* **Note 7**). Use the timer to add the assay mix to each sample at regular intervals, and stop the reaction after 5–10 min at room temperature in a similarly timed manner by addition of 30 L 2X SDS sample buffer (*see* **Note 8**).
10. Boil the samples and proceed with SDS-PAGE and autoradiography as described in **Subheading 3.2.1.** (*see* **Notes 9** and **10**).

4. Notes

1. Only small amounts of acrylamide–PDA stock solution should be prepared each time because it appears to lose resolving power with age. Similarly, the ammonium persulfate solution should be no older than 1 wk.
2. We find that a 9% separating gel is optimal for the separation of CDK7 isoforms, whereas10–12% is ideal for CDK1.
3. It is preferable to use a cyclin antibody for immunoprecipitation because the bulk of CDK1 is present in its monomeric, and hence unphosphorylated, form in the cell. In each case it is also necessary to probe with an antibody that recognizes the CDK irrespective of phosphorylation state to control for differences in the efficiency of immunoprecipitation. Also note that cyclin A immunoprecipitates will contain some CDK1, and therefore an antibody specific for CDK2 should be used (i.e., not the PSTAIRE antibody that recognizes both CDKs 1 and 2).
4. If the quantity of starting material is not limiting, starting with large amounts of protein (0.5–1 mg per assay) will reduce effects of sample-to-sample variations in the amounts of kinase complexes isolated.
5. Dilute lysates to a volume that will allow for good mixing with the Protein G beads during immunoprecipitation (300 L for a 0.5-mL tube or 0.7 mL for a 1.5-mL tube).
6. Some autophosphorylation on the CDK subunit may occur during this reaction. It is essential to include a control omitting any CAK-containing extracts or fractions and to subtract any resulting background autophosphorylation from the signals measured in the presence of CAK. This problem can be minimized or eliminated by using a catalytically inactive CDK as a substrate.
7. Cut off the pipet tips to facilitate mixing of the beads with the kinase assay cocktail. It is also important to make sure that beads do not stick to the wall of the tip during mixing.
8. Do not incubate the substrates with the immunoprecipitated enzyme for too long. Sensitivity is rarely an issue, and it is easy to surpass the linear range of the reaction with prolonged incubation. (This will yield similar signals even when significantly different

amounts of CDK activity are present.) Also, avoid saturation that can be caused by an insufficient quantity of substrate; we typically include at least 3–5 g of protein substrate (e.g., histone H1 or the carboxyl terminal domain (CTD) of RNA polymerase II) and a soluble, calibrated, and active CDK/cyclin complex as a positive control to ensure that we are in the linear range of the assay.

9. A portion of the kinase assay can also be used for immunoblotting to ensure that an apparent difference in activity does not simply reflect varying amounts of CDK7 in the final immunoprecipitate. We find it more reliable to take an aliquot of the kinase assay itself after denaturation in sample buffer for immunoblot analysis, instead of splitting the precipitate prior to the kinase reaction. As little as 1/5 of each reaction should be sufficient for immunoblotting, leaving some sample in case a problem occurs while one is running and/or transferring the gels. If GST-CTD is the radiolabeled substrate, we actually cut the gel in two, using the bottom part containing CDK7 (~40 kDa) for immunoblotting while subjecting the top part of the gel containing the phosphorylated GST-CTD (~90 kDa) to autoradiography.
10. Because monomeric CDK7 is virtually inactive, it is important to verify that equal amounts of both its regulatory subunits, cyclin H and MAT1, are also present in similar amounts in each immunoprecipitate.

References

1. Morgan, D. O. (1995) Principles of CDK regulation. *Nature* **374,** 131–134.
2. Gould, K. L., Moreno, S., Owen, D. J., Sazer, S., and Nurse, P. (1991) Phosphorylation at Thr167 is required for *Schizosaccharomyces pombe* p34^{cdc2} function. *EMBO J.* **10,** 3297–3309.
3. Kaldis, P., Sutton, A., and Solomon, M. J. (1996) The Cdk-activating kinase (CAK) from budding yeast. *Cell* **86,** 553–564.
4. Thuret, J.-Y., Valay, J.-G., Faye, G., and Mann, C. (1996) Civ1 (CAK in vivo), a novel Cdk-activating kinase. *Cell* **86,** 565–576.
5. Espinoza, F. H., Farrell, A., Erdjument-Bromage, H., Tempst, P., and Morgan, D. O. (1996) A cyclin-dependent kinase-activating kinase (CAK) in budding yeast unrelated to vertebrate CAK. *Science* **273,** 1714–1717.
6. Larochelle, S., Pandur, J., Fisher, R. P., Salz, H. K., and Suter, B. (1998) Cdk7 is essential for mitosis and for in vivo Cdk-activating kinase activity. *Genes Dev.* **12,** 370–381.
7. Lee, K. M., Saiz, J. E., Barton, W. A., and Fisher, R. P. (1999) Cdc2 activation in fission yeast depends on Mcs6 and Csk1, two partially redundant Cdk-activating kinases (CAKs). *Curr. Biol.* **9,** 441–444.
8. Wallenfang, M. R. and Seydoux, G. (2002) *cdk-7* is required for mRNA transcription and cell cycle progression in *Caenorhabditis elegans* embryos. *Proc. Natl. Acad. Sci. USA* **99,** 5527–5532.
9. Murray, A. W. and Marks, D. (2001). Can sequencing shed light on cell cycling? *Nature* **409,** 844–846.
10. Saiz, J. E. and Fisher, R. P. (2002) A CDK-activating kinase network is required in cell cycle control and transcription in fission yeast. *Curr. Biol.* **12,** 1100–1105.
11. Chen, J., Larochelle, S., Li, X., and Suter, B. (2003) Xpd/Ercc2 regulates CAK activity and mitotic progression. *Nature* **424,** 228–232.
12. Kaldis, P., Russo, A.A., Chou, H.S., Pavletich, N.P., and Solomon, M.J. (1998). Human and yeast cdk-activating kinases (CAKs) display distinct substrate specificities. *Mol. Biol. Cell* **9,** 2545–2560.

13. Desai, D., Wessling, H.C., Fisher, R.P., and Morgan, D.O. (1995). The effect of phosphorylation by CAK on cyclin binding by CDC2 and CDK2. Mol. Cell. Biol. 15, 345-350.
14. Kumagai, A. and Dunphy, W.G. (1995) Control of the Cdc2/Cyclin B complex in *Xenopus* egg extracts arrested at a G2/M checkpoint with DNA synthesis inhibitors. *Mol. Biol. Cell* **6,** 199–213.
15. Makowski, G. S. and Ramsby, M. L. (1993) pH modification to enhance the molecular sieving properties of sodium dodecyl sulfate-10% polyacrylamide gels. *Anal. Biochem.* **212,** 283–285.
16. Garrett, S., Barton, W. A., Knights, R., Jin, P., Morgan, D. O., and Fisher, R. P. (2001) Reciprocal activation by cyclin-dependent kinases 2 and 7 is directed by substrate specificity determinants outside the T-loop. *Mol. Cell. Biol.* **21,** 88–99.
17. Larochelle, S., Chen, J., Knights, R., et al. (2001) T-loop phosphorylation stabilizes the CDK7-cyclin H-MAT1 complex *in vivo* and regulates its CTD kinase activity. *EMBO J.* **20,** 3749–3759.
18. Ross, K. E., Kaldis, P., and Solomon, M. J. (2000) Activating phosphorylation of the *Saccharomyces cerevisiae* cyclin-dependent kinase, Cdc28p, precedes cyclin binding. *Mol. Biol. Cell* **11,** 1597–1609.
19. Fisher, R. P., Jin, P., Chamberlin, H. M., and Morgan, D. O. (1995) Alternative mechanisms of CAK assembly require an assembly factor or an activating kinase. *Cell* **83,** 47–57.
20. Devault, A., Martinez, A.-M., Fesquet, D., et al. (1995) MAT1 (‘*ménage à trois*), a new RING finger protein subunit stabilizing cyclin H-cdk7 complexes in starfish and *Xenopus* CAK. *EMBO J.* **14,** 5027–5036.
21. Tassan, J.-P., Jaquenod, M., Fry, A. M., Frutiger, S., Hughes, G., and Nigg, E. A. (1995) *In vitro* assembly of a functional human cdk7/cyclin H complex requires MAT1, a novel 36 kD RING finger protein. *EMBO J.* **14,** 5608–5617.
22. Desai, D., Gu, Y., and Morgan, D. O. (1992) Activation of human cyclin-dependent kinases in vitro. *Mol. Biol. Cell* **3,** 571–582.
23. Rosenblatt, J., De Bondt, H., Jancarik, J., Morgan, D. O., and Kim, S.-H. (1993) Purification and crystallization of human cyclin-dependent kinase 2. *J. Mol. Biol.* **230,** 1317–1319.
24. Fesquet, D., Labbé J.-C., Derancourt, J., et al. (1993) The MO15 gene encodes the catalytic subunit of a protein kinase that activates cdc2 and other cyclin-dependent kinases (CDKs) through phosphorylation of Thr161 and its homologues. *EMBO J.* **12,** 3111–3121.
25. Poon, R. Y. C., Yamashita, K., Adamczewski, J. P., Hunt, T., and Shuttleworth, J. (1993) The cdc2-related protein $p40^{MO15}$ is the catalytic subunit of a protein kinase that can activate $p33^{cdk2}$ and $p34^{cdc2}$. *EMBO J.* **12,** 3123–3132.
26. Solomon, M. J., Harper, J. W., and Shuttleworth, J. (1993) CAK, the $p34^{cdc2}$ activating kinase, contains a protein identical or closely related to $p40^{MO15}$. *EMBO J.* **12,** 3133–3142.
27. Nagahara, H., Ezhevsky, S.A., Vocero-Akbani, A.M., Kaldis, P., Solomon, M. J., and Dowdy, S. F. (1999) Transforming growth factor beta targeted inactivation of cyclin E:cyclin-dependent kinase 2 (Cdk2) complexes by inhibition of Cdk2 activating kinase activity. Proc. Natl. Acad. Sci. USA 96, 14961-14966.
28. Kaldis, P. and Solomon, M.J. (2000) Analysis of CAK activities from human cells. *Eur. J. Biochem.* **267,** 4213–4221.
29. Ukomadu, C. and Dutta, A. (2003) Inhibition of cdk2 activating phosphorylation by mevastatin. *J. Biol. Chem.* **278,** 4840–4846.

17

Cyclins, Cyclin-Dependent Kinases, and Cyclin-Dependent Kinase Inhibitors

Detection Methods and Activity Measurements

Gavin Brooks

Summary

The cyclin/cyclin-dependent kinase (Cdk) complexes and the Cdk inhibitors (CDKI) are crucial regulators of cell cycle progression in all eukaryotic cells. Using rat cardiac myocytes as a model system, this chapter provides a detailed account of methods that can be employed to measure both cyclin/Cdk activity in cells and the extent of CDKI inhibitory activity present in a particular cell type.

Key Words

Cardiac myocyte; CDK kinase assay; CDKI inhibitory activity assay; Cyclin; Immunoprecipitation; p21; p27.

1. Introduction

Normal cellular proliferation is under the tight control of both positive and negative regulators that determine whether a particular cell can progress through the cell cycle (*see* **refs. *1–3*** for reviews). This carefully ordered progression of the mammalian cell cycle is controlled by the sequential formation, activation, and deactivation of a series of cell cycle regulatory molecules (**Fig. 1**) that exist as a series of complexes consisting of a catalytic kinase subunit (known as the cyclin-dependent kinase [CDK]) and a regulatory cyclin subunit. These cyclin/CDK complexes form part of the positive regulatory machinery that drives the cell through the cycle. Most cyclin mRNAs and proteins show a dramatic fluctuation in their expression during the cell cycle. For example, cyclin A and B expression accumulates transiently at the onset of S-phase and in late

From: *Methods in Molecular Biology, vol. 296, Cell Cycle Control: Mechanisms and Protocols*
Edited by: T. Humphrey and G. Brooks © Humana Press Inc., Totowa, NJ

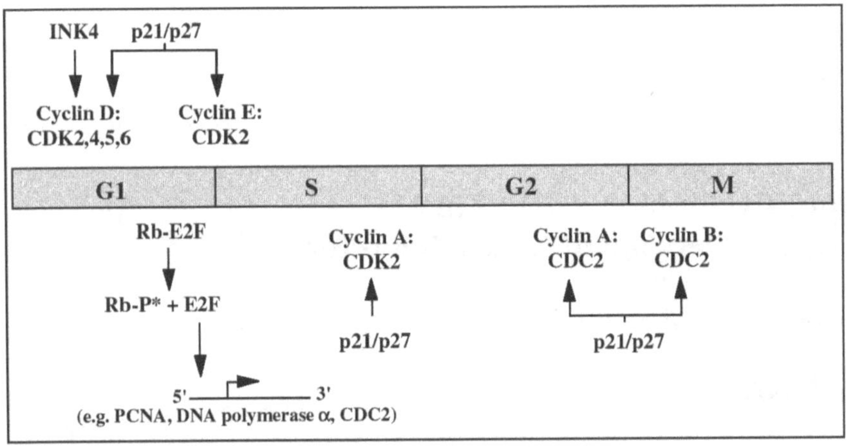

Fig. 1. The mammalian cell cycle showing where various cell cycle regulatory molecules act. PCNA, proliferating cell nuclear antigen.

G_2, respectively, followed rapidly by their degradation via the ubiquitin/proteosome pathway *(1,2)*. In contrast, expression of the various CDK molecules remains relatively constant throughout the cell cycle.

Specific CDK molecules interact with specific cyclin partners at various stages of the cell cycle, such that each complex displays its kinase activity at a particular point in the cell cycle (**Fig. 1**). These cyclin/CDK complexes are able to bind to other regulatory proteins to form large, heterologous multi-complexes that can act as regulatory elements in the control of cellular events. For example, cyclin D/CDK4 can bind proliferating cell nuclear antigen (PCNA) and the cyclin-dependent kinase inhibitor (CDKI) protein $p21^{CIP1}$ (*see* next paragraph; *4,5*). The binding of other proteins to these cyclin/CDK complexes, such as the CDKIs, can regulate the kinase activity of the complexes throughout the cell cycle such that different cell cycle-regulatory molecules are activated at specific times in the cell cycle to "drive" the cell through successive checkpoints i.e., G_1-, S-, G_2-, and M-phase of the cycle (**Fig. 1**).

The CDKIs e.g., $p21^{CIP1}$ *(6,7)* $p27^{KIP1}$ *(8,9)*, and $p16^{INK4}$ *(10,11)*, exert a negative regulatory effect on the cell cycle machinery by binding to, and inhibiting the activities of, specific cyclin/CDK complexes. Primarily, these proteins cause arrest in the G_1-phase of the cell cycle, and it is likely that they may play a role in the development of the terminally differentiated phenotype.

This chapter outlines the purification of various CDK and CDKI molecules from cellular lysates and describes the subsequent measurement of their enzyme activities in vitro.

2. Materials

1. Chemicals to be used in these procedures should be of the best grade available commercially. Solutions should be prepared with double-distilled sterile water, unless otherwise indicated.

2. Cyclin, CDK, and CDKI antipeptide antibodies and the glutathione-S-transferase-retinoblastoma protein (GST-pRb) fusion protein may be obtained from various sources, including BD Biosciences Pharmingen (Oxford, UK), New England Biolabs (Herts, UK), Santa Cruz Biotechnology (Santa Cruz, CA), Serotec (Oxford, UK), and Upstate Group (Lake Placid, NY). The various immunizing peptides for each antibody are also available for use in competition studies.
3. Adenosine trisphosphate (ATP): 10 m*M* stock solution in water. Store at –20°C in aliquots (Sigma-Aldrich, Poole, Dorset, UK);
4. 3000 Ci/mmol [γ-^{32}P]ATP redivue (Amersham International, Bucks, UK). Alternatively, [γ-^{32}P]ATP of the same specific activity may be obtained from other suppliers, e.g., ICN Biomedicals (Thame, Oxon, UK or Sigma-Aldrich).
5. Histone H1 substrate protein (Boehringer Mannheim, Lewes, East Sussex, UK).
6. Protein-A Sepharose beads (Pharmacia Biotech, St. Albans, Herts, UK).
7. 0.1% (w/v) Coomassie brilliant blue solution: 50% methanol, 10% acetic acid, 0.1% (w/v) Coomassie brilliant blue R-250, and 40% water. Store at room temperature. Can be used many times (between 20 and 40 times) before replacing.
8. Destain Buffer: 5% methanol, 7% acetic acid, and 88% water. Store at room temperature.
9. CDK lysis buffer (*see* **Note 1**): 50 m*M* Tris-HCl, pH 7.4, 0.25 *M* sodium chloride, 0.1% (v/v) Nonidet P-40, 0.005 *M* EDTA, 0.05 *M* sodium fluoride, 0.001 M sodium orthovanadate, 0.001 *M* sodium pyrophosphate, 0.01 *M* benzamidine, 50 g/mL phenylmethylsulphonylfluoride (PMSF), 10 g/mL *N*-toysl-phenylalanine chloromethyl ketone (TPCK), 10 g/mL soybean trypsin inhibitor (STI), 1 mg/mL aprotinin, and 1 g/mL leupeptin. Prepare fresh every time and store on ice.
10. Wash buffer: 0.05 *M* Tris-HCl, pH 7.4, 0.01 *M* $MgCl_2$, and 1 mmol/L ditiothreitol (DTT). Store at 4°C; stable for up to 1 mo.
11. ATP buffer (*see* **Note 1**): wash buffer plus 10 *M* ATP and 1 Ci/ L [γ-32P]ATP. Prepare fresh every time and store on ice.
12. 2X Sample buffer: 0.12 *M* Tris-HCl, pH 6.8) 4%v/v sodium dodecylsulphate (SDS), 20%v/v glycerol, 10% (v/v) β-mercaptoethanol (or 200 m*M* DTT), and 0.002% Bromophenol Blue. Stable for many months at 4°C.
13. Rb buffer (*see* **Note 1**): 20 m*M* sodium-β-glycerophosphate, pH 7.3, 15 m*M* $MgCl_2$, 5 g/mL leupeptin, 1 m*M* benzamidine, 0.5 m*M* PMSF, 0.1 m*M* sodium orthovanadate, and 1 m*M* DTT. Prepare fresh every time and store on ice.
14. Rb assay buffer (*see* **Note 1**): Rb buffer plus, 50 *M* ATP, 10 Ci [γ-^{32}P]ATP, 0.5 g GST-pRb fusion protein. Prepare fresh every time and store on ice.

3. Methods

3.1. CDK Activity Assays

The following procedure is a modification of the method of Draetta et al. *(12)* and has been used routinely in my laboratory for determining CDK activities in CDC2-, CDK2-, CDK4-, CDK5-, and CDK6-containing complexes in cardiac myocytes *(13)*. Each kinase has a different specificity for a particular substrate protein, although the individual kinases can be placed into one of two separate groups depending on their abilities to phosphorylate either histone H1 or a GST-pRb fusion protein. Thus, the G_1-acting CDKs, CDK4 and CDK6, phosphorylate a GST-pRb fusion protein efficiently, whereas the G_1/S and G_2/M kinases, CDK2 and CDC2, and the G_1-acting kinase CDK5, efficiently phosphorylate histone H1. It should be noted that these substrates are not totally specific for any one group of kinases since it is possible for CDC2, CDK2, and CDK5 to phosphorylate a GST-pRb fusion protein and similarly for CDK4 and CDK6 to phosphorylate histone H1. However, the extent of phosphorylation in these latter cases is much less than that observed with their more specific substrates.

3.1.1. Preparation of Cellular Lysates Prior to Immunoprecipitation

1. Freshly isolated or trypsinized cells are initially lysed on ice in ice-cold CDK lysis buffer. Use the minimum volume of buffer necessary to obtain lysis of cells, e.g., approx 100–200 L buffer per 1×10^6 cells for large cells (e.g., adult rat cardiac myocytes) or 100–200 L buffer per $0.5–1 \times 10^7$ cells for smaller cells (e.g., ovarian carcinoma cells).
2. Sonicate lysed samples on ice (three pulses of 5 s each on maximum setting) and incubate on ice for 30 min. Centrifuge sonicated samples at 4*g* in a benchtop refrigerated centrifuge set at 4°C. Supernatants then are snap-frozen on dry ice and stored overnight in liquid nitrogen prior to assaying for CDK complex activity (*see* **Note 2**). Prior to freezing, remove a 5 L volume of supernatant containing the sample of interest for protein analysis.

The above method can also be used for the preparation of tissue lysates prior to measurement of CDK activities.

3.1.2. Immunoprecipitations and CDK Assays

1. Preswell protein-A sepharose beads (0.1 g/mL) in ice-cold phosphate buffered saline (PBS), pH 7.4 and then equilibrate in CDK lysis buffer for 1 h at 4°C with gentle mixing on a rotating blood wheel.
2. Add 40 L preswollen protein A sepharose beads to 250–750 g of cell lysate protein in an Eppendorf tube. Make up to a final volume of 750 L using CDK lysis buffer, and mix gently on a rotating blood wheel for 1 h at 4°C. Pellet the beads using a 15-s pulse in a benchtop microfuge, and remove the supernatant to a clean Eppendorf tube ("precleared cell lysate").
3. Immunoprecipitate the precleared cell lysate with the appropriate CDK antibody (e.g., 1 g/sample as determined by titration) for 1 h at 4°C (*see* **Note 3**). Add 50 L of preswollen protein-A sepharose beads, and mix gently for 1 h at 4°C on a rotating blood wheel.

The protocol then differs slightly depending on which CDK is being assayed, as follows.

3.1.2.1. CDC2, CDK2, and CDK5 Assays

1. Wash the immunocomplex bound beads four times with 1 mL CDK lysis buffer each and once in 1 mL wash buffer. Pellet beads using a 15-s pulse in a microfuge, resuspend the pellet in 20 L wash buffer containing 125 g/mL histone H1 substrate protein, and incubate for 5 min at 30°C.
2. Add 5 L ATP buffer to each reaction tube, incubate for 10 min at 30°C, and then terminate the reaction by the adding of 25 L 2X sample buffer. Then move to **Subheading 3.1.2.2., step 3** below.

3.1.2.2. CDK4 and CDK6 Assays

1. Wash the immunocomplex bound beads four times with 1 mL CDK lysis buffer each and then once in 1 mL of Rb buffer. Pellet beads using a 15-s pulse in a microfuge, and resuspend the pellet in 30 L of Rb assay buffer per sample.

2. Incubate the reaction at 30°C for 1 h and terminate by the addition of 30 L 2X sample buffer.
3. Boil samples for 3 min, and separate proteins by 12.5% SDS-polyacrylamide gel electrophoresis (PAGE). Separated proteins can be visualized by staining the gel in 0.1% (w/v) Coomassie Brilliant Blue solution (*see* **Note 4**). The gel then is dried and exposed to photographic film overnight at –80°C using intensifying screens. The resultant autoradiograph can then be scanned using a laser densitometer, or the phosphorylated bands can be excised directly from the gel and counted in a scintillation counter. **Figure 2A** and **B** shows representative autoradiographs of CDK2 and CDK6 kinase activities in developing rat cardiac myocytes (*see* **Note 5**).

It is also possible to use this method to determine specific cyclin-associated kinase activities. For instance, if we wished to measure the kinase activity associated with cyclin E in a cellular extract, we would immunoprecipitate cyclin E-containing complexes from a cellular lysate using a specific anti-cyclin E antibody at the appropriate concentration, e.g., 1 g/sample as determined by titration. The resultant immunoprecipitated protein would then be analyzed for CDK2 activity (cyclin E associates specifically with CDK2), using a similar protocol to that just described, with histone H1 as a substrate. This method can be used to determine the kinase-associated activity of any cyclin molecule, providing the appropriate substrate protein for the CDK molecule that complexes with the immunoprecipitated cyclin is used in the in vitro kinase assay.

3.2. CDKI Inhibitory Activity Assays

The following procedure is a modification of the methods of Draetta et al. ***(12)*** and Guo et al. ***(14)*** and has been used successfully in my laboratory to determine the inhibitory activity of $p21^{CIP1}$ present in adult rat cardiac myocytes against neonatal rat cardiac myocyte CDK2 activity ***(15,16)***.

1. Prepare cell lysates as described above by lysing in ice-cold CDK lysis buffer. Boil lysates to obtain a heat-stable protein preparation (*see* **Note 6**), snap-freeze samples on dry ice, and store in liquid nitrogen prior to use (*see* **Note 2**). Prior to freezing, remove a 5 L vol of supernatant containing the sample of interest for protein analysis.
2. Thaw samples and incubate 50 g of neonatal myocyte lysate protein (containing activated CDK2) with 50 g of boiled adult myocyte lysate (*see* **Note 7**) overnight at 4°C with mixing on a rotating blood wheel.
3. Immunoprecipitate CDK2 from the mix using an anti-CDK2 antibody exactly as described in **Subheading 3.1.2.** above.
4. Wash immunocomplex bound beads four times with 1 mL CDK lysis buffer each and once with 1 mL wash buffer. Pellet beads using a 15-s pulse in a microfuge, resuspend the pellet in 20 L wash buffer containing 125 g/mL histone H1 substrate protein, and incubate for 5 min at 30°C.
5. Add 10 L of ATP buffer to each sample tube, and incubate for 10 min at 30°C.
6. Terminate the reaction by the addition of 30 L 2X sample buffer.
7. Boil samples for 3 min, and separate proteins by 12.5% SDS-PAGE. Separated proteins can be visualized by staining the gel in 0.1% w/v Commassie brilliant blue solution (*see* **Note 4**). The gel then is dried and exposed to photographic film overnight at –80°C using intensifying screens. The phosphorylated histone H1 bands are then excised from the gel and counted in a liquid scintillation counter.

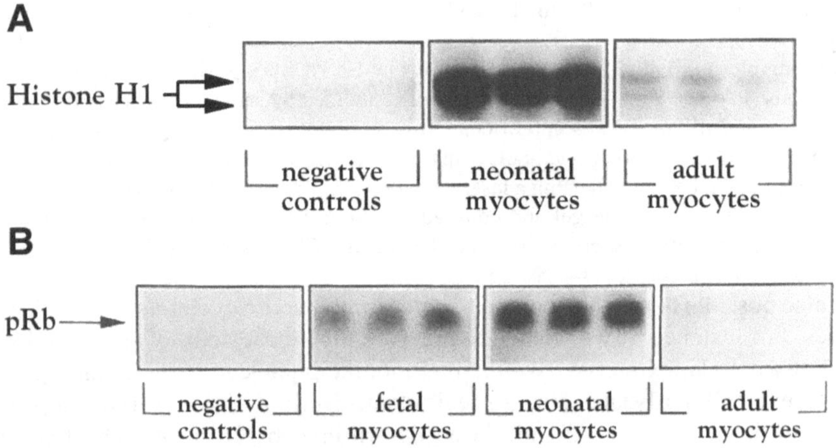

Fig. 2. Representative autoradiographs of (A) CDK2-mediated phosphorylation of histone H1 and (B) CDK6-mediated phosphorylation of GST-pRb fusion protein in rat cardiac myocytes prepared from hearts of animals at different developmental ages. The negative control lanes represent no antibody, cellular lysate (left-hand lane); CDK antibody, no cellular lysate (middle lane); and CDK antibody plus immunizing peptide, cellular lysate (right-hand lane).

4. Notes

1. All assay buffers should be prepared fresh on the day of assay. We have found that storage of such solutions overnight at 4°C leads to significant loss of measurable kinase activity.
2. CDK activities in frozen lysates remain measurable for 2–3 d following preparation when stored in liquid nitrogen. However, the degree of measurable activity deteriorates rapidly such that only 50–75% of original activity (i.e., that present in freshly prepared cell lysates) remains after one day of storage and 25–50% after 2 d.
3. It is advisable to include a series of control immunoprecipitations in all kinase assays to ensure that the results obtained are specific for the CDK of interest. Suggested controls include no CDK, antipeptide antibody, cellular lysate; CDK antipeptide antibody plus immunizing peptide (mix together for 5–10 min at room temperature prior to addition to lysate), cellular lysate; CDK antibody, no cellular lysate; and no antibody, no lysate. All immunoprecipitations and kinase assays should be carried out in triplicate, and each individual experiment should be repeated at least three times to ensure good statistical results.
4. It is advisable to stain all SDS-PAGE gels with Coomassie brilliant blue solution for approx 30–60 min at room temperature followed by destaining until blue bands and a clear background are obtained (typically 1–2 h). Gels should then be dried down prior to autoradiography. This approach enables clear observation of the band(s) of interest directly on the gel and also reduces significantly the amount of background radioactivity that could affect the final results.
5. Phosphorylated histone H1 protein migrates as two distinct bands with M_r 32 and 34 kDa by 12.5% SDS-PAGE. Phosphorylated GST-pRb fusion protein (e.g., from Santa Cruz Biotechnology) migrates with an M_r of 40 kDa under the same conditions.

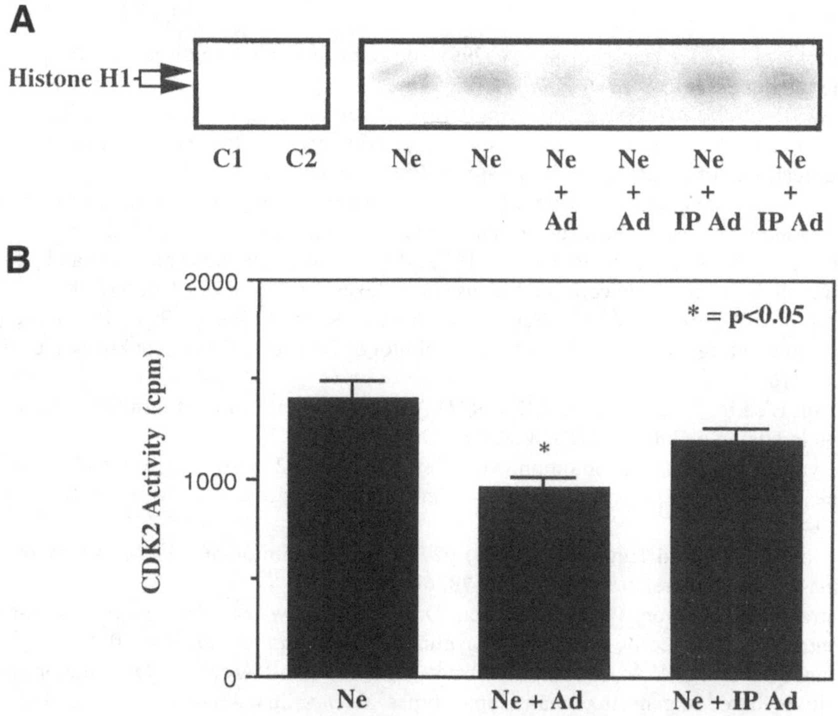

Fig. 3. **(A)** Representative autoradiograph of p21^{CIP1}-mediated inhibition of CDK2 activity in rat cardiac myocytes showing histone H1 phosphorylation by immunoprecipitated CDK2 complexes from neonatal myocytes in the presence or absence of adult myocyte lysates that contain p21^{CIP1}. **(B)** Graph showing means ± SD of histone H1 phosphorylation obtained from three separate experiments. C1, no CDK2 antibody, cellular lysate; C2, CDK2 antibody, no cellular lysate; Ne, neonatal myocyte lysate; Ne + Ad, 50 g neonatal and 50 g heat-stable adult myocyte lysate; Ne + IP Ad, neonatal and p21^{CIP1} immunodepleted heat-stable adult myocyte lysate.

6. Most CDKI molecules described to date are heat-stable proteins. The use of a heat-stable cellular extract in these studies enriches for those CDKI molecules that may be expressed at relatively low levels compared with other heat-labile proteins. Thus, if a heat stable extract was not used, it might prove difficult to demonstrate inhibitory activity of CDKI molecules in a particular cellular lysate.
7. To demonstrate that a particular CDKI molecule is present and is responsible for CDK inhibitory activity in a cellular lysate, it is necessary to compare the CDK activity measured in mixes of CDK-containing lysate with CDKI-containing lysate that has been (1) depleted of a particular CDKI molecule by prior immunoprecipitation with a CDKI-specific antibody, and (2) not depleted of any CDKI molecule (*see* **Fig. 3** for an example).
8. A typical time frame, from start to finish, for the above assays is 1–2 d.

References

1. Brooks, G. and La Thangue, N. B. (1999) The cell cycle and drug discovery: the promise and the hope. *Drug Discov. Today* **4,** 455–463.
2. Nurse P. (2000) A long twentieth century of the cell cycle and beyond. *Cell* **100,** 71–78.
3. Bicknell, K. A., Surry, E. L., and Brooks, G. (2003) Targeting the cell cycle machinery for the treatment of cardiovascular disease. *J. Phar. Pharmacol.* **55,** 571–591.
4. Xiong, Y., Zhang, H., and Beach, D. (1993) Subunit rearrangement of the cyclin-dependent kinases is associated with cellular transformation. *Genes Dev.* **7,** 1572–1583.
5. Zhang, H., Xiong, Y., and Beach, D. (1993) Proliferating cell nuclear antigen and p21 are components of multiple cell cycle kinase complexes. *Mol. Cell. Biol.* **4,** 897–906.
6. Harper, J. W., Adami, G. R., Wei, N., Keyomarsi, K., and Elledge, S. J. (1993) The p21 Cdk-interacting protein Cip1 is a potent inhibitor of G1 cyclin-dependent kinases. *Cell* **75,** 805–816.
7. Jiang, H., Lin, J., and Fisher, P.B. (1994) A molecular definition of terminal differentiation in human melanoma cells. *Mol. Cell Differ.* **2,** 221–239.
8. Polyak, K., Kato, J-Y., Solomon, M. J., et al. (1994) p27Kip1, a cyclin-Cdk inhibitor, links transforming growth factor-β and contact inhibition to cell cycle arrest. *Genes Dev.* **8,** 9–22.
9. Toyoshima, H. and Hunter, T. (1994) p27, a novel inhibitor of G1 cyclin-Cdk protein kinase activity, is related to p21. *Cell* **78,** 67–74.
10. Serrano, M., Hannon, G. J., and Beach, D. (1993) A new regulatory motif in cell-cycle control causing specific inhibition of cyclin D/CDK4. *Nature* **366,** 704–707.
11. Kamb, A., Grus, N. A., Weaver-Feldhaus, J., et al. (1994) A cell cycle regulator potentially involved in genesis of many tumor types. *Science* **264,** 436–440.
12. Draetta, G., Brizuels, L., Potashkin, J., and Beach, D. (1987) Identification of p34 and p13, human homologs of the cell cycle regulators of fission yeast encoded by $cdc2^+$ and $suc1^+$. *Cell* **50,** 319–325.
13. Brooks, G., Poolman, R. A., McGill, C. J., and Li, J-M. (1997) Expression and activities of cyclins and cyclin-dependent kinases in developing rat ventricular myocytes. *J. Mol. Cell. Cardiol.* **29,** 2261–2271.
14. Guo, K., Wang, J., Andres, V., Smith, R. C., and Walsh, K. (1995) MyoD-induced expression of p21 inhibits cyclin-dependent kinase activity upon myocyte terminal differentiation. *Mol. Cell. Biol.* **15,** 3823–3829.
15. Poolman, R. A., Gilchrist, R., and Brooks G. (1998) Cell cycle profiles and expression of $p21^{CIP1}$ and $p27^{KIP1}$ during myocyte development. *Int. J. Cardiol.* **67,** 133–142.
16. Poolman, R. A., Li, J-M., Durand, B. and Brooks, G. (1999) Altered expression of cell cycle proteins and prolonged duration of cardiac myocyte hyperplasia in $p27^{KIP1}$ knockout mice. *Circ. Res.* **85,** 117–127.

18

Measurement of Wee Kinase Activity

Paul R. Mueller and Walter F. Leise III

Summary

The Wee kinases (Wee1, Wee2, and Myt1) are major regulators of mitotic entry. They function by phosphorylating Cdc2 and related Cdks on conserved tyrosine and threonine residues. This phosphorylation blocks the activity of the Cdc2 and prevents entry into mitosis. The abundance and activity of the Wee kinases are regulated during the cell cycle and development. In this chapter, we describe several procedures to measure the activity of the Wee kinases found either in crude extracts or in purified preparations. Specific protocols include the production and purification of recombinant Cdc2/Cyclin B substrate, the production of crude subcellular extract fractions, the purification of endogenous or recombinant Wee kinases, Wee kinase assays, and the Histone H1 kinase assay to measure Cdc2 activity. In addition, support protocols are provided that describe the use and production of Ni-IDA beads for the purification of Histidine-tagged proteins, and the use of the baculovirus expression system to produce recombinant proteins.

Key Words

Wee1; Wee2; Myt1; Cdc2; baculovirus; Xenopus; kinase assay; phosphorylation; Histone H1; His6; immunoprecipitation.

1. Introduction

Entry into mitosis is dependent on the activity of a highly conserved kinase, Cdc2 (reviewed in **ref.** ***1***). By itself, monomeric Cdc2 is inactive. Instead, the activation of Cdc2 is dependent on the accumulation of a binding partner, Cyclin B, and on the cyclin-dependent kinase (CDK)-activating kinase (CAK)-mediated phosphorylation of Cdc2 on Thr161. Together, the cyclin binding and Thr161 phosphorylation change the structure of Cdc2, enabling its activity. In the absence of other modifications, these structural changes would generate active Cdc2/Cyclin B, but during interphase, two additional phosphorylations on Thr14 and Tyr15 of Cdc2 suppress Cdc2 activity (reviewed in **refs.** ***2*** and ***3***). These phosphorylations dominantly inhibit the Cdc2/Cyclin B complex and block entry into mitosis. At the G_2/M transition, the Cdc25 protein

From: *Methods in Molecular Biology, vol. 296, Cell Cycle Control: Mechanisms and Protocols*
Edited by: T. Humphrey and G. Brooks © Humana Press Inc., Totowa, NJ

dephosphorylates Thr14 and Tyr15, thereby allowing maturation-promoting factor (MPF) to phosphorylate its mitotic substrates (reviewed in **ref.** *4*) (**Fig. 1**).

Collectively, the Wee family of kinases is responsible for phosphorylating Thr14 and/or Tyr15 of Cdc2 and thus inhibiting the activity of the Cdc2/Cyclin B complex during interphase. The first Wee kinase to be identified was Wee1 of *Schizosaccharomyces pombe* when it was isolated as a critical negative regulator of mitosis *(5)*. Fission yeast deficient in this kinase divide at a small size (hence the name Wee) owing to their premature entry into mitosis. Subsequently, one or more Wee1 homologs have been found in all other eukaryotes examined including *Saccharomyces cerevisiae*, humans, *Xenopus*, *Drosophila*, and *Caenorhabditis elegans* *(6–11)*. The Wee1 and Wee1 homologs phosphorylate Cdc2 exclusively on Tyr15. Myt1, a related member of the Wee family of kinases, also phosphorylates Cdc2 on Tyr15 as well as the adjacent Thr14 *(12–14)*. Because of their common function, we will use the generic phrases "Wee kinases" or "Wee family of kinases" to include both the Myt1 kinases and the Wee1/Wee1 homologs *(15)*.

At the amino acid level, Wee kinases share the greatest level of similarity in their kinase domains. A diagnostic feature found in all members of the Wee kinase family is that they contain five conserved residues in the kinase domain. These are a Trp in subdomain IV, a Glu and Asp in subdomain VIII, and a Trp and Arg in subdomain X. These residues are not commonly found in other, non-Wee kinases at these positions and therefore serve as a "signature" of Wee kinase family identity *(8)*. Whether these conserved residues are responsible for any unique function of the Wee kinase remains to be determined. Outside the kinase domain, the Wee kinases can be quite divergent, but there is some similarity between putative orthologs *(15)*. These noncatalytic regions of the protein are thought to play important roles in the regulation, protein–protein interaction, and subcellular localization of the Wee kinases.

One of the more divergent members of the Wee kinase family is Myt1. Not only is Myt1 a dual-specificity kinase (phosphorylating both Thr14 and Tyr15 of Cdc2), but it is also a membrane-associated kinase that colocalizes with the cytoplasmic endoplasmic reticulum and Golgi apparatus *(12–14)*. This differs from Wee1 and Wee1-like kinases, which phosphorylate Tyr15 exclusively and are soluble and predominantly nuclear in their subcellular location *(8,16–19)*. Finally, unlike the Wee1 homologs, Myt1 is found only in metazoans. Despite these unique properties, Myt1 has the other characteristics that designated it as a member of the Wee kinase family. For example, Myt1 will only phosphorylate Cdc2 that is prebound to Cyclin B, and the Myt1 kinase domain shows strong similarity to other Wee kinases including the five signature residues *(12)*.

In addition to Myt1, vertebrates have two distinct Wee1-like kinases. These are classified as distinct subgroups of the Wee kinase family based on their sequence similarity outside the kinase domain, their developmental expression patterns, and their levels of activity toward Cdc2/Cyclin B complexes *(15,20,21)*. One subgroup is found exclusively as a maternal gene transcript (designated Wee1 and Wee1A in *Xenopus* and Wee 1B in mammals); the other is found predominately as a zygotic transcript (designated Wee2 and Wee1B in *Xenopus* and Wee1 and Wee1A in mammals) *(7,8,15,20–22)*. To minimize the nomenclature confusion, we will refer to the maternally expressed subgroup as "Wee1" and the mostly zygotic subgroup as "Wee2" *(15)*.

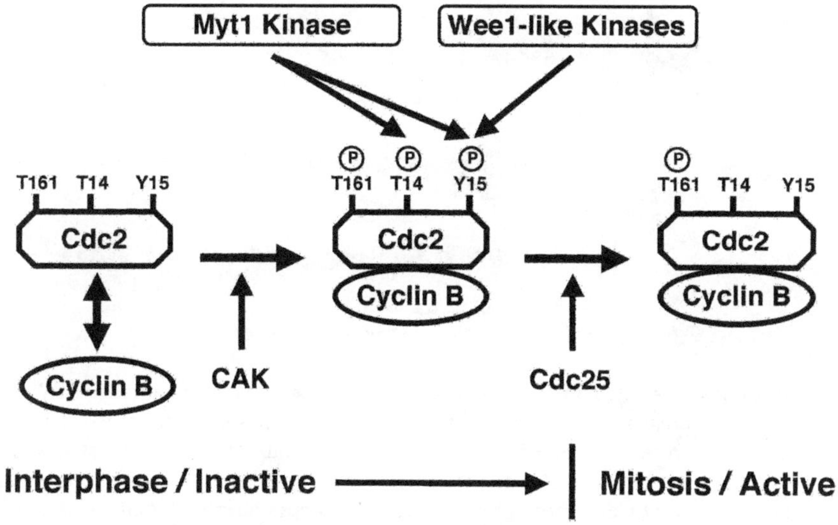

Fig. 1. Control of Cdc2 activity through inhibitory phosphorylation. During interphase, Cdc2 associates with the accumulating Cyclin B protein and is phosphorylated by CAK on Thr161. Concurrently, the Myt1 and Wee1-like kinases phosphorylate the Cdc2/Cyclin B complex on two conserved residues in the active site of Cdc2, Thr14, and Tyr15. These phosphorylations dominantly block Cdc2/Cyclin B activity. At the G_2/M phase transition, the Cdc25 phosphatase removes these inhibitory phosphates.

Biochemically, Wee1 and Wee2 function similarly. Both phosphorylate Cdc2 exclusively on Tyr15 in a Cyclin B-dependent manner, and both require the same purification and assay conditions. One important difference that has been reported in *Xenopus* is that Wee2 is about 10-fold more active than Wee1 ***(15)***. It remains to be determined whether this activity difference exists in other vertebrates.

This chapter describes several methods for measuring Wee kinase activity both in crude cellular extracts and in defined biochemical systems. Depending on the goals of the experiment, one method may be more appropriate than the other. If the goal is simply to determine the presence of Wee activity, a Western blot of total cell lysates with an antibody directed against the phospho-Tyr15 form of Cdc2 might be most appropriate (**Fig. 2**). However, because the in vivo Tyr15 phosphorylation status of Cdc2 is a reflection of the ratio of Wee kinase to Cdc25 phosphatase activity, this approach cannot give a complete measure of Wee activity. Furthermore, this antibody cannot discriminate between Wee1/2 and Myt1 activity.

A more direct approach is to provide a defined substrate (recombinant Cdc2/Cyclin B) for Wee kinase activity. The production of this substrate is described in the first part of this chapter (**Fig. 3**). This is a variation of the method described by Kumagai and Dunphy ***(23)*** and uses the robust baculovirus expression system (reviewed in **refs. *24–26***). One of the advantages of using a recombinant substrate is that several mutant forms of Cdc2 can be produced. These, in turn, can be used to define better the enzyme

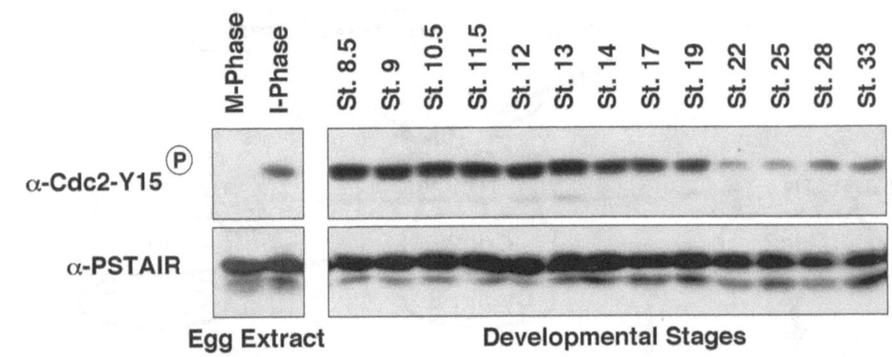

Fig. 2. Cdc2 phosphorylated on Tyr15 can be detected in crude lysates. Top panels: Western analysis showing the presence of phospho-Tyr15 Cdc2 in interphase, but not mitotic, *Xenopus* egg extracts, and the presence of phospho-Tyr15 Cdc2 in various developmental stages of *Xenopus* (*see* **Note 41**). Bottom panels: Western analysis of the same samples used on top with an antibody that detects total Cdc2 (both phosphorylated and nonphosphorylated; *see* **Note 39**). The developmental stages range from midblastula (stage 8.5) to early tadpole (stage 33).

responsible for the Wee kinase activity (**Fig. 4**). The recombinant substrate can be used in two general ways. First, it can be used to detect and measure Wee kinase activity in crude cellular extracts (**Fig. 5**). This method is based on the gel mobility shift that Cdc2 undergoes when it is phosphorylated on Thr14 and Tyr15 ***(27,28)***. As a supporting method for this shifting assay, we describe procedures for the production and fractionation of cellular extracts. Second, the recombinant Cdc2/Cyclin B complex can be used to measure the activity of purified Wee kinases from endogenous or recombinant sources. We describe two measurement assays, one based on the Wee-mediated incorporation of ^{32}P into Cdc2 (**Fig. 6**), and the other based on the Wee-mediated reactivity of Cdc2 to a phospho-tyrosine antibody (**Fig. 4**).

As supporting methods for these more defined measurement assays, we describe methods for immunoprecipitating the endogenous Wee kinases from cellular extracts and the production of recombinant Wee kinases from the baculovirus expression system. These purified Wee proteins may be tested directly for Wee kinase activity (**Figs. 4** and **6**) or may be used to test for Wee kinase regulators found in cellular extracts or more biochemically defined systems. Finally we describe the Histone H1 kinase assay (**Fig. 4**). This method measures the activity of the Cdc2/Cyclin B complex before or after Wee modification and can be used to prove that a candidate Wee kinase has the ability to inhibit Cdc2.

Throughout this chapter, we make extensive use of a baculovirus expression system to produce recombinant Wee kinase substrates and recombinant versions of Wee proteins. Although this system is somewhat more laborious than the production of recombinant proteins in bacteria, we find it to be very robust and reliable. Our attempts to produce active forms of cell cycle proteins in bacteria have met with mixed results at best, and we do not recommend this approach for routine use. We also make exten-

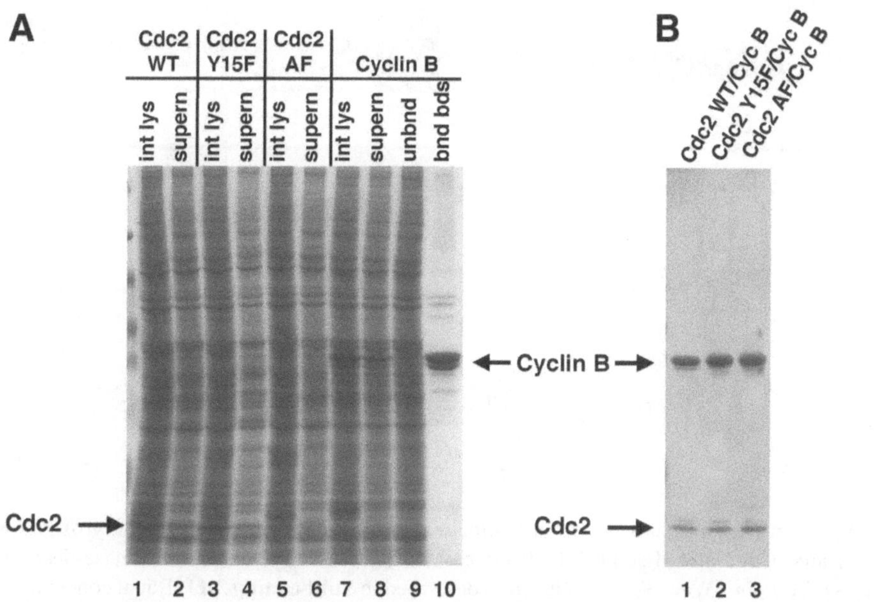

Fig. 3. Production of recombinant Cdc2/Cyclin B complex. **(A)** Production of Cdc2 lysates and Cyclin B beads from Sf9 cells as described in **Subheadings 3.4.** and **3.5.** Lanes 1–6, initial lysates (int lys) and supernatant (supern) of Cdc2 WT-, Y15F-, or AF-infected cells. Note the low production of Cdc2 AF in lanes 5 and 6. Lanes 7–10, initial lysates (int lys), supernatant (supern), unbound flowthrough (unbnd), and Cyclin-B bound beads (bnd bds) of Cyclin B-infected cells. **(B)** Production of recombinant Cdc2/Cyclin B complexes as described in **Subheading 3.6.**, except that for Cdc2 AF/Cyclin B a 500- L Cdc2 AF lysate was first prebound as described in **Note 24**. WT, Y15F, and AF complexes were eluted in 120, 60, and 80 L elution buffer, respectively. In all three cases, 36 L of the eluted complex was loaded on the SDS-PAGE gel. Note that similar, but not equal, concentrations of Cdc2 are present in the three samples. In the next experiment, 90, 50, and 80 L of elution buffer was used to elute the WT, Y15F, and AF complexes, respectively, 36 L was loaded, and equal concentrations of Cdc2 were obtained (data not shown). Gels in both **(A)** and **(B)** are stained with Coomassie blue.

sive use of nickel-iminodiacetic acid (Ni-IDA) beads to purify His6-tagged recombinant proteins (**Fig. 3**). We find that we can get relatively pure protein preparations using this tag and bead system, particularly with our recommended wash buffers (*see* **Note 1**). Supporting methods for producing Ni-IDA beads and infecting of Sf9 (*Spodoptera frugiperda*) cells with recombinant baculovirus are described.

2. Materials

2.1. General Solutions and Equipment

These reagents and materials are used in two or more of the procedures.

1. dH_2O: deionized (Millipore quality), autoclaved water.

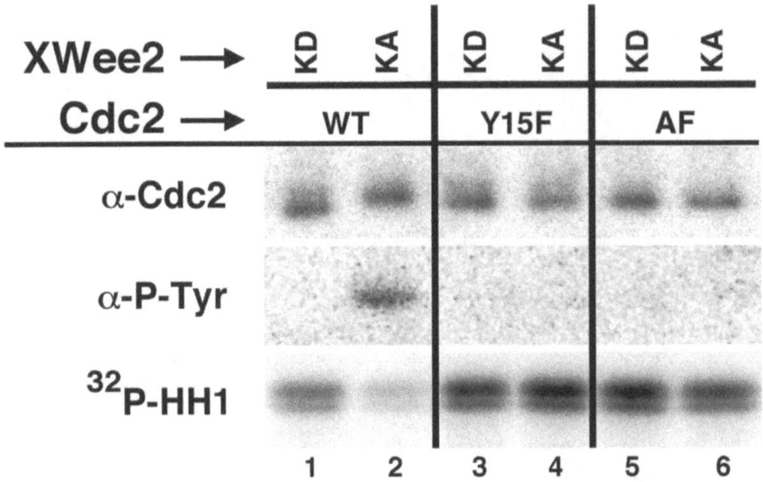

Fig. 4. In vitro measurement of Wee kinase activity via Western analysis for tyrosine-phosphorylated Cdc2 and Histone H1 kinase assays. Kinase-active (KA) and kinase-dead (KD) forms of *Xenopus* Wee2 were produced as described in **Subheading 3.11.** Equal concentrations (data not shown) of these recombinant forms of Wee2 were then used in a Cdc2 anti-phosphotyrosine antibody kinase assay to detect Wee kinase activity as described in **Subheading 3.12.2.** The top panel shows both phosphorylated and nonphosphorylated Cdc2 at the completion of the assay. The middle panel shows that kinase-active Wee2 phosphorylates wild-type Cdc2 (WT) but not Cdc2 (Y15F) or Cdc2 (AF; *see* **Subheading 3.5.**). A small portion of the Cdc2 kinase reaction was then used in the Histone H1 kinase assay as described in **Subheading 3.13.** (bottom panel). The lack of ^{32}P incorporation into Histone H1 (lane 2) shows that kinase-active, but not kinase-dead Wee2 can inhibit Cdc2/Cyclin B activity as long as a Tyr15 residue is present in the Cdc2. Together, these results show that Wee2 is a Wee1-like kinase because it both phosphorylates and inhibits Cdc2 in a Tyr15-dependent manner.

2. 5 *M* NaCl. Autoclave and store room temperature.
3. 1 *M* Tris-HCl, pH 7.5. Autoclave and store room temperature.
4. 10X TBS: 1.4 *M* NaCl, 100 m*M* Tris-HCl, pH 7.5, 50 m*M* KCl. Autoclave and store room temperature.
5. 10X PBS: 1.37 *M* NaCl, 10 m*M* Na_2HPO_4, 2.7 m*M* KCl, 2 m*M* KH_2PO_4, pH 7.4. Autoclave and store room temperature.
6. 200 m*M* Phenylmethylsulfonyl fluoride (PMSF; Sigma, cat. no. P 7626). Make fresh, keep on ice. Make approx 1 mL of 200 m*M* PMSF in 100% ethanol in a 1.5-mL microcentrifuge tube. Mix by inversion. Use caution, and use within approx 12 h (*see* **Note 2**).
7. 1 *M* HEPES, pH 7.5. Autoclave and store room temperature.
8. 0.5 *M* EGTA, pH 8.0. Autoclave and store room temperature.
9. 10% Nonidet P40 (NP-40; w/v in dH_2O). Make from high-grade Nonidet P40 (Roche, cat. no. 1-754-599) and store 10% stock in the dark for ≤6 mo at room temperature.
10. 1000X PCL: a mixture of pepstatin (Roche, cat no. 1-359-053), chymostatin (Roche, cat no. 1-004-638), and leupeptin (Sigma, cat. no. L 2884) (10 g/mL each) in dimethyl

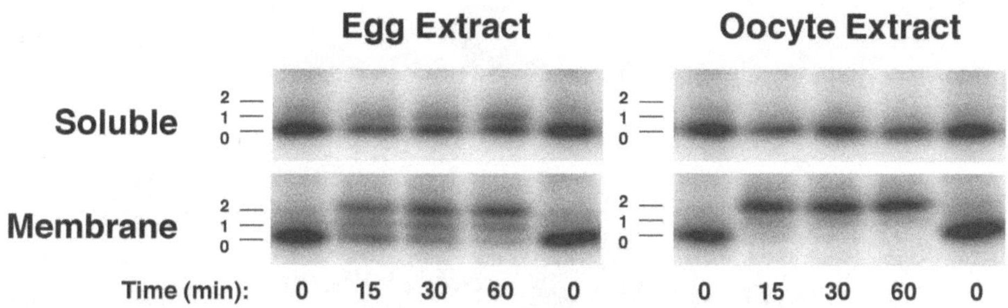

Fig. 5. Cdc2 shifting assay to measure Wee kinase activity in cellular extracts. Soluble and membrane fractions were made from *Xenopus* oocyte and interphase egg extracts as described in **Subheading 3.7.** and **Note 27**. These were subject to the Cdc2 shifting assay to detect Wee activity as described in **Subheading 3.8.** Note that the egg, but not oocyte, soluble fractions have Wee1 kinase activity (indicated by the single shift). This corresponds to the absence of Wee1 kinase protein in *Xenopus* oocytes ***(34)***. In contrast, both egg and oocyte membrane fractions have Myt1 activity (indicated by the double shift), again corresponding to the presence of Myt1 in both cell types. 0, unshifted; 1, single-shifted; 2, double-shifted.

sulfoxide (DMSO). Use caution and store in 10- L aliquots at –20°C for ≤6 mo (*see* **Note 3**).

11. Low-salt wash buffer with PMSF (LSWB-PMSF; make fresh, and make 100 mL): 10 m*M* HEPES, 150 m*M* NaCl. Adjust pH to 7.4, and chill to 4°C on ice. Shortly before use, make 0.2 m*M* PMSF (*see* **Note 2**).
12. Low-salt wash buffer with PMSF and PCL (LSWB-PMSF-PCL; make fresh): transfer 10 mL of LSWB-PMSF to a tube on ice. Shortly before use, make 1X PCL (*see* **Note 3**).
13. 7 mL Dounce Tissue Grinder with approx 0.05-mm clearance.
14. Liquid nitrogen (LN_2; preferred) or a dry-ice/ethanol bath.
15. 13-mL tube with matching snap-cap (Starstedt, tube cat. no. 55.518; cap cat. no. 65.816).
16. 1 *M* Imidazole, pH 7.4. Autoclave and store room temperature.
17. 100 m*M* ATP, pH 7.5. Store –20°C.
18. 1 *M* $MgCl_2$. Autoclave and store room temperature.
19. 1 *M* NaF (*see* **Note 4**). Filter-sterilize, and store at room temperature.
20. 100 m*M* Sodium pyrophosphate (NaPPi). Autoclave and store at room temperature.
21. 1 *M* Dithiothreitol (DTT) (Store –20°C).
22. 10% Triton X-100 (w/v in dH_2O). Make from high-grade Triton X-100 (Roche, cat. no. 789-704), store 10% stock in the dark for ≤6 mo at room temperature.
23. 500 *M* Sodium orthovanadate (Na_3VO_4; Sigma, cat. no. S 6508). Make fresh and *see* **Note 4**.
24. 200 m*M* Creatine phosphate (CP; Roche, cat. no. 621-714). Make in dH_2O, and store at –20°C.
25. 5 g/mL Creatine phosphokinase (CPK; Sigma, cat. no. C 3755). Make in dH_2O, and store at –80°C.
26. 4X Sodium dodecyl sulfate (SDS) sample buffer: 250 m*M* Tris-HCl, pH 6.8, 40% (w/v) glycerol, 8% SDS, 65 m*M* DTT, 0.004% bromophenol blue. (Store at –20°C for ≥6 mo).
27. Dry ice.

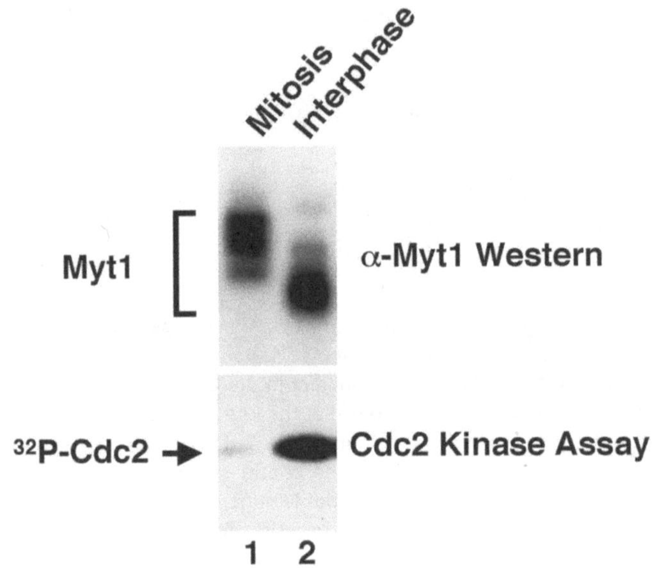

Fig. 6. In vitro measurement of Wee kinase activity via radiolabeling of Cdc2. Myt1 was immunoprecipitated from *Xenopus* egg extracts that were either in mitosis or interphase as described in **Subheading 3.9.** The immunoprecipitated Myt1 was then used in a Cdc2 ^{32}P labeling kinase assay to detect Wee kinase activity as described in **Subheading 3.12.1.** Note that the mitotic form of Myt1 has less activity than the interphase form (bottom panel). This corresponds to mitotic modifications of Myt1 that change its gel mobility, but not its abundance, as detected in the Western analysis (top panel).

28. EB buffer: 80 m*M* β-glycerol phosphate, 20 m*M* EGTA, 15 m*M* $MgCl_2$. Adjust pH to 7.3 with KOH, filter-sterilize, and store at 4℃.
29. 100 mg/mL Ovalbumin (albumin, Chicken Egg Grade V, Sigma, cat. no. A 5503). Make in dH_2O, and store at –20℃.
30. [γ-^{32}P]ATP (*see* **Note 5**), specific activity >4000 mCi/mL (i.e., ICN, cat. no. 38101X, IsoBlue Stabilized; or ICN #35001X).

2.2. Making Nickel-Iminodiacetic Acid Beads

1. 100 m*M* NiCl. Make fresh, ~50 mL, room temperature.
2. 1X TBS: make 250 mL from 10X stock (room temperature).
3. IDA-sepharose (iminodiacetic acid coupled to Sepharose, Fast Flow, Amersham Pharmacia, cat. no. 17-0575-01). Store at 4℃, but warm to room temperature before using.
4. Vacuum apparatus: 500-mL vacuum flask, 30-mL Kimax Buchner funnel with (10–15 m) fritted disc (Kimble cat. no. 28400 32) with neoprene adapter for filtering flask (i.e., Fisher, cat. no. 10-184-4).
5. 10% Sodium azide. **Caution:** Use extreme care when working with sodium azide, as it is highly toxic. Store at 4℃.

2.3. Infection of Sf9 (S. frugiperda) Cells With Recombinant Baculovirus

1. Tissue culture hood, 27°C tissue culture incubator (without CO_2), 10-cm coated tissue culture plates (Falcon, cat. no. 3003, or equivalent), spinner plate, and flask for tissue culture use.
2. Sf9 media: Grace's Insect Media (Invitrogen, cat. no. 11605-102) containing 10% fetal bovine serum (FBS), 1% Fungizone® antimycotic (Invitrogen, cat. no. 15290018), and 0.05 mg/mL gentamicin (Invitrogen, cat. no. 15710072). Make 0.1% (1X) Pluronic® F-68 (Invitrogen, cat. no. 24040032, 10% [100X] stock) if cells are growing in suspension.
3. P3 virus stock of specific protein to be produced. The multiplicity of infection (MOI) of the P3 stocks should be $\geq 10^8$/mL.

2.4. Collection and Washing of Infected Sf9 Cells

1. 1X PBS-PMSF (make fresh, and make 200 mL). Make 1X PBS from 10X stock and chill to 4°C on ice for several hours/overnight. Shortly before use, make 0.2 m*M* PMSF (*see* **Note 2**).

2.5. Purification of Cyclin B on Ni-IDA Sepharose Beads

1. Cyclin B lysis buffer-PMSF (make fresh, and make 100 mL): 10 m*M* HEPES, 150 m*M* NaCl, 5 m*M* EGTA, 0.5% NP-40. Adjust pH to 7.4, and chill to 4°C on ice. Shortly before use, make 0.2 m*M* PMSF (*see* **Note 2**).
2. Cyclin B lysis buffer-PMSF-PCL (make fresh, and make 10 mL): transfer 10 mL of Cyclin B lysis buffer-PMSF to a tube on ice. Shortly before use, make 1X PCL (*see* **Note 3**).
3. LSWB-PMSF (*see* **Subheading 2.1.**; make fresh).
4. LSWB-PMSF-PCL (*see* **Subheading 2.1.**; make fresh).
5. Washed pellet of Sf9 cells infected with His6 Cyclin B virus (*see* **Subheading 3.3.**).
6. Ni-IDA beads (*see* **Subheading 3.1.**).
7. 20-mL Chromatography column with a porous bed support (Bio-Rad Econo-Pac Column, cat. no. 732-1010, or equivalent) and two-way stopcock (Bio-Rad, cat. no. 732-8102, or equivalent).

2.6. Production of Crude Lysate Containing Cdc2 Protein

1. Cdc2 hypotonic lysis buffer-PMSF-PCL (make fresh, and make 25 mL): 10 m*M* HEPES, 10 m*M* NaCl. Adjust pH to 7.4 and chill to 4°C on ice. Shortly before use, make 0.2 m*M* PMSF (*see* **Note 2**), and then take 6 mL and make 1X PCL (*see* **Note 3**).
2. Washed pellet of Sf9 cells infected with untagged Cdc2 virus (*see* **Subheading 3.3.**).

2.7. Making Recombinant Cdc2/Cyclin B Complex

1. Bound Cyclin B beads and Cdc2 lysate (*see* **Subheadings 3.4.** and **3.5.**).
2. Cdc2 complex wash buffer A (make fresh): 150 m*M* NaCl, 10 m*M* HEPES, 5 m*M* EGTA, 0.1% NP-40. Adjust pH to 7.4, and chill on ice. Before use, take 5 mL, and make 1X PCL (*see* **Note 3**).
3. Cdc2 complex wash buffer B (make fresh): 150 m*M* NaCl, 10 m*M* HEPES. Adjust pH to 7.4, and chill on ice. Before use, take 5 mL and make 1X PCL (*see* **Note 3**).
4. Cdc2 complex elution buffer (make fresh). Add 75 L 1 *M* imidazole, pH 7.4, to 475 L Cdc2 complex wash buffer B. The final imidazole concentration will be 150 m*M*. Chill on ice.
5. [γ-^{32}P]ATP (*see* **Note 5**), specific activity >4000 mCi/mL (such as ICN, cat. no. 35001X; do not use "stabilized" versions of ATP such as ICN, cat. no. 38101X IsoBlue, as we have

found that they work poorly for this specific procedure). This reagent is optional and only needed if one is labeling Cdc2 on Thr161 (*see* **Note 6**).

2.8. Preparation of Total and Fractionated Cellular Extract

1. 1X PBS (make fresh, and make 200 mL). Make from 10X stock, and chill to 4°C on ice for several hours/overnight.
2. Hypotonic lysis buffer (HLB; make fresh, and make 25 mL): 20 m*M* HEPES, 5 m*M* NaF, 1 m*M* NaPPi, 1 m*M* DTT. Adjust pH to 7.6, and chill to 4°C on ice. Shortly before use, take 5 mL and make 1X PCL (*see* **Note 3**).
3. Membrane wash buffer (MWB; make fresh, and make 25 mL): 500 m*M* NaCl, 20 m*M* HEPES, 5 m*M* NaF, 1 m*M* NaPPi, 1 m*M* DTT. Adjust pH to 7.6, and chill to 4°C on ice. Shortly before use, take 5 mL and make 1X PCL (*see* **Note 3**).
4. Extract dilution buffer (EDB; make fresh, and make 25 mL): 100 m*M* NaCl, 20 m*M* HEPES, 5 m*M* NaF, 1 m*M* NaPPi, 1 m*M* DTT. Adjust pH to 7.6, and chill to 4°C on ice. Shortly before use, take 5 mL and make 1X PCL (*see* **Note 3**).
5. Micro-ultra-centrifuge (Sorvall, cat. no. RC M120EX or equivalent) with swinging bucket rotor (Sorvall, cat. no. RP55S or equivalent) and 1.4-mL high-speed tubes (Sorvall polycarbonate, cat. no. 45237 or equivalent that are capable of ~260,000*g*). This equipment is optional and required only if one is fractionating the extract into membrane and soluble fractions.

2.9. Shifting Assay to Detect Wee Kinase Activity

1. 75 m*M* Na_3VO_4. Make fresh from 500 *M* stock (*see* **Note 4**).
2. 750 m*M* $MgCl_2$. Make from 1 *M* stock.
3. 2.5% Triton X-100. Make from 10% stock.
4. Shifting assay 11X ATP/regeneration mix (make fresh, and keep on ice): 10 L 100 m*M* ATP, 10 L 5 mg/mL CPK, 25 L 200 m*M* CP.
5. ^{32}P-T161-labeled Cdc2/Cyclin B complex (*see* **Subheading 3.6.** and **Note 6**).
6. Various extracts to be measured, for example, total, soluble, membrane (*see* **Subheading 3.7.**).
7. 1.25X Cdc2 shifting stop buffer: 1.89 mL dH_2O, 1.0 mL 4X SDS loading buffer, 310 L 1 *M* DTT sample buffer (*see* **Note 7**). Can be made fresh or stored at approx 20°C for ~1 mo. Warm to room temperature, and mix well before use.

2.10. Immunoprecipation of Endogenous Wee1, Wee2, or Myt1

1. Crude cell lysate that is between 10 and 50 g/ L protein (*see* **Note 8**).
2. Specific, species reactive antibody against Myt1, Wee1, or Wee2. This antibody should not change the activity of the kinase.
3. Okadaic acid (*see* **Note 4**).
4. Sepharose CL-4B Protein A beads (Sigma, cat. no. P 3391).
5. Bead wash: EB that is 0.5% NP-40 for Wee1/Wee2 *or* 0.5% Triton X-100 for Myt1. Keep chilled on ice, and shortly before use; make 1X PCL (*see* **Note 3**). Make fresh, and make 4 mL for four reactions (*see* **Note 9**).
6. IP wash A: EB that is 1 mg/mL ovalbumin, 25 m*M* NaF, 1 m*M* Na_3VO_4, 1 *M* okadaic acid, and 0.1% NP-40 for Wee1/Wee2 *or* 0.1% Triton X-100 for Myt1. Keep chilled on ice, and shortly before use make 1X PCL (*see* **Note 3**). Make fresh, and make 2 mL for four reactions (*see* **Note 9**).

7. IP wash B: EB that is 1 mg/mL ovalbumin, 25 m*M* NaF, 1 m*M* Na_3VO_4, and 0.1% NP-40 for Wee1/Wee2 *or* 0.1% Triton X-100 for Myt1. Keep chilled on ice, and shortly before use make 1X PCL (*see* **Note 3**). Make fresh, and make 4 mL for four reactions (*see* **Note 9**).
8. IP wash C: EB that is 1 mg/mL ovalbumin, 25 m*M* NaF, 1 m*M* Na_3VO_4. Keep chilled on ice, and shortly before use make 1X PCL (*see* **Note 3**). Make fresh, and make 5.5 mL for four reactions (*see* **Note 9**).
9. Kinase wash buffer: 50 m*M* Tris-HCl, pH 7.5, 10 m*M* $MgCl_2$, 1 m*M* DTT, 1 mg/mL ovalbumin. Keep chilled on ice, and shortly before use make 1X PCL (*see* **Note 3**). Make fresh and keep on ice; make 4 mL for four reactions (*see* **Note 9**).

2.11. Purification of Recombinant Myt1

1. Myt1 lysis buffer-PMSF (make fresh, and make 100 mL): 10 m*M* HEPES, 150 m*M* NaCl, 5 m*M* EGTA, 0.5% Triton X-100. Adjust pH to 7.4, and chill to 4°C on ice. Shortly before use, make 0.2 m*M* PMSF (*see* **Note 2**).
2. Myt1 lysis buffer-PMSF-PCL (make fresh, and make 10 mL). Transfer 10 mL of Myt1 lysis buffer-PMSF to a tube on ice. Shortly before use, make 1X PCL (*see* **Note 3**).
3. High-salt wash buffer (make fresh, and make 25 mL): 10 m*M* HEPES, 500 m*M* NaCl, 5 m*M* EGTA, 0.1% TX-100. Adjust pH to 7.4, and chill to 4°C on ice. Shortly before use, make 0.2 m*M* PMSF (*see* **Note 2**).
4. High-salt wash buffer-PMSF + PCL (make fresh, and make 5 mL): transfer 5 mL of high-salt wash buffer-PMSF to a tube on ice. Shortly before use, make 1X PCL (*see* **Note 3**).
5. LSWB-PMSF (*see* **Subheading 2.1.**; make fresh).
6. LSWB-PMSF-PCL (*see* **Subheading 2.1.**; make fresh).
7. Myt1 elution buffer (make fresh): add 60 L 1 *M* imidazole, pH 7.4, to 340 L LSWB-PMSF-PCL. The final imidazole concentration will be 150 m*M*. Chill on ice.
8. Washed pellet of Sf9 cells infected with His6 Myt1 virus (*see* **Subheading 3.3.**).
9. Ni-IDA beads (*see* **Subheading 3.1.**).

2.12. Purification of Recombinant Wee1 or Wee2

1. Wee1/2 lysis buffer-PMSF (make fresh, and make 50 mL): 50 m*M* Na_2HPO_4, 300 m*M* NaCl, 10 m*M* imidazole, 0.5% NP-40. Adjust pH to 7.8, and chill to 4°C on ice. Shortly before use, make 15 m*M* B-ME (β-mercaptoethanol) and 0.2 m*M* PMSF (*see* **Note 2**).
2. Wee1/2 lysis buffer-PMSF-PCL (make fresh, and make 10 mL): transfer 10 mL of Wee1/2 lysis buffer-PMSF to a tube on ice. Shortly before use, make 1X PCL (*see* **Note 3**).
3. Wee1/2 high-salt wash buffer-PMSF without B-ME (make fresh, and make 50 mL): 50 m*M* Na_2HPO_4, 750 m*M* NaCl, 25 m*M* imidazole, 5 m*M* EGTA. Adjust pH to 7.8, and chill to 4°C on ice. Shortly before use, make 0.2 m *M* PMSF (*see* **Note 2**).
4. Wee1/2 high-salt wash buffer-PMSF-PCL with 10, 5, 2.5, or 0 m*M* of B-ME. These are made from the Wee1/2 high-salt wash buffer-PMSF without B-ME; make four 5-mL aliquots of Wee1/2 wash buffer-PMSF that have 10, 5, 2.5, or 0 m*M* B-ME each. Store on ice. Shortly before use, make 1X PCL (*see* **Note 3**).
5. Washed pellet of Sf9 cells infected with His6 Wee1 or Wee2 virus (*see* **Subheading 3.3.**).
6. Sonic cell disruptor (Branson model 450 Sonifier®, cat. no. 33995-310 with a tapered 1/8-inch microtip, cat. no. 33996-185 or equivalent).

7. Ni-IDA beads (*see* **Subheading 3.1.**).
8. Wee1/2 elution buffer-PMSF-PCL (make fresh, and make 25 mL): 50 m*M* Na_2HPO_4, 300 m*M* NaCl, 250 m*M* imidazole. Adjust pH to 7.2, and chill to 4°C on ice. Shortly before use, make 0.2 m*M* PMSF (*see* **Note 2**), and then take 1 mL and make 1X PCL (*see* **Note 3**).
9. Dialysis buffer: 10 m*M* HEPES, 150 m*M* NaCl, pH 7.4. Make 3 L, autoclave, and then transfer to three 1-L beakers and chill to 4°C.
10. Dialysis membrane tubing (12–14 kDa MWCO, Spectra/Por 2, cat. no. 132676).

2.13. Cdc2 Kinase Assay to Detect Wee Kinase Activity

These reagents are used in both Cdc2 kinase assays (*see* **Subheadings 3.12.1.** and **3.12.2.**).

1. 10X kinase buffer (make fresh, and keep chilled on ice): 500 m*M* Tris-HCl, pH 7.5, 100 m*M* $MgCl_2$, 10 m*M* DTT.
2. Kinase dilution buffer (make fresh, and keep chilled on ice): 1X kinase buffer with 1 mg/mL ovalbumin. Make from 10X and 100 mg/mL stocks, respectively.
3. 5X Kinase buffer (make fresh, and keep chilled on ice): make from 10X stock.
4. 5X Kinase buffer + 1% Triton X-100 (make fresh, and keep chilled on ice): make from 10X and 10% stocks, respectively. Only needed if one is measuring Myt1 activity.
5. Purified Wee1, Wee2, or Myt1 kinase (*see* **Subheadings 3.9.**, **3.10.**, or **3.11.**). These preparations of kinase will probably need to be diluted with kinase dilution buffer. If the concentration of recombinant kinase is known, start with 1–20 ng kinase per reaction (*see* **Note 10**). If working with immunoprecipitated kinase from extracts, start with kinase immunoprecipitated from 0.5–10 L of extract (*see* **Note 11**). Hopefully, these amounts of kinase will be in the linear range of the assay, but this will vary with the activity and concentration of the kinase (*see* **Notes 12** and **13**).

2.13.1. Cdc2 ^{32}P Labeling Kinase Assay to Detect Wee Kinase Activity

1. Recombinant, kinase-dead Cdc2/Cyclin B substrate(s) (*see* **Subheading 3.6.**).
2. 100 *M* ATP. Make fresh, make from 100 m*M* stock, and keep chilled on ice.
3. Hot ATP cocktail. Make fresh, and keep chilled on ice (*see* **Note 5**):
   Per reaction, the following amounts are needed: 3.8 L dH_2O, 0.2 L 100 *M* cold ATP, and 1.0 L [γ-^{32}P]ATP, for 5 L total/reaction.
   Multiply these volumes by the number of reactions to be performed, plus one to account for pipeting loss.

2.13.2. Cdc2 Antiphospho-Tyrosine Antibody Kinase Assay to Detect Wee Kinase Activity

1. Recombinant Cdc2/Cyclin B substrate(s) (*see* **Subheading 3.6.**).
2. Kinase assay 12.4X ATP/regeneration mix (make fresh, and keep on ice): 5 L 100 m*M* ATP, 10 L 5 mg/mL CPK, 25 L 200 m*M* CP. This is sufficient for approx 11 reactions (*see* **Note 14**).

2.14. Histone H1 Kinase Assay

1. 5 mg/mL Histone H1. Make from lyophilized powder (Roche, cat. no. 223549) in dH_2O. Store in 5- L aliquots at –80°C. Do not refreeze aliquots after thawing.
2. 2X HH1 buffer: 40 m*M* HEPES, 10 m*M* EGTA, 20 m*M* $MgCl_2$. Adjust pH to 7.3. Store in 500- L aliquots at –20°C.

3. Samples for assay: Wee kinase-modified Cdc2/Cyclin B complex (*see* **Subheading 3.12.2., step 5**).

3. Methods

3.1. Making Nickel-Iminodiacetic Acid Beads

Several of the procedures in this chapter use metal-chelate affinity purification to isolate and immobilize recombinant proteins that are tagged with a 6-histidine (His6) tag (reviewed in **ref. *29***). The His6 tag will bind to a variety of metal-chelated beads. Although these beads are available commercially, it is both easy and cost-effective to make Ni-IDA beads *(23)*. In our hands, these 'home-made'beads work as well if not better than premade beads. This can be quite a savings since some of the wash steps we recommend use agents (EGTA, 2-mercaptoethanol) that degrade the bead linkage, making it unwise to reuse the Ni-IDA beads (*see* **Note 1**).

1. While waiting for the IDA-sepharose to warm to room temperature, make the 1X TBS and 100 m*M* NiCl, and set up the fritted funnel with a neoprene adapter on the 500-mL vacuum flask. Using low vacuum, rinse the funnel three times with 30 mL dH_2O.
2. Resuspend the IDA-sepharose by gentle inversion. Use a 10-mL pipet to transfer 6.5 mL of the sepharose slurry (equivalent to 5 mL of beads) to the fritted funnel. Use a slight vacuum to remove the liquid from beads, and then rinse the beads twice with 30 mL of dH_2O under vacuum. Do not let the sepharose dry out~~as~~ soon as the liquid is gone, either release the vacuum or add more liquid. After the last rinse, release the vacuum.
3. Use sequential, approx 10 mL aliquots of 100 m*M* NiCl to resuspend the rinsed sepharose from **step 2**. (The liquid will not flow through the funnel since no vacuum is applied). Transfer and pool the aliquots of resuspended sepharose to a 50-mL conical tube. Adjust the final volume to approx 50 mL with 100 m*M* NiCl. Rotate NiCl/sepharose slurry for 10 min at room temperature.
4. Set up the fritted funnel again, and apply a slight vacuum. Pour the NiCl/sepharose slurry into a funnel, and drain the liquid with low vacuum. Once again, do not let the beads dry out. The beads will have acquired a light blue-green color owing to the absorption of Nickel.
5. Using a low vacuum, wash the beads two or three times with approx 30 mL of dH_2O each time, and then wash six or seven times with approx 30 mL 1X TBS each time. When the last liquid is gone, immediately remove the vacuum.
6. Use three approx 6 mL aliquots of 1X TBS to resuspend and transfer Ni-IDA sepharose beads to a fresh 50-mL conical tube. Adjust the final volume to 25 mL with 1X TBS, and make 0.02% sodium azide. This makes a 20% slurry. Store at 4°C for $\leq$6 mo.

3.2. Infection of Sf9 (S. frugiperda) Cells with Recombinant Baculovirus

This procedure works for most recombinant proteins and usually yields a sufficient amount of biologically active protein for analysis. As outlined, this procedure is for 10 plates, but it may be scaled up or down as needed. All growth of Sf9 cells is at 27°C and all work should be done in a tissue culture hood to maintain aseptic technique (reviewed in **refs. *24–26***). Viral stocks for the expression of specific proteins can be made by the use of several commercial kits. We have used the Bac-to-Bac kit (Invitrogen, cat. no. 10584-027) with a high degree of success. A His6 tag is used in the purification of many recombinant proteins in this chapter. In most cases, this small tag on the amino-terminus will not interfere with the function of the protein. The one exception is Cdc2. For Cdc2, an amino terminal tag slightly weakens the association of Cdc2 with Cyclin B. Therefore, we produce an untagged Cdc2 protein but then purify it by subsequent binding to Cyclin B (*see* **Subheading 3.6.**).

1. At 72 h before the desired harvest time, plate Sf9 cells at a density of 1×10^7 per 10-cm plate (*see* **Notes 15** and **16**).
2. Infect Sf9 cells with recombinant viral stock 48 h before desired harvest time. Aspirate the media off after the cells, and then add 3 mL of a diluted virus stock to each plate (1 mL of a P3 virus stock to 2 mL of fresh media [no Pluronic® F-68]). Gently rock the plates a few times to mix. After 1 h, add 6 additional mL of fresh media, and gently rock the plates a few times to mix. Let infection continue for 48 h (*see* **Notes 17** and **18**) before harvest (*see* **Subheading 3.3.**).

3.3. Collection and Washing of Infected Sf9 Cells

This simple cell collection/wash procedure is designed to remove contaminating serum proteins and to collect the infected cells with minimal damage to the cells. This same wash procedure is used for all proteins, but individual cell lysis procedures are used for each specific protein (*see* **Subheadings 3.4.**, **3.5.**, **3.10.**, and **3.11.**). Important: be ready to proceed immediately to the specific lysis/purification procedure.

1. After 48 h of infection, collect the Sf9 cells (*see* **Notes 18** and **19**). While cells are still in Sf9 media, dislodge any attached cells by gently scraping with a rubber policeman. Transfer and distribute the suspended cells to three or four 50-mL conical tubes. Wash the plate with 3 mL of 1X PBS-PMSF, and pool this wash with cells in the media. From this point on, keep cells on ice.
2. Wash infected Sf9 cells. Pellet cells by centrifugation for 5 min at 4°C at 800 *g*. Resuspend each cell pellet in approx 5 mL of 1X PBS-PMSF. Do this by gently pipeting up and down. The cells should be very easy to resuspend. Once the cells are resuspended, add approx 40 mL of 1X PBS-PMSF to each tube, mix by gentle inversion, and respin. Resuspend the cell pellet in 5 mL 1X PBS-PMSF as before, then pool all resuspended pellets into one 50-mL tube, bring the volume up to approx 50 mL with 1X PBS-PMSF, mix by gentle inversion, and respin. Remove as much of the PBS wash from the pellet as possible with this last wash. Proceed immediately to protein-specific cell lysis and protein harvest.

3.4. Purification of Cyclin B on Ni-IDA Sepharose Beads

In this procedure, recombinant, His6-tagged Cyclin B is purified from infected Sf9 cells and then left immobilized on Ni-IDA sepharose beads for future use, as described in **Subheading 3.6.** The results of the procedure can be seen in **Fig. 3A**, lanes 7–10. Important: the purification of Cyclin B from Sf9 infected cells should follow immediately from **Subheading 3.3.**

1. Lyse Sf9 cells infected with Cyclin B virus by douncing (*see* **Note 20**).
2. Remove insoluble material (*see* **Note 21**). While waiting for this centrifuge run, equilibrate 1 mL of packed Ni-IDA sepharose beads into Cyclin B lysis buffer-PMSF (*see* **Note 22**). Prepare these beads in a 15-mL conical tube.
3. Bind the approx 6 mL of His6-tagged Cyclin B protein supernatant (*see* **Note 21**) to the equilibrated Ni-IDA beads (*see* **Note 22**). Rotate end over end for 1 h (*see* **Note 23**).
4. Pour the Cyclin B-bound bead slurry into a 20-mL chromatography column, and adjust the flow to approx 1 drop per second with a two-way stopcock (*see* **Note 23**). Collect the flowthrough as the beads settle. Pass the flowthough over the column one more time.

Drop-freeze 5 L of the unbound Cyclin B supernatant (flowthrough) in LN_2 for a test gel. (Store the test sample at –80°C until use.) Immediately proceed to washing the column so that the beads do not dry out.

5. Wash the column sequentially with 50 mL of Cyclin B lysis buffer-PMSF, 50 mL of LSWB-PMSF, and 5 mL of LSWB-PMSF-PCL (*see* **Note 23**).
6. Collect the Cyclin B beads (*see* **Note 23**). Seal the bottom of the chromatography column containing the washed Cyclin B beads with parafilm. Using a total of 5 ml of LSWB-PMSF-PCL, resuspend the beads, and transfer them to a 50-mL conical tube. This will make a 20% slurry.
7. Make 50 L aliquots of the 20% Cyclin B bead slurry in 600 L microcentrifuge tubes (*see* **Note 23**). While making these aliquots, keep the beads suspended by frequent gentle swirling of the 50-mL conical tube containing the bound Cyclin B. Do not use a narrow-bore pipet tip to pipet bead solutions. Immediately drop-freeze these aliquots in LN_2, and store at –80°C. In addition, drop-freeze 25 L of the 20% Cyclin B bead slurry in LN_2 for a test gel. (Store at –80°C until use.)

3.5. Production of Crude Lysate Containing Cdc2 Protein

In this procedure, a crude lysate containing recombinant Cdc2 protein is prepared from infected Sf9 cells. The Cdc2 in this lysate will be subjected to further purification, modification, and binding to Cyclin B when the Cdc2/Cyclin B complex is made in **Subheading 3.6.** In addition to wild-type Cdc2, many forms of Cdc2 can be made by this procedure *(8,12,15,23)*. To list a few examples: kinase-dead Cdc2 (Cdc2 N133A, Asp133 changed to Ala), Cdc2 that cannot be phosphorylated by Wee1 or Wee2 (Cdc2 Y15F, Tyr15 changed to Phe), and Cdc2 that cannot be phosphorylated by Wee1, Wee2, or Myt1 (Cdc2 AF, a double mutant with Thr14 changed to Ala and Tyr15 changed to Phe). Each of these is useful in measuring Wee kinase specificity and activity. The results of the procedure can be seen in **Fig. 3A**, lanes 1–6. Important: the preparation of the Cdc2 lysate from Sf9 infected cells should follow immediately from **Subheading 3.3.**

1. Lyse Sf9 cells infected with Cdc2 virus by hypotonic swelling and douncing (*see* **Note 20**) with the following modifications:
   a. Before douncing, let cells sit in Cdc2 hypotonic lysis buffer-PMSF-PCL for 10–15 min on ice to permit cell swelling in the hypotonic solution.
   b. After rinsing and pooling the lysate from the dounce, add 168 mL 5 M NaCl to make a final concentration of 150 mM NaCl in the lysate.
   c. Then mix by inversion, drop-freeze the test sample, and then remove the insoluble material (step 2).
2. Remove insoluble material (*see* **Note 21**).
3. Make 80 L aliquots of the Cdc2-containing supernatant into 600- L microcentrifuge tubes (*see* **Note 23**). Immediately drop-freeze these aliquots in LN_2, and store at –80°C. In some cases, larger aliquots may be desired (*see* **Note 24**).

3.6. Making Recombinant Cdc2/Cyclin B Complex

In this procedure, the Cdc2/Cyclin B complex is made *(23)*. This is the functional form of Cdc2, and it is used to measure Wee kinase activity in several procedures (*see* **Subheadings 3.8., 3.12.1., 3.12.2.**, and **3.13.**). Once made, this complex is moderately unstable. It should be used within approx 1 h and cannot be frozen. Four key events take place during this procedure. First, Cdc2 is bound to Cyclin B, which is immobilized in Ni-IDA beads. Second, CAK activity that is present in the crude Cdc2 lysate phosphorylates Cdc2 on Thr161. This phosphorylation is required for Cdc2 activity but will only happen after Cdc2 is bound to Cyclin B. Furthermore, some Wee kinases, such as *Xenopus* Myt1, can phosphorylate Cdc2 only after it is phosphorylated on Thr161. Third, the complex is washed relatively free of contaminating insect cell proteins. Fourth, the Cdc2/Cyclin B complex is eluted from the Ni-IDA beads. The results of the procedure can be seen in **Fig. 3B**.

1. Thaw aliquots of Cdc2 lysate and Cyclin B bead slurry on ice. Pack the Cyclin B beads by centrifugation in a swinging bucket rotor at 4°C for 10 s at 1600 *g*. Discard supernatant, but do not lose any beads. Return the pelleted beads to ice; this will be 10 L of packed beads. Centrifuge the aliquot of Cdc2 at 4°C for 2 min at 18,000 *g*. Transfer 70 L of the cleared Cdc2 lysate to the tube containing the 10 L of packed cyclin beads (*see* **Note 24**). Discard the remaining Cdc2 lysate and any pellet.
2. Bind and activate the Cdc2/Cyclin B complex (*see* **Note 6**). To the lysate/bead mixture, add 0.8 L 1 *M* $MgCl_2$ and 0.8 L 100 m*M* ATP. Rotate this at room temperature for 30 min, and then return samples to ice.
3. Wash the Cdc2/Cyclin B complex on the beads (*see* **Note 25**). Centrifuge the bead slurry for 10 s at 1600*g* in a swinging bucket rotor, and discard the supernatant. Wash the beads four times with 400 L each time of wash buffer A, and then four times with 400 L each of Wash buffer B (*see* **Note 26**). For the last spin, remove as much buffer as possible, but no beads.
4. Elute complex from beads: Add 70 L Cdc2 complex elution buffer to the tube and rotate for 10–15 min at 4°C (*see* **Note 24**). Centrifuge beads at 4°C for 20 s at 1600 *g* in a swinging bucket rotor. Carefully remove, and save the supernatant into a fresh, chilled tube; avoid taking any beads. The Cdc2/Cyclin B complex should be used as soon as possible, typically within an hour. If desired, drop-freeze a large aliquot (~35 L) of the Cdc2/Cyclin B complex in LN_2 for analysis on a Coomassie blue-stained gel (**Fig. 3B**). This is needed if one is comparing different forms of Cdc2 (*see* **Note 24**).

3.7. Preparation of Total and Fractionated Cellular Extract

This procedure is designed to make a cellular extract that can be used to measure Wee kinase activity in a Cdc2 shifting assay (*see* **Subheading 3.8.**). In some cases, only the total cellular extract will be needed and the procedure may be stopped at **step 2**. Alternatively, if the goal is to examine the specific contributions of the membrane associated kinase (Myt1) or the soluble kinases (Wee1 or Wee2) to the total Wee kinase activity that is found in the extract *(12)*, the cell extract can be further fractionated into soluble and membrane fractions as described in **steps 3–7** of this procedure. This procedure describes the production of extract from tissue culture cells, but extracts from *Xenopus* eggs and oocytes may also be used (*see* **Fig. 5** and **Note 27**). Although the procedure described here works well, any of several cell lysis procedures may be used. However, note the suggested use of phosphatase inhibitors in the dilution and wash buffers, and avoid the intitial use of detergents if the total extract is to be fractionated into membrane and soluble fractions. If detergents are used, Triton X-100 is suggested, as it does not inhibit Myt1 activity (*see* **Note 28**).

1. Collect and wash approx 5×10^8 tissue culture cells with ice-cold 1X PBS as described (*see* **Subheading 3.3.**), except leave PMSF out of the 1X PBS solution.
2. Obtaining the total cell extract.
   a. Lyse the tissue culture cells by hypotonic swelling and douncing.
   b. Resuspend cells in 2 ml HLB (*see* **Note 14**), and let cells sit for 10–15 min on ice to permit swelling.
   c. Transfer the cells to an ice-cold Dounce tissue grinder with a tight pestle, and disrupt the cells on ice. The cells should break open in 20 or 40 strokes.
   d. Microscopically examine the cells to be sure they are disrupted. It is helpful to look at the cells before and after douncing.
   e. Recover and save the lysate in a chilled tube, rinse the Dounce/pestle with 0.5 mL hypotonic lysis buffer, and pool this with the lysate.
   f. Make the final concentration of NaCl in the pooled lysate cell lysate 100 m*M* by adding 50 L of 5 *M* NaCl and mixing gently. This is the total cell extract.
   g. If the total cell extract is to be fractionated, fill an ice-cold, high-speed centrifuge tube with 1.4 mL of the total cell extract for use in **step 3** (*see* **Note 14**).
   h. Make 59- L aliquots of the remaining total cell extract in ice-cold 600- L microcentrifuge tubes, immediately drop-freeze these in LN_2, and store at –80€.
3. To fractionate the extract, centrifuge it at 260,000g for 1 h in a swinging bucket rotor at 4€.
4. Carefully remove, and save the supernatant into a chilled tube. This is the soluble fraction and may be set aside (on ice) until **step 6**. When collecting the soluble fraction (supernatant), do not disturb the brownish cloudy membrane pellet. This is best accomplished by removing and saving only approx 90% of the supernatant. Then, if the pellet is still intact, remove and discard the remaining supernatant. Otherwise, leave the small amount of supernatant with the pellet. Important: while handling the tubes, keep them on ice as much as possible.
5. Resuspend the membrane pellet from **step 4**. Measure out, and keep chilled, 1.3 mL of MWB. This will be used to resuspend the membrane (see Notes 14 and 29). Using a 1000-mL pipetor, take approx 300 mL of the measured-out MWB, and resuspend some of the membrane by pipeting up and down several times. Do not resuspend all the membrane in this small volume. After 10–20 pipeting strokes, transfer the resuspended membrane to a fresh, ice-cold high-speed centrifuge tube, but avoid transferring any "chunks" of membrane that are not yet resuspended. Repeat this resuspension procedure several times until all the membrane is resuspended and the entire 1.3 mL of MWB that was measured out is used up. If any insoluble material remains in the centrifuge tube, discard it. When done correctly, the second high-speed centrifuge tube will be full of resuspended membrane. Mix this by gently pipeting up and down with a 1000-mL pipetor a few times. Remember, it is important to keep all the tubes on ice as much as possible.
6. Centrifuge the resuspended membrane at 260,000g for 1 h in a swinging bucket rotor at 4€. While the membrane is being centrifuged, aliquot the soluble fraction from **step 4**. Make 59-mL aliquots of the soluble fraction in ice-cold 600-mL microcentrifuge tubes, immediately drop-freeze these in LN_2, and store at –80€.

7. Discard the supernatant from the MWB centrifuge run (**step 6**). Use extreme caution so as not to disturb or lose any of the pellet. Once the supernatant is removed, resuspend the washed membrane pellet in 1.3 mL of EDB (*see* **Note 14**) per the method described in **step 5**, except transfer the resuspended membrane into an ice-cold microcentrifuge tube. This is the membrane fraction. Make 59-mL aliquots of the membrane fraction in ice-cold 600-mL microcentrifuge tubes, immediately drop-freeze these in LN_2, and store at –80°C. Remember to keep everything on ice.

3.8. Shifting Assay to Detect Wee Kinase Activity

In this procedure, the activity of the various Wee kinases that are present in crude cellular extracts can be measured *(12)*. This measurement is based on a gel mobility shift of Cdc2 when Cdc2 is phosphorylated by any of the Wee kinases *(27,28)*. The degree to which Cdc2 is shifted helps identify the Wee kinase(s) responsible. Specifically, Wee1 and Wee2 phosphorylate Cdc2 on Tyr15 (causing a single shift), whereas Myt1 phosphorylates Cdc2 on Thr14 and/or Tyr15 (causing a single or double shift) *(12)*. Another key difference between the Wee kinases is that Wee1 and Wee2 are soluble, whereas Myt1 is membrane-associated. Finally, the rate at which Cdc2 is shifted indicates the relative activity of the Wee kinase(s). Thus, depending on the degree of shifting (single or double), on which extract fraction shifts Cdc2 (membrane or soluble), and on the rate of shifting over time, it is possible to measure and compare relative Wee1/Wee2 and Myt1 kinase activities in crude cellular extracts (for an example, *see* **Fig. 5**).

1. Make ^{32}P-T161–labeled Cdc2/Cyclin B complex as described in **Subheading 3.6.** (and *see* **Note 6**).
2. Thaw (on ice) the 59 L extract samples that are to be measured (*see* **Subheading 3.7.**).
3. Aliquot 1.25X Cdc2 shifting stop buffer into microcentrifuge tubes (*see* **Note 30**). The 0-min time-point will need 56 L stop buffer; all other time-points will need 28 L (*see* **Note 14**). Keep these tubes at room temperature.
4. Keeping the thawed extract samples (from **step 2**) chilled on ice, add 3.2 L 2.5% Triton-X 100, 1 L 750 mM $MgCl_2$, and 1 L 75 mM Na_3VO_4 to each tube. Mix by gently flicking the tube, and then quick-spin the samples at 4°C to collect the mixture at the bottom of the tube. Keep samples on ice.
5. The reaction is started by addition of the ATP/regeneration mix and the Cdc2/Cyclin B substrate and by moving the sample to room temperature. To process several samples in one experiment, it is useful to stagger/stop the reactions (*see* **Note 31**).
   a. Working with one sample at a time, add 6.8 L of 11X ATP/regeneration mix, then immediately add 4 L of labeled Cdc2/Cyclin B complex (from **step 1**), and mix the sample by gently pipeting up and down. Keep the sample chilled on ice for now.
   b. Next, immediately take 14 L of this sample and add it to 56 L of the prealiquoted stop buffer (from **step 3**). This is the 0-min time-point, and the sample should be quickly frozen on dry ice.
   c. The remainder of the reaction should then be moved to room temperature to start the reaction.
   d. At this point, repeat this starting procedure with the next sample to be assayed.
6. At the desired time (*see* **Note 30**), transfer 7 L of the sample to 28 L of the prealiquoted stop buffer, mix, and quickly freeze on dry ice. Allow the reaction to continue to incubate

at room temperature. At the next time-point, repeat the procedure (*see* **Note 31**). The samples frozen on dry ice may be stored at –80°C until use.

7. Boil the samples, and then load and resolve them on a 12 or 15% SDS-PAGE gel. The various phosphorylated forms of Cdc2 will run between 30 and 35 kDa. The gel can be transferred to membrane or dried down to reveal the shifted forms of Cdc2 with a PhosphorImager or X-ray film.

3.9. Immunoprecipitation of Endogenous Wee1, Wee2, or Myt1

This procedure uses immunoprecipitation to recover any of the Wee kinases from cellular extracts. This allows isolation of the Wee kinase from other cell constituents and makes direct measurement of its activity relatively straightforward (*see* **Subheadings 3.12.1.** and **3.12.2.**). The main challenge of this approach is that it requires a specific, noninterfering antibody to immunoprecipitate the Wee kinase. In most cases, an antibody directed toward the carboxyl-terminal region of the kinase works well for this purpose ***(8,12)***. Unfortunately, most commercially available antibodies are specific to human or mammalian kinases. Therefore, for studies in other species, custom-made antibodies must be obtained or produced. During the isolation of the Wee kinases, it is important to preserve the native activity of the kinase. For example, both Myt1 and Wee1 are less active during mitosis owing to phosphorylation by mitotic kinases (reviewed in **ref. *4***). The use of phosphatase inhibitors (*see* **Note 4**) during the wash steps helps preserves these modifications and permits the accurate, subsequent measurement of kinase activity in vitro. **Figure 6** shows an example of immunoprecipitated *Xenopus* Myt1.

1. On ice, add 5 L 10% NP-40 for Wee1/Wee2 *or* 10% Triton X-100 Myt1 to 95 L cell extract (*see* **Notes 8** and **14**). Add approx 2–4 g of Myt1-, Wee1-, or Wee2-specific antibody (*see* **Note 32**). Rotate at 4°C for approx 1 h. During this incubation, prepare protein A beads as described in **step 2**.
2. Swell and wash the protein A beads. For four reactions, weigh out 12.5 g of dried beads (*see* **Notes 14** and **33**). Add 1 mL of bead wash, and allow the beads to soak for 15 min with occasional flicking of the tube to mix. Centrifuge beads for 15 s at 1600*g* in a swinging bucket rotor, and discard the supernatant. Wash the beads two more times with 1-mL bead wash each time. Resuspend the beads with 450 L of bead wash (final volume, 500 L). Aliquot 100 L of the bead slurry into four different tubes, and discard the remaining unused beads (*see* **Note 34**). Centrifuge aliquoted beads as before, and discard supernatant. There should be 10 L of swelled beads per tube. Keep these beads on ice.
3. Transfer the antibody/cell extract mixture from **step 1** to the beads prepared in **step 2**. Rotate at 4°C for approx 1 h.
4. Centrifuge beads for 30 s in a swinging bucket rotor at 1600g at 4°C (*see* **Note 23**). Carefully remove and discard the supernatant without disturbing the beads. Batch-wash beads once with 400 L of IP wash A (*see* **Note 26**).
5. Batch-wash beads two times with 400 L of IP wash B (*see* **Notes 23** and **26**).
6. Batch-wash beads three times with 400 L of IP wash C (*see* **Notes 23** and **26**).
7. Batch-wash beads two times with 400 L of kinase wash buffer (*see* **Notes 23** and **26**). After the last wash, remove as much of the liquid as possible from the beads. Then resuspend the beads in kinase wash buffer so that the final volume is equal to that of the extract volume used in **step 1**. (In this example, use 85 L kinase buffer so the final volume with

the beads is 95 L; *see* **Note 14**). Make 10 to 20 L aliquots of the bead slurry in 600- L microcentrifuge tubes, drop-freeze these in LN_2, and store at –80°C until use, (*see* **Subheadings 3.12.1.** and **3.12.2.** and **Note 11**).

3.10. Purification of Recombinant Myt1

In this procedure, recombinant, His6-tagged Myt1 is purified from infected Sf9 cells using Ni-IDA beads ***(12)***. The Myt1 is then eluted from these beads for subsequent use in measurement assays (*see* **Subheadings 3.12.1.** and **3.12.2.**) or as a substrate for Myt1 regulators such as p90 Rsk ***(30)***. The yield of *Xenopus* Myt1 by this procedure is not particularly good, but it is enough for biochemical analysis (*see* **Note 35**). Importantly, the recovered Myt1 is active. The purification of Myt1 from Sf9-infected cells should follow immediately from **Subheading 3.3.**

1. Lyse Sf9 cells infected with Myt1 virus by douncing (*see* **Note 20**).
2. Remove insoluble material (*see* **Note 21**). While waiting for this centrifuge run, equilibrate 50 L of packed Ni-IDA sepharose beads into Myt1 lysis buffer-PMSF (*see* **Note 22**). Prepare these beads in a 15-mL conical tube.
3. Bind the approx 6 mL of His6-tagged Myt1 protein supernatant (see **Note 21**) to the equilibrated Ni-IDA beads (*see* **Note 22**). Rotate end over end for 1 h (*see* **Note 23**).
4. Recover beads with bound Myt1 protein (*see* **Note 23**). Centrifuge beads with bound protein for 1 min at 4°C at 1600 *g*. Remove the unbound supernatant, but drop-freeze 5 L of the Myt1 unbound supernatant in LN_2 for a test gel. (Store the test sample at –80°C until use.) Save the beads for **step 5** on ice.
5. Wash the Myt1-bound beads (*see* **Note 23**). Using two sequential 400- L aliquots of high-salt wash buffer-PMSF-PCL, transfer the beads to a 1.5-mL microcentrifuge tube. Centrifuge beads for 20 s in a swinging bucket rotor at 1600*g*. Carefully remove and discard the supernatant without disturbing the beads. Batch-wash beads with approx 800 L of Myt1 high salt wash buffer-PMSF-PCL three more times (*see* **Note 26**).
6. Batch-wash beads with approx 800 L LSWB-PMSF-PCL four times (*see* **Notes 23** and **26**). After the last wash, remove as much liquid as possible.
7. Elute the Myt1 protein (*see* **Note 23**). Add 75 L of Myt1 elution buffer to the beads. Rotate end over end for 20 min at 4°C. Spin for 20 s at 1600 *g* in swinging bucket rotor, remove, and save eluted Myt1 protein in a fresh microcentrifuge tube. Repeat elution procedure with a fresh 75- L aliquot of elution buffer. Pool elutes. Drop-freeze 40 L of the pooled elute in LN_2 for a test gel. To the remaining pooled elute, add ovalbumin to make 1 mg/mL ovalbumin, then make 10–20- L aliquots, drop-freeze these in LN_2, and store at –80°C until use (*see* **Subheadings 3.12.1.** and **3.12.2.** and **Note 35**).

3.11. Purification of Recombinant Wee1 or Wee2

In this procedure, recombinant, His6-tagged Wee1 or Wee2 is purified from infected Sf9 cells using Ni-IDA beads ***(15)***. The Wee1 or Wee2 is then eluted from these beads for subsequent use in measurement assays (*see* **Subheadings 3.12.1.** and **3.12.2.**) or as a substrate for Wee1/2 regulators such as Cdc2 or Nim1 ***(8,31)***. The yields of both *Xenopus* Wee1 and *Xenopus* Wee2 by this procedure are very good; commonly we isolate ≥1 g per 10-cm plate of infected Sf9 cells. Importantly, the recovered Wee1 or Wee2 is active. The purification of Wee1 or Wee2 from Sf9-infected cells should follow immediately from **Subheading 3.3.**

1. Resuspend the cell pellet from **Subheading 3.3.** in 10 mL of Weel/2 lysis buffer-PMSF-PCL by gently pipeting up and down. Transfer the cell suspension to a 13-mL Sarstedt tube, and disrupt the cells by sonication (*see* **Note 36**).
2. Remove insoluble material (*see* **Note 21**). While waiting for this centrifuge run, equilibrate 100 L of packed Ni-IDA Sepharose beads into Weel/2 lysis buffer-PMSF (*see* **Note 22**). Prepare these beads in a 15-mL conical tube.
3. Bind the 10 mL of His6-tagged Weel or Wee2 protein supernatant (*see* **Note 21**) to the equilibrated Ni-IDA beads (*see* **Note 22**). Rotate end over end for 1 h (*see* **Note 23**).
4. Recover beads with bound Weel or Wee2 protein (*see* **Note 23**). Centrifuge beads with bound protein for 1 min at 4C at 1600 *g*. Remove the unbound supernatant, but drop-freeze 5 L of the Weel or Wee2 unbound supernatant in LN_2 for a test gel. (Store the test sample at –80C until use.) Save the beads for **step 5** on ice.
5. Wash the Weel or Wee2-bound beads (*see* **Note 23**). Batch-wash beads four times with 5 mL Weel/2 lysis buffer-PMSF each time. For each wash, mix by inversion and spin for 10 s in a swinging bucket rotor at 1600*g*. Then batch-wash beads four times with 5 mL of Weel/2 wash buffer-PMSF containing sequentially decreasing amounts of B-ME (10, 5, 2.5, and 0 m*M*), once each. Use the last wash to transfer the beads to a 1.5-mL microcentrifuge tube. Spin this tube for 10 s in a swinging bucket rotor at 1600*g*, and remove as much liquid as possible.
6. Elute the Weel or Wee2 protein (*see* **Note 23**). Add 150 L of Weel/2 elution buffer to the beads. Rotate end over end for 20 minutes at 4C. Spin for 20 s at 1600 *g* in a swinging bucket rotor, remove, and save the eluted Weel or Wee2 protein in a fresh microcentrifuge tube. Repeat the elution procedure with a fresh 150- L aliquot of elution buffer. Pool the elutes. Drop-freeze 5 L of the pooled elution in LN_2 for a test gel. (Store the test sample at –80C until use.)
7. Dialyze the eluted Weel or Wee2 against three 1-L changes of dialysis buffer at 4C (*see* **Note 23**). Collect the dialyzed protein, drop-freeze 5 L in LN_2 for a test gel. To the remaining dialyzed elute, make 10–20- L aliquots, and drop-freeze these in LN_2, and store at –80C until use (*see* **Subheadings 3.12.1.** and **3.12.2.**).

3.12. Cdc2 Kinase Assay to Detect Wee Kinase Activity

3.12.1. Cdc2 ^{32}P Labeling Kinase Assay to Detect Wee Kinase Activity

In this procedure, the activity of purified Wee kinases is measured in vitro using recombinant Cdc2/Cyclin B complex as a substrate. The reaction is performed in the presence of [γ-^{32}P]ATP, and the degree of ^{32}P incorporation into Cdc2 is an indicator of Wee activity ***(8,12,15)***. The advantage of this method is that it is simple and quite sensitive. The disadvantage is that the signal can be obscured by background caused by contaminating kinases. For this reason, it requires relatively pure preparations of Wee kinases and the use of kinase-dead Cdc2 substrates such as the N133A mutant (*see* **Subheading 3.5.**). The use of a mutant of Cdc2 that cannot be phosphorylated by the Wee kinases (i.e., Cdc2 Y15F, N133A or Cdc2–AF, N133A) is a useful control to include (*see* **Fig. 6** for an example of this assay).

1. Make unlabeled, kinase-dead Cdc2/Cyclin B complex(es) (*see* **Subheading 3.6.**).
2. Thaw purified kinase (Weel, Wee2, or Mytl) to be measured on ice (*see* **Subheadings 3.9., 3.10.**, or **3.11.**).

3. Set up the reaction by adding components to microcentrifuge tubes in the order listed below. Keep tubes on ice throughout. When setting up the reaction, add all of one reagent to all of the samples, and then move on to the next reagent. At each addition, mix the reaction by slowly pipeting up/down a couple of times with the tip used to add the reagent.
   a. First: dilute kinase in kinase dilution buffer so that the volume of kinase is 31 L (*see* **Notes 10–13** and **34**).
   b. Second: add 2 L of 5X kinase buffer if measuring Wee1/Wee2 *or* 2 L of 5X kinase buffer + 1% Triton X-100 if measuring Myt1.
   c. Third: add 2 L Cdc2/Cyclin B complex.
   d. Fourth: add 5 L Hot ATP cocktail to each tube (*see* **Note 5**).

   The final concentrations in the 40 L reaction will be: 50 m*M* Tris-HCl, pH 7.5, 10 m*M* $MgCl_2$, 1 m*M* DTT, 1 mg/mL ovalbumin, 0.05% Triton X-100 (for assays with Myt1), 500 n*M* ATP, 55 n*M* [γ-^{32}P]ATP; 5 Ci/reaction, plus Cdc2/Cyclin B complex and Myt1, Wee1, or Wee2 kinase.
4. After the final addition, mix by gently flicking the tubes, quick-spin 4°C, and incubate for 15 min at room temperature. If the kinase is on beads, rotate the reaction tubes, or mix by flicking occasionally.
5. To stop the reaction, add 13.5 L 4X sample buffer, flick the tube to mix, and either place on ice (if planning to boil immediately and load samples on a 10% SDS-PAGE gel), or freeze on dry ice and store at –80°C (if planning to run samples on a gel in future). Cdc2 runs between 30 and 35 kDa (*see* **Notes 13** and **37**).

3.12.2. Cdc2 Antiphospho-Tyrosine Antibody Kinase Assay to Detect Wee Kinase Activity

In this procedure, the activity of purified Wee kinases is measured in vitro using recombinant Cdc2/Cyclin B complex as a substrate ***(8,12,15)***. All Wee kinases phosphorylate a conserved tyrosine residue (Tyr15) on Cdc2. Several commercially available antibodies recognize phosphorylated tyrosine. In this procedure, Western blotting of Wee-treated Cdc2 samples with these antibodies is used to detect and measure Wee kinase activity (*see* **Note 38**). The advantages of this method are that it gives very clean results and that it can be performed using active forms of Cdc2. This is because the only tyrosine kinase present in this semipurified system is the added Wee kinase being measured. In addition, because active forms of Cdc2 may be used, this assay can be combined with the Histone H1 assay to show that the Wee kinase actually inhibits the activity of the Cdc2/Cyclin B complex (*see* **Subheading 3.13.**). The disadvantages of this procedure are that it is not as sensitive and that it is a bit more laborious than measuring Wee activity through the use of incorporated ^{32}P (*see* **Subheading 3.12.1.**). The use of a mutant of Cdc2 that cannot be phosphorylated by the Wee kinases (i.e., Cdc2 Y15F or Cdc2–AF) is a useful control to include (*see* **Subheading 3.5.** and **Fig. 4**, for an example of this assay).

1. Make unlabeled Cdc2/Cyclin B complex(es) (*see* **Subheading 3.6.**).
2. Thaw purified kinase (Wee1, Wee2, or Myt1) to be measured on ice (*see* **Subheadings 3.9., 3.10.**, or **3.11.**).
3. Set up the reaction by adding components to microcentrifuge tubes in the order listed below. Keep tubes on ice throughout. When setting up the reaction, add all of one reagent to all of the samples and then move on to the next reagent. At each addition,

mix the reaction by slowly pipeting up/down a couple of times with the tip used to add the reagent.

a. First: dilute kinase in kinase dilution buffer so that the volume of kinase is 31 L (*see* **Notes 10–13** and **34**).
b. Second: add 2 L of 5X kinase buffer if measuring Wee1/Wee2 *or* 2 L of 5X kinase buffer + 1% Triton X-100 if measuring Myt1.
c. Third: add 3.2 L kinase assay 12.4X ATP/regeneration mix.
d. Fourth: add 3.8 L Cdc2/Cyclin B complex.

The final concentrations in the 40 L reaction will be: 50 m*M* Tris-HCl, pH 7.5, 10 m*M* $MgCl_2$, 1 m*M* DTT, 1 mg/mL ovalbumin, 0.05% Triton X-100 (for assays with Myt1), 1 m*M* ATP, 100 g creatine phosphokinase, 10 m*M* creatine phosphate, plus Cdc2/Cyclin B complex and Myt1, Wee1, or Wee2 kinase.

4. After final addition, mix by gently flicking the tubes, quick-spin 4°C, and incubate for 30 min at room temperature. If the kinase is on beads, rotate reaction tubes, or mix by flicking occasionally.
5. Stop the reaction by placing samples on ice. If desired, drop-freeze 3 L of the reaction in LN_2 for a future Histone H1 kinase assay (*see* **Subheading 3.13.**). To the remaining material, add 1/3 vol 4X sample buffer, flick the tube to mix, and either place on ice (if planning to boil immediately and load samples on 10% SDS-PAGE gels), or freeze on dry ice and store at –80°C (if planning to run samples on gels in future). These samples should be loaded onto two SDS-PAGE gels. On one gel, load approx 85% of the sample. The bottom portion of this gel will be used for detecting the phospho-tyrosine signal and, if desired, the top portion can be used to detect the added Wee kinase. Cdc2 runs between 30 and 35 kDa, and most Wee kinases run at ≥60 kDa. On the other gel, load approx 15% of the sample. This will be used to detect all Cdc2 (phosphorylated or not). At the end of the run, transfer the gels to membranes and perform western analysis with anti-phosphotyrosine (*see* **Notes 13** and **38**), anti-Cdc2 (*see* **Note 39**), and anti-Wee kinase antibodies (if desired).

3.13. Histone H1 Kinase Assay

This procedure measures the activity of Cdc2/Cyclin B complexes *(32)*. This assay is very sensitive and relatively easy to perform. In the cell, this activity of the Cdc2/Cyclin B complex depends on the balance of Wee kinase and the Cdc25 phosphatase activity. However, in the case of recombinant assays in which only the Wee kinases are added (*see* **Subheading 3.12.2.**), the activity of Cdc2/Cyclin B becomes an indirect measure of Wee activity. Although the samples measured in this procedure contain all the reaction components from the Cdc2 kinase assay (*see* **Subheading 3.12.2.**), these components will not interfere with the Histone H1 measurement (*see* **Fig.4** for an example).

1. Thaw samples containing Wee-modified Cdc2/Cyclin B complex (*see* **Subheading 3.12.2.**) on ice. Make serial dilutions of these samples in ice-cold EB buffer (*see* **Note 40**). Keep the samples on ice.
2. Prepare reaction buffer cocktail (make fresh and keep on ice). For each reaction, the following amounts will be needed: 9.08 L 2X HH1 buffer, 0.02 L 100 mM ATP, 0.5 L ^{32}P gATP (*see* **Note 5**), and 0.4 L Histone H1, for 10 L total per reaction.
Multiply these volumes by the number of reactions to be performed, plus one to account for pipeting loss.

3. Aliquot 10 L of the reaction buffer cocktail prepared in step 2 into pre-chilled microcentrifuge tubes. To each, add 10 L of the diluted Cdc2/Cyclin B samples prepared in **step 1**. Mix the reaction by slowly pipeting up/down a few times with the tip used to add the diluted samples (*see* **Note 31**).
4. After exactly 10 min at room temperature (*see* **Note 31**), stop the reaction by adding 8 L of 4X SDS sample buffer to each tube, flick the tube to mix, and then place on ice. The samples should be either immediately boiled and loaded on a 10% SDS-PAGE for analysis or frozen at –80C for future analysis. Histone H1 will run below 30 kDa but above the bromphenol blue marker (*see* **Notes 13** and **37**).

4. Notes

1. The inclusion of low levels of EGTA or imidazole in the lysis and/or wash buffers minimizes the binding of nonspecific proteins to Ni-IDA beads.
2. PMSF is a serine protease inhibitor. **Caution:** use extreme care when working with PMSF, since it is highly toxic. PMSF is poorly soluble in aqueous solutions. Therefore, a concentrated stock of PMSF is made in ethanol first. The concentrated PMSF should be added to a vigorously stirred solution to prevent the PMSF from "crashing out." Since PMSF has a short half-life in aqueous solutions, PMSF should be made fresh and added to all buffers shortly before they are used, and these buffers should be used within 1 or 2 h.
3. PCL is a mixture of protease inhibitors dissolved in DMSO. Use extreme caution when working with PCL, since these compounds are toxic. Before use, warm PCL to room temperature to thaw, then add to solutions, and mix immediately.
4. NaF, Na_3VO_4, and okadaic acid are used in several procedures as phosphatase inhibitors.
5. When using radioactivity, make sure to use appropriate caution, shielding, and monitoring equipment to prevent injury to self or others.
6. Cdc2 is activated by Cyclin B binding and CAK phosphorylation of Cdc2. When making the Cdc2/Cyclin B complex (*see* **Subheading 3.6.**), the CAK is supplied in the crude Sf9 cell lysate containing the Cdc2. If desired, the Cdc2 can be ^{32}P-labeled on Thr161 during this step. Prepare the beads as in **step 1**, but then in **step 2**, add 20 L [γ-^{32}P]ATP (*see* **Note 5**), 1 L 1 *M* $MgCl_2$, and 1 L 25 m*M* "cold" ATP. Cold ATP is required to maintain sufficient ATP for the CAK.
7. We observe better resolution of the various phosphorylated forms of Cdc2 when the loading buffer contains additional DTT.
8. We commonly make cell lysates in an HBS-based buffer (150 m*M* NaCl, 10 m*M* HEPES, pH 7.5), but just about any crude cell extract can be used for this purpose. This includes variations on the lysis procedures used to make recombinant proteins from Sf9 cells (*see* **Subheadings 3.4.**, **3.5.**, **3.10.**, and **3.11.**), crude lysates for the shifting assay (*see* **Subheading 3.7.**), or egg extracts from *Xenopus* ***(33)***. However, there are several factors to consider. If you are attempting to immunoprecipitate Myt1, note the sensitivity of Myt1 to detergents (*see* **Note 28**). If you are attempting to immunoprecipitate Wee1 or Wee2, note that these enzymes are predominantly nuclear. Therefore, it is useful either to sonicate the cells during the lysis procedure or to use a nonionic detergent such as NP-40 to permeablize the nucleus.
9. These wash buffers can be conveniently made by adding concentrated stocks (*see* **Subheading 2.1.**) to chilled EB; the slight dilution of EB will not matter; use 100 mg/mL ovalbumin, 1 *M* NaF, 500 *M* Na_3VO_4, 10% Triton X-100 or 10% NP-40, 500 *M* okadaic acid, and 1000X PCL. Volumes may be scaled up or down as needed, but the ratio of components should be maintained. Note the use of specific detergents for immunoprecipitation of Wee1/Wee2 or Myt1.

10. Concentrations of imidazole of about 45 m*M* or more will interfere with the Cdc2 kinase assay. Imidazole is added to the Cdc2 kinase assay with the Wee kinase (if not dialyzed) and the Cdc2/Cyclin B substrate. Therefore, for example, if the Wee kinase was eluted in 150 m*M* imidazole, up to approx 8 L (before dilution) of kinase may be added to a 40- L reaction. A good control is to use an equivalent volume of elution or dialysis buffer in a "negative-control" reaction.
11. We have not attempted to elute the immunoprecipitated Wee kinase from the antibody or protein A beads but instead use the slurry of beads/kinase in the kinase assays (*see* **Subheadings 3.12.1.** and **3.12.2.**). However, once this slurry is frozen, do not assume that the interaction of antibody/beads/kinase is stable. Therefore, when diluting the immunoprecipitated Wee kinase for the kinase assay, treat the slurry as a whole, and do not repellet the beads. Once the samples are boiled in SDS sample buffer, the Wee kinase is released, and the beads may be discarded.
12. To confirm that the kinase assay is within the linear range (*see* **Note 13**), make serial dilutions of the kinase (two- or threefold steps). The kinase should be diluted in kinase dilution buffer. As described in the procedure, 31 L of each dilution will be needed per reaction. If using kinase that is immobilized on beads, *see* **Note 34**.
13. When in the linear range of the assay, the signal will increase according to the amount of enzyme added. If it does not, determine the linear range empirically.
14. Volumes used in this procedure may be scaled up or down as needed, but the ratio of components and volumes should be maintained.
15. Sf9 cells can be grown both on plates and in suspension. We commonly maintain our noninfected stocks of Sf9 cells in suspension in spinner flasks but perform protein production from adherent cells on plates. In most cases, this results in a better yield of protein. For suspension stock cultures, cells are grown in spinner flasks in the presence of an anticlumping agent such as Pluronic F-68. To keep the suspension stock healthy, cells should be passaged when the density reaches approx 2×10^6/mL. Allowing the stocks to reach densities much higher than approx 3×10^6/mL adversely affects subsequent protein production. Stocks in the spinner flasks can be passaged for up to 4 mo, at which point they should be replaced with a fresh thaw from frozen stocks.
16. Sf9 cells from the suspension stock will settle within 30–60 min and adhere loosely to plates. There is no need to continue growing cells in the presence of Pluronic F-68, but there is no reason to remove the Pluronic F-68 purposely from the cells either; just do not add any more Pluronic F-68 with subsequent media changes. Once plated, cells are allowed to grow for 24 h. This will permit one cell doubling, and the cells will be at a density of 2×10^7 per 10-cm plate at the time of viral infection.
17. Because Sf9 cells are easily disrupted by a jet of media from a 10-mL pipet, add media slowly to adherent Sf9 cells. Initially infecting the cells for 1 h in a minimal volume of media as described increases the yield of recombinant protein in our hands.
18. Do not be alarmed if by the end of this infection (48 h), most of the cells have detached from the plates. It is not recommended that the infection be allowed to proceed beyond 48 h, since the yield of active recombinant protein may decrease.
19. There is no need to maintain aseptic tissue culture technique while collecting the injected cells. However, with the exception of the initial scraping of cells, all these steps should be performed on ice with ice-cold solutions.
20. Resuspend the cell pellet from the washed, infected cells (*see* **Subheading 3.3.**) in 3 mL of protein-specific lysis buffer-PMSF-PCL by gently pipeting up and down. Transfer to an ice-cold Dounce tissue grinder. The cells should break open in 20 or 40 strokes. Microscopically examine the cells to be sure they are disrupted. It is helpful to look at the

cells before and after douncing, but note that Sf9 cells have a large nucleus that may remain intact. Once the cells are disrupted, transfer the mixture to an ice-cold 13-mL Starstedt tube. Rinse the Dounce and pestle with a fresh, 3-mL aliquot of protein-specific lysis buffer-PMSF-PCL, and then pool this rinse with the original cell lysate. Mix by inversion, and drop-freeze 5 L of this initial lysate in LN_2 for a test gel. (Store the test sample at –80°C until use.)

21. Spin the lysate at 16,000*g* for 20 min at 4°C. A Sorvall HB4 or HB6 rotor (10,000 rpm) with a rubber adapter (Sorvall, cat. no. 00363) works well for this purpose. Note: do not completely fill the Starstedt tubes during this spin (~10 mL is the upper limit), or they may break at the *g* force and temperature used. At the completion of this spin, transfer and save the supernatant to a tube chilled on ice; drop-freeze 5 L of this supernatant in LN_2 for a test gel. (Store the test sample at –80°C until use.)
22. Resuspend the 20% Ni-IDA sepharose beads (*see* **Subheading 3.1.**) by gentle inversion to make a slurry, and then transfer five times the volume desired for the packed beads to a clean centrifuge tube (i.e., if 1 mL of packed beads is needed, transfer 5 mL of the 20% slurry). Pack the beads by spinning in a centrifuge at 1,600*g* for 30 s. Batch-wash the beads a total of three times with 10 vol of protein-specific lysis buffer-PMSF each. After the last wash, remove as much liquid as possible, and then chill the washed beads on ice.
23. This step should be done at 4°C in a cold room or equivalent, all material (tubes, pipet tips, and so on) should be prechilled to 4°C, and all solutions and samples should be kept on ice.
24. The volumes listed here are a good starting point. However, for each form of Cdc2 and each preparation of Cdc2 (*see* **Subheading 3.5.**), the optimal amount of Cdc2 added to the Cyclin B beads and the optimal volume of the Cdc2 complex elution buffer will need to empirically determined (*see* **Subheading 3.6.**). This is particularly important when comparing different forms of Cdc2 in subsequent kinase assays (*see* **Subheadings 3.12.1.** and **3.12.2.**). For some forms of Cdc2, the level of Cdc2 expressed in the Sf9 cells is particularly low (for example, Cdc2-AF; *see* **Fig. 3A**, lanes 5 and 6). In these cases, a prebinding step is useful when making the complex (*see* **Subheading 3.6.**). Briefly, a 500 L aliquot of the cleared Cdc2 lysate is pre-bound to the standard amount of Cyclin B beads (10 L packed) by rotation at 4°C, however this is performed in the absence of $MgCl_2$ or ATP. The immobilized Cdc2/Cyclin B complex is recovered by centrifugation, the supernatant is discarded, and then the prebound complex is incubated with the standard 70- L aliquot of matching Cdc2, $MgCl_2$, and ATP as described in **step 2, Subheading 3.6.**
25. For convenience, we commonly do this step at room temperature, but keep solutions and samples on ice when they are not in the centrifuge.
26. To batch-wash beads in microcentrifuge tubes, add wash buffer, resuspend beads by flicking the tube with finger and inverting, and then immediately respin (10 s, ~1600*g*, 4°C)
27. Prepare egg or oocyte extracts as described ***(33)***, but then dilute these with 2 vol of EDB to make them the equivalent to the total cell extract (**step 2**). The extract can then be aliquoted and/or separated into membrane and soluble fractions as described (*see* **Subheading 3.7.**).
28. Triton X-100 is required both to solublize Myt1 from the membrane and to enhance Myt1 activity. The use of other detergents including CHAPS, Zwittergent 3-14, deoxycholic Acid, and NP-40 has been shown to inhibit *Xenopus* Myt1 activity (P. R. Mueller and W. G. Dunphy, unpublished data).
29. MWB uses high salt to wash off proteins that are nonspecifically associated with the membrane.

30. The desired time-course will determine the number of tubes to prepare. As a start, we suggest the following time course: 0, 2, 4, 8, 16, 32, and 64 min. In addition, for the 0-min time-point, it is useful to prepare a double reaction. This way, the 0-min sample can be loaded on each side of the timed progression (*see* **Fig. 5**).
31. To control the timing of the reaction, it is helpful to stagger the start/stop time of each reaction by 30 s. This will ensure that each sample is treated equally while providing sufficient time to handle the samples.
32. The optimal amount of antibody required to immunoprecipitate the Wee kinase might need to be empirically determined. If the extract is particularly viscous, it helps to dilute the extract in the appropriate lysis buffer. The addition of detergent aids in the accessibility of the antibody to the antigen and helps to minimize nonspecific background. However, note the use of specific detergents for immunoprecipitation of Wee1/Wee2 or Myt1. A useful control to include is an immunoprecipitation with a nonspecific antibody that is raised in the same species as the primary antibody. For example, when working with a rabbit polyclonal antibody, use an IgG, whole molecule rabbit polyclonal antimouse (ICN, cat. no. 55480). This should give no reaction in subsequent Western blotting or kinase assays.
33. The beads swell by a factor of approx 4. Therefore, if 12.5 g of dried beads are weighed out, they will swell to 50 L. This is sufficient for four reactions, leaving 20% extra for pipeting loss.
34. When pipeting samples that contain beads, gently mix beads by pipeting up and down with a pipetman before each pipeting. This will ensure that equal amounts of slurry are pipeted. Do not use a narrow-bore pipet tip to pipet bead solutions.
35. The yield of *Xenopus* Myt1 protein is particularly poor, only approx 10–20 ng per 10-cm plate. We have not tried Myt1 from other species, so we do not know whether this is a universal problem. From this size prep, we have not been able to recover enough Myt1 protein to perform Bradford protein determinations. Therefore use antibody and/or gel standards to determine the concentration/yield of Myt1. We typically find that much of the Myt1 protein does not bind to the beads, and some does not elute from the beads. In our hands, increasing the amount of beads or imidazole concentration does not improve either of these problems. Fortunately, the recombinant Myt1 has abundant activity. Because the concentration of Myt1 is low, we typically add ovalbumin to the experimental stock. This helps prevent loss owing to nonspecific sticking or proteolysis.
36. Although douncing works (*see* **Note 20**), we get a greater yield of Wee1 and Wee2 protein by using a sonicator to disrupt the cells. This is probably because Wee1 and Wee2 are nuclear proteins. Avoid cavitation, and keep the cell suspension ice-cold during the sonication. The following has worked well in our hands using a Branson model 450 Sonifer® with a microtip: sonicate in a 4°C cold room with two sets of 20 bursts at 45% power. Between each set of bursts, keep cells on ice for 2 min. Also, during the burst cycles, the tube containing the cells should be surrounded by ice. At the end of the sonication, microscopically examine the cells to be sure they are disrupted; the nucleus should not be intact. Once the cells are disrupted, drop-freeze 5 L of this initial lysis in LN_2 for a test gel (store the test sample at –80°C until use).
37. Do not allow the bromophenol blue marker to run off the gel or the free [γ-^{32}P]ATP will run into the lower buffer chamber. Instead, stop the run when the bromophenol blue is 1–4 cm from the bottom of the gel. Carefully cut the gel with a razor blade approx 1 mm above the bromophenol blue. Discard the lower portion of the gel containing the bromophenol blue and a majority of the unincorporated radioactive ATP into radioactive

waste. The top portion of the gel can be transferred to membrane or dried down to reveal the labeled substrate with a PhosphorImager or X-ray film.

38. We have found the anti-phospho-Tyr antibody (4G10, Upstate, cat. no. 05-321) to work well for detecting tyrosine-phosphorylated Cdc2 in this assay. We use the antibody at a 1:1000 dilution in 3% BSA-TSBT. We do not recommend the use of a Cdc2-specific, anti-phospho-Tyr15 antibodies (i.e., Cell Signaling, cat. no. 9111; *see* **Fig. 1** and **Note 41**), because they are not as sensitive, and their specificity is not needed in this semipurified in vitro assay.
39. A variety of antibodies may be used to detect all forms of Cdc2. We routinely use an antibody that detects the PSTAIR sequence (for example, Santa Cruz Biotech, cat. no. SC-53) that is conserved in Cdc2 from yeast to human. Alternatively, a more specific Cdc2 antibody may be used.
40. A suggested series of dilutions is: 1:25 (2 L of the sample from **step 5**, **Subheading 3.12.2.** + 48 L EB buffer), 1:50 (50% dilution of 1:25 dilution in EB buffer), and 1:100 (50% dilution of 1:50 dilution in EB buffer). In most cases, this will be in the linear range of the assay (*see* **Note 13**).
41. We have found that the phospho-Cdc2 (Tyr15) antibody from Cell Signaling (cat. no. 9111) works well for detecting phospho-Tyr15 Cdc2 in a crude lysate. One of its advantages is that it recognizes phosphorylated Cdc2 from a wide variety of eukaryotes. However, it is critical to use the antibody exactly as described by the manufacturer when performing Western blots. Deviation from the recommended wash and incubation times or the buffer conditions will result in suboptimal results. We use this antibody at a 1:500 dilution.

Acknowledgments

We would like to thank Thomas Coleman, William Dunphy, Akiko Kumagai, and Marc Lake for their help and input in developing these methods and Reshma Relwani and Troy McSherry for their critical comments on the manuscript. This work is supported in part by a grant from the National Institutes of Health (1 RO1 CA84007 to P. R. M.).

References

1. Morgan, D. O. (1997) Cyclin-dependent kinases: engines, clocks, and microprocessors. *Annu. Rev. Cell Dev. Biol.* **13,** 261–291.
2. Dunphy, W. G. (1994) The decision to enter mitosis. *Trends Cell. Biol.* **4,** 202–207.
3. Lew, D. J. and Kornbluth, S. (1996) Regulatory roles of cyclin dependent kinase phosphorylation in cell cycle control. *Curr. Opin. Cell Biol.* **8,** 795–804.
4. OFarrell, P. H. (2001) Triggering the all-or-nothing switch into mitosis. *Trends Cell. Biol.* **11,** 512–519.
5. Russell, P. and Nurse, P. (1987) Negative regulation of mitosis by wee1+, a gene encoding a protein kinase homolog. *Cell* **49,** 559–567.
6. Booher, R. N., Deshaies, R. J., and Kirschner, M. W. (1993) Properties of *Saccharomyces cerevisiae* weel and its differential regulation of p34CDC28 in response to G1 and G2 cyclins. *EMBO J.* **12,** 3417–3426.
7. Igarashi, M., Nagata, A., Jinno, S., Suto, K., and Okayama, H. (1991) Wee1(+)-like gene in human cells. *Nature* **353,** 80–83.

8. Mueller, P. R., Coleman, T. R., and Dunphy, W. G. (1995) Cell cycle regulation of a *Xenopus* Wee1-like kinase. *Mol. Biol. Cell* **6,** 119–134.
9. Watanabe, N., Broome, M., and Hunter, T. (1995) Regulation of the human WEE1Hu CDK tyrosine 15-kinase during the cell cycle. *EMBO J.* **14,** 1878–1891.
10. Campbell, S. D., Sprenger, F., Edgar, B. A., and OFarrell, P. H. (1995) *Drosophila* Wee1 kinase rescues fission yeast from mitotic catastrophe and phosphorylates *Drosophila* Cdc2 in vitro. *Mol. Biol. Cell* **6,** 1333–1347.
11. Wilson, M. A., Hoch, R. V., Ashcroft, N. R., Kosinski, M. E., and Golden, A. (1999) A *Caenorhabditis elegans* wee1 homolog is expressed in a temporally and spatially restricted pattern during embryonic development. *Biochim. Biophys. Acta* **1445,** 99–109.
12. Mueller, P. R., Coleman, T. R., Kumagai, A., and Dunphy, W. G. (1995) Myt1: a membrane-associated inhibitory kinase that phosphorylates Cdc2 on both threonine-14 and tyrosine-15. *Science* **270,** 86–90.
13. Liu, F., Stanton, J. J., Wu, Z., and Piwnica-Worms, H. (1997) The human Myt1 kinase preferentially phosphorylates Cdc2 on threonine 14 and localizes to the endoplasmic reticulum and Golgi complex. *Mol. Cell. Biol.* **17,** 571–583.
14. Booher, R. N., Holman, P. S., and Fattaey, A. (1997) Human Myt1 is a cell cycle-regulated kinase that inhibits Cdc2 but not Cdk2 activity. *J. Biol. Chem.* **272,** 22300–22306.
15. Leise, W. F. 3rd and Mueller, P. R. (2002). Multiple Cdk1 inhibitory kinases regulate the cell cycle during development. *Dev. Biol.* **249,** 156–173.
16. Parker, L. L. and Piwnica-Worms, H. (1992) Inactivation of the p34cdc2-cyclin B complex by the human WEE1 tyrosine kinase. *Science* **257,** 1955–1957.
17. McGowan, C. H. and Russell, P. (1993) Human Wee1 kinase inhibits cell division by phosphorylating p34cdc2 exclusively on Tyr15. *EMBO J.* **12,** 75–85.
18. Kornbluth, S., Sebastian, B., Hunter, T., and Newport, J. (1994) Membrane localization of the kinase which phosphorylates p34cdc2 on threonine 14. *Mol. Biol. Cell* **5,** 273–282.
19. McGowan, C. H. and Russell, P. (1995) Cell cycle regulation of human WEE1. *EMBO J.* **14,** 2166–2175.
20. Nakanishi, M., Ando, H., Watanabe, N., et al. (2000) Identification and characterization of human Wee1B, a new member of the Wee1 family of Cdk-inhibitory kinases. *Genes Cells* **5,** 839–847.
21. Okamoto, K., Nakajo, N., and Sagata, N. (2002) The existence of two distinct Wee1 isoforms in *Xenopus*: implications for the developmental regulation of the cell cycle. *EMBO J.* **21,** 2472–2484.
22. Murakami, M. S. and Vande Woude, G. F. (1998) Analysis of the early embryonic cell cycles of *Xenopus*; regulation of cell cycle length by Xe-wee1 and Mos. *Development* **125,** 237–248.
23. Kumagai, A. & Dunphy, W. G. (1995) Control of the Cdc2/cyclin B complex in *Xenopus* egg extracts arrested at a G2/M checkpoint with DNA synthesis inhibitors. *Mol. Biol. Cell* **6,** 199–213.
24. Murphy, C. I., Piwnica-Worms, H., Grüwald, S., and Romanow, W. G. (1997) Maintenance of insect cell cultures and generation of recombinant baculoviruses, in *Current Protocols in Molecular Biology* (Ausubel, F. M., Brent, R., Kingston, R. E., et al., eds.), John Wiley & Sons, New York, pp. 16.10.1–16.10.17.
25. Murphy, C. I., Piwnica-Worms, H., Grüwald, S., and Romanow, W. G. (1997) Expression and purification of recombinant proteins using the baculovirus system, in *Current Protocols in Molecular Biology* (Ausubel, F. M., Brent, R., Kingston, R. E., et al., eds.), John Wiley & Sons, New York, pp. 16.11.1–16.11.12.

26. Murphy, C. I., Piwnica-Worms, H., Grüwald, S., and Romanow, W. G. (1997) Overview of the baculovirus expression system, in *Current Protocols in Molecular Biology* (Ausubel, F. M., Brent, R., Kingston, R. E., et al., eds.), John Wiley & Sons, New York, pp. 16.9.1–16.9.10.
27. Solomon, M. J., Lee, T., and Kirschner, M. W. (1992) Role of phosphorylation in p34cdc2 activation: identification of an activating kinase. *Mol. Biol. Cell* **3,** 13–27.
28. Edgar, B. A., Sprenger, F., Duronio, R. J., Leopold, P., and OFarrell, P. H. (1994) Distinct molecular mechanism regulate cell cycle timing at successive stages of *Drosophila* embryogenesis. *Genes Dev.* **8,** 440–452.
29. Petty, K. J. (1996) Metal-chelate affinity chromatography, in *Current Protocols in Protein Science* (Coligan, J. E., Dunn, B. M., Ploegh, H. L., Speicher, D. W., and Wingfield, P. T., eds.), John Wiley & Sons, New York, pp. 9.4.1–9.4.16.
30. Palmer, A., Gavin, A. C., and Nebreda, A. R. (1998) A link between MAP kinase and p34(cdc2)/cyclin B during oocyte maturation: p90(rsk) phosphorylates and inactivates the p34(cdc2) inhibitory kinase Myt1. *EMBO J.* **17,** 5037–5047.
31. Coleman, T. R., Tang, Z., and Dunphy, W. G. (1993) Negative regulation of the wee1 protein kinase by direct action of the nim1/cdr1 mitotic inducer. *Cell* **72,** 919–929.
32. Dunphy, W. G. and Newport, J. W. (1989) Fission yeast p13 blocks mitotic activation and tyrosine dephosphorylation of the *Xenopus* cdc2 protein kinase. *Cell* **58,** 181–191.
33. Murray, A. W. (1991) Cell cycle extracts. *Methods Cell Biol.* **36,** 581–605.
34. Nakajo, N., Yoshitome, S., Iwashita, J., et al. (2000) Absence of Wee1 ensures the meiotic cell cycle in *Xenopus* oocytes. *Genes Dev.* **14,** 328–338.

19

CDC25 Dual-Specificity Protein Phosphatases

Detection and Activity Measurements

Sorab N. Dalal and Melanie Volkening

Summary

Most cyclin-dependent kinases are negatively regulated by phosphorylation of two residues, a threonine at residue 14 and a tyrosine at residue 15. These residues are dephosphorylated by the cdc25 family of dual-specificity phosphatases leading to cell cycle progression. These phosphatases are inactivated by cellular checkpoint pathways in response to DNA damage leading to cell cycle arrest. Checkpoint pathways regulate the function of these phosphatases by regulating their stability, localization, association with substrate, and their activity. Hence, determining these properties for the cdc25 family of phosphatases becomes crucial for understanding how checkpoint pathways regulate the function of the cdc25 family members and, hence, cell cycle progression. This chapter describes methods to determine the activity, levels, phosphorylation status, and localization of both endogenous and overexpressed cdc25 proteins.

Key Words

cdc25; cell cycle progression; phosphatase; phosphorylation; RNA interference.

1. Introduction

Most cyclin dependent kinases (cdks) are negatively regulated by phosphorylation at two residues: a threonine at position 14 (T14) and a tyrosine at position 15 (Y15) (reviewed in **refs.** ***1*** and *2*). These residues are dephosphorylated by the cdc25 family of dual-specificity phosphatases. Three cdc25 family members have been cloned in mammalian cells. cdc25C was the first human phosphatase cloned on the basis of its homology to the yeast and *Drosophila* cdc25 proteins ***(3,4)***. Two other human cdc25 homologs, cdc25A and cdc25B, were cloned later based on their ability to rescue a temperature-sensitive cdc25 mutant in fission yeast ***(5)***. cdc25A can

From: *Methods in Molecular Biology, vol. 296, Cell Cycle Control: Mechanisms and Protocols*
Edited by: T. Humphrey and G. Brooks © Humana Press Inc., Totowa, NJ

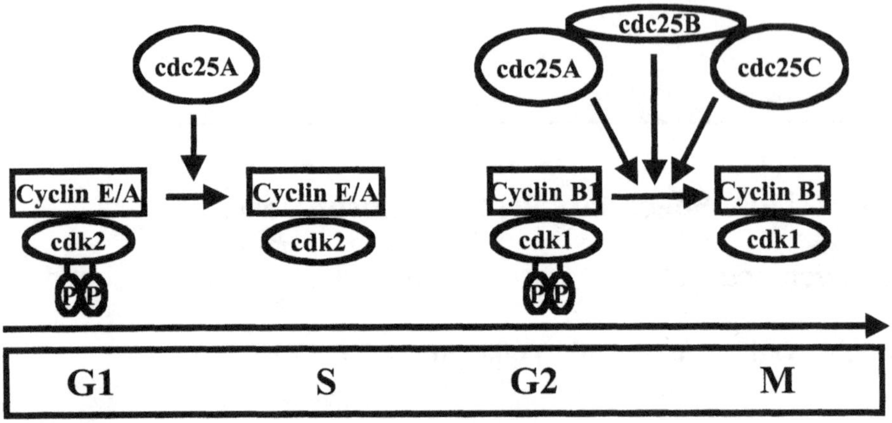

Fig. 1. Regulation of cell cycle transitions by the different cdc25 family members. cdc25A has been shown to activate both cyclin E/cdk2 and cyclin A/cdk2 complexes, leading to S-phase progression, whereas all cdc25 family members can activate cyclin B1/cdk1, thus leading to mitotic progression.

dephosphorylate and activate cdk2 in vitro ***(6)***, which correlates with cdc25A's ability to promote the G_1/S transition, when overexpressed in mammalian cells ***(6–8)*** (**Fig. 1**). All three cdc25 family members may play distinct roles in promoting progression from G_2- to M-phase, as they can all dephosphorylate and activate cdk1/cyclin B1 in vitro ***(4,6,9,10)*** (**Fig. 1**). The depletion of different cdc25 family members from cell lysates leads to a decrease in the ability of the extracts to activate cyclin B1/cdk1 in vitro ***(10)***. The downregulation of cdc25A using RNA interference delays mitotic progression ***(10)***. Overexpression of cdc25B can lead to premature mitotic progression and centrosomal abnormalities ***(11–13)***, whereas overexpression of cdc25C leads to premature mitotic progression during S-phase and the override of a DNA damage-induced G_2 arrest ***(14,15)***.

The cdc25 family members are regulated by cell cycle-dependent phosphorylation events. Hyperphosphorylation of cdc25C prior to mitosis results in a stimulation of its phosphatase activity ***(16–18)***. In vitro cdc25C can be phosphorylated by its substrate cyclin B1/cdk1 ***(16,17,19)***, and in vivo it is phosphorylated on Ser198 by Plk1, which regulates its nuclear translocation during prophase ***(20)***. Cdc25B protein accumulates and is phosphorylated prior to mitosis. The mitotic phosphorylation of cdc25B is thought to be required for the stimulation of cdc25B phosphatase activity ***(12)***. The cyclin B1/cdk1 complex phosphorylates cdc25A on two serine residues, S17 and S115, resulting in the stabilization of cdc25A and mitotic progression ***(10)***.

All eukaryotic cells have evolved mechanisms to prevent cell cycle progression in the presence of DNA damage or incomplete S-phase (reviewed in **ref. *21***). DNA damage activates a series of responses, one of which is cell cycle arrest. The cell cycle arrest that occurs in response to DNA damage is dependent on the inhibition of cdc25 function. One mechanism by which cell cycle checkpoint pathways regulate cdc25 function is by inducing cell cycle-dependent phosphorylation of the cdc25 proteins, which result in an inhibition of cdc25 function ***(6,15,22,23)***. cdc25A can be phosphorylated on a serine residue at position 123 (S123) by the checkpoint kinases chk1 and chk2 in mammalian cells in response to either UV light or γ-radiation. The DNA damage-mediated phosphorylation of cdc25A at S123 leads to its ubiquitin-mediated proteolysis and cell cycle arrest ***(6,22)***, which can be abrogated by the ectopic overexpression of cdc25A ***(10)***. Checkpoint pathways also inhibit cdc25C function. Both chk1 ***(15)*** and chk2 ***(23)*** can phosphorylate cdc25C on a serine residue at position 216, which results in an inhibition of cdc25C function. Arsenite-induced DNA damage can induce the turnover of cdc25C by the ubiquitin pathway and is dependent on the presence of a KEN-box in cdc25C ***(24)***. cdc25B, like other cdc25 family members, is a target of the DNA damage checkpoint. Interestingly, cdc25B is not phosphorylated by the checkpoint kinases chk1 and chk2 in response to UV irradiation but by the mitogen-activated protein (MAP) kinase p38 ***(25)***. Phosphorylation of cdc25B by p38 in response to UV irradiation seems to be a critical event in the initiation of a G_2 delay. Finally, irradiation of cells with UV also blocks the premitotic accumulation and phosphorylation of cdc25B ***(12)***.

Phosphorylation of the S216 residue in cdc25C results in the generation of a consensus binding site for 14-3-3 proteins ***(15,26,27)***. cdc25C is a cytoplasmic protein during interphase and enters the nucleus just prior to mitosis ***(14)***, a phenotype that is very similar to its substrate cyclin B1/cdk1 ***(28,29)***. 14-3-3 proteins regulate cdc25C function in part by regulating its cellular localization, as 14-3-3 binding mutants of cdc25C show a pan-cellular localization, which is different from the cytoplasmic localization observed for the wild-type protein ***(14)***. In addition to regulating the cytoplasmic localization of cdc25C, 14-3-3 proteins may also inhibit complex formation between cdc25C and cyclin B1, thereby preventing activation of cyclin B/cdk1 ***(30)***. Cdc25B also binds to 14-3-3 proteins ***(31,32)***. 14-3-3 binding mutants of cdc25B show an altered cellular localization ***(33)***, and 14-3-3 binding inhibits the association of cdc25B with the mitotic cyclin/cdk complex ***(34)***. The phosphorylation of cdc25B by p38 in response to UV irradiation also results in the formation of a 14-3-3 binding site on cdc25B ***(25)***. chk1 phosphorylates cdc25A on S178 and T507 to facilitate 14-3-3 binding. 14-3-3 binding to the C-terminal cycin B1 binding domain of cdc25A blocks functional interaction between cdc25A and cyclin B1-cdk1 ***(35)***.

As described above, cdc25 family members are regulated by phosphorylation, which dramatically affects their activity, stability, and intracellular localization. In this chapter we have tried to cover techniques to measure cdc25 phosphatase activity as well as techniques used to determine cdc25 levels, phosphorylation status, and subcellular localization.

2. Materials

2.1. Measuring cdc25 Phosphatase Activity

1. pGEX expression vectors (Amersham).
2. *E. coli* host strains: XL1-blue for plasmid propagation and BL21 (DE3) for protein expression (Stratagene).

3. Luria Bertani (LB) medium.
4. Ampicillin (Sigma).
5. Isopropyl-D-thiogalactopyranoside (IPTG; Sigma), 1 *M* stock solution stored at –20°C).
6. Glutathione sepharose CL-4B (Amersham).
7. Single-use columns (PolyPrep columns, Bio-Rad).
8. *E. coli* lysis buffer: 50 m*M* Tris-HCl, pH 7.5, 250 m*M* NaCl, 1 m*M* EDTA, 0.1% Triton X-100, 0.1 m*M* phenylmethylsulfonlyl fluoride (PMSF), 10 mg/mL soybean trypsin inhibitor, 1 mg/mL aprotinin.
9. Sodium dodecyl sulfate polyacrylamide gel electrophoresis (SDS-PAGE) equipment.
10. Coomassie brilliant blue R 250 (Sigma).
11. Fluorescein-diphosphate (FDP; Molecular Probes), 10 m*M* stock solution stored at –20°C).
12. Fluorescence spectrophotometer (F-2000, Hitachi).
13. FDP buffer: 20 mM Tris-HCl, pH 8.5, 75 m*M* NaCl, 0.57 m*M* EDTA, 0.033% bovine serum albumin (BSA), 1 m*M* dithiothreitol (DTT).
14. Eukaryotic cell lysis buffer: 50 m*M* Tris-HCl, pH 7.4, 0.25 *M* NaCl, 50 m*M* NaF, 0.1% Triton X-100, 5 m*M* EDTA, 1 m*M* DTT, 1 m*M* Na_3VO_4, 2 m*M* PMSF, 10 mg/mL soybean trypsin inhibitor, 10 mg/mL L-1-chlor-3-(4-tosylamido)-4-phenyl-2-butanon [TPCK], 5 mg/mL L-1-chlor-3-(4-tosylamido)-7-amino-2-heptanon-hydrochloride (TLCK), 1 mg/mL aprotinin. (All protease and phosphatase inhibitors are added freshly before use.)
15. Cl-4B Sepharose (Amersham).
16. Protein A or G-Sepharose (Amersham).
17. Phosphatase assay buffer (prepared freshly): 40 m*M* Tris-HCl, pH 8.0, 300 m*M* NaCl, 10 m*M* EDTA, 10 m*M* DTT.
18. Histone H1 kinase buffer (prepared freshly): 50 m*M* Tris-HCl, pH 7.5, 10 m*M* $MgCl_2$, 1 m*M* DTT, 50 μ*M* ATP, 5 mCi [γ-^{32}P]ATP, 10 μg Histone H1 (Roche) per reaction.
19. Coomassie blue stain solution: 2.5 g/L Coomassie brilliant blue R-250 (Sigma), 400 mL methanol, 100 mL acetic acid, 500 mL H_2O.
20. Coomassie blue destain solution: 400 mL methanol, 100 mL acetic acid, 500 mL H_2O.
21. Antibodies.
    a. Anti-cdc25A (F6, Santa Cruz Biotechnology, cat no. #7389).
    b. Anti-cdc25B (H-85, Santa Cruz Biotechnology, cat. no. #5619).
    c. Anti-cdc25C (H-6, Santa Cruz Biotechnology, cat. no. sc #13138).
    d. Anti-cyclinB1 (GNS1, Santa Cruz Biotechnology, cat. no. #245).

2.2. Detection and Localization of Human cdc25 Isoforms in Human Cells

1. The human osteosarcoma cell line, U-2OS, and the primary diploid human fibroblast strains, MRC-5 and WI-38, were obtained from the ATCC and are cultured in Dulbecco's modified Eagle's medium (DMEM; Cellgro) supplemented with 10% Fetal Clone-I serum (Hyclone), 100 U penicillin per mL, and 100 μg streptomycin per mL. The cells are grown in the presence of 10% CO_2.
2. Transfection reagents. All transfection reagents should be tissue culture grade from Sigma or any other manufacturer.
    a. 2X BBS (BES-buffered solution), pH 6.95: 50 m*M* BES, 280 m*M* NaCl, 1.5 m*M* Na_2HPO_4 (*see* **Note 1**).
    b. 0.5 *M* $CaCl_2$.
3. Antibodies (*see* **Note 2**).
    a. Anti-cdc25C: monoclonal antibodies TC14, TC15, TC19, and TC113 (Upstate Biotechnology and Neomarkers).

   b. Anti-cdc25A and cdc25B (*see* **Subheading 2.1.**).
   c. Anti-phospho-histone H3 antibody (Upstate Biotechnology).
   d. Anti-myc monoclonal antibody 9E-10 and anti-myc polyclonal antibody A-14 (Santa Cruz Biotechnology).
   e. Anti-HA polyclonal antibody 12CA5 (Roche).
   f. MPM2 monoclonal antibody (Upstate Biotechnology).
   g. Anti-rabbit/mouse horseradish peroxidase (HRP) conjugates (Pierce or Amersham).
   h. Anti-rabbit/mouse fluorescent conjugates (Boehringer Mannheim).
   i. Anti-cyclin B1 monoclonal CB169 (Upstate Biotechnology).
   j. Anti-Rb MAb 245 (Pharmingen).
4. Protein A-sepharose (Amersham).
5. EBC buffer: 50 m*M* Tris-HCl, pH 8.0, 120 m*M* NaCl, 0.5% Nonidet P-40 [NP-40], 10 mg/mL of aprotinin, 10 mg/mL of leupeptin, 0.1 m*M* PMSF, 50 m*M* NaF, 1 m*M* sodium orthovanadate, 1 m*M* EDTA. (All protease and phosphatase inhibitors are added freshly before use.)
6. Whole cell extract (WCE) buffer: 50 m*M* Tris-HCl, pH 8.0, and 2% SDS.
7. 1X SDS-PAGE sample buffer (1X SB): 50 m*M* Tris-HCl, pH 6.8, 10% glycerol, 2% SDS, 0.1% bromophenol blue, 10 m*M* DTT. Make aliquots and freeze at –20°C.
8. NET-N buffer: 20 m*M* Tris-HCl, pH 8.0, 100 m*M* NaCl, 1 m*M* EDTA, pH 8.0, 0.5% NP-40.
9. 3X SDS-PAGE sample buffer (3X SB): 150 m*M* Tris-HCl, pH 6.8, 30% glycerol, 6% SDS, 0.3% bromophenol blue, 30 m*M* DTT. Make aliquots and freeze at –20°C.
10. TBS-T: 10 m*M* Tris-HCl pH 8.0, 150 m*M* NaCl, 0.5% Tween-20.
11. 1X Transfer buffer: 12.11 g Tris base, 57.6 g glycine, 800 mL methanol, 4 mL 10% SDS. Add dH_2O to 4 L.
12. 1X Running buffer for SDS-PAGE gels: 3.03 g Tris base, 18.77 g glycine, 1 g SDS. Add dH_2O to 1 L.
13. ECL detection reagent (Pierce or Amersham).
14. 30% Acrylamide (Protogel or Bio-Rad).
15. Buffer A for cellular fractionation: 10 mM HEPES, pH 7.9, 10 m*M* KCl, 1 mM EDTA, 0.1 m*M* EGTA, 1 m*M* DTT, 10 µg/mL of aprotinin, 10 µg/mL of leupeptin, 0.1 m*M* PMSF, 50 m*M* NaF, 1 m*M* sodium orthovanadate (all protease and phosphatase inhibitors added fresh before use).
16. Additional phosphatase inhibitors for fractionation: 10 µ*M* cypermethrin, 200 µ*M* dephostatin, 200 n*M* okadaic acid, 25 n*M* tautomycin.
17. 1X Phosphate buffered saline (PBS) for immunofluorescence: 8 g NaCl, 0.2 g KH_2PO_4, 0.2 g KCl, 2.18 g Na_2HPO_4.
18. 4% Paraformaldehyde in PBS: add the paraformaldehyde (Sigma) to PBS, and heat to 65°C to dissolve the paraformaldehyde. Cool and store at 4°C.
19. 0.3% Triton X-100 in PBS
20. 0.1% NP-40 in PBS
21. N-PER fractionation kit from Pierce: add 10 mg/mL of aprotinin, 10 mg/mL of leupeptin, 0.1 m*M* PMSF, 50 m*M* NaF, 1 m*M* sodium orthovanadate, 1 m*M* EDTA. (All protease and phosphatase inhibitors are added fresh before use).
22. 5 µg/mL 4',6-Diamidino-2-phenylindone (DAPI) in PBS.
23. BSA.
24. Citifluor mounting fluid (Calbiochem).

25. Slides and cover slips.
26. Bio-Rad protein assay reagent.
27. Prestained markers (Bio-Rad, Invitrogen, Sigma).
28. Nitrocellulose membrane (Amersham, Bio-Rad).
29. Protein gel and transfer apparatus, power packs.
30. Skim milk powder.
31. Goat serum (Invitrogen).
32. Confocal microscope with the appropriate filters.

3. Methods

3.1. Measuring cdc25 Phosphatase Activity

3.1.1. Expression of Recombinant GST-cdc25A,B,C in E. coli

1. 1 µg of pGEX-cdc25A, B, or C plasmid, respectively, is transformed in chemically or electrocompetent *E. coli* BL21 DE3.
2. One single colony of the transformants is used to inoculate 50–100 mL of LB medium supplemented with 100 µg/mL ampicillin.
3. The *E. coli* culture is grown overnight at 37°C and shaken vigorously.
4. The following day, the overnight culture is diluted 1:20 with fresh LB medium (containing 100 mg/mL ampicillin) and further incubated at 37°C for approx 3 h until OD_{600} = 0.4–0.5.
5. The culture is then cooled down to 30°C, and IPTG (Sigma) is added to a final concentration of 0.5 m*M*.
6. The culture is further incubated for 3 h at 30°C under shaking and then harvested by centrifugation for 30 min at 3000 rpm (Sorvall, GS3 rotor) at 4°C
7. The bacterial pellets can be stored at –20°C until protein isolation.

3.1.2. Purification of GST-cdc25A,B,C from E. coli

Recombinant GST-cdc25A, B, C fusion proteins can easily be purified by coupling them to glutathione *S*-transferase (GST), a 26-kDa protein that efficiently binds to glutathione sepharose ***(36)***.

1. The bacterial pellets containing the induced GST fusion proteins are resuspended in 10 ml lysis buffer (per liter over day culture) and incubated on ice for 15 min.
2. The cells are then lysed by sonification four times 30 s, on ice.
3. The lysate is further incubated on ice for 5 min, and Triton X-100 is added to a final concentration of 1%. Cell debris is removed by centrifugation for 30 min at 14,000 rpm in the Sorvall SS-34 rotor at 4°C.
4. Glutathione sepharose beads (2 mL; Amersham) are washed with 10 mL lysis buffer.
5. The beads are centrifuged at 3000 rpm in a Megafuge R 1.0 (Heraeus) for 2 min at 4°C, and the supernatant is removed.
6. The bacterial supernatant is added to the beads and incubated on a wheel either at 4°C for a minimum of 30 min or for 10 min at room temperature.
7. Beads with bound GST-cdc25 fusion proteins are centrifuged as above, the supernatant is removed, and the beads are washed twice with 30 mL of ice-cold lysis buffer.
8. After the last wash, the beads are resuspended in 10 mL of lysis buffer and loaded on a single-use column (Bio-Rad).

9. The column is washed with 40 mL cold lysis buffer followed by 10 mL of ice-cold 50 m*M* Tris-HCl, pH 7.5.
10. The fusion proteins are eluted at room temperature with 2 mL of 50 m*M* Tris-HCl, pH 7.5, and 20 m*M* reduced glutathione (Sigma).
11. The eluates are collected in 500-μL aliquots.
12. The protein content of the aliquots is analyzed by 10% SDS-PAGE (10 μL is normally enough) followed by Coomassie blue staining.
13. Protein concentrations are determined, and aliquots are prepared and frozen at –70°C.

*3.1.3. cdc25 Phosphatase Assay Using Fluorescein-Diphosphate as Substrate (see **Note 3**)*

1. Recombinant or immunoprecipitated cdc25 protein is incubated with 20 μ*M* FDP in FDP buffer for 1 h at 37°C. The immunoprecipitate is washed twice in FDP buffer prior to the assay.
2. The reactions are stopped through addition of 1 *M* NaOH (to a final concentration of 0.1 *M*).
3. Reactions with immunoprecipitates containing sepharose beads are filtered through a filter Eppendorf tube (Millipore) to remove all beads from the solution.
4. The fluorescence at 530 nm is measured with a fluorescence spectrophotometer (F-2000, Hitachi), and the excitation wavelength is 490 nm.

3.1.4. Measuring cdc25 Phosphatase Activity In Vivo

Owing to the low abundance of cdc25 proteins in a variety of human cell lines, it is not easy to measure cdc25 activity. Therefore a two-step assay has been developed *(8)*. The assay procedure is depicted in **Fig. 2**. In the first step an inactive cyclinB1/cdk1 complex is immunoprecipitated from S-phase extracts. cdc25 is immunoprecipitated from cell extracts from which the cdc25 activity has to be determined using cdc25-specific antibodies. Then the two immunoprecipitates are combined, and a phosphatase assay is performed. In the second step a histone H1 assay is performed to determine the activity of the cyclinB1/cdk1 complex. The amount of active cyclinB1/cdk1 is a direct measure of the initial cdc25 activity. An example is shown in **Fig. 3**.

3.1.4.1. Preparing S-Phase Cell Extracts

Adherent HeLa cells are treated with 10 m*M* hydroxyurea (HU) for 18 h under standard cell culture conditions. If large amounts of cellular extracts are needed, a suspension of HeLa S3 cells may be used. Extracts are prepared as described below in **Subheading 3.1.4.2.** (*see* **Note 4**).

3.1.4.2. Preparing Cell Extracts From Which the cdc25 Activity Has to be Determined

1. The cell monolayer is washed twice with PBS and the cells are harvested by trypsinization followed by centrifugation.
2. Resuspend the pellet in 3–5 vol of lysis buffer.
3. Incubate for 30 min on ice.
4. The lysates are cleared by centrifugation for 10 min at 13,000 rpm. The total protein concentration is determined by standard assay procedures (e.g., Bio-Rad assay system).

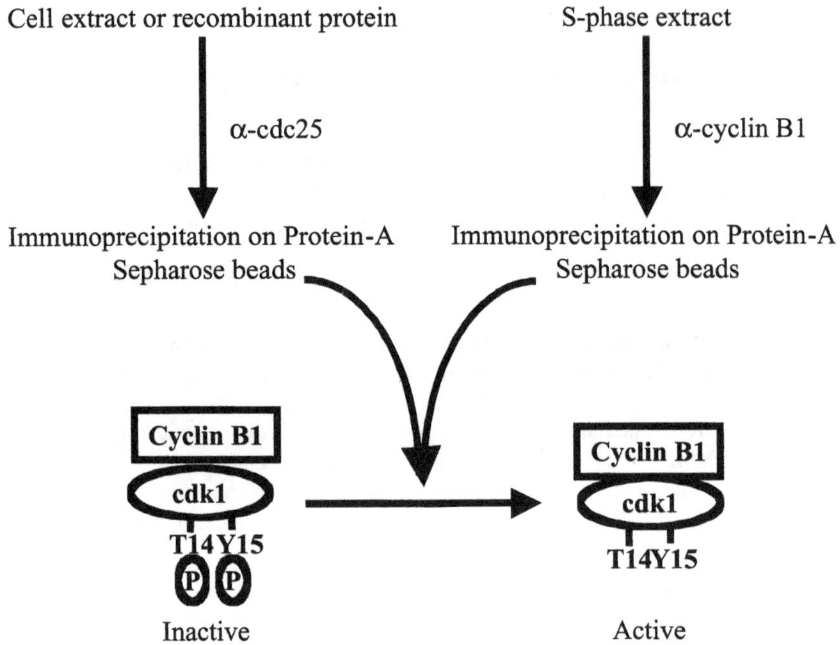

Fig. 2. Scheme for assaying endogenous cdc25 activity. cdc25 is immunoprecipitated from cell extracts and incubated with an inactive cyclin B1/cdk1 complex. The cyclin B1/cdk1 activity is subsequently measured using a Histone H1 kinase assay.

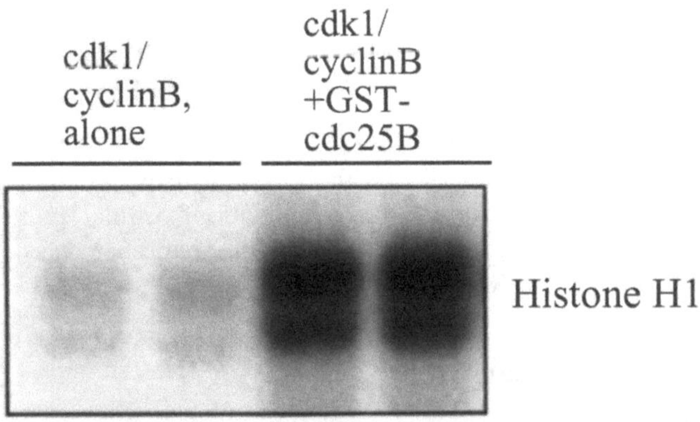

Fig. 3. cdk1/cyclin B complex from S-phase cells is activated by GST-cdc25B. cdk1/cyclin B was immunoprecipitated from S-phase cells and incubated with bacterially expressed GST-cdc25B (right lanes) or left untreated (left lanes) to measure background activity of the cdk1/cyclin B complex. The activity of the cdk1/cyclin B complex was measured in a kinase assay using Histone H1 as a substrate. The 12.5% SDS-PAGE gel was stained with Coomassie brilliant blue, and radioactively labeled Histone H1 was visualized by autoradiography.

3.1.4.3. Preclearing the Protein Extracts (*see* Note 5)

1. Each sample for measuring the cdc25 activity requires one sample of immunoprecipitated cyclinB1/cdk1 complex.
2. Wash 100 µL Cl-4B sepharose of a 25% slurry with lysis buffer, and pellet the beads at 13,000 rpm for 1 min.
3. Add 500 µg S-phase cell extract or 2 mg cell extract for cdc25 activity. Take the volume to 1 mL with lysis buffer.
4. Incubate on a rotating wheel for 30 min at 4°C.
5. Pellet the beads by centrifugation for 1 min at 13,000 rpm.

3.1.4.4. Immunoprecipitations

1. Transfer the supernatants to a new Eppendorf tube containing washed protein A or G sepharose beads (according to the antibody used).
2. Add 5 µL antibody for cyclin B1 IP or 10 µL antibody for cdc25 IP, respectively.
3. Incubate for 2 h at 4°C on a rotating wheel.
4. Collect the beads by centrifugation for 1 min at 13,000 rpm.
5. Remove the supernatant, and wash the beads three times with 1 mL lysis buffer, collecting the beads by centrifugation after each wash.

3.1.4.5. Phosphatase Assay

1. Add 500 µL phosphatase assay buffer to the cdc25 IP, resuspend, and transfer to the tube containing the cyclin B1 IP. Mix well.
2. Spin down and remove the supernatant completely.
3. Add 100 µL phosphatase buffer, and incubate at 30°C for 30 min with constant shaking (200 rpm).

3.1.4.6. Histone H1 Kinase Assay

1. After incubation spin down and remove the supernatant.
2. Add 50 µL kinase buffer containing 10 µg histone H1 (Roche) per reaction.
3. Incubate at 30°C for 15 min with constant shaking (200 rpm).
4. Stop the kinase reaction by adding 15 µL 4X Laemmli buffer. Boil for 4 min.
5. Mix well, and load supernatant completely on a 12.5% SDS-PAGE gel.
6. Run at 50 V overnight in SDS gel running buffer.

3.1.4.7. Coomassie Staining and Autoradiography

1. Cut off and discard the bromophenol blue band since it comigrates with free [γ-^{32}P]ATP.
2. Stain the gel in Coomassie brilliant blue solution for about 0.5–1 h.
3. Destain until the histone H1 bands at around 33 kDa are clearly visible; the background stain should be low.
4. Dry the gel in a gel dryer for 2 h at 80°C under vacuum.
5. Expose to X-ray films as necessary.
6. The signal should appear within 2 h.

3.2. Detection and Localization of Human cdc25 Isoforms in Human Cells

All the experiments described in this section can be performed in cells that have been synchronized as described *(37)* or in cells treated with UV or γ-radiation to induce a DNA damage response.

3.2.1. Detection of Endogenous cdc25 Family Members

1. Grow U-20S cells (or the cell line of interest) to approx 70% confluence.
2. Harvest the cells by trypsinization or scraping.
3. Boil the cell pellet in WCE buffer (1 mL buffer per 100-mm dish) for 10 min.
4. Clear the extracts by centrifugation at 12,000*g* for 15 minutes at 4°C.
5. Estimate total protein in the extract using the Bio-Rad Protein Assay kit (*see* **Note 6**). Add 0.5 vol of 3X SB, and boil for 4 min.
6. Load 25–100 μg of extract onto a 7.5% SDS-PAGE gel. Run the gel until the dye is off the gel.
7. Transfer to nitrocellulose membrane for either 3 h at 55 V or overnight at 8 V in a Bio-Rad chamber.
8. Block the membrane in 5% milk, 1% goat serum in TBS-T for 1 h at room temperature (RT).
9. Rinse the blot in TBS-T to remove all residual milk. Incubate the blot with primary antibody diluted in TBS-T containing 1% BSA and 0.01% sodium azide overnight on a rocking platform at 4°C (*see* **Note 7**).
10. Wash three times with TBS-T for 5 min at RT.
11. Incubate the blot with the appropriate secondary antibody diluted in 2.5% milk, 0.5% goat serum in TBS-T for 45 min to an hour at RT on a rocking platform.
12. Wash three times with TBS-T for 10 min at RT.
13. Develop the blots with ECL reagents per the manufacturers instructions (*see* **Note 8**).
14. Typically a 1-min exposure is enough to detect most proteins under these conditions. The Kodak gel emulsion-treated films are much more sensitive than regular Kodak film.

3.2.2. Immunoprecipitation of cdc25 Family Members

1. Grow U-20S cells (or the cell line of interest) to approx 70% confluence.
2. Harvest the cells by trypsinization or scraping.
3. Lyse the cells in 1 mL EBC buffer per 100-mm dish. Leave the extracts on ice for 15 min.
4. Clear the extracts by centrifugation at 12,000*g* for 15 minutes at 4°C.
5. Estimate the protein in the extract using a Bio-Rad assay kit (*see* **Note 6**).
6. To 2 mg of EBC extract, add either 200 μL of the tissue culture supernatants or 2 μL of purified antibody. Add 50 μL of a 50% slurry of protein-A-sepharose, and rock it overnight at 4°C. An immunoprecipitation with an isotype-matched antibody should be included as a negative control in each experiment.
7. Wash the immune complexes three times with NET-N. Boil in 50 μL 1X SB, and resolve the immune complexes on a 7.5% SDS-PAGE gel, followed by Western blotting, as described above.
8. To determine whether the cdc25 proteins are phosphorylated, the IP can be treated with λ-phosphatase (NEB) before being resolved on an SDS-PAGE gel. To do the phosphatase assay, wash the immune complexes three times with NET-N and once with phosphatase buffer.
9. Resuspend the beads in a 50 μL reaction volume containing 1 mL NEB λ-phosphatase and phosphatase buffer. Do two reactions for each sample and to one add sodium orthovanadate to a final concentration of 1 m*M* and sodium fluoride to 50 m*M* to inhibit

the λ-phosphatase completely. This reaction serves as a negative control. Incubate at 37°C for 30 min. Add 25 mL of 3X SB to each reaction, boil, and resolve on a 7.5% SDS-PAGE gel followed by Western blotting as described in **Subheading 3.2.1.**

3.2.3. Immunoprecipitation of Epitope Tagged cdc25 Family Members

1. Split U-2OS cells so that they are approx 30–50% confluent the next day.
2. Change the media on the cells the next morning (*see* **Note 9**).
3. Each 100-mm dish is transfected with a total of 25 μg of plasmid DNA. Some amount of this will be the plasmid encoding the cDNA that you wish to express plus carrier DNA (generally a plasmid such as bluescript [pBSK$^-$]). For cdc25C we have transfected 15 μg of the cdc25C, plasmid per 100-mm dish transfected, with the total amount of DNA adjusted to 25 μg with pBSK$^-$. To determine empirically the amount of cdc25 construct to transfect into cells, refer to **Note 10**.
4. Take the volume of DNA to 250 μL with dH_2O. Add 250 μL of 0.5 *M* $CaCl_2$. To this mix add 500 μL of 2X BBS dropwise. Mix by pipeting up and down a few times, and let the precipitate form at RT for 15–20 min.
5. Mix the precipitate, and then add it dropwise to the tissue culture dish. A fine precipitate should be observed under the microscope. Leave the precipitate on the cells overnight in a 5% CO_2 incubator.
6. The next morning, remove the precipitate from the dish and wash the cells twice with PBS. Feed with fresh media.
7. Harvest the cells 40–44 h post transfection, and perform immunoprecipitations with antibody to the epitope tag followed by Western blots with antibody to the cdc25 family member or the epitope tag as described above.

3.2.4. Determining the Localization of cdc25 Proteins Using Cellular Fractionation

The Pierce N-PER assay kit is a convenient way to make cytoplasmic and nuclear extracts without much fuss. The protocol described here, which is a modification of a protocol developed by Schreiber et. al. ***(38)***, provides similar results, and either technique will provide good-quality cytoplasmic and nuclear extracts.

1. Harvest U-2OS cells at approx 50% confluence by trypsinization.
2. Extract some of the cells in EBC (500 μL of EBC for each 50% confluent 100-mm dish). This is your whole cell extract.
3. To prepare cytoplasmic extracts, lyse cells in buffer A on ice for 10 min (*see* **Note 11**).
4. Spin the extracts for 30 s in a microfuge at 4°C.
5. Remove the supernatant to a fresh tube on ice. This is your cytoplasmic extract.
6. Extract the pellet with an equal amount of EBC (as described in **step 2**) to generate nuclear extracts.
7. Equivalent volumes of each extract can be used either for direct Western Blotting or immunoprecipitations followed by Western blotting.

3.2.5. Localization of Endogenous cdc25 by Immunofluorescence

1. Cells are grown on a glass cover slip in a 35-mm dish to no more than 50% confluence.
2. Remove the medium, and wash the cells three times with 2 mL of PBS.
3. Fix the cells in 2 mL of 4% paraformaldehyde in PBS for 20 min at RT.
4. Wash three times with PBS, and permeabilize the cells with 0.3% Triton X-100 in PBS for 10 min at RT.

5. Wash the cells twice in PBS with 0.1% NP-40. After the second wash, carefully remove all the excess liquid, and place the cover slip (cell side up) on Parafilm (*see* **Note 12**).
6. Incubate the fixed permeabilized cells with primary antibody to the cdc25 isoform of your choice, diluted in PBS containing 3% BSA and 0.1% NP-40, overnight at 4°C (*see* **Note 13**). To identify cells that are entering prophase, the cells can be costained with a phospho-specific antibody to Histone H3 ***(39)***. Cells can be costained with as many antibody combinations as you like. The only issue is having enough secondary antibodies with different fluorescent conjugates and a microscope with the appropriate filters.
7. Wash the cover slips alternately with PBS containing 0.1% NP40 and PBS 6 times.
8. Incubate the cover slips with the appropriate secondary antibody in PBS containing 3% BSA and 0.1% NP-40 for 30 min at RT.
9. Wash the cover slips alternately with PBS containing 0.1% NP-40 and PBS six times.
10. Counterstain the cells with the DNA dye DAPI at a concentration of 5 μg/mL.
11. Wash the cells three times with PBS. Mount on a slide using Citifluor mounting fluid (or equivalent product), and seal the cover slips with nail polish. The slides can be maintained for up to 2 d at 4°C without significant loss of signal.
12. Confocal images can be obtained using a Bio-Rad MRC-1024/2P instrument interfaced with a Zeiss Axiovert S100 microscope. A krypton–argon laser with emission lines at 488 nm and 568 nm is used for conventional excitation of fluorescein and rhodamine with bandpass emission filters at 522 and 598 nm. A Spectra-Physics Tsunami femtosecond pulsed laser tuned to 770 nm is used for the multiphoton excitation of DAPI.

3.2.6. Localization of Exogenously Expressed cdc25 by Immunofluorescence

1. Split U-2OS cells so that they are approx 30–50% confluent the next day into a 35 mm dish containing a cover slip.
2. Change the media on the cells the next morning.
3. We transfect 3 μg of the cdc25 plasmid with 2 μg of pBSK$^-$ per 35-mm dish. The amount of plasmid to be transfected will have to be determined empirically, as described in **Note 10**.
4. Take the volume of DNA to 50 μL with dH_2O. Add 50 μL of 0.5 *M* $CaCl_2$. To this mix, add 100 μL of 2X BBS dropwise. Mix by pipeting up and down a few times, and let the precipitate form at RT for 15–20 min.
5. Mix the precipitate, and then add it dropwise to the tissue culture dish. Leave the precipitate on the cells overnight in a 5% CO_2 incubator.
6. The next morning remove the precipitate from the dish and wash the cells twice with PBS. Feed with fresh media.
7. At 40–44 h post transfection, fix and permeabilize the cells as described in **Subheadings 3.2.5., step 2–5**. Stain the cells with antibody to the epitope tag for 2 h at RT. Other than the time required for staining with the primary antibody the rest of the protocol is identical to the one described in **Subheading 3.2.5., steps 7–12** for the endogenous cdc25 protein.

4. Notes

1. The pH of the transfection buffer is absolutely critical. If you are making buffers for the first time, the best idea is to make four batches of 2X BBS in which the pH differs by 0.01 pH units. The 2X BBS should be filter-sterilized and stored at –20°C. The transfection efficiency of the different batches can then be tested by transfection of a plasmid encoding green fluorescent protein (GFP). The solution can be freeze-thawed a number of times

and will last for long periods in the freezer; 2X BBS will last for about 2 wk at 4°C. Either cesium chloride-purified DNA or Qiagen DNA can be used for the transfections.

2. All cdc25 antibodies should be tested for cross-reactivity with the other family members. It is best to do this yourself to confirm the manufacturer's claims. The easiest way to do this is to synthesize GST-cdc25 proteins in bacteria (as described in **Subheading 3.1.**). The bacterial pellets can be boiled in 1X sample buffer and resolved on SDS-PAGE gels, and Western blots can be performed with either a GST antibody or the individual cdc25 antibodies.
3. Among the exogenous substrates that can be used in cdc25 phosphatase assays, fluorescein diphosphate (FDP) is the most sensitive. It has been shown previously that FDP can be dephosphorylated by tyrosine phosphatases ***(40)***.The product of this reaction is a molecule that exhibits fluorescence at 530 nm when excited at 490 nm. A phosphatase assay that uses FDP as a substrate is simpler and more direct than measuring the activation of dephosphorylated cdk1/cyclin B1. The FDP assay, however, is not as sensitive as endogenous tyrosine-phosphorylated cdk1/cyclin B1. Therefore this assay is only recommended when recombinant GST/cdc25 fusion protein is used and is not recommended for measuring small amounts of endogenous cdc25 phosphatase activity observed in immunoprecipitates from mammalian cells (as described in **Subheading 3.1.**).
4. During the whole assay it is crucial to keep the extracts and the immunoprecipitated proteins on ice to prevent protein degradation and especially to inhibit dephosphorylation events. All buffers, reaction tubes, and centrifuges must be cooled down prior to use.
5. To measure the background activity of the cdk1/cyclin B complex, it is recommended to include at least one additional sample of immunoprecipitated cdk1/cyclin B that is not incubated with recombinant or immunoprecipitated cdc25. As a positive control, you can use M-phase cell extract, from which you immunoprecipitate cdk1/cyclin B and perform the Histone H1 assay without cdc25 incubation or perform the kinase assay with recombinant cdk1/cyclin B.
6. Bio-Rad makes several protein assay kits. Please ensure that the assay kit you use is compatible with the buffer constituents.
7. The antibody solution stored at 4°C and reused until it gives no signal. We have typically used blotting solutions for anywhere from 6 mo to 1 yr.
8. The Pierce Restore solution is a wonderful way of stripping and reprobing ECL blots. It is very quick (15 min maximum for antibodies with high affinity for the antigen) and can be used to strip a blot up to three times. This is of great importance if your sample size is limited.
9. We begin thawing the 2X BBS at this point at 4°C. The 2X BBS is allowed to come to room temperature after it has thawed. The $CaCl_2$ and the dH_2O used for the transfection should also be at RT, as the formation of the precipitate depends on the pH of the solution.
10. The amount of DNA transfected will have to be determined empirically for each expression construct used. To determine the optimum amount of DNA to be used for the transfection, do a titration transfecting 1, 2, 5 10, 15, 20, and 25 μg of DNA into cells, and determine the amount that gives you the highest expression of the epitope-tagged protein. When doing this titration, and indeed any other transfection, the amount of expression vector should be the same in all the transfections, e.g., in the transfection described above, if you transfect in 1 μg of cdc25 expression vector the transfection should contain 24 μg of expression vector (not $pBSK^{-}$) as carrier so as to be comparable to the transfection in which 25 μg of expression plasmid is used. Cells transfected with the vector alone should always be included in each transfection.

11. The amount of buffer A depends on the number of cells. Add 500 μL buffer A for each 50% confluent 100-mm dish that you use. If this ratio is not maintained, the nuclear extracts will be contaminated with cytoplasmic proteins. For the N-Per assay reagent, follow the manufacturer's instructions. We had previously observed that cdc25C was dephosphorylated in cytoplasmic extracts. To prevent the dephosphorylation of cdc25C in the cytoplasmic extracts, additional phosphatase inhibitors (10 m*M* cypermethrin, 200 m*M* dephostatin, 200 n*M* okadaic Acid, and 25 n*M* tautomycin) were added to buffer A.
12. Do not aspirate the liquid off the cover slip after the second wash in PBS 0.1% NP-40. Instead, lift the cover slip up with a surgical forceps, aspirate any excess liquid, place the cover slip on Parafilm, and immediately add antibody to the cover slip. This prevents the cover slip from drying. It is extremely important to ascertain that the cover slips do not go dry at any point during the staining procedure, as that will result in high background and poor signal quality.
13. It takes 350 μL of staining solution to cover the surface of the average cover slip completely. If your antibody is limiting, you can stain with as low as 25 μL of staining solution by putting the staining solution onto Parafilm and flipping the cover slip cell side down onto the staining solution. We recommend that all staining done overnight be done in a box that can be converted into a humidified chamber as follows: wet three to four sheets of Whatman paper, and use them to line the lid of the box. This will prevent evaporation and drying of the antibody solution, resulting in better signals.

References

1. Nurse, P. (1994) Ordering S phase and M phase in the cell cycle. *Cell* **79,** 547–550.
2. Nurse, P. (1998) Checkpoint pathways come of age. *Cell* **91,** 865–867.
3. Sadhu, K., Reed, S.I., Richardson, H., and Russell, P. (1990) Human homolog of fission yeast cdc25 mitotic inducer is predominantly expressed in G2. *Proc. Natl. Acad. Sci. USA* **87,** 5139–5143.
4. Strausfeld, U., Labbe, J. C., Fesquet, D., et al. (1991) Dephosphorylation and activation of a p34^{cdc2}/cyclin B complex in vitro by human cdc25 protein. *Nature* **351,** 242–245.
5. Galaktionov, K. and Beach, D. (1991) Specific activation of cdc25 tyrosine phosphatases by B-type cyclins: evidence for multiple roles of mitotic cyclins. *Cell* **67,** 1181–1194.
6. Mailand, N., Falck, J., Lukas, C., Syljuasen, R. G., Welcker, M., Bartek, J., and Lukas, J. (2000) Rapid destruction of human cdc25A in response to DNA damage. *Science* **288,** 1425–1429.
7. Jinno, S., Suto, K., Nagata, A., Igarashi, M., Kanaoka, Y., Nojima, H., and Okayama, H. (1994) Cdc25A is a novel phosphatase functioning early in the cell cycle. *EMBO J.* **7,** 1549–1556.
8. Blomberg, I. and Hoffmann, I. (1999) Ectopic expression of cdc25A accelerates the G1/S transition and leads to premature activation of cyclinE- and cyclinA-dependent kinases. *Mol. Cell. Biol.* **1999,** 6183–6194.
9. Honda, R., Ohba, Y., Nagata, A., Okayama, H., and Yasuda, H. (1993) Dephosphorylation of human p34cdc2 kinase on both Thr-14 and Tyr-15 by human cdc25B phosphatase. *FEBS Lett.* **318,** 331–334.
10. Mailand, N., Podtelejnikov, A. V., Groth, A., Mann, M., Bartek, J., and Lukas, J. (2002) Regulation of G2/M events by cdc25A through phosphorylation-dependent modulation of its stability. *EMBO J.* **21,** 5911–5920.
11. Gabrielli, B. G., DeSouza, C. P. C., Tonks, I. D., Clark, J. M., Hayward, N. K., and Ellem, K. A. O. (1996)Cytoplasmic accumulation of cdc25B phosphatase in mitosis triggers centrosomal microtubule nucleation in HeLa cells. *J. Cell Sci.* **109,** 1081–1093.

12. Gabrielli, B. G., Clark, J. M., McCormack, A. K., and Ellem, K. A. O. (1997) Ultraviolet light-induced G2 phase cell cycle checkpoint blocks cdc25-dependent progression into mitosis. *Oncogene* **15,** 749–758.
13. Karlsson, C., Katich, S., Hagting, A., Hoffmann, I., and Pines, J. (1999) Cdc25B and cdc25C differ markedly in their properties as initiators of mitosis. *J. Cell Biol.* **146,** 573–583.
14. Dalal, S. N., Schweitzer, C. M., Gan, J., and DeCaprio, J. A. (1999) Cytoplasmic localization of human cdc25C during interphase requires an intact 14-3-3 binding site. *Mol. Cell. Biol.* **19,** 4465–4479.
15. Peng, C.-Y., Graves, P. R., Thoma, R. S., Wu, Z., Shaw, A. S., and Piwnica-Worms, H. (1997) Mitotic and G2 checkpoint control: regulation of 14-3-3 protein binding by phosphorylation of cdc25C on serine-216. *Science* **277,** 1501–1505.
16. Hoffmann, I., Clarke, P. R., Marcote, M. J., Karsenti, E., and Draetta, G. (1993) Phosphorylation and activation of human cdc25C by cdc2-cyclin B and its involvement in the self amplification of MPF at mitosis. *EMBO J.* **12,** 53–63.
17. Izumi, T. and Maller, J. L. (1993) Elimination of cdc2 phosphorylation sites in the cdc25 phosphatase blocks initiation of M-phase. *Mol. Biol. Cell* **4,** 1337–1350.
18. Izumi, T. and Maller, J. L. (1995) Phosphorylation and acitvation of the *Xenopus* cdc25 phosphatase in the absence of cdc2 and cdk2 kinase activity. *Mol. Biol. Cell* **6,** 215–226.
19. Strausfeld, U., Fernandez, A., Capony, J. -P., et al. (1994) Activation of $p34^{cdc2}$ protein kinase by microinjection of human cdc25C into mammalian cells. Requirement for prior phosphorylation of cdc25C by $p34^{cdc2}$ on sites phosphorylated at mitosis. *J. Biol. Chem.* **269,** 5989–6000.
20. Toyoshima-Morimoto, F., Taniguchi, E., and Nishida, E. (2002) Plk1 promotes nuclear translocation of human Cdc25C during prophase. *EMBO Rep.* **3,** 341–348
21. Zhou, B.-B. S. and Elledge, S. J. (2000) The DNA damage response: putting checkpoints in perspective. *Nature* **408,** 433–439.
22. Falck, J., Mailand, N., Syljuasen, R.G., Bartek, J., and Lukas, J. (2001) The ATM-chk2-cdc25A checkpoint pathway guards against radio-resistant DNA synthesis. *Nature* **410,** 842–847.
23. Matsuoka, S., Huang, M., and Elledge, S. J. (1998) Linkage of ATM to cell cycle regulation by the chk2 protein kinase. *Science* **282,** 1893–1897.
24. Chen, F., Zhang, Z., Bower, J., et al. (2002) Arsenite-induced cdc25C degradation is through the KEN-box and ubiquitin-proteasome pathway. *Proc. Natl. Acad. Sci. USA* **99,** 1990–1995.
25. Bulavin, D. V., Higashimoto, Y., Popoff, I. J., et al. (2001) Initiation of a G2/M checkpoint after ultraviolet radiation requires p38 kinase. *Nature* **411,** 102–107.
26. Muslin, A. J., Tanner, J. W., Allen, P. M., and Shaw, A. S. (1996) Interaction of 14-3-3 with signaling proteins is mediated by recognition of phosphoserine. *Cell* **84,** 889–897.
27. Yaffe, M. B., Rittinger, K., Volinia, S., et al. (1998) The structural basis for 14-3-3 phosphopeptide binding specificity. *Cell* **91,** 961–971.
28. Hagting, A., Karlsson, C., Clute, P., Jackman, M., and Pines, J. MPF localization is controlled by nuclear export. EMBO J. 17, 4127-4138 (1998).
29. Toyoshima, F., Moriguchi, T., Wada, A., Fukuda, M., and Nishida, E. (1998) Nuclear export of cyclin B1 and its possible role in the DNA damage-induced G2 checkpoint. *EMBO J.* **17,** 2728–2735.
30. Morris, M. C., Heitz, A., Mery, J., Heitz, F., and Divita, G. (2000) An essential phosphorylation-site domain of cdc25C interacts with both 14-3-3 and cyclins. *J. Biol. Chem.* **275,** 28849–28857.

31. Mils, V., Baldin, V., Goubin, F., et al. (2000) Specific interaction between 14-3-3 isoforms and the human CDC25B phosphatase. *Oncogene* **19,** 1257–1265.
32. Conklin, D.S., Galaktionov, K., and Beach, D. (1995) 14-3-3 proteins associate with cdc25 phosphatases. *Proc. Natl. Acad. Sci. USA* **92,** 7892–7896.
33. Davezac, N., Baldin, V., Gabrielli, B., Forrest, A., Theis-Febvre, N., Yashida, M. and Ducommun, B. (2000) Regulation of cdc25B phosphatases subcellular localization. *Oncogene* **19,** 2179–2185.
34. Forrest, A. and Gabrielli, B. (2001) Cdc25B activity is regulated by 14-3-3. *Oncogene* **20,** 4393–4401.
35. Chen, M.-S., Ryan, C. E., and Piwnica-Worms, H. (2003) Chk1 kinase negatively regulates mitotic function of cdc25A phosphatase through 14-3-3 binding. *Mol. Cell. Biol.* **23,** 7488–7497.
36. Asubel, F. M., Brent, R., Kingston, R. E., Moore, D. D., Seidman, J. G., Smith, J. A. and Struhl, K. (eds.) (1987) *Current Protocols in Molecular Biology*, John Wiley & Sons, New York.
37. Krek, W. and DeCaprio, J. A. (1995) Cell synchronization. *Meth. Enzymol.* **254,** 114–124.
38. Schreiber, E., Matthias, P., Muller, M. M., and Schaffner, W. (1989) Rapid detection of octamer binding proteins with 'mini-extracts', prepared from a small number of cells. *Nucleic Acids Res.* **17,** 6419.
39. Hendzel, M. J., Wei, Y., Mancini, M. A., et al. (1997) Mitosis-specific phosphorylation of histone H3 initiates prmarily within pericentromeric heterochromatin during G2 and spreads in an ordered fashion coincident with mitotic chromosome condensation. *Chromosoma* **106,** 348–360.
40. Huang, Z., Olson, N. A., You, W., and Houglang, R. P. (1992) A sensitive and competitive ELISA for 2, 4-dinitrophenol using 3, 6-fluorscein diphosphate as a fluorogenic substrate. *J. Immunol. Methods* **149,** 261–266.

20

Assaying Cell Cycle Checkpoints

Activity of the Protein Kinase Chk1

Carmela Palermo and Nancy C. Walworth

Summary

Eukaryotic cells regulate progression through the cell cycle in response to DNA damage. Cell cycle checkpoints are the signal transduction pathways that couple the detection of DNA damage to the proteins that control transitions in the cell cycle. The protein kinase Chk1, originally discovered in fission yeast, but conserved in humans, is essential for preventing mitotic entry in the presence of DNA damage or blocks to DNA replication that cannot be reconciled. Chk1 is phosphorylated in response to DNA damage. Phosphorylation depends on the activity of conserved components of the checkpoint pathway including Rad3, a member of the ATM/ATR family of kinases. Phosphorylation leads to activation of Chk1 kinase activity. In this chapter, we describe an assay for monitoring the activity of Chk1 isolated.

Key Words

Chk1; checkpoint; DNA damage; cell cycle; protein kinase.

1. Introduction

Mitosis is one of many essential processes in eukaryotic cells that depend on protein kinase cascades to control signaling events. At the core of the mitotic machinery is the highly conserved family of cyclin-dependent protein kinases (CDKs) *(1)*. Cell cycle progression is mediated by CDK kinase activity, which is sharply periodic as cells cycle between DNA synthesis (S-phase) and mitosis (M-phase). The intervening G_1- and G_2-phases of the cell cycle ensure the timing and accuracy of cell division by imposing interdependency between S- and M-phase, such that the onset of mitosis depends on the completion of DNA replication in the previous S-phase *(2)*. However,

From: *Methods in Molecular Biology, vol. 296, Cell Cycle Control: Mechanisms and Protocols*
Edited by: T. Humphrey and G. Brooks © Humana Press Inc., Totowa, NJ

a significant problem for cells is the common occurrence of DNA damage as a consequence of both environmental stresses and normal metabolic processes.

Eukaryotic cells have evolved a sophisticated approach to maintain genomic stability when they are challenged with DNA damage. The DNA damage checkpoint is a signal transduction pathway that coordinates cell division by sensing DNA damage and delaying entry into mitosis by relaying a signal that ultimately influences the timing of CDK activation ***(3–5)***. Just as protein kinase cascades drive cell cycle progression, checkpoint activation and cellular survival following DNA damage depend on activated kinases. In the fission yeast *Schizosaccharomyces pombe*, the protein kinase Chk1 has an essential role in the checkpoint to transduce a delay signal to the cell cycle machinery when DNA damage is sensed ***(6)***. Specifically, Chk1 is a serine/threonine kinase that phosphorylates particular components of the cell cycle machinery, which will ultimately prevent entry into mitosis until genomic integrity is restored ***(7)***. Chk1, like many of the kinases involved in regulating the cell cycle, is evolutionarily conserved from yeast to humans. In fact, a comparison of the "kinomes" of yeast, flies, and worms with that of humans reveals that many kinases have been functionally conserved across these diverse species during evolution ***(8)***. Thus, simpler model organisms are ideal to study kinase functions in protein phosphorylation pathways or, alternatively, may be used for cross-species analysis, particularly of human kinases.

S. pombe has been a valuable organism for studying the cell cycle owing to the ease with which genetic, biochemical, and morphological experiments can be performed. Our lab focuses on the role of Chk1 in the DNA damage checkpoint pathway and how regulation of its activity affects cell cycle progression. In response to DNA damage, a significant fraction of Chk1 is phosphorylated, which results in a form of the protein that exhibits decreased mobility on sodium dodecyl sulfate-polyacrylamide gel electrophoresis (SDS-PAGE) ***(9)***. The appearance of a Chk1 mobility shift has been a valuable tool to indicate that the checkpoint is activated when studying functional mutants of Chk1 or mutated proteins that may interact with Chk1 in the pathway. Recently, we have developed an assay for Chk1 kinase activity to detect the biochemical changes that are induced in response to DNA damage. Although Chk1 possesses a basal kinase activity, this activity increases twofold in response to DNA damage ***(10)***. Analysis of nonfunctional mutants of Chk1 indicates that the increase in Chk1 kinase activity is necessary for cell cycle delay and cellular survival following DNA damage ***(10)***. As a result, we have at hand an additional method to correlate the response to DNA damage with a quantifiable induction of kinase activity. Although we are successfully exploiting this kinase assay to investigate the checkpoint pathway and have optimized it for analysis of fission yeast Chk1, we believe the principles of the assay are generally applicable.

2. Materials

2.1. Culture Growth

1. Yeast.
2. YEA medium: 0.5% yeast extract, 3% dextrose, supplemented with adenine (150 mg/L).
3. Camptothecin lactone (CPT), 40 m*M* stock prepared in dimethyl sulfoxide (DMSO) and stored at –20°C.

2.2. Protein Extraction

1. Immunoprecipitation buffer (IP buffer): 10 m*M* phosphate buffer pH 7.0, 0.15 *M* NaCl, 1% NP-40, 10 m*M* EDTA, 50 m*M* NaF, 2 m*M* dithiothreitol (DTT), 50 m*M* PMSF, and Complete Protease Inhibitor Tablet (Roche Diagnostics).
2. 425–600-mm Glass beads (Sigma, cat. no. G 9268).
3. Fast Prep vortexing machine (Bio 101).
4. Bradford reagent (Bio-Rad).

2.3. Immunoprecipitation

1. Protein A sepharose beads (50% slurry beads:IP buffer).
2. Antibody to protein of interest (F-7 anti-HA mouse monoclonal).
3. IP buffer (*see* **Subheading 2.2.**).
4. Rotating wheel at 4℃.

2.4. Washes

1. IP buffer.
2. 2X Kinase Buffer: 50 m*M* Tris-HCl, pH 7, 1 m*M* DTT, 5 m*M* $MgCl_2$, 0.4 m*M* $MnCl_2$, 25% Sigma glycerol, 0.1% Triton-X, 100 m*M* ATP.

2.5. Kinase Assay

1. 2X Kinase buffer.
2. 2 mg/mL Substrate peptide (Chk1 substrate peptide: RIARAASMAAALARK).
3. [γ-^{32}P]ATP (3000 Ci/mmol).
4. Platform rocker at 30℃.
5. P81 Phosphocellulose paper (Whatman).
6. 1-L Plastic beaker with holes punched through.
7. 2-L Glass beaker with stir bar.
8. 3 L 0.5% Phosphoric acid.
9. 100% Ethanol.
10. Hair dryer.
11. Liquid scintillation fluid.
12. Scintillation vials.
13. Liquid scintillation counter.

3. Methods

For the kinase assays described herein, Chk1 is isolated by immunoprecipitation using a strain in which the *chk1* gene is tagged at the C-terminus with three copies of the HA epitope and either integrated into the genome at the *chk1* locus ***(9)*** or encoded by a plasmid. The strain is grown in the absence or presence of CPT, a topoisomerase inhibitor, to induce DNA damage and thus activate the kinase ***(10,11)***. The cells are then lysed in IP buffer and the protein fraction is collected. Chk1 is immunoprecipitated by incubating the protein extract with a monoclonal antibody directed to the HA epitope tag of Chk1, followed by incubation with protein A sepharose beads. The Chk1/antibody complex will remain bound to the beads, and extensive washes are then carried out to clear away the lysate (*see* **Note 1**). The kinase reaction, consisting of the phosphorylatable substrate peptide [γ-^{32}P]ATP and the kinase buffer, is then added to each IP. Hence, the kinase assays are performed with the kinase of interest bound to the protein A beads, while the peptide substrate is freely soluble. Following 10 min of incubation at 30°C, the reactions are centrifuged, and the supernatant, containing the phosphorylated peptide, is spotted onto P81 phosphocellulose paper. The filter papers are analyzed by liquid scintillation counting, with counts per minute reflecting the level of phosphorylation of the peptide and hence corresponding to the activity of the kinase toward the substrate. The specific methods are detailed below.

3.1. Culture Growth and Exposure to DNA Damage

1. Grow a 200 mL YEA culture to a density of 5×10^6- to 1×10^7 (*see* **Note 2**).
2. Split the culture into two 100-mL cultures in separate flasks.
3. Treat one culture with 40 μ*M* CPT, and leave the other as an untreated control. Continue to grow the cells for an additional 2.5 h, with continuous shaking at 30°C.
4. Harvest the cells by centrifugation once, and wash with 50 mL IP buffer.

3.2. Protein Extraction

1. Resuspend the cell pellet in 1 mL of IP buffer, and aliquot 200 μL to approx 10 Eppendorf tubes. The kinase assay will be performed in triplicate. Thus, a sufficient volume of protein must be extracted.
2. Add acid-washed glass beads to the meniscus.
3. Lyse the cells in a Fast Prep vortex machine, with the following settings: Speed 6.5, Time 20 s. The Fast Prep is maintained at 4°C.
4. Place tubes on ice until all samples have gone through the Fast Prep.
5. Repeat the Fast Prep lysis, with the following settings: Speed 6.5, Time 10 s.
6. Use a syringe equipped with an 18-gage needle to pierce a small hole in the tube containing the cellular lysate and glass beads. Insert this tube into a second Eppendorf tube.
7. Centrifuge the two tubes for 5 min at 325*g* and 4°C in a microfuge to collect the cell lysate (*see* **Note 3**).
8. Discard the upper Eppendorf tube containing only the glass beads. Pool the cell lysate collected in the lower tubes into a single Eppendorf tube.
9. Centrifuge the pooled lysate for 10 min, 8160*g*, at 4°C. This step will serve to pellet insoluble debris and unlysed cells.
10. Transfer the supernatant to a fresh Eppendorf tube. Utilize approx 1 μL for the Bradford assay to quantify the amount of protein contained in the extract (*see* **Fig. 1**).

3.3. Immunoprecipitation of Chk1 Kinase

1. Once the protein concentration of the extract is determined, aliquot approx 5 mg of protein to four Eppendorf tubes. Each aliquot will be subject to IP. Three will be used for the kinase assay, and one (the control) will be used to assess the amount of kinase in the IP. The control IP will go through all the following steps but will not be used in the kinase reaction. It will instead be analyzed by SDS-PAGE and Western blot, serving as a control to demonstrate that the kinase was efficiently immunoprecipitated in the parallel reactions (*see* **Subheading 3.6.**). An additional 100–500 g of lysate should also be retained for SDS-PAGE and Western blot analysis of the whole cell lysate.
2. Aliquot 20 L of F-7 anti-HA antibody (Santa Cruz) to the 5 mg of protein extract. Bring the volume of each IP to 500 mL, and incubate on a rocker at 4°C overnight (*see* **Note 4**).
3. To each IP add 20 L of protein A sepharose beads equilibrated with IP buffer (*see* **Note 4**). Incubate the IPs for 1 h at 4°C on a rocker. During this incubation begin to prepare the 2X kinase buffer (*see* **Note 5**).

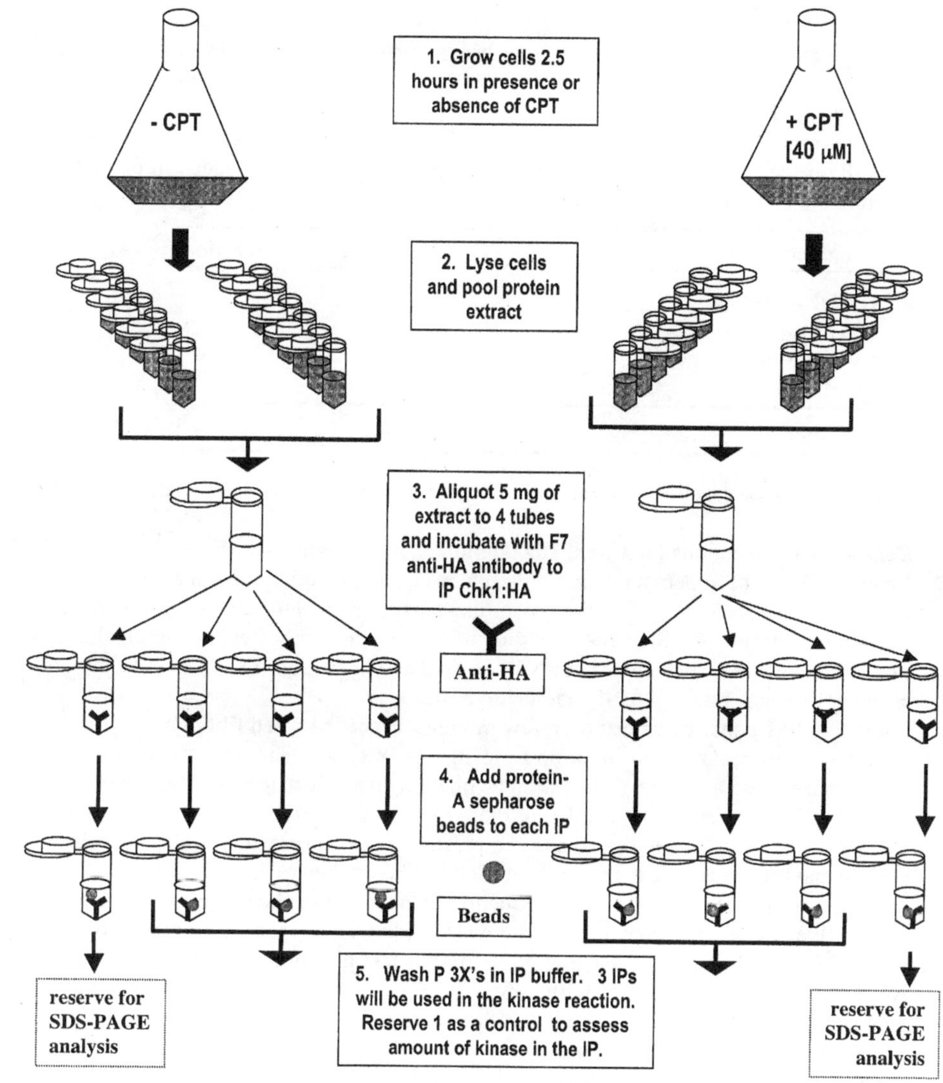

Fig. 1. Culture growth and immunoprecipitation of Chk1 from *S. pombe*. Cultures containing chk1::HA are grown in the presence or absence of camptothecin lactone (CPT) for 2.5 h, after which the cells are harvested by centrifugation. The cell pellet is resuspended in immunoprecipitation (IP) buffer and aliquoted to multiple Eppendorf tubes for cell lysis, and the protein extract is pooled. To immunoprecipitate Chk1, 5 mg of protein is incubated with an anti-HA antibody followed by incubation with protein A sepharose beads.

4. Centrifuge for 1 min, 82*g*, in a microfuge at 4C to collect the beads, and aspirate the supernatant. Proceed to wash the beads 3 times with IP buffer (*see* **Fig. 1**).

3.4. The Kinase Reactions

1. Subsequent to the IP buffer washes, the beads must be washed 3 times in 1X kinase buffer. For each wash, the IPs are centrifuged for 1 min, 16,000 *g*, in a microfuge at 4°C (*see* **Note 5**). Following the final wash, the beads are aspirated dry.
2. A master mix of the kinase reaction is prepared, and 24 L of the reaction is aliquoted to each IP on ice. The kinase reaction components and volume to be added in one 24 L reaction are as follows (*see* **Note 6** and **Fig. 2**):

Reaction components	Volume for one 24- L reaction
2X Kinase buffer 1	2.0 L
γ-^{32}P[ATP]	0.72 L
[2 mg/mL] Substrate peptide	6 .00 L
ddH_2O	5.28 L
Total	24.0 L

3. Place each kinase reaction on a rocker at 30°C for 10 min, and then immediately transfer on ice.
4. Centrifuge the reactions for 1 min, 16,000*g* to pellet the beads.
5. Pipet 20 L of the supernatant, avoiding the beads, and carefully spot the reaction onto prelabeled and prefolded P81 phosphocellulose paper (*see* **Note 6** and **Fig. 3**). A filter paper must be spotted with a 20- L aliquot of kinase reaction buffer containing all reaction components *except* isotope. This will be used as a blank control and must go through all the following washes with the radioactive filters.
6. Allow the P81 paper to air dry for a few minutes. Place the dried filter papers in a 1-L polypropylene beaker with three pouring lips (VWR) that has been punctured with holes (resembling a strainer). Holes can be made with an electric drill. Place the beaker in a 2-L glass beaker containing 1 L of 0.5% phosphoric acid, vigorously stirring with a magnetic stir bar. Manually, shake the plastic beaker submerged in the phosphoric acid so that the papers are in motion and each can be washed of any free isotope. Allow the filters to wash for 5 min and subsequently dispose of the phosphoric acid wash in a radioactive waste container. Repeat the wash with 1 L of phosphoric acid two more times, for a total of three washes. Following the last phosphoric acid wash, utilize a wash bottle to rinse the papers with 100% ethanol, and collect the ethanol in the glass beaker. Remove the plastic beaker, containing the washed filter papers, from the glass beaker and cover with a cardboard beaker cap (VWR). Thoroughly dry the papers with a hair dryer (*see* **Note 6** and **Fig. 3**).
7. Use forceps to extract each filter from the beaker, and lay them out on the glass plate in the order you wish to analyze them. Place each paper in a scintillation vial containing 5 mL of liquid scintillant, and gently mix.
8. Count each sample in a liquid scintillation counter for 30 s.

3.5. Quantifying the Results

The counts per minute for each set of triplicate IPs can be averaged and displayed in a bar graph. The standard deviation of each set of IPs should also be depicted.

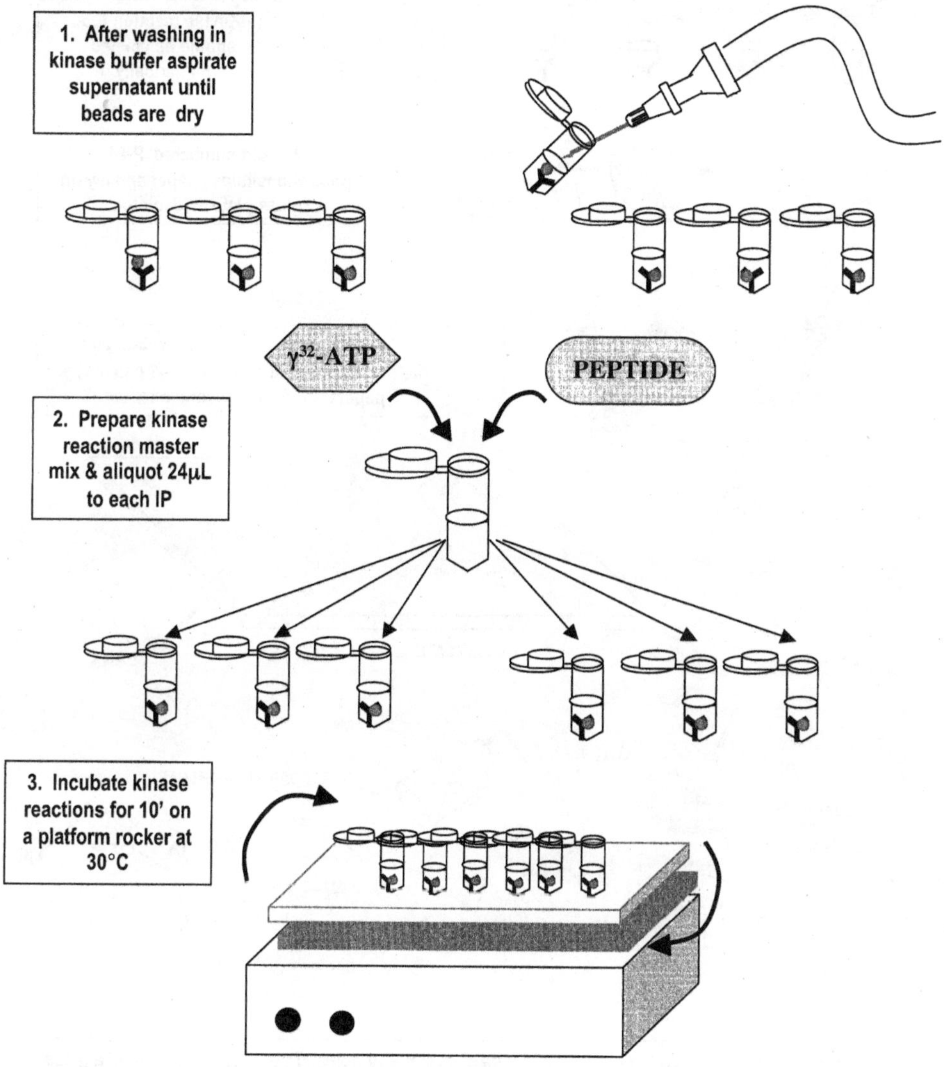

Fig. 2. The kinase reaction. The protein-A sepharose beads with the Chk1/antibody complex attached to the beads are washed in kinase buffer and aspirated dry using a vacuum apparatus equipped with a 27 1/2-gage needle. A master mix of the kinase reaction containing kinase buffer, substrate peptide, and γ-^{32}P[ATP] is prepared and aliquoted to each immunoprecipitation (IP). The kinase reactions are incubated for 10 min at 30C on a platform rocker.

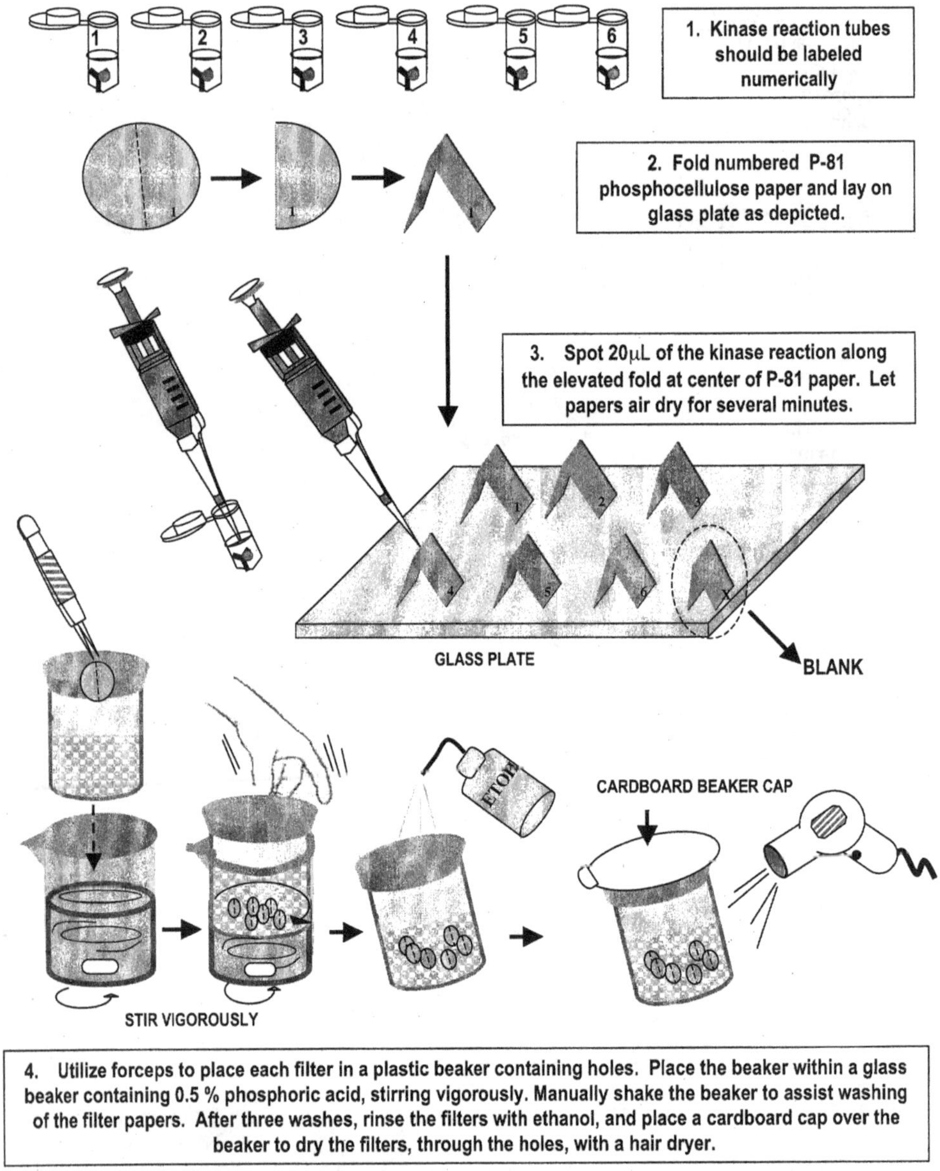

Fig. 3. Spotting the kinase reactions on P-81 phosphocellulose paper and phosphoric acid washes.

3.6. SDS-PAGE Analysis of Nonradioactive Samples

To determine indirectly if the kinase was efficiently immunoprecipitated in the assay, boil the reserved IP reaction for 5 min. Analyze the supernatant as well as the whole cell lysate by SDS-PAGE.

4. Notes

1. Theoretically, contaminating kinases may be coprecipitated with Chk1. These contaminating kinases may affect the results of the assay and contribute to residual noise.
2. A control that may be used in this assay is a strain that possesses a mutation within the conserved catalytic domain of the kinase. For example, a catalytically inactive allele of Chk1 (D155A) possesses a mutation in a conserved residue that helps orient the γ-phosphate of ATP by chelating the activating Mg^{2+} ions that bridge the β- and γ-phosphates ***(12)***. In our assays, the Chk1 D155A protein possesses a negligible level of kinase activity and serves as a control when one is testing the level of kinase activity of other Chk1 mutants ***(10)***.
3. Protein extraction: an 18-gage needle is utilized to puncture the Eppendorf tube. The hole that is inserted in the tube containing the cellular extract is small enough so that during centrifugation at low speed the lysate will be collected in the lower tube, while the beads will remain in the upper tube.
4. Immunoprecipitation of the kinase: incubation of the protein extract with the antibody may be executed for a minimum of 1–2 h. The protein A beads are stored in 70% ethanol at 4€. To utilize the beads in IP reactions, an appropriate volume of beads is transferred to an eppendorf tube and centrifuged for 1 min at 82*g*. It is critical to be gentle with the beads and centrifuge only at 82*g*. The ethanol is extracted by vacuum suction using a 30 1/2-gage needle, which allows the supernatant to be extracted with minimal bead loss. The beads are then washed three times with IP buffer and resuspended in an equal volume of IP buffer. The beads must be washed in a similar manner following immunoprecipitation. However, each time the supernatant is aspirated, a residual amount (~100 mL) of the buffer is left behind. Following the final wash, aspirate most of the IP buffer from the beads by allowing the needle to pierce through the bead pellet to the bottom of the Eppendorf tube.
5. Kinase buffer: when preparing the kinase buffer, the ATP and DTT must be thawed on ice and added last. It is also critical that the glycerol utilized be Sigma 99+ % grade (G-5516). The kinase buffer must be well mixed, with no apparent gradations, before the ATP and DTT can be added. Upon addition of the final two components, the kinase buffer must remain on ice. To wash the IPs with kinase buffer, the reactions must be centrifuged at 4€ and at a speed of 16,000 *g*. This top speed must be utilized at this point of the assay, because it is difficult to pellet the beads in the glycerol-containing buffer. To ease removal of the viscous buffer, a P200 pipet tip is used to aspirate most of the buffer and then switched to the 30 1/2-gage needle to get closer to the bead pellet. Subsequent to the final wash, the 30 1/2-gage needle will be used to aspirate the beads dry. The kinase reaction must follow immediately, once the beads are dry.
6. Kinase reactions: when calculating the volume of the kinase reaction master mix, prepare enough of the mixture to be aliquoted to each IP plus an additional five reactions. This is because the glycerol in the reaction introduces pipeting error. The P81 phosphocellulose paper can be labeled with a pen ahead of time and should be folded in half. When you are ready to spot the kinase reactions, lay each paper on a glass plate so that the fold at the

center of the paper is not touching the glass. When spotting, carefully pipet the reaction along the center fold, and allow it to slowly soak across the paper. If the reaction is pipeted hastily, it will not be absorbed and will run off the sides onto the glass. During the filter paper washes, be sure to wipe the glass plate clean. Often some of the kinase reactions will inadvertently contaminate the plate, and you want to avoid spreading the free isotope onto the filters in a later step. When drying the filters, it is important to hold the cardboard beaker lid down as you direct the hair dryer through the holes in the beaker to avoid having the filters fly out of the beaker.

Acknowledgments

The assay described in this paper is derived from one that was established for recombinant Chk1 in the laboratory of Dr. Matthew OConnell, Peter MacCallum Cancer Centre. The assistance of the OConnell laboratory was invaluable in the development of this assay for use with immunoprecipitated Chk1, and its contribution is gratefully acknowledged. The authors also thank Dr. Holly Capasso for her work in initially establishing this assay in our lab and Drs. Matthew OConnell and Nicole den Elzen, Peter MacCallum Cancer Centre, for their advice in refining the assay and for helpful discussions. Work in the authors'laboratory is supported by a grant from the NIH, GM53194.

References

1. Pines, J. (1995) Cyclins and cyclin-dependent kinases: a biochemical view. *Biochem J.* **308,** 697–711.
2. Nurse, P. (1994) Ordering S phase and M phase in the cell cycle. *Cell* **79,** 547–550.
3. Hartwell, L. H. and Weinert, T. A. (1989) Checkpoints: controls that ensure the order of cell cycle events. *Science* **246,** 629–634.
4. Zhou, B. B. and Elledge, S. J. (2000) The DNA damage response: putting checkpoints in perspective. *Nature* **408,** 433–439.
5. Nyberg, K. A., Michelson, R. J., Putnam, C. W., and Weinert, T. A. (2002) Toward Maintaining the Genome: DNA damage and replication checkpoints. *Annu. Rev. Genet.* **36,** 617–656.
6. Walworth, N., Davey, S., and Beach, D. (1993) Fission yeast chk1 protein kinase links the rad checkpoint pathway to cdc2. *Nature* **363,** 368–371.
7. OConnell, M. J., Walworth, N. C., and Carr, A. M. (2000) The G2-phase DNA-damage checkpoint. *Trends Cell Biol.* **10,** 296–303.
8. Manning, G., Plowman, G. D., Hunter, T., and Sudarsanam, S. (2002) Evolution of protein kinase signaling from yeast to man. *Trends Biochem. Sci.* **27,** 514–520.
9. Walworth, N. C. and Bernards, R. (1996) rad-Dependent response of the chk1-encoded protein kinase at the DNA damage checkpoint. *Science* **271,** 353–356.
10. Capasso, H., Palermo, C., Wan, S., et al. (2002) Phosphorylation activates Chk1 and is required for checkpoint-mediated cell cycle arrest. *J. Cell Sci.* **115,** 4555–4564.
11. Wan, S., Capasso, H., and Walworth, N. C. (1999) The topoisomerase I poison camptothecin generates a Chk1-dependent DNA damage checkpoint signal in fission yeast. *Yeast* **15,** 821–828.
12. Hanks, S. K. and Hunter, T. Protein kinases 6. The eukaryotic protein kinase superfamily: kinase (catalytic) domain structure and classification. *FASEB J.* **9,** 576–596.

21

Polo-Like Kinase-1

Activity Measurement and RNAi-Mediated Knockdown

Marcel A. T. M. van Vugt and René H. Medema

Summary

Polo-like kinase-1 (Plk-1) is an important cell cycle regulatory kinase that has been implicated in a multitude of cell cycle events. In this chapter we review those multiple functions of Plk-1 and describe the methods routinely used in our laboratory to purify Plk-1 from cellular lysates and measure Plk-1 kinase activity in vitro. In addition, we describe a method to analyze cell cycle progression after depletion of Plk-1 by RNA-interference in tissue culture cells.

Key Words

Mitosis; synchronization; small interfering RNA; kinase assay.

1. Introduction

Originally identified as the *Polo* gene in *Drosophila melanogaster*, Polo-like kinases were shown to be required for proper mitotic progression in various species (*see* **refs. *1–3*** for reviews). Whereas only a single Polo-like kinase has been identified in *Saccharomyces cerevisiae, Schizosaccharomyces pombe*, and *Drosophila melanogaster*, mammalian cells appear to express up to four distinct Polo-like kinases (Plk-1–3 and SAK) ***(3)***.

Polo-like kinases regulate multiple processes in mitosis (**Fig. 1**). At the G_2/M transition, a role for Polo-like kinases was suggested in activation of the cyclin B/Cdk1 complex. This complex is essential for the onset of mitosis. On the one hand, Polo-like kinases can activate the Cdc25C phosphatase by direct phosphorylation and thereby contribute to activation of cyclin B/Cdk1 complexes ***(4–6)***. In addition, Plk-1 promotes the nuclear translocation of Cdc25C and cyclin B, suggesting that Plk-1 not only contributes to activation of cyclin B/Cdk1 but also regulates the redistribution of the complex to the nucleus at the onset of mitosis ***(7,8)***.

From: *Methods in Molecular Biology, vol. 296, Cell Cycle Control: Mechanisms and Protocols*
Edited by: T. Humphrey and G. Brooks © Humana Press Inc., Totowa, NJ

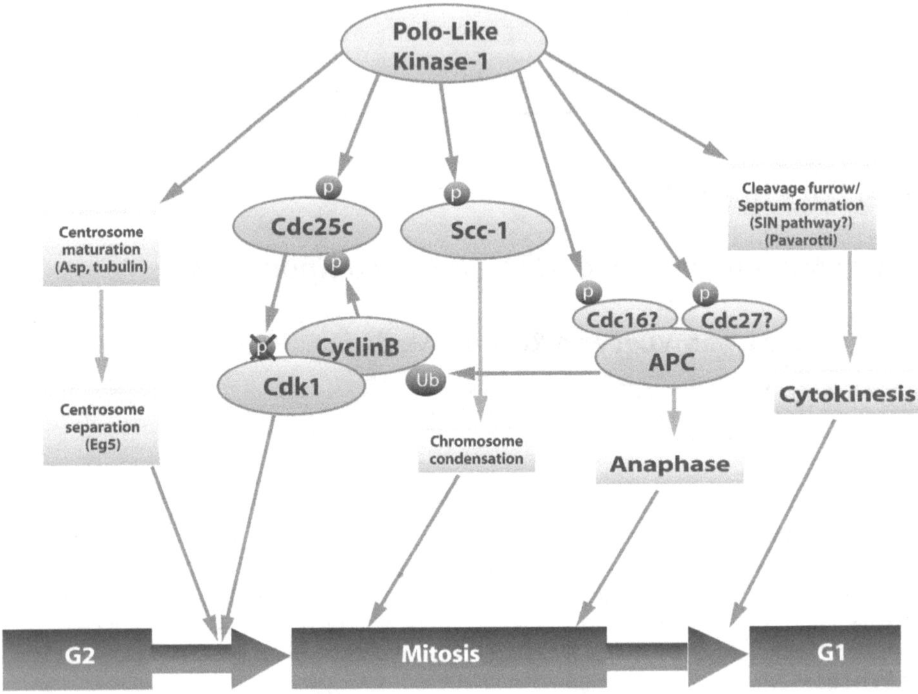

Fig. 1. Multiple roles for Polo-like kinases during cell division. Polo-like kinases (Plks) have been shown to play a role in centrosome maturation and separation by stimulating the centrosomes microtubule nucleating activity (which is suggested to involve direct regulation of Asp and tubulin subunits by Polo/Plk-1). At the entry of mitosis, Plks play a role in a positive feedback loop of cyclin B/cdc2 activation that triggers the G_2/M transition, by direct phosphorylation and (partial) activation of Cdc25C. In prophase, cohesin is displaced from the chromosome arms, and this has been shown to involve direct phosphorylation of Scc1 by Plk. During mitosis, Plks have been shown to play a role in activation of the anaphase-promoting complex (APC), which directs the degradation of a variety of proteins (cohesin, cyclin B1, and so on) that inhibit the onset of anaphase and mitotic exit. This could involve direct phosphorylation of APC components such as Cdc16/Cdc27 shown to occur at least in vitro. Finally, Plks were demonstrated to play an essential role during cytokinesis, at least in *S. pombe* and *Drosophila*. In *S. pombe*, Plo1 is required for septum formation by regulating the septum-inducing network (SIN), whereas in *Drosophila* the role of Polo in cytokinesis could be mediated through an effect on Pavarotti, a motor protein that is essential for restructuring the mitotic spindle in telophase to allow formation of the contractile ring.

During the early stages of mitosis, Polo-like kinases have been shown to mediate the displacement of the majority of cohesin from the chromosomes ***(9,10)***. In mammals, cohesin is displaced from the chromosome arms in prophase and prometaphase, whereas it is specifically retained at the centromeric region ***(11)***. Cohesin displacement in early mitosis does not involve anaphase-promoting complex/cyclosome (APC/

C)-dependent cleavage of cohesin, but rather involves loss of binding to chromosome arms through Plk-1-mediated phosphorylation ***(9,10)***. The centromeric fraction of cohesin that remains present until chromosomes are aligned is degraded in an APC/C-dependent fashion (for review, *see* **ref.** ***12***). Again, Plk-1 was suggested to play a regulatory role by phosphorylation of three subunits of the APC/C, Cdc16, Cdc27, and Tsg24 ***(13,14)***.

Polo-like kinases have also been shown to perform essential roles during the later stages of mitosis, particularly in *S. cerevisiae*, *S. pombe*, and *Drosophila* (reviewed in **ref.** ***1***). Exit from mitosis in yeast has been shown to depend on a mitotic exit network (MEN; *S. cerevisiae*) or septum initiation network (SIN; *S. pombe*), which are regulated by the Polo-like kinases Cdc5 and Plo1, respectively ***(1)***.

In addition to the described roles for Polo-like kinases in chromosome division, regulatory roles for Polo-like kinases in the centrosome cycle and formation of the mitotic spindle have also been reported (reviewed in **refs.** ***1–3***). Mutations in the *Drosophila Polo* gene result in defects in the architecture of the mitotic spindle, characterized by the appearance of monopolar spindles, or bipolar spindles with one of the poles broadened ***(15,16)***. Indeed, *Polo* can directly phosphorylate Asp, a microtubule-associated protein that binds to the centrosome and is required to organize microtubule asters ***(17,18)***. A role for Polo-like kinases in the centrosome cycle appears to be conserved in other organisms, as interference with the function of Plk-1 in HeLa cells impaired centrosome maturation and subsequently resulted in a mitotic arrest ***(19)***.

At present, little is known about how the multiple functions of Polo-like kinases are separated in space and time to allow orderly progression through the different stages of mitosis. Most of what we know about this comes from studies on Plk-1. Phosphorylation of Plk-1 is a means of regulating its kinase activity ***(20)***, but this appears to be rather complex ***(21)***. It is very well possible that subsequent functions of Plk-1 throughout cell division depend on progressive phosphorylation–dephosphorylation events. Subcellular localization of Plk-1 is also subject to extensive regulation and could direct and separate its actions. Plk-1 is cytoplasmic in interphase, localizes to centrosomes in late G_2 ***(22)***, and translocates into the nucleus just prior to cyclin B1 at the onset of mitosis ***(23)***. In mitosis Plk-1 is found on a variety of mitotic structures, such as the centrosomes, the centromeres, and the spindle midzone ***(22,24)***. Targeting of Plk-1 to these locations is mediated by the conserved C-terminal polo-box motifs, which appear to be sufficient for localization ***(25)***. So far, the polo-box motifs have been shown to have at least two clear roles in regulation of polo kinase activity. These two roles are in remarkable contrast. The first role of the polo-box motifs is binding and inactivating the Plk-1 kinase domain ***(25)***. The polo-box motifs specifically recognize, bind, and inactivate the unphosphorylated form of the kinase domain, whereas phosphorylation of Thr210 of the kinase domain abolishes binding and restores kinase activity ***(25)***. The second role of the polo-box motifs was shown to be the binding to peptides containing phosphorylated serine or threonine residues ***(26)***. Therefore the polo-box motifs may play an important auxiliary role in substrate recognition and subcellular targeting of Plk-1 during mitosis.

Plk-1 is overexpressed in a variety of human tumors ***(27)***, and ectopic expression of Plk-1 in NIH3T3 cells causes malignant transformation ***(28)***. Moreover, deregulation of Plk-1 affects the cellular response to DNA damage ***(29)***. These observations have led to the proposal that Plk-1 could be a bona fide target for antitumor therapies, and several lines of evidence suggest that interference with the function of Plk-1 may indeed result in tumor-selective apoptosis ***(19,30,31)***. Thus, understanding (de)regulation of Plk-1 function in normal vs transformed cells may help identify novel strategies to target tumor outgrowth.

In this chapter we describe methods to purify Plk-1 from cellular lysates and subsequent measurement of Plk-1 kinase activity in vitro. In addition, we describe a method to analyze cell cycle progression after interference with the function of Plk-1 through RNA interference (RNAi)-mediated depletion in tissue culture cells.

2. Materials

1. Plk-1, phospho-histone H3, and phospho-MPM2 antipeptide antibodies can be obtained from Upstate Biotechnology (Lake Placid, NY). Cy5-coupled donkey antirabbit IgG can be obtained from Jackson Immunoresearch (West Grove, PA). Protein A/G sepharose can be obtained from Santa Cruz Biotechnology (Santa Cruz, CA).
2. Dephosphorylated α-casein (Sigma-Aldrich): prepare a fresh 10 mg/mL stock solution in water. Do not store this stock, as α-casein easily forms aggregates in solution.
3. Adenosine triphosphate (ATP; Sigma-Aldrich) and guanosine triphosphate (GTP; Sigma-Aldrich): prepare as a 10 m*M* stock solution in water, and store at –20°C in aliquots of 100 L.
4. [γ-^{32}P]ATP redivue (3000 Ci/mmol; Amersham Life Sciences). Other suppliers that offer [γ-^{32}P]ATP of the same specific activity can be used.
5. Plk-1 lysis buffer: 50 m*M* HEPES, pH 7.4, 1% NP-40, 100 m*M* sodium chloride, 25 m*M* sodium fluoride, 25 m*M* β-glycerophosphate, 1 m*M* sodium orthovanadate, 1 g/mL leupeptin, 1 g/mL aprotinin, 10 g/mL soybean trypsin inhibitor, 50 g/mL phenyl methylsulphonylfluoride (PMSF). Prepare fresh and store on ice (*see* **Note 1**).
6. Plk-1 wash buffer: 20 m*M* HEPES, pH 7.4, 150 m*M* potassium chloride, 10 m*M* magnesium chloride, 1 m*M* EGTA, 0.5 m*M* dithiothreitol, 5 m*M* sodium fluoride. Prepare fresh and store on ice (*see* **Note 1**).
7. Plk-1 kinase buffer: Plk-1 wash buffer supplemented with 10 *M* ATP, 0.15 Ci/mL [γ-^{32}P]ATP and 0.33 g/mL dephosphorylated α-casein. Prepare fresh and store on ice (*see* **Note 1**).
8. 2X Sample buffer: 120 m*M* Tris-HCl, pH 6.8, 4% sodium dodecyl sulfate (SDS), 20% glycerol, 10% β-mercaptoethanol, 0.005% bromophenol blue. Prepare this solution in distilled water, and store in 500- L aliquots at –20°C (*see* **Note 2**).
9. Thymidine (Sigma-Aldrich): prepare a 100X stock of thymidine (250 m*M*) by dissolving 0.6055 g thymidine in 10 mL distilled water, and store at room temperature. This solution is passed over a sterile filter (0.22 m, Millipore) prior to use in tissue culture. After prolonged storage (>2 wk), crystals will form, and the solution should be discarded.
10. Puromycin (Sigma-Aldrich): dissolve 10 mg of puromycin in 1 mL distilled water. Pass the solution over a sterile filter (0.22 m, Millipore), and store small aliquots (100 L) at –20°C.
11. pBabePuro ***(32)***, green fluorescent protein (GFP)-spectrin ***(33)*** (*see* **Note 3**), and pSuper ***(34)*** have been described elsewhere and can be requested/purchased from the appropriate

sources (*see* www.molbio.princeton.edu/facility/flowcyt/GFPPI.html for GFP-spectrin and www.oligoengine.com for pSuper).

12. DNA staining buffer: prepare a solution of 10 g/mL propidium iodide and 250 g/mL RNase A in phosphate-buffered saline (PBS). This solution is freshly prepared from a 10 mg/mL stock of propidium iodide (Sigma-Aldrich) and a 10 mg/mL stock of RNase A (Boerhinger Mannheim; *see* **Note 1**). The propium iodide stock is stored at 4°C. To prepare the RNAse A stock, dissolve 100 mg RNAse A in 10 mL TN buffer (10 m*M* Tris-HCl, pH 7.5, 10 m*M* NaCl). This solution is incubated for 15 min at 100°C, gradually cooled down to room temperature, and stored in 250- L aliquots at –20°C.
13. PBS/0.05% Tween-20: 250 L Tween-20 (Sigma-Aldrich) in 500 mL PBS; store at 4°C.
14. 2X HEPES-buffered saline (HBS): 50 m*M* HEPES, 10 m*M* KCl, 12 m*M* glucose, 1.5 m*M* Na_2HPO_4, 280 m*M* NaCl, pH 7.05. It is important that the pH of this solution be exactly 7.05 at room temperature, as the pH is an important determinant of transfection efficiency. To prepare this solution, weigh the appropriate amount of each compound to prepare a 500-mL batch, and dissolve in 450 mL distilled water. After calibrating the pH to 7.05 with 5 N KOH, the volume is brought to 500 mL. Filter-sterilize (0.22- m filter, Millipore) the solution, and store at –20°C in 5-mL aliquots. Before use, aliquots are thawed and brought to room temperature. Stocks should not be frozen repeatedly.
15. $CaCl_2$ (Sigma-Aldrich): prepare a 2.5 *M* stock of $CaCl_2$ in distilled water. Filter-sterilize (0.22- m filter, Millipore) the solution, and store at –20°C. Notably, this solution stays liquid at –20°C and should be filtered again when it freezes at this temperature.
16. Lowry assay buffers. Lowry buffer A: 2% Na_2CO_3 is dissolved in 0.1 *M* NaOH. Lowry buffer B: 0.5%$CuSO_4 \cdot 5H_2O$ is dissolved in 1% NaCitrate. Both buffer A and B are stored at room temperature and are stable for months. Buffer B is stored protected from light. Lowry buffer C is prepared freshly by mixing buffers A/B in a 50:1 ratio. Lowry buffer D is prepared freshly by adding 1 mL of Folin & Ciolateu's phenol reagent (BDH laboratory Supplies, England) to 2 mL H_2O.
17. Culture medium: Dulbecco's modified eagle's medium (DMEM) with glutamax (Gibco-BRL) supplemented with 8% fetal calf serum (FCS; Sigma) and 50 U/mL penicillin/streptomycin (Gibco-BRL).
18. Paclitaxel and monastrol (Sigma): prepare a 1 m*M* solution of paclitaxel in distilled water. Pass the solution over a sterile filter (0.22 m, Millipore), and store small aliquots (100 L) at –20°C. This stock is diluted 1000X into culture medium when used. Prepare a 200 m*M* solution of monastrol in dimethylsulfoxide (DMSO), and store at –20°C.

3. Methods

3.1. Plk-1 Kinase Activity Assay

The following procedure is a modification of the method of Golsteyn et al. *(22)*. We routinely use this method to determine Plk-1 activity in cell lysates obtained from (synchronized) cultures of human cells, but it can also be used to determine Plk-1 activity from tissue homogenates or from cell lysates obtained from mouse cells.

3.1.1. Preparation of Cellular Lysates

1. To prepare the cellular lysate, harvest 0.3–1 × 10^7 cells by trypsinization. As Plk-1 kinase activity is highest in mitotic cells, care should be taken not to lose the mitotic cells during harvesting of the cells. Therefore, pool the culture medium, PBS wash buffer, and trypsinized cells for each sample in a single 15-mL tube, and pellet the cells by centrifugation (500g for 5 min at 4°C).

2. Discard the supernatant, and wash the cells once with 5 mL of ice-cold PBS. Resuspend the cells in 500 L PBS, and transfer the samples to Eppendorf tubes. Pellet the cells by centrifugation in a benchtop refrigerated centrifuge (500*g* for 5 min at 4°C).
3. Remove the supernatant (care should be taken at this point to remove as much of the supernatant as possible), and lyse the cells in 500 L Plk-1 lysis buffer on ice. Lysis is allowed to reach completion by incubating the samples on ice for 30 min and occasionally inverting the Eppendorf tubes.
4. Centrifuge the lysates at 12,000*g* in a benchtop refrigerated centrifuge (10 min at 4°C), and transfer the supernatant to a clean Eppendorf tube. At this point, small aliquots (5–10 L) of each sample are taken for protein concentration determination (*see* **Note 4**). We routinely continue with immunoprecipitation of Plk-1 directly after lysis, but samples can be stored for a limited amount of time at 4°C or –20°C (*see* **Note 5**).

3.1.2. Immunoprecipitation and Plk-1 Kinase Assay

1. Transfer 150 g of total cell lysate to a clean Eppendorf tube, and bring the final volume to 500 L with Plk-1 lysis buffer.
2. In a separate Eppendorf tube, prepare the protein A/G sepharose beads. (Use 15 L of the protein A/G sepharose stock per sample.) Wash the beads three times with excess Plk-1 lysis buffer (~1 mL buffer per 100 L bead volume) before use. Resuspend the washed beads in Plk-1 lysis buffer at a ratio of 50 L lysis buffer per 15 L bead volume.
3. Add 50 L of washed protein A/G sepharose beads (15 L bead volume) to each sample. To add protein A/G sepharose beads, it is important to resuspend the washed beads carefully to ensure that the beads are equally distributed over the different samples. Also, it is necessary to cut off the ends of the pipet tips to resuspend the beads.
4. Add 5 L of anti-Plk-1 (0.5 g) antibody, and tumble the samples on a rotating incubator for 3–18 h at 4°C.
5. Pellet the beads using a 5-s pulse at 12,000*g* in a refrigerated benchtop centrifuge at 4°C. It is important not to spin the beads much longer than 10 s per run, as this will lead to rupture of the beads and results in high background in the activity measurements. Wash the beads three times with excess ice-cold Plk-1 lysis buffer (1 mL per sample).
6. Wash the beads once with 1 mL Plk-1 wash buffer, and remove the supernatant with a Gilson pipet, leaving approx 50 L of wash buffer on the beads. Carefully remove the remaining supernatant with a 1-mL syringe equipped with a 25-gage needle. Leave just enough liquid not to let the beads dry out and quickly proceed to the next step.
7. To each sample add 30 L kinase buffer, and incubate at 30°C for 30 min. Terminate the reactions by adding 30 L of 2X sample buffer. At this point, samples can be stored at –20°C for several days.
8. Samples are heated to 95°C in a heating block for 5 min, and proteins are separated on regular 12% SDS-polyacrylamide gels. Gels are fixed in methanol/acetic acid/water (30:10:60), dried, and exposed to X-ray film or PhosphoImager screens for the appropriate exposure time. Exposure times will vary depending on the type of cells used and the method of detection. **Figure 2A** shows a representative autoradiograph of a Plk-1 kinase assay from synchronized U2OS osteosarcoma cells (*see* **Note 6**).

3.2. Plk-1 Depletion by Vector-Driven siRNA Expression

The method we describe here allows depletion of Plk-1 in human cells in tissue culture. It utilizes the recently described pSuper vector for expression of small interfering RNAs (siRNAs) *(34)*. It goes beyond the scope of this chapter to describe a

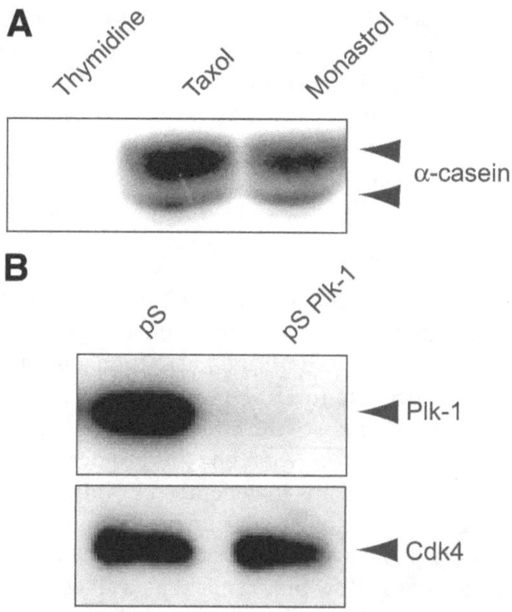

Fig. 2. **(A)** Plk-1 kinase activity measurement. U2OS cell were treated with 1 *M* of the microtubule-stabilizing agent paclitaxel or 100 *M* of the Eg5 inhibitor monastrol, in order to arrest cells in mitosis. Alternatively, cells were treated with thymidine for 24 h to arrest cells at the G_1/S transition. Plk-1 was immunoprecipitated from 150 g protein lysate. Plk-1 activity was determined in kinase asssays with dephosphorylated α-casein as a substrate. **(B)** Biochemical analysis of Plk-1–depleted cells. U2OS cells were transfected with 10 g pSuper or pSuper-Plk-1 in combination with 1 g pBabePuro. At 18 h after transfection, cells were incubated with thymidine and puromycin for 24 h, in order to arrest cells at the G_1/S border and select for transfected cells. At 18 h after thymidine release, Plk-1 levels were analyzed by Western blotting. Cdk4 protein levels were detected by Western blotting and served as a loading control. (Rabbit anti-Cdk4 was from Santa Cruz Biotechnology, Santa Cruz, CA.).

detailed procedure for cloning of a pSuper-derived targeting construct, and we will only briefly discuss the cloning procedure that was followed to construct the targeting vector for Plk-1. A detailed description of pSuper and the cloning procedures to construct a targeting vector are available at www.oligoengine.com. pSuper-derived targeting vectors allow transient as well as stable depletion from cultured cells. Since stable depletion of Plk-1 is not compatible with sustained cell proliferation (our own unpublished observations), we have optimized our RNAi approaches for transient depletion in synchronized cultures of cells.

The approach described here is applicable for any mitotic regulatory protein as long as an efficient targeting vector can be constructed (*see* **Note 7**). Following introduction of the targeting construct, cells are synchronized at the G_1/S-phase transition by the addition of thymidine to the culture medium. This synchronization procedure permits depletion of a mitotic kinase such as Plk-1 without causing the deleterious effects one would normally see if the cells are allowed to continue to cycle. In addition, it allows one to monitor carefully (synchronous) cell cycle progression of Plk-1-depleted cultures through G_2 and into mitosis upon release of the cells from the thymidine block. We find this method particularly useful to combine single cell-based assays (time-lapse microscopy, immunofluorescence) with population-based assays (flow cytometry, kinase assays, Western blotting) to determine various aspects of cell cycle progression.

3.2.1. Construction of a pSuper-Based Targeting Vector

In the method described here, vector-driven RNA interference is used to deplete a protein of interest from mammalian cells. This technology was first described by Brummelkamp et al. ***(34)***. The method described here is an adaptation of that original description to optimize the procedure for depletion of mitotic proteins by utilizing cell synchronization following transfection of the targeting construct(s). Although here we describe a method for depletion of Plk-1, we found this protocol very useful for siRNA-mediated knockdown of mitotic proteins in general (*see* **Note 7**). Following depletion of the protein of interest, cells can be released from the thymidine block, and cell cycle progression can be followed using a variety of assays.

Here we describe the assay to allow biochemical as well as flow cytometric analysis. To prepare the pSuper-derived targeting construct, the target site is selected from the mRNA of choice, and oligos that contain an inverted repeat of this sequence separated by a short loop are cloned into the pSuper vector. For Plk-1, our most effective sequence is cggcagcgtgcagatcaac, corresponding to nucleotides 1581–1599 in the human Plk-1 gene. The pSuper vector contains a polymerase-III H1-RNA gene promoter that drives transcription of this sequence (**Fig. 3**). Specific cleavage of this transcript at the termination site results in the synthesis of short, single-stranded mRNAs that form a stem–loop structure, whereas RNA processing within the host cell subsequently results in the formation of double-stranded RNAs that can function as siRNAs to remove the target mRNA specifically.

3.2.2. Transfection, Synchronization, and Selection of Cells for RNAi-Mediated Depletion

1. Plate U2OS cells on 10-cm dishes so that on the day of transfection they will have reached a confluency of appromimately 40–60%. Remove the culture medium 2 h before transfection, replace it with prewarmed fresh medium, and place the cells back in the CO_2 incubator. This step will ensure a neutral pH value of the medium at the moment the calcium–phosphate/DNA precipitate is added.
2. To prepare the calcium/DNA solution, combine 10 g pSuper-Plk DNA with 1 g pBabePuro DNA (for biochemical analysis, *see* **Subheading 3.2.4.**) or 1 g GFP-spectrin DNA (for cell cycle analysis, *see* **Subheading 3.2.5.**) in a sterile Eppendorf tube. Add sterile H_2O to a final volume of 450 L.
3. Add 50 L $CaCl_2$ to each tube, and vortex this mixture briefly. Collect the solution to the bottom of the tube by a short pulse in a benchtop centrifuge.
4. Remove a single culture dish from the CO_2 incubator, and immediately prepare the calcium–phosphate/DNA precipitate for this dish (**step 5**). If multiple dishes are transfected,

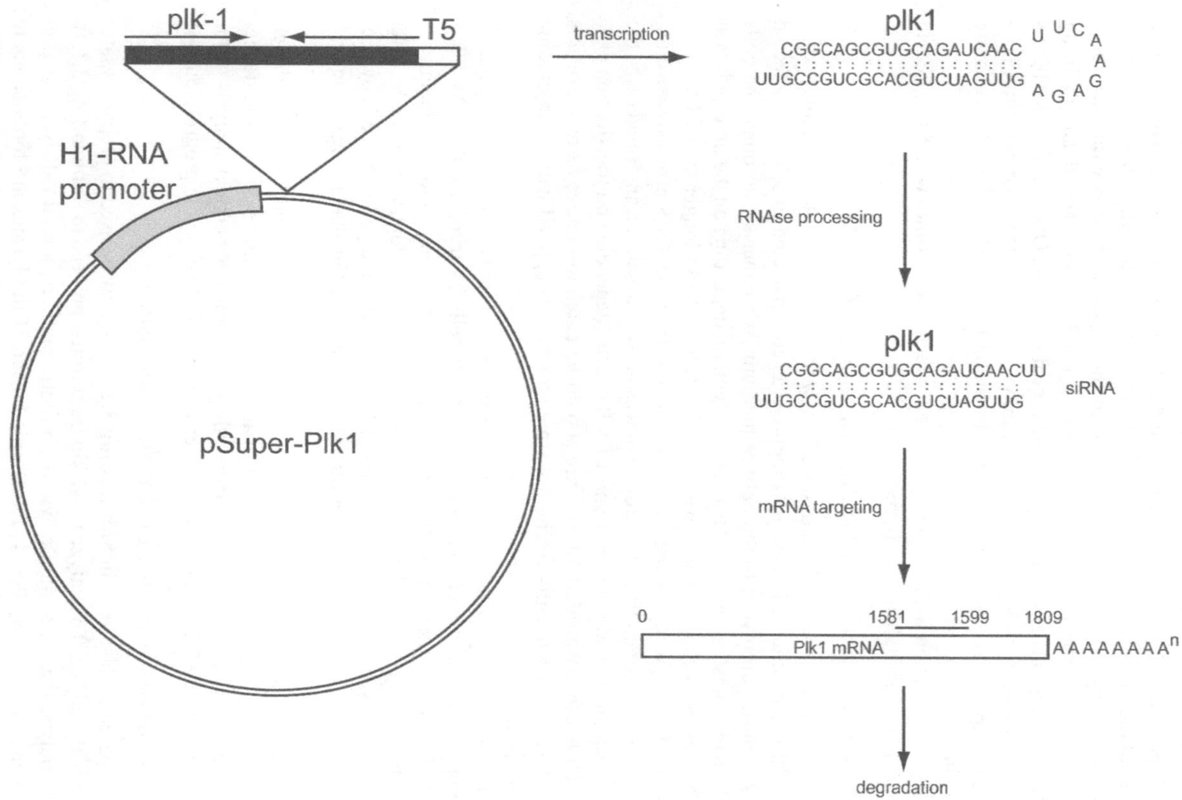

Fig. 3. Schematic representation of the mode of action of a pSuper-derived targeting construct. A polymerase III H1-RNA gene promoter drives transcription of a palindromic Plk-1 targeting sequence, followed by a T5 termination signal. The palindromic sequence results in the formation of a stem–loop structure. RNAse processing within the host results in the production of 19-mer siRNAs, which specifically target Plk-1 mRNA for degradation.

do not take more than four dishes at a time, as the pH of the medium will go off, affecting the transfection efficiency.

5. Slowly add 500 L 2X HBS to the calcium/DNA solution from **step 3**. This is the most critical step of the procedure, as high local concentrations of phosphate may trigger formation of excessively large crystals that will not be taken up by the cells. Therefore layer the 2X HBS into the calcium/DNA solution. To do this, bring the pipet tip with 2X HBS to the bottom of the tube, and slowly eject the 2X HBS into the calcium/DNA solution. At the same time, slowly move the pipet tip upward through the solution while making a swirling motion. If the surface is reached before all the 2X HBS is added, simply bring the pipet tip to the bottom of the tube again, and repeat the cycle. Once all the 2X HBS is added, mix the solution by pipeting it up and down once or twice. The calcium-phosphate/DNA precipitate is immediately added to the culture medium and the cells are placed back into the CO_2-incubator.
6. After overnight incubation with the precipitate cells are washed 3 times with 5 mL PBS to remove the calcium-phosphate/DNA precipitate.

3.2.3. Synchronization and Selection of Transfected Cells

1. To synchronize the cells, add 100 L of the 250 m*M* thymidine stock to each culture dish (containing 10 mL medium). In case pBabePuro was cotransfected to select transfected cells, add 2 g/mL puromycin to the culture medium. We either synchronize the cells directly following transfection or at 8 h after transfection, depending on the time-points at which we harvest the cells. This way we avoid thymidine blocks longer than 24 h.
2. At 24 h after addition of thymidine, cells are released from the G_1/S arrest by washing three times with prewarmed PBS. If puromycin was present to select transfected cells, the medium can contain considerable amounts of cell debris (depending on transfection efficiency, which is usually around 60%). Care should be taken to remove as much of this debris as possible. Subsequently fresh prewarmed medium is applied without thymidine and/or puromycin.
3. Typically at 18 h after release from the thymidine block, cells are harvested by trypsinization. To prepare the cell pellets, harvest $0.3–1 \times 10^7$ cells by trypsinization. As Plk-1 depletion leads to enrichment in mitotic cells, care should be taken not to lose the mitotic cells during harvesting of the cells. Therefore, pool the culture medium, PBS wash buffer and trypsinized cells for each sample in a single 15-mL tube, and pellet the cells by centrifugation (500*g* for 5 min at 4°C). Other time-points can be taken, depending on the cell cycle phase of interest.
4. After centrifugation of the cells, carefully aspirate the supernatant and wash the cell pellet with 5 mL PBS. Pellet the cells by centrifugation (500*g* for 5 min at 4°C). Aspirate the supernatant carefully, and proceed to **Subheading 3.2.4.** for biochemical analysis of the cells or to **Subheading 3.2.5.** for flow cytometric analysis of cell cycle progression.

3.2.4. Biochemical Analysis of Cell Cycle Progression

1. The cells prepared as described in **Subheading 3.2.3.** can now be used to prepare lysates for biochemical analysis. To analyze Plk-1 kinase activity, proceed to **Subheading 3.1.1.**, **step 3**. To analyze the expression of various mitotic markers, a total cell lysate is prepared by lysing the cells in 500 L 2X sample buffer. If the lysates are too viscous to pipet properly, lysates can be sheared with a 1-mL syringe equipped with a 2-gage needle.
2. Transfer the lysates to Eppendorf tubes, and make a small hole in the lid of the tube with the 25-gage needle. Heat the closed tubes for 5 min at 95°C. At this point, cell lysates can be stored at –20°C for several weeks.

3. Determine the protein concentration of the lysates with the Lowry assay (*see* **Note 2**).
4. Run 50 g of total protein lysates per lane on an SDS-polyacrylamide gel (percentage depending on the size of the protein to be analyzed), and transfer the proteins to a nitrocellulose membrane, using a regular Western blotting procedure.
5. Proteins are detected with the appropriate antibodies. We routinely check expression of Plk-1 to check the efficiency of RNAi-mediated depletion and analyze for the presence/absence of useful mitotic markers such as phospho-histone H3, cyclin A, or cyclin B. An example is shown in **Fig. 2B**.

3.2.5. Flow Cytometric Analysis of Cell Cycle Progression

1. Resuspend the cell pellet prepared in **Subheading 3.2.3.** carefully in 100 L PBS.
2. Fix the cells in 70% ethanol, by adding 5 mL of ice-cold 70% ethanol dropwise to the cell suspension.
3. Incubate the cells at 4°C for at least 30 min. At this point, fixed cells can be stored in the 70% ethanol for up to 2 wk at 4°C.
4. Spin down the cells (500*g* for 5 min at 4°C), and carefully remove the supernatant.
5. Resuspend the cell pellet in the little bit of liquid that is left behind, add 5 mL PBS, and collect the cells by by centrifugation (500*g* for 5 min at 4°C).
6. Remove as much as possible of the supernatant, and resuspend the cells in 100 L PBS, containing 1 L anti-phospho-histone H3. Incubate for 3 h at room temperature.
7. Wash the cells with 5 mL of PBS/0.05% Tween-20, and collect the cells by centrifugation (500*g* for 5 min at 4°C).
8. Remove as much as possible of the supernatant, and resuspend the cells in 100 L PBS, containing 1 L Cy5-conjugated donkey antirabbit antibody. Incubate for 1 h at room temperature; protect the samples from light.
9. Wash the cells with 5 mL of PBS/0.05% Tween-20, and collect the cells by centrifugation (500*g* for 5 min at 4°C).
10. Remove as much as possible of the supernatant, resuspend the cells in 500 L DNA staining buffer, incubate at 37°C for 15 min.
11. DNA profiles and phospho-histone H3 positivity can be analyzed by trivariate flow cytometry using a fluorescence-activated cell sorter (FACS), equipped with a 488 nm and a 633-nm light source, to excitate GFP, propidium iodide, and Cy5, respectively. (We use a Becton Dickinson FACS-Calibur, equipped with a Argon 488 laser, and a helium/neon diode to produce 633-nm light). To analyze the data, Cell Quest software (Becton Dickinson) is used. A typical result is shown in **Fig. 4**.

4. Notes

1. All buffers should be prepared fresh on the day of the assays. We have found that longer term storage of such solutions often leads to formation of precipitates and a reduction in the measurable kinase activity.
2. To determine protein concentrations in lysates prepared with 2X sample buffer, cells should be lysed with 2X sample buffer prepared without bromophenol blue and β-mercaptoethanol. Then 5–10 L of the lysates is transferred to clean Eppendorf tubes to determine protein concentration using the Lowry assay. The remainder of the lysate is supplemented with bromophenol blue and β-mercaptoethanol from a 20X stock. For the Lowry assay, add 90 L of distilled water in Eppendorf tubes. (Samples containing 10, 20, 50, or 100 g of BSA in 5–10 L of 2X sample buffer are included for calibration.) Add 1 mL of Lowry buffer C, vortex, and incubate for 10 min at room temperature. Add

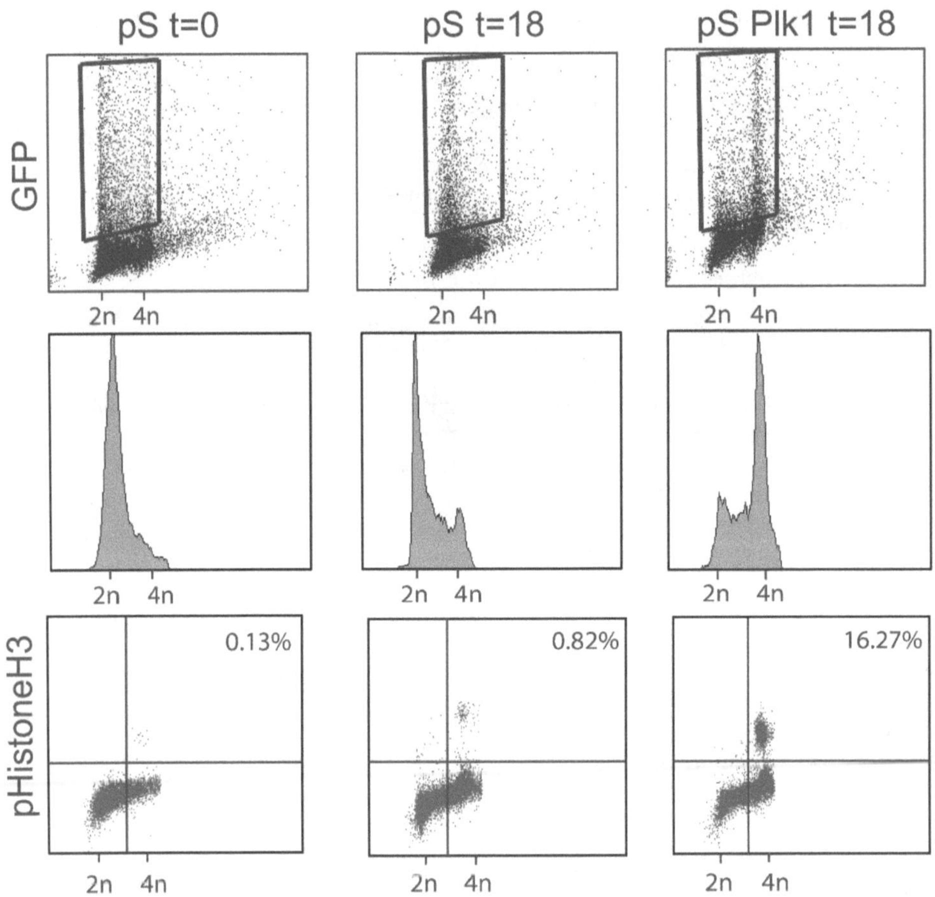

Fig. 4. Analysis of cell cycle progression of Plk-1–depleted cells by flow cytometry. U2OS cells were transfected with 10 g pSuper or pSuper-Plk-1 in combination with 1 g green fluorescent protein (GFP)-spectrin. At 18 h after transfection, cells were incubated in thymidine for 24 h, to arrest cells at the G_1/S border. At indicated time-points after thymidine release, cells were harvested and fixed in ethanol. Mitotic cells were stained with rabbit anti-phospho-histone H3 (pHistone H3) in combination with Cy5-conjugated donkey antirabbit antibody. DNA was counterstained with propidium iodide, and 10^4 cells were analyzed on a Becton Dickinson FACS-Calibur using CellQuest software.

100 L of Lowry buffer D, vortex, and incubate at room temperature for 30 min. Protein concentration is measured at 750 nm.

3. We use spectrin-GFP, rather than normal GFP, to analyze cell cycle profiles of transfected cells, because regular GFP will leak out of the cells during fixation.
4. For protein determination of Plk kinase assay lysates, Bio-Rad assays are performed. To this end, 10 L of Plk kinase assay lysate is added to 790 L distilled water. Samples are vortexed, and 200 L Bio-Rad Protein Assay solution is added. Samples are incubated at

room temperature for 5 min. For calibration, 10 g BSA in 10 L Plk kinase buffer is diluted in 790 L of distilled water, and 200 L Bio-Rad Protein Assay solution is added. Protein concentration is determined at 595 nm.

5. We routinely immunoprecipitate Plk-1 directly following preparation of cell lysates in Plk-1 lysis buffer. However, lysates can be stored overnight at 4°C before immunoprecipitation, or immunoprecipitation can be carried out for periods up to 48 h. Do not freeze the lysates prior to kinase measurement. Both procedures result in suboptimal kinase activities and should be avoided if possible.
6. As controls for Plk-1 kinase assays, it is advisable to carry out a number of additional immunoprecipitation reactions, one leaving out the anti-Plk-1 antibody, one leaving out the lysate, and one with another unrelated antibody. Also, as a means to rule out contamination of the immunoprecipitate with casein kinase II, one can do the kinase reaction in the presence of 500 g/mL heparin or 10 m*M* unlabeled GTP. Casein kinase II is inhibited by heparin, and can substitute ATP for GTP as a phosphate donor in the kinase reaction, whereas Plk-1 cannot ***(22)***.
7. At present, not much is known about the determinants of a good targeting sequence for RNAi. We find that target site selection is basically a question of trial and error. Some of our most efficient targeting constructs do not adhere to any published rules, and some that do fail to downregulate their target efficiently.

References

1. Glover, D. M., Hagan, I. M. and Tavares, A. A. (1998) Polo-like kinases: a team that plays throughout mitosis. *Genes Dev.* **12,** 3777–3787.
2. Donaldson, M. M., Tavares, A. A., Hagan, I. M., Nigg, E. A., and Glover, D. M. (2001) The mitotic roles of Polo-like kinase. *J. Cell Sci.* **114,** 2357–2358.
3. Nigg, E. A. (1998) Polo-like kinases: positive regulators of cell division from start to finish. *Curr. Opin. Cell Biol.* **10,** 776–783.
4. Abrieu, A., Brassac, T., Galas, S., Fisher, D., Labbe, J. C., and Doree, M. (1998) The Polo-like kinase Plx1 is a component of the MPF amplification loop at the G2/M-phase transition of the cell cycle in *Xenopus* eggs. *J. Cell Sci.* **111,** 1751–1757.
5. Kumagai, A. and Dunphy, W. G. (1996) Purification and molecular cloning of Plx1, a Cdc25-regulatory kinase from *Xenopus* egg extracts. *Science* **273,** 1377–1380.
6. Qian, Y. W., Erikson, E., and Maller, J. L. (1998) Purification and cloning of a protein kinase that phosphorylates and activates the polo-like kinase Plx1. *Science* **282,** 1701–1704.
7. Toyoshima-Morimoto, F., Taniguchi, E., Shinya, N., Iwamatsu, A., and Nishida, E. (2001) Polo-like kinase 1 phosphorylates cyclin B1 and targets it to the nucleus during prophase. *Nature* **410,** 215–220.
8. Toyoshima-Morimoto, F., Taniguchi, E., and Nishida, E. (2002) Plk1 promotes nuclear translocation of human Cdc25C during prophase. *EMBO Rep.* **3,** 341–348.
9. Sumara, I., Vorlaufer, E., Stukenberg, P. T., et al. (2002) The dissociation of cohesin from chromosomes in prophase is regulated by Polo-like kinase. *Mol. Cell.* **9,** 515–525.
10. Losada, A., Hirano, M., and Hirano, T. (2002) Cohesin release is required for sister chromatid resolution, but not for condensin-mediated compaction, at the onset of mitosis. *Genes Dev.* **16,** 3004–3016.
11. Waizenegger, I. C., Hauf, S., Meinke, A., and Peters, J. M. (2000) Two distinct pathways remove mammalian cohesin from chromosome arms in prophase and from centromeres in anaphase. *Cell* **103,** 399–410.

12. Orr-Weaver, T. L. (1999) The ties that bind: localization of the sister-chromatid cohesin complex on yeast chromosomes. *Cell* **99,** 1–4.
13. Golan, A., Yudkovsky, Y., and Hershko, A. (2002) The cyclin-ubiquitin ligase activity of cyclosome/APC is jointly activated by protein kinases Cdk1-cyclin B and Plk. *J. Biol. Chem.* **277,** 15552–15557.
14. Kotani, S., Tugendreich, S., Fujii, M., et al. (1998) PKA and MPF-activated polo-like kinase regulate anaphase-promoting complex activity and mitosis progression. *Mol. Cell.* **1,** 371–380.
15. Llamazares, S., Moreira, A., Tavares, A., et al. (1991) polo encodes a protein kinase homolog required for mitosis in *Drosophila. Genes Dev.* **5,** 2153–2165.
16. Sunkel, C. E. and Glover, D. M. (1988) polo, a mitotic mutant of *Drosophila* displaying abnormal spindle poles. *J. Cell Sci.* **89,** 25–38.
17. do Carmo Avides, M., Tavares, A., and Glover, D. M. (2001) Polo kinase and Asp are needed to promote the mitotic organizing activity of centrosomes. *Nat. Cell Biol.* **3,** 421–424.
18. Gonzalez, C., Sunkel, C. E., and Glover, D. M. (1998) Interactions between mgr, asp, and polo: asp function modulated by polo and needed to maintain the poles of monopolar and bipolar spindles. *Chromosoma* **107,** 452–460.
19. Lane, H. A. and Nigg, E. A. (1996) Antibody microinjection reveals an essential role for human polo-like kinase 1 (Plk1) in the functional maturation of mitotic centrosomes. *J. Cell Biol.* **135,** 1701–1713.
20. Lee, K. S. and Erikson, R. L. (1997) Plk is a functional homolog of *Saccharomyces cerevisiae* Cdc5, and elevated Plk activity induces multiple septation structures. *Mol. Cell. Biol.* **17,** 3408–3417.
21. Wind, M., Kelm, O., Nigg, E. A., and Lehmann, W. D. (2002) Identification of phosphorylation sites in the polo-like kinases Plx1 and Plk1 by a novel strategy based on element and electrospray high resolution mass spectrometry. *Proteomics* **2,** 1516–1523.
22. Golsteyn, R. M., Mundt, K. E., Fry, A. M., and Nigg, E. A. (1995) Cell cycle regulation of the activity and subcellular localization of Plk1, a human protein kinase implicated in mitotic spindle function. *J. Cell Biol.* **129,** 1617–1628.
23. Jackman, M., Lindon, C., Nigg, E. A., and Pines, J. (2003) Active cyclin B1-Cdk1 first appears on centrosomes in prophase. *Nat. Cell Biol.* **5,** 143–148.
24. Arnaud, L., Pines, J., and Nigg, E. A. (1998) GFP tagging reveals human Polo-like kinase 1 at the kinetochore/centromere region of mitotic chromosomes. *Chromosoma* **107,** 424–429.
25. Jang, Y. J., Lin, C. Y., Ma, S., and Erikson, R. L. (2002) Functional studies on the role of the C-terminal domain of mammalian polo-like kinase. *Proc. Natl. Acad. Sci. USA* **99,** 1984–1989.
26. Elia, A. E., Cantley, L. C., and Yaffe, M. B. (2003) Proteomic screen finds pSer/pThr-binding domain localizing Plk1 to mitotic substrates. *Science* **299,** 1228–1231.
27. Yuan, J., Horlin, A., Hock, B., et al. (1997) Polo-like kinase, a novel marker for cellular proliferation. *Am. J. Pathol.* **150,** 1165–1172.
28. Smith, M. R., Wilson, M. L., Hamanaka, R., et al. (1997) Malignant transformation of mammalian cells initiated by constitutive expression of the polo-like kinase. *Biochem. Biophys. Res. Commun.* **234,** 397–405.
29. Smits, V. A., Klompmaker, R., Arnaud, L., Rijksen, G., Nigg, E. A., and Medema, R. H. (2000) Polo-like kinase-1 is a target of the DNA damage checkpoint. *Nat. Cell Biol.* **2,** 672–676.
30. Elez, R., Piiper, A., Giannini, C. D., Brendel, M., and Zeuzem, S. (2000) Polo-like kinase1, a new target for antisense tumor therapy. *Biochem. Biophys. Res. Commun.* **269,** 352–356.

31. Yuan, J., Kramer, A., Eckerdt, F., Kaufmann, M., and Strebhardt, K. (2002) Efficient internalization of the polo-box of polo-like kinase 1 fused to an *Antennapedia* peptide results in inhibition of cancer cell proliferation. *Cancer Res.* **62,** 4186–4190.
32. Morgenstern, J. P. and Land, H. (1990) Advanced mammalian gene transfer: high titre retroviral vectors with multiple drug selection markers and a complementary helper-free packaging cell line. *Nucleic Acids Res.* **18,** 3587–3596.
33. Kalejta, R. F., Shenk, T., and Beavis, A. J. (1997) Use of a membrane-localized green fluorescent protein allows simultaneous identification of transfected cells and cell cycle analysis by flow cytometry. *Cytometry* **29,** 286–291.
34. Brummelkamp, T. R., Bernards, R., and Agami, R. (2002) A system for stable expression of short interfering RNAs in mammalian cells. *Science* **296,** 550–553.

22

The Ipl1/Aurora Kinase Family

Methods of Inhibition and Functional Analysis in Mammalian Cells

Claire Ditchfield, Nicholas Keen, and Stephen S. Taylor

Summary

The Ipl1/Aurora family of protein kinases are required for accurate chromosome segregation. Because members of this family are often overexpressed in human tumors, they have recently received much attention, both from the academic community and the pharmaceutical industry. Indeed, two small molecule Aurora kinase inhibitors have recently been described. In this chapter, we describe several methods for investigating the function of the Aurora kinases, focusing on Aurora B. We describe the use of the small-molecule inhibitor ZM447439, RNA interference, and overexpression of a catalytic mutant. All of these methods have proved useful in studying Aurora B as well as validating it as a potential anticancer drug target. However, while all three methods are useful for probing the function of Aurora B, each has inherent advantages and disadvantages. Furthermore, because the mechanism underlying the inhibition is different in each case, caution must be taken when interpreting the data.

Key Words

Mitosis; kinetochore; spindle checkpoint; Ipl1; Aurora B.

1. Introduction

Accurate chromosome segregation is essential for the maintenance of genomic stability. Several families of protein kinases play distinct roles throughout mitosis and cytokinesis to ensure that each daughter cell inherits one complete copy of the genome. Members of the Ipl1/Aurora family of protein kinases are key regulators of both chromosome segregation and cytokinesis ***(1–4)***. The founding members of this family are Ipl1p from *Saccharomyces cerevisiae* ***(5)*** and aurora from *Drosophila melanogaster*

From: *Methods in Molecular Biology, vol. 296, Cell Cycle Control: Mechanisms and Protocols*
Edited by: T. Humphrey and G. Brooks © Humana Press Inc., Totowa, NJ

(6). Although the yeasts contain a single Aurora kinase, higher eukaryotes such as worms, flies, and frogs contain at least two family members. Three kinases, Aurora A, B, and C, have been identified in humans. These three proteins range in size from 309 to 403 amino acids and share 67–76% sequence identity in their catalytic domains, located at the C-terminus. In contrast, the N-terminal domains, which vary in size, do not share any obvious sequence similarity *(2)*.

The three human Aurora kinases are cell cycle-regulated, with both their protein levels and their activity peaking during G_2 and mitosis. However, the subcellular localization patterns of each protein are distinct (**Fig. 1**). Aurora A associates mainly with the centrosome and mitotic spindle throughout mitosis, whereas Aurora B is a chromosome passenger protein, localizing to centromeres in early mitosis, then transferring to the spindle midzone upon the onset of anaphase, and finally accumulating at the midbody in telophase *(1–3)*. In many organisms Aurora A has been shown to play a role in centrosome separation and/or maturation *(6–8)*. Aurora B functions in a complex with two other passenger proteins, inner centromere protein (INCENP) and Survivin, and plays multiple roles in chromosome segregation and cytokinesis *(1)*. Aurora B has been shown to be required for the phosphorylation of histone H3, chromosome alignment, and spindle checkpoint control. The function of Aurora C is unknown, but it is expressed primarily in the testis and localizes to centrosomes only at the very latter stages of mitosis *(9)*.

Interest in the Aurora family of protein kinases rose following the discovery that they are overexpressed in many human tumors *(10–12)*. The *Aurora A* gene maps to chromosome 20q13, a region often amplified in human cancers, and overexpression of Aurora A in Rat-1 fibroblasts leads to transformation, centrosome amplification, and genomic instability. Furthermore, cells overexpressing wild-type Aurora A, but not a kinase dead mutant, form tumors when injected into nude mice *(10)*. More recently, quantitative trait loci mapping has identified *Aurora A* as a low-penetrance cancer susceptibility gene in both mouse and humans *(13)*. In addition, overexpression of Aurora A induces taxol resistance *(14)*. As a result, the Aurora kinases are being considered as potential targets for novel anticancer drugs.

Although the Auroras have been extensively studied over the last few years, it is still unclear how exactly the different kinases function during mitosis. One informative way to investigate the role of a particular protein is to inhibit its function in cultured cells and then analyze the resulting phenotypes. This should then provide insight into the processes for which the protein is normally required. Several different strategies can be used to inhibit protein function in human cells including gene targeting, RNA interference (RNAi), antibody injection, or ectopic overexpression of mutants. When analyzing the role of a protein kinase, it is also possible to overexpress catalytic mutants or inhibit catalytic activity directly through the use of small molecules. Each technique inhibits protein function by a different mechanism, and therefore care must be taken when interpreting the phenotypes observed. For example, small molecules such as ZM447439 and Hesperadin directly inhibit Aurora kinase activity without altering protein localisation or expression level *(15,16)*.

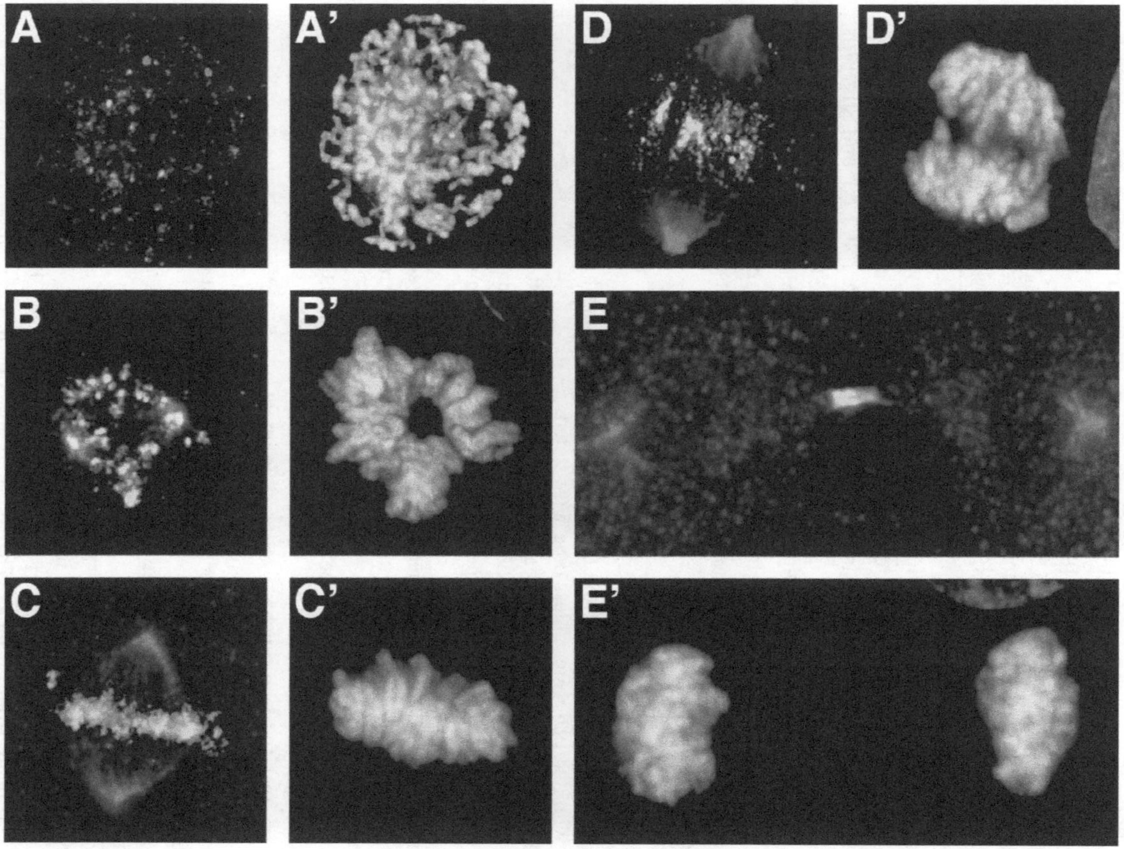

Fig. 1. Subcellular localization of Aurora A and B during **(A)** prophase, **(B)** prometaphase, **(C)** metaphase, **(D)** anaphase, and **(E)** telophase. Human DLD-1 cells were fixed and stained (A'–E') to detect Aurora A (red), Aurora B (green), and DNA (blue). Image stacks were taken, deconvolved and projected onto a single plane.

There are several additional advantages to using small molecules. First, they are readily absorbed by the cell, and therefore Aurora kinase activity is inhibited across the entire population of cells. Second, their action is rapid, allowing the drugs to be added at specific time-points during the cell cycle. In contrast, gene knockouts, RNAi-mediated repression, or overexpression of kinase mutants take longer to act and rely on good transfection efficiency or the generation of stable cell lines. Moreover, these techniques alter the amount of the protein present in the cell, which may have other consequences in addition to simply reducing kinase activity. This is exemplified by recent Aurora B studies in which inhibition by either small molecules, RNAi, antibody injection and overexpression of a kinase mutant yielded similar but not identical phenotypes ***(15–18)***. These differences probably arise because inhibition of Aurora B with a small molecule does not mislocalize Aurora B or Survivin, suggesting that the Aurora B/INCENP/Survivin complex is still intact ***(15)***. In contrast, repression of Aurora B or overexpression of a kinase mutant does mislocalize INCENP and Survivin ***(15,19)***, indicating that disruption of the Aurora B/INCENP/Survivin complex has consequences additional to simple inhibition of Aurora B kinase activity. Following on from these recent observations, this chapter focuses on three different strategies to inhibit Aurora B in cultured cells and then describes several assays that can be used to analyze the resulting phenotypes.

2. Materials

1. Chemicals used should be of the best grade available, and all solutions should be prepared with distilled water unless otherwise stated.
2. ZM447439 (AstraZeneca): prepare a 10 m*M* solution in sterile dimethylsulfoxide (DMSO), and store in small aliquots at –20°C to avoid freeze–thaw cycles. It is stable for up to 6 mo. On the day of the experiment dilute ZM447439 in media, and use at a final concentration of 2 *M*.
3. Culture media: Dulbecco's modified Eagle's medium (DMEM) supplemented with 10% fetal boving serum (FBS), 100 U/mL penicillin, 100 g/mL streptomycin, and 2 m*M* L-glutamine (all from Invitrogen).
4. TA-HeLa cells ***(20)*** and DLD-1 cells (American Type Culture Collection).
5. Thymidine (Sigma): prepare a 200 m*M* solution in water, filter-sterilize, and store at 4°C. Make up fresh for each synchronization experiment.
6. Nocodazole (Sigma): prepare a 5 mg/mL solution in DMSO, and store at –20°C in aliquots. Dilute in media on the day of the experiment to a final concentration of 0.2 g/mL.
7. Optimem (Invitrogen).
8. Oligofectamine (Invitrogen): store at 4°C for up to 12 mo. Mix gently before use to ensure the lipid is evenly dispersed.
9. siRNA oligos can be purchased from various manufacturers as purified and annealed duplexes. Reconstitute, and store in small aliquots at –20°C.
10. 2X $CaCl_2$ and 2X HBS are part of the Profection Mammalian Transfection System Calcium Phosphate from Promega. The solutions are aliquoted and stored at –20°C. Use a fresh aliquot each time to avoid freeze–thaw cycles.
11. Transfection serum: FBS from Hyclone (Pierce and Warriner, cat. no. SH30070.03).
12. PBST: phosphate-buffered saline (PBS) containing 0.1% Triton X-100.
13. Blocking solution: 5% nonfat milk in PBST.

14. MPM-2 FSE conjugate (Upstate Biotechnology): store at –20°C in the dark, and use at a final concentration of 1.25 g/mL.
15. Antiphospho-Histone H3 (Upstate Biotechnology): the antibody is aliquoted and stored at –20°C, with a working aliquot stored at 4°C to avoid freeze–thaw cycles. Use at 1:200.
16. Sheep polyclonal antibodies against Bub1 and BubR1 have been made in the laboratory and are routinely used for immunofluorescence analysis ***(21)***. Store current aliquot at 4°C, and use at 1:1000. Mouse monoclonal antibodies are now available commercially from Chemicon and give good kinetochore staining when used at a dilution of 1:1000 (unpublished data).
17. Anti-Aim1, a mouse monoclonal antibody that detects Aurora B (Transduction Laboratories): store at –20°C and use at 1:200. We also use a sheep polyclonal antibody against human Aurora B that was made in the laboratory (1:1000) ***(15)***.
18. Cy2-, Cy3-, and Cy5-conjugated secondary antibodies (Jackson ImmunoResearch): reconstitute, and add an equal volume of 100% glycerol, store at –20°C, and use at 1:500.
19. Mounting media: 90% glycerol, 20 m*M* Tris-HCl, pH 8.0.

3. Methods

3.1. Approaches to Inhibit Aurora Function

The following section outlines three different strategies for effective inhibition of Aurora B function in cultured cells. In particular, it describes how the small molecule ZM447439 can be used to inhibit Aurora kinase activity in cells synchronized either at the G_1/S boundary or in mitosis. By analyzing synchronous populations of cells that have been released into the inhibitor, it is possible to determine the effect of inhibiting Aurora kinase activity on progression through the cell cycle and the exit from mitosis, respectively.

3.1.1. Cell Cycle Analysis Following Release from G_1/S

This method has been routinely used in the laboratory to synchronize cells at the G_1/S boundary following a double thymidine block (*see* **Note 1**).

1. Plate HeLa cells at a density of 3.5×10^5 cells per 100-mm dish in 10 ml culture media, and grow overnight.
2. Add 100 L thymidine (200 m*M*) to each dish to give a final concentration of 2 m*M*, and incubate for 16 h overnight.
3. Wash the cells twice with 5 mL PBS and return to 10 mL normal culture media for at least 8 h. After 8 h, re-add 100 L thymidine (200 m*M*) to each dish and incubate for a further 14-16 hours overnight.
4. Wash the cells twice with 5 mL PBS, and return to 8 mL normal culture medium containing 2 *M* ZM447439 or 0.02% DMSO to control for the solvent. The cells can be harvested at various time points as they continue through the cell cycle and then analyzed by fluorescence-activated cell sorting (FACS), Western blot, or cytospin/immunofluorescence as described in **Subheading 3.2.**

3.1.2. Cell Cycle Analysis Following a Nocodazole Block and Release

This method has been routinely used in the laboratory for analyzing cells as they exit mitosis. Cells are treated with nocodazole for 12–14 h overnight to accumulate cells in mitosis (*see* **Note 2**). Mitotic cells are then harvested by selective detachment or “shake-off” and replated into media containing different drug combinations, allowing cells to be analyzed at different time-points following release from the nocodazole block.

1. Plate HeLa cells at a density of 2×10^6 cells per 150-mm dish in 20 mL culture media, and grow overnight. The following evening, add nocodazole to a final concentration of 0.2 g/mL, and incubate for 12–14 h overnight.
2. To harvest the mitotic cells by “shake-off,” transfer the media to a 50-mL Falcon tube, gently tap the dish to dislodge any rounded-up mitotic cells, and transfer them to the tube of media. Pellet the cells at approx 110*g* for 3 min, aspirate the supernatant, and resuspend at a density of approx $0.5–1 \times 10^6$ cells/mL.
3. Add 1 mL cells per 100-mm dish containing 9 ml culture media plus drug at the appropriate concentration, and incubate at 37°C in a CO_2 incubator. Harvest the cells at the time-points of interest and then analyze by FACS, Western blot, or cytospin/immunofluorescence as described in **Subheading 3.2.**

3.1.3. Repression of Aurora B by RNAi

The process of RNAi results in the specific and efficient posttranscriptional silencing of gene expression in mammalian cells through the transfection of short interfering RNAs (siRNAs). The following procedure is adapted from Elbashir et al. ***(22)*** and results in effective repression in both HeLa and DLD-1 cells, 24–48 h post transfection.

1. Plate 500 L DLD-1 cells at a concentration of 1×10^5 cells/mL into 24-well plates in culture media without antibiotics, and grow overnight to achieve 30–50% confluency on the day of transfection.
2. Set up two Eppendorf tubes for each well to be transfected. In tube A, dilute 3 L of a 20 M stock of oligonucleotide (*see* **Note 3**) into 39 L Optimem (*see* **Note 4**). In tube B, dilute 3 L Oligofectamine reagent into 5 L Optimem. Incubate for 10 minutes at room temperature. Combine the contents of tubes A and B, mix by gentle pipeting, and incubate at room temperature for a further 20 min.
3. Wash the cells once with Optimem, and add 200 L per well. Overlay 50 L of the oligonucleotide–lipid complex onto the cells, and incubate for 4 h at 37°C in a humidified 5% CO_2 incubator.
4. Following incubation, add 500 L growth media containing 20% FCS to each well.
5. Assay for gene expression 24–48 h after transfection.

3.1.4. Calcium Phosphate Transfection of Wild-Type and Kinase Mutant (K106R) Aurora B

The method of calcium phosphate transfection used in our laboratory routinely results in excellent transient transfection of the TA-HeLa cell line. Typically, about 50% of cells express Myc-Aurora B when transfected with the pcDNA3 wild-type or K106R Aurora B plasmids.

1. All small-scale transfections are performed on cells grown on 19-mm glass cover slips. Float a piece of Parafilm on water in a tissue culture dish, and aspirate away the water so that the film adheres. Place coverslips onto drops of water on the parafilm, and again aspirate away the water until the cover slips are secure. Wash the cover slips twice with distilled water, and sterilize under a UV light. Plate 450 L HeLa cells at a concentration of 6×10^4 cells/mL onto each cover slip, and incubate overnight.

2. One hour prior to performing the transfection, prefeed the cells with 450 L fresh culture media.
3. Set up one Eppendorf tube for each transfection and add 30 L 2X $CaCl_2$ and 2 g DNA (*see* **Note 5**). Add 30 L 2X HBS to each tube while gently mixing with the pipet tip. Ensure that the contents are thoroughly mixed by swiping the tube along the grid at the front of the tissue culture hood.
4. Incubate at room temperature for 20 min to allow the precipitates to form. Add 400 L prewarmed fresh media containing 10% serum (*see* **Note 6**). Aspirate the media from the cells, and immediately add 450 L of the transfection mix. Return the cells to the incubator, and leave for approx 16 h overnight.
5. The following morning, wash each cover slip four times with prewarmed culture media, and then incubate for a further 24 h before analyzing for protein expression.

3.2. Analyzing the Effect of Inhibiting Aurora B Function

Once Aurora B function has been effectively inhibited by any of the methods described in **Subheading 3.1.**, the cells can be processed and analyzed by a variety of techniques. The methods described here have been extensively used in the laboratory to look at cell cycle progression, the phosphorylation status of histone H3, and the localization of various spindle checkpoint proteins by immunofluorescence.

3.2.1. Flow Cytometry and MPM2 Staining

Flow cytometry analysis of cells that have been stained with propidium iodide and the MPM-2 antibody enables both DNA content and mitotic index to be determined, respectively.

1. Transfer the medium from the culture dish into a 15 mL Falcon tube to collect any floating cells, rinse the dish once with PBS, and transfer to the same Falcon tube. To harvest the remaining adherent cells, add 1 mL trypsin/EDTA to the cell monolayer, and incubate at 37°C for 5 min.
2. Meanwhile, centrifuge the contents of the tube at 110*g* for 3 min, and aspirate the supernatant. Collect the trypsinized cells in 5 mL PBS, transfer to the appropriate Falcon tube, and centrifuge at 110*g* for a further 3 min. Aspirate the supernatant, and resuspend the cell pellet in 200 L PBS.
3. Fix the cells by adding 500 L –20°C ethanol dropwise while gently vortexing, and store at –20°C overnight.
4. Add 5 mL PBS to the ethanol-fixed samples, and centrifuge at approx 440*g* for 5 min. Aspirate the supernatant, and wash once more with 5 mL PBS.
5. Dilute the MPM-2 antibody 1:500 in PBS, and resuspend the cell pellet in 200 L of antibody solution. Incubate for 1 h on ice in the dark.
6. Wash the cells twice with 5 mL PBS. Make up the propidium iodide/RNase A solution by adding 8 L propidium iodide (5 mg/mL stock) and 5 L RNase A (5 mg/mL stock) per 1 ml PBS, and resuspend the cell pellet in 500 L of this solution. Incubate for 30 min at room temperature in the dark.
7. Place on ice until ready to analyze by flow cytometry to determine DNA content and MPM-2 reactivity.

3.2.2. Immunofluorescence Analysis

Generally we use DLD-1 cells for immunofluorescence analysis as these cells remain relatively flat during mitosis. It is important to consider which protein is to be studied before fixing the cells, as different antigens require different fixation conditions. Most of the centromeric and kinetochore proteins are analyzed following fixation with 1% formaldehyde in PBST. However, we routinely use methanol fixation for preserving microtubules.

3.2.2.1. Formaldehyde Fixation

1. Place 19-mm glass cover slips on a piece of Parafilm in a tissue culture dish as described in **Subheading 3.1.4. (step 1)**, wash twice with distilled water, and sterilize under a UV light. Plate 450 L DLD-1 cells, at a concentration of 12×10^4 cells/mL, onto the sterilized cover slips, and grow for 48 h.
2. Prepare a 1% formaldehyde solution in PBS. Aspirate the media from the cells, and fix in 1% formaldehyde for 5 min at room temperature.
3. Wash the coverslips three times with PBST. Add 200 L blocking solution (5% nonfat milk in PBST), and block at room temperature for 20 min.
4. Dilute the primary antibodies in blocking solution and apply 100 L to each coverslip. Incubate for 30 minutes at room temperature.
5. Wash the cover slips three times with PBST. Dilute the secondary antibodies in blocking solution, apply 100 L to each cover slip, and incubate for a further 30 min at room temperature.
6. Wash three times with PBST. Add 100 L Hoechst solution (stock 10 mg/mL in water, diluted 1:10,000 in PBST) to stain the DNA, and incubate for 2 min at room temperature. Wash once with PBST.
7. Apply a drop of mounting media in the center of a microscope slide, and carefully invert the cover slip onto the slide. Aspirate any excess mounting media and seal the edges with nail varnish. Clean the back of the cover slips with water to remove any PBS salts, and then view under a fluorescence microscope.

3.2.2.2. Methanol Fixation

1. Place 19-mm glass cover slips on a piece of parafilm in a tissue culture dish as described in **Subheading 3.1.4. (step 1)**, wash twice with distilled water and sterilise under a UV light. Plate 450 L Dld-1 cells, at a concentration of 12×10^4 cells/mL onto the sterilised coverslips and grow for 48 h.
2. Immerse the cover slip in a beaker of PBS to remove serum proteins, drain on a piece of filter paper and immediately immerse in a beaker of –20℃ methanol for 20 s. Transfer the coverslip into a 6-well plate (cell side upward) containing 3 mL –20℃ methanol, and incubate the plate at –20℃ for 10 min to fix the cells.
3. Remove the cover slip from the methanol, rinse in a beaker of PBS, and return to the tissue culture dish. Aspirate any remaining PBS from the cover slip, and wash twice with 500 L PBST.
4. Block and incubate with the relevant antibodies as in **Subheading 3.2.2.1. (steps 4–7)**.

3.2.3. Cytospin and Phospho-Histone H3 Staining

The method of cytospin immunofluorescence is particularly useful for analyzing cells in suspension or when a large proportion of the cells are mitotic. The mitotic cells round up and become detached from the culture dish; hence they are easily lost during normal fixation procedures. In this method both floating and adherent cells are collected and then spun onto glass slides, allowing the total cell population to be analyzed.

1. Transfer the medium from the culture dish into a 15-mL Falcon tube to collect any floating cells, rinse the dish once with PBS, and transfer to the same Falcon tube. To harvest the remaining adherent cells, add 1 mL trypsin/EDTA to the cell monolayer, and incubate at 37℃ for 5 min.
2. Meanwhile, centrifuge the contents of the tube at approx 110g for 3 min, and aspirate the supernatant. Collect the trypsinized cells in 5 mL PBS, transfer to the appropriate Falcon tube, and centrifuge at approx 110g rpm for a further 3 min. Aspirate the supernatant and resuspend the cell pellet in 200 mL PBS.
3. Dilute the cell suspension in PBS to a concentration of approx 1×10^6 cells/mL. Load 50 mL into a cytospin centrifuge, and centrifuge at approx 250g for 2 min to spin the cells onto a glass microscope slide. Fix the cells by immersing the slide in a Coplin jar containing 40 mL of 1% formaldehyde in PBS, and incubate for 10 min at room temperature. Wash the slides in a Coplin jar of PBST, and then immerse in blocking solution and incubate at room temperature for 20 min.
4. Remove the slides from the Coplin jar, and drain any excess liquid. Dilute the phospho-histone H3 antibody (1:200) in blocking solution, and then apply 100 mL onto the cells. Use a small piece of Parafilm as a flexible cover slip to prevent the cells from drying out, and incubate at room temperature for 30–45 min.
5. Wash by immersing in PBST, and then repeat step 4 above, applying 100 mL of Cy3 donkey antirabbit secondary antibody, diluted 1:500 in blocking solution.
6. Wash in PBST, incubate in Hoechst solution (1:1000 diluted in PBST) for 2 min, and then wash in PBST.
7. Apply a drop of mounting media over the cells on the microscope slide, and carefully mount a cover slip. Aspirate any excess mounting media, and seal the edges with nail varnish. Carefully clean the slides with water to remove any PBS salts, and then view under a fluorescence microscope.

4. Notes

1. The critical stage in a double thymidine block is the release period following the first overnight block with thymidine. For HeLa cells, an 8-h release period is sufficient to allow all the cells to pass through S-phase into G_2, but this may vary between cell lines. It is possible to perform a single thymidine block; however, a second thymidine treatment overnight will cause most cells to accumulate at the G_1/S boundary.
2. For a nocodazole block and release experiment, the cells should only be treated with nocodazole for 12–14 h because following prolonged periods the cells will eventually exit mitosis and then commit to apoptosis ***(20)***. Following a 12-h nocodazole block, approx 60–70% of the cells are in mitosis, and these cells are capable of maintaining mitotic arrest for a further 4 h during the analysis period.
3. We have found that successful repression of Aurora B is obtained using the siRNA oligo 5'-AACGCGGCACUUCACAAUUGA-3'. Repression of Aurora A has been successfully achieved using the oligo 5'-AAGCACAAAAGCUUGUCUCCA-3'. A 21-mer scrambled oligonucleotide should be included as a negative control. It is possible to scale up the reaction into 6-well plates if required, but cell density, the amount of oligo, and the volume of oligofectamine should be re-optimized. We plate 1.5×10^5 cells into each well of a 6-well plate and grow overnight. For the transfection, 10 L oligo is diluted in 175 L transfection media, and 6 L oligofectamine is diluted in 9 L transfection media. After the 4-h incubation period with the transfection mix, the cells are overlaid with 2 mL growth media containing 20% FCS.

4. In general, most transfections are carried out in serum-free media without antibiotics (Optimem). However, for HeLa cells, the transfection efficiency is increased in the presence of serum, so we use normal culture media (DMEM) without the addition of antibiotics.
5. DNA from Qiagen maxipreps (2 g at a concentration of 0.5 g/ L) or from Promega Wizard Plus minipreps (8–10 L of plasmid DNA if a 1.5 mL culture is eluted in 50 L TE) gives good results, although the efficiency can vary with different plasmids, so it is worth testing both methods of purification.
6. The pH of the media and source of the serum are critical to the success of calcium phosphate transfections. We use fresh DMEM (frozen aliquots) and FBS from Hyclone (Pierce and Warriner, cat. no. SH30070.03) The color of the DNA/$CaPO_4$/media mix should be 'peachy' in color, not pink/purple.

References

1. Adams, R.R., Carmena, M., and Earnshaw, W. C. (2001) Chromosomal passengers and the (aurora) ABCs of mitosis. *Trends Cell Biol.* **11,** 49–54.
2. Bischoff, J. R. and Plowman, G. D. (1999) The Aurora/Ipl1p kinase family: regulators of chromosome segregation and cytokinesis. *Trends Cell Biol.* **9,** 454–459.
3. Giet, R. and Prigent, C. 1999. Aurora/Ipl1p-related kinases, a new oncogenic family of mitotic serine-threonine kinases. *J. Cell Sci.* **112,** 3591–3601.
4. Nigg, E. A. (2001) Mitotic kinases as regulators of cell division and its checkpoints. *Nat. Rev. Mol. Cell. Biol.* **2,** 21–32.
5. Francisco, L. and Chan, C. S. (1994) Regulation of yeast chromosome segregation by Ipl1 protein kinase and type 1 protein phosphatase. *Cell. Mol. Biol. Res.* **40,** 207–213.
6. Glover, D. M., Leibowitz, M. H., McLean, D. A., and Parry, H. (1995) Mutations in aurora prevent centrosome separation leading to the formation of monopolar spindles. *Cell* **81,** 95–105.
7. Roghi, C., Giet, R., Uzbekov, R., et al. (1998) The *Xenopus* protein kinase pEg2 associates with the centrosome in a cell cycle-dependent manner, binds to the spindle microtubules and is involved in bipolar mitotic spindle assembly. *J. Cell Sci.* **111,** 557–572.
8. Hannak, E., M. Kirkham, A.A. Hyman, and K. Oegema, 2001. Aurora-A kinase is required for centrosome maturation in *Caenorhabditis elegans. J. Cell Biol.* **155,** 1109–1116.
9. Kimura, M., Matsuda, Y. ,Yoshioka,T., and Okano, Y. (1999) Cell cycle-dependent expression and centrosome localization of a third human aurora/Ipl1-related protein kinase, AIK3. *J. Biol. Chem.* **274,** 7334–7340.
10. Bischoff, J.R., Anderson, L., Zhu, Y., et al. (1998) A homologue of *Drosophila* aurora kinase is oncogenic and amplified in human colorectal cancers. *EMBO J.* **17,** 3052–3065.
11. Tatsuka, M., Katayama, H., Ota, T., et al. (1998) Multinuclearity and increased ploidy caused by overexpression of the aurora- and Ipl1-like midbody-associated protein mitotic kinase in human cancer cells. *Cancer Res.* **58,** 4811–4816.
12. Zhou, H., Kuang, J., Zhong, L., et al. (1998) Tumour amplified kinase STK15/BTAK induces centrosome amplification, aneuploidy and transformation. *Nat. Genet.* **20,** 189–193.
13. Ewart-Toland, A., Briassouli, P., De Koning, J. P., et al. (2003) Identification of Stk6/STK15 as a candidate low-penetrance tumor-susceptibility gene in mouse and human. *Nat Genet.* **34,** 403–412.
14. Anand, S., Penrhyn-Lowe, S., and Venkitaraman, A. R. (2003) AURORA-A amplification overrides the mitotic spindle assembly checkpoint, inducing resistance to Taxol. *Cancer Cell* **3,** 51–62.

15. Ditchfield, C., Johnson, V. L., Tighe, A., et al. (2003) Aurora B couples chromosome alignment with anaphase by targeting BubR1, Mad2, and Cenp-E to kinetochores. *J. Cell Biol.* **161,** 267–280.
16. Hauf, S., Cole, R. W., LaTerra, S., et al. (2003) The small molecule Hesperadin reveals a role for Aurora B in correcting kinetochore-microtubule attachment and in maintaining the spindle assembly checkpoint. *J. Cell Biol.* **161,** 281–294.
17. Kallio, M.J., McCleland, M. L., Stukenberg, P.T., and Gorbsky, G. J. (2002) Inhibition of aurora B kinase blocks chromosome segregation, overrides the spindle checkpoint, and perturbs microtubule dynamics in mitosis. *Curr. Biol.* **12,** 900–905.
18. Murata-Hori, M. and Wang, Y. L. (2002) The kinase activity of aurora B is required for kinetochore-microtubule interactions during mitosis. *Curr. Biol.* **12,** 894–899.
19. Honda, R., Korner, R., and Nigg, E. A. (2003) Exploring the functional interactions between Aurora B, INCENP, and survivin in mitosis. *Mol. Biol. Cell.* **14,** 3325–3341.
20. Taylor, S. S. and McKeon, F. (1997) Kinetochore localization of murine Bub1 is required for normal mitotic timing and checkpoint response to spindle damage. *Cell* **89,** 727–735.
21. Taylor, S.S., Hussein, D., Wang, Y., Elderkin, S., and Morrow, C. J. (2001) Kinetochore localisation and phosphorylation of the mitotic checkpoint components Bub1 and BubR1 are differentially regulated by spindle events in human cells. *J. Cell Sci.* **114,** 4385–4395.
22. Elbashir, S. M., Harborth, J., Lendeckel, W., Yalcin, A., Weber, K., and Tuschl, T. (2001) Duplexes of 21-nucleotide RNAs mediate RNA interference in cultured mammalian cells. *Nature* **411,** 494–498.

23

Purification of the Ndc80 Kinetochore Subcomplex From *Xenopus* Eggs

Mark L. McCleland and P. Todd Stukenberg

Summary

The identification of protein binding partners often facilitates understanding of protein complex function. However, identifying binding partners has proven difficult because proteins are often bound to insoluble structures or are only present during certain stages of the cell cycle. Fortunately, *Xenopus* eggs stockpile many proteins, which are typically insoluble, as soluble subcomplexes to facilitate rapid early embryonic divisions. We exploited this by developing a purification scheme using *Xenopus* egg extracts to isolate Coomassie-stainable amounts of the xNdc80 kinetochore complex. In this scheme *Xenopus* eggs are directly made into a mitotic high-speed supernatant and then flowed over three chromatographic columns: heparin, Mono-Q, and Superose 6 gel filtration columns. A final immunoprecipitation is then performed from the peak Superose 6 column to yield Ndc80 complex purified to homogeneity. With minor modification and manipulation of *Xenopus* egg extracts, this protocol can easily be adapted for purification of other protein complexes.

Key Words

Mitosis; kinetochore fractionation; Ndc80 complex; protein complex purification; Spc24; Spc25; *Xenopus* extract.

1. Introduction

Kinetochores are key regulators of mitosis that link chromosomes to the mitotic spindle (reviewed in **ref. *1***). The kinetochore is a large insoluble complex composed of over 30 proteins that assembles on centromeric heterochromatin from prometaphase to telophase. Although many kinetochore components have been identified, there are still many kinetochore proteins that have yet to be discovered. One reason for this is that whole kinetochores have not been biochemically purified. Biochemical purification of kinetochores has been difficult for two reasons. First, functional kinetochores

From: *Methods in Molecular Biology, vol. 296, Cell Cycle Control: Mechanisms and Protocols*
Edited by: T. Humphrey and G. Brooks © Humana Press Inc., Totowa, NJ

only exist during mitosis, and many of the identified components are synthesized in late G_2 and degraded at the end of mitosis. This means that the components are very rare in most tissues and asynchronous cell lysates. Second, throughout mitosis most of the components are tightly associated on mitotic chromosomes and therefore not amenable to chromatographic separation. Recently it has become clear from work in *Xenopus* and yeast that many kinetochore components are found in subcomplexes with other kinetochore proteins that, upon entry into mitosis, assemble onto a constitutive inner kinetochore anchor ***(2,3)***. Unlike whole kinetochores, these subcomplexes are soluble and can be biochemically fractionated. We and others have exploited the subassembly nature of the kinetochore to purify kinetochore subcomplex components and identify novel proteins.

In this chapter we will describe our purification of the Ndc80 kinetochore subcomplex from *Xenopus* egg extracts, a procedure that identified two novel kinetochore proteins, which we have named Spc24 and Spc25. The largest member of the Ndc80 complex, Hec1/Ndc80, was first identified in a two-hybrid interaction as a protein that interacts with human retinoblastoma protein ***(4)***. The Lee group produced high-quality antibodies that localized the protein to the kinetochore, and injections of the antibodies showed phenotypes consistent with important kinetochore functions. This work was highlighted when the Kilmartin group purified the spindle pole complex from budding yeast and identified a homolog to Hec1 ***(5)***. After characterization they named the protein Ndc80, as the phenotypes were reminiscent of a core budding yeast kinetochore component, Ndc10. Purification of budding yeast Ndc80 complex identified a four-protein complex with three other presumptive spindle pole proteins Nuf2, Spc24, and Spc25 ***(6–8)***. All four proteins were characterized in budding yeast and shown to be kinetochore components that have important roles in microtubule attachment, spindle checkpoint signaling, and chromosome segregation.

Sequence homology identified vertebrate and worm homologs to Hec1/Ndc80 and Nuf2, and both proteins were subsequently shown to be key kinetochore components with roles in microtubule attachment, spindle checkpoint signaling, and kinetochore assembly ***(3,9,10)***. Metazoan homologs of Spc24 or Spc25 could not be found using sequence alignment algorithms, so we set out to purify the complex from *Xenopus* egg extracts by following Ndc80 and Nuf2 through multiple chromatographic steps with antibodies against each protein. We developed a procedure that started with a clarification spin followed by three conventional steps (heparin, Mono-Q, and Superose 6), which we estimate gave us 1000-fold purification. This was followed by an immunoprecipitation that purified the complex to homogeneity. Mass spectrometry analysis of the four Coomassie stained bands identified xNdc80, xNuf2, and two proteins that were encoded by expressed sequence tags of unknown function. Since the novel proteins had limited sequence homology to yeast Spc24 and Spc25, and similar coiled-coil structure, bound both Nuf2 and Ndc80, and localized to kinetochores, we designated them as homologs of Spc24 and Spc25.

The following protocol generated Coomassie-stainable amounts of protein for protein identification from the eggs of only eight frogs. We feel that an important reason for the success of the project was the choice of *Xenopus* extracts as the protein source for purification. In somatic cell cycles the complex is only present from late G_2 through mitosis. However, in the *Xenopus* embryo the complex is stockpiled in a soluble form to facilitate rapid early embryonic cell divisions. Moreover, since the extracts have no chromatin, the subcomplexes cannot assemble into kinetochores and therefore are amendable to chromatography. Many chromatin and signaling components are similarly stockpiled in solubilized forms in *Xenopus* extracts, so it is likely that the general methodology described here would be amenable to purify other chromatin-bound complexes.

2. Materials

2.1. Clarified Mitotic Egg Lysate

1. 1X MMR: 100 m*M* NaCl, 2 m*M* KCl, 1 m*M* $MgCl_2$, 2 m*M* $CaCl_2$, 0.1 m*M* EDTA, 5 m*M* HEPES, pH 7.8. A 20X stock is prepared, titrated with NaOH to pH 7.8, and stored at room temperature.
2. The protocols for handling and care of *Xenopus* laevis have been well documented and will not be covered here. For purification, we started with the eggs of eight frogs that were primed with pregnant mare serum gonadotropin (PMSG). Eggs were laid into 8 L of 20°C MMR overnight.
3. Dejelly buffer: 2% cysteine (Sigma), 100 m*M* KCl, 0.1 m*M* $CaCl_2$, 1 m*M* $MgCl_2$ pH 7.7. This buffer must be made just before use.
4. EB: 80 m*M* β-glycerol phosphate (Sigma), 20 m*M* EGTA, 1 m*M* $MgCl_2$.
5. EB+++: EB supplemented with 1 m*M* dithiothreitol (DTT), 1 *M* ATP (pH 7.5), 1 m*M* microcystin LR (Alexis), 10 g/mL leupeptin, 10 g/mL pepstatin, and 10 g/mL chymostatin. The leupeptin, pepstatin, and chymostatin should be dissolved in dimethyl sulfoxide at 10 mg/mL and stored in small aliquots at –20°C.

2.2. Chromatography

1. Slyde-A-Lyzer Dialysis cassettes, 10,000 MWCO 3–15-mL capacity (Pierce, cat. no. 66410), 0.5–3-mL capacity (cat. no. 66425).
2. Microcon Centrifugal Filter Concentrators (Amicon, cat. no. 42407).
3. 25 mL Affi-Gel Heparin Gel (Bio-Rad, cat. no. 153-6173) bought in bulk and packed in a C 10/40 column (Amersham-Pharmacia),1 mL Mono-Q (Amersham-Pharmacia), 24 mL Superose 6 HR 10/30 (Amersham-Pharmacia).
4. Heparin buffer: 20 m*M* MES, pH 6.8, 0.1 m*M* EDTA, 1 m*M* DTT, and 10 m*M* NaCl.
5. Q buffer: 20 m*M* Tris-HCl, pH 8.0, 0.1 m*M* EDTA, 1 m*M* DTT, and 10 m*M* NaCl.
6. Superose buffer: 10 m*M* K-HEPES, pH 7.7, 300 m*M* NaCl, 1 m*M* $MgCl_2$, and 50 m*M* sucrose.

2.3. Immunoprecipitation

1. 100 L of Affi-prep (Bio-Rad) protein-A beads covalently coupled using dimethylpimylimidate (Pierce) to 100 g of rabbit IgG (Sigma) or 100 g anti-Ndc80 antibodies according to the methods in Harlow and Lane ***(11)***.
2. IP wash buffer 1: 10 m*M* K-HEPES, pH 7.7, 350 m*M* NaCl, 1 m*M* $MgCl_2$, 0.05% Triton X-100, and 50 m*M* sucrose.
3. IP wash buffer 2: 10 m*M* Tris-HCl, pH 7.0, 50 m*M* NaCl.
4. IP elution buffer: 0.1 *M* glycine pH 2.5 made fresh.
5. 100% Trichloroacetic acid w/v (TCA) is made up by adding 227 mL H_2O to 500 g TCA (Fisher, cat. no. 322-500) and stored at 4°C.

6. Na-deoxycholate is diluted to 125 g/mL from a 10 mg/mL stock stored at –20°C.
7. 5–20% gradient gels for SDS-PAGE (Bio-Rad) were purchased, and all buffers for SDS-PAGE and Commassie stain were 0.2- m-filtered to make samples clean for subsequent mass spec analysis.

3. Methods

3.1. Clarified Mitotic Egg Lysate

The following is a protocol for a diluted, clarified lysate (clarified mitotic egg lysate [MOE]) that is quite different from the classic cytostatic factor (CSF) arrested egg extracts developed by Lohka and Masui and then improved by Murray and Kirschner ***(12–14)***. This protocol was developed by Jian Kuang (M.D. Anderson) as a method to generate lysate rapidly for biochemical fractionation of proteins and maintain the mitotic phosphorylation status of proteins found in *Xenopus* eggs ***(15)***. Its major advantages are its relative ease and speed for obtaining clarified protein for chromatography. Moreover, the mitotic kinases are much more stable than in standard CSF extracts and maintain full kinase activity after quick freezing and long-term storage at –80°C.

1. Five to 7 d before lysate preparation, eight frogs are primed with 50 U of PMSG injected into the dorsal lymph sac with a 27-gage needle.
2. Ovulation is induced by injecting 100 U of human chorionic gonadotropin (Sigma) into the dorsal lymph sac of primed frogs with a 25- or 27-gage needle. Injections are performed around 6:00 PM the night before lysate preparation, and the frogs are stored in groups of two in 2 L of MMR at 20°C. Also, at least 4 L of MilliQ water is chilled to 20°C overnight, and all buffers are made with the chilled water the next day.
3. Laid eggs are collected, and debris and poor quality eggs are removed.
4. The jelly coats of the eggs are removed by pouring off as much MMR as possible and adding at least 4 vol of dejelly solution. The eggs are watched closely with occasional gentle swirling (~3 min). Once the jelly coat is removed, the eggs will pack closely together.
5. As soon as tight packing is detected, the eggs are washed at least four times in at least 6 egg volumes of MMR until the eggs are relatively clear of debris, jelly coats, and lysed eggs.
6. The eggs are washed two times in 6 egg vol of EB.
7. As much EB as possible is removed, and 1 egg volume of ice cold EB+++ is added. The eggs are placed on ice and lysed by pipeting up and down in a 25-mL pipet approx 20 times.
8. The lysate is centrifuged in a Beckman 50.2-Ti rotor at 184,000*g* for 1 h at 4°C.
9. After the spin the lysate has four separated layers. At the top is a thick yellow lipid layer; the clear lysate (which contains most cytoplasmic proteins), a fluffy membrane layer, sits on top of the dense gray/black yolk platelets. The clear lysate layer is carefully removed using a peristaltic pump. Tubing is carefully placed through the lipid layer and the clear lysate is pumped into an ice-cold 50-mL conical tube. If the lysate looks cloudy because it is contaminated with either the lipid or membranes, it should be centrifuged again.
10. The lysate is quick-frozen in liquid N_2 in 4-mL aliquots in 5 mL-Nalgene cryogenic vials and stored at –80°C until use.

3.2. Chromatographic Purification of xNdc80 Complex

Purification of complexes should be carried out as quickly as possible to prevent degradation or dissociation of potential complex members. Therefore, before we started the purification we ran small-scale test columns to determine how the Ndc80 complex would migrate over various columns. We found that the Ndc80 complex bound to a heparin column and a Q column and ran as a large 1-MDa complex on a Superose 6 gel filtration column. We ran the heparin column first because most proteins flow through this negatively charged resin, allowing us to use a relatively small column (16 ml), yet decrease the amount of total protein by approximately 10 fold. This significant reduction in total protein allowed us to next run the heparin fractions containing Ndc80 complex over a small 1-mL mono-Q column. The elutions off the Q column were in a relatively small volume, which allowed for easy sample concentration with a Microcon concentrator and direct loading of complex containing fractions on the Superose 6 column. Ndc80 and Nuf2 ran as the expected 1-MDa complex after the Superose 6 column, confirming that the complex did not dissociate through the purification. Superose 6 fractions containing Ndc80 and Nuf2 were precleared with rabbit IgG and then incubated with anti-Ndc80 beads. Both sets of beads were subsequently washed and eluted with low pH. Elutions were TCA-precipitated and then run on an SDS-PAGE gradient gel. Coommassie-stained proteins specific to the anti-Ndc80 immunoprecipitation were excised and identified by mass spectrophotometry. Depending on how your complex migrates over chromatographic columns, different columns or different order should be considered.

1. 50 mL of MOE (~500 mg of protein) is thawed and centrifuged in a Beckman 50.2-Ti rotor at 184,000*g* for 30 min at 4€ to ensure that the extract is clarified (*see* **Note 5**).
2. Slowly dilute the MOE with 200 mL of water containing 1 m*M* Microcystin to reduce conductivity. To remove any protein precipitate that may have occurred during dilution, spin the extract at 26,900*g* for 20 min in an SS-34 rotor at 4€. All chromatography steps are performed at 4€ to prevent protein degradation.
3. Equilibrate the 16-mL heparin column with 32 mL of heparin buffer at 2.5 mL/minute.
4. Use a peristaltic pump to load the 250 mL of extract directly onto the heparin column at 2.5 mL/min. Collect the flowthrough in case the complex does not bind the column. Wash the heparin column with 80 mL (5 column vol) of heparin buffer. Elute bound proteins with a 160 mL (10 column vol) linear gradient of heparin buffer and heparin buffer containing 500 m*M* NaCl. Collect the heparin elutions in 5-mL fractions. Assay the fractions by immunoblot analysis using antibodies to Ndc80 and Nuf2 (*see* **Notes 2** and **3**).
5. Pool fractions containing Ndc80 and Nuf2 and dialyze overnight using a Slyde-A-Lyzer cassette into Q buffer.
6. Equilibrate a 1-mL mono-Q column with 5 mL of Q buffer at 1 mL/min.
7. Use a Dynal-loop or equivalent to load the dialyzed heparin fractions onto the mono-Q column at 1 mL/min. Collect the flowthrough in case the complex does not bind the column. Wash the mono-Q column with 5 mL (5 column vol) of Q buffer. Elute bound proteins with a 10 mL (10 column vol) linear gradient of Q buffer and Q buffer containing 500 m*M* NaCl. Collect the Q elutions in 0.3-mL fractions. Assay fractions by immunoblot analysis using antibodies to Ndc80 and Nuf2.
8. Pool fractions containing Ndc80 and Nuf2, and concentrate the sample to approx 300 L by centrifugation in a Microcon device at 14,000*g*.
9. Equilibrate the Superose 6 column in 24 mL of Superose buffer at 0.4 mL/min.
10. Load the Q fractions onto the Superose 6 column with Superose buffer at a flow rate of 0.4 mL/min. Collect the Superose 6 column elution in 0.5-mL fractions. Assay fractions by immunoblot analysis using antibodies to Ndc80 and Nuf2.

3.3. Immunoprecipitation of the Ndc80 Complex

1. Pool the Superose 6 fractions containing Ndc80 and Nuf2. Preclear the pooled fractions with 100 L of Affi-prep protein A beads covalently coupled to 100 g of rabbit IgG for 1 h at 4°C with rotation.
2. Remove the precleared beads by centrifugation at 2700*g* in a microcentifuge, and transfer the supernatant to a tube containing 100 L of Affi-prep protein A beads covalently coupled to 100 g of anti-Ndc80 antibody. Rotate the supernatant for 2 h with anti-Ndc80 beads at 4°C to allow maximal immunoprecipitation (*see* **Notes 4** and **5**).
3. Wash the precleared beads and anti-Ndc80 beads eight times with 1.5 mL of IP wash buffer.
4. Wash both sets of beads two times in IP wash buffer 2.
5. Elute each set of beads four times with 300 mL of IP elution buffer (*see* **Note 6**). Combine the elutions, and immediately add 60 L (1/5 vol) of 1 *M* Tris-HCl, pH 8.0 to each elution.
6. Precipitate the 1.5 mL elutions by adding 18.75 L of Na-deoxycholate (10 mg/mL) and 112.5 L of 100% TCA. Incubate this mixture on ice for 30 min, and then centrifuge the sample in a microcentrifuge at full speed for 15 min at 4°C.
7. Carefully remove the supernatant, and then wash pellets with 0.5 mL of –20°C acetone. Centrifuge the sample again at full speed for 10 min at 4°C. Remove the acetone, and allow the pellet to air-dry.
8. Resuspend the pellet in 40 L of 1.5X SDS-PAGE sample buffer. Load 30 L of the sample on a Bio-Rad 5–20% gradient gel. Use running buffers that are 0.2 *M*-filtered (*see* **Note 7**).

4. Notes

A few general points that are important to consider when adapting this protocol to purifications of other complexes.

1. Before starting the prep, we estimated the concentration of the complex using Western analysis with recombinant proteins as standards. Therefore we knew that we started with about 20 g of Ndc80 and Nuf2 protein in 50 mL of lysate, and all we needed was a 5% total recovery to give Coomassie-stainable material.
2. When one follows a protein complex through a purification by immunoblot, a number of cross-reacting bands can appear that are not seen upon blotting a whole cell lysate. These cross-reacting bands appear because proteins become highly concentrated during purification (**Fig. 1**). Our ability to follow two proteins simultaneously in the complex was critical to allow us to follow the complex correctly throughout the entire purification. Moreover, we were careful not to overload SDS-PAGE gels with protein in order to reduce the number of cross-reacting bands.
3. The Western blot is rate-limiting in most steps of the procedure. We minimized the time of many steps in the Western protocol by using minigels and testing to determine the shortest time that we could block and incubate with primary and secondary antibodies.
4. The critical reagent for this complex purification was an antibody that both immunoprecipitation and western blots. We knew before our purification that our rabbit polyclonal antibody could not only immunoprecipitate, but also immunodeplete the complex from *Xenopus* extracts.
5. The major contaminant of many purifications with final IP steps is IgG. This can eliminate the ability to detect proteins of 52–58 kDa and 22–28 kDa. Therefore, we thoroughly prewashed our antibody beads with IP elution buffer just before using them. In the end

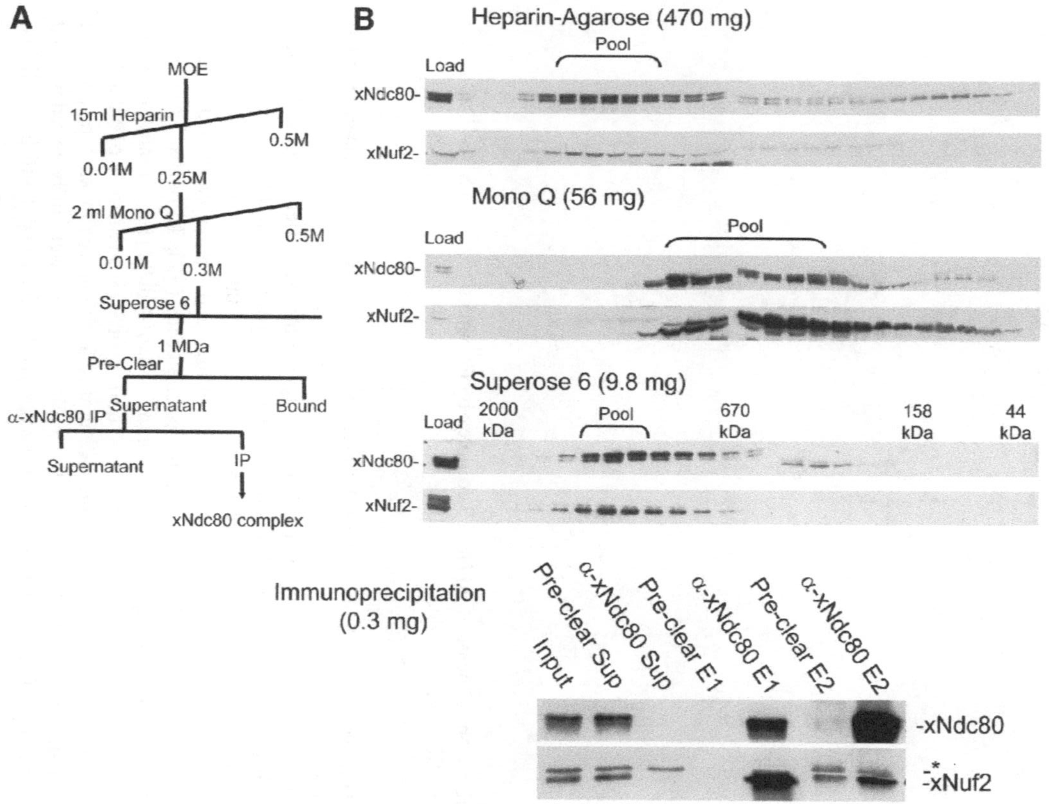

Fig. 1. Immunoblot analysis of column fractionation of the Ndc80 complex. **(A)** Scheme of the Ndc80 complex purification. **(B)** Anti-Ndc80 and Anti-Nuf2 immunoblot analysis of column profiles from each step of the purification. The indicated step is followed by the total milligrams of protein loaded on the column. The fractions that were collected and further chromatographed are indicated by the "pooled" fractions. *See* **Subheading 3.** for details. For the immunoprecipitation step, *see* **Note 6** for details on the elutions. MOE, mitotic egg lysate.

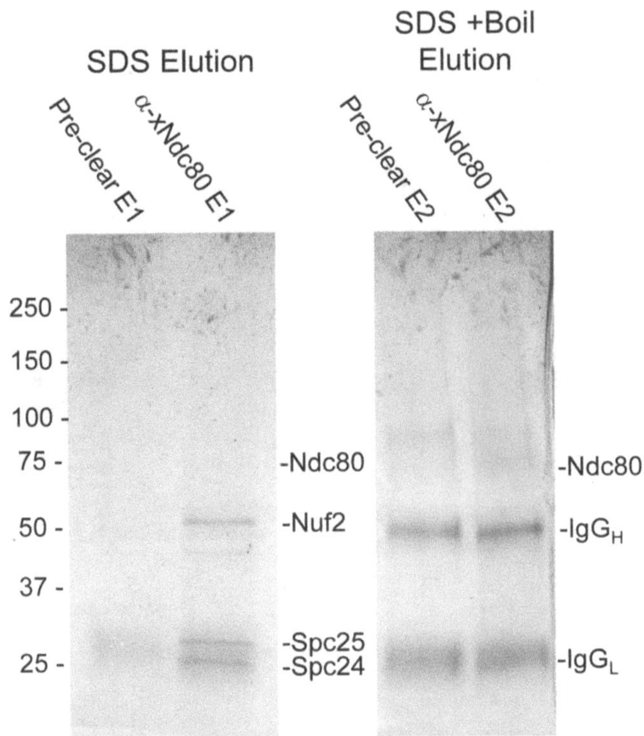

Fig. 2. Visualization of the purified Ndc80 complex by Coomassie-stained SDS-PAGE. *See* **Note 6** for details. SDS, sodium dodecyl sulfate.

the IgG beads were not detectable by Coomassie stain, but they were detected by mass spectrometry.

6. We actually went through this purification twice with almost identical results. The step that we modified in the second purification was the elution of the anti-Ndc80 immunoprecipitation. In the first purification we eluted the beads with 100 L of SDS-PAGE sample buffer for 10 min at room temperature, spun down the beads, removed the supernatant (E1), then added fresh sample buffer, and boiled the beads (E2). To our surprise in E1 contained the three interacting proteins but not the bulk of Ndc80 (**Figs. 1** and **2**). Breaking of the antibody–antigen interaction required boiling in SDS, so the bulk of Ndc80 was found in E2. This boiling also eluted a great deal of IgG. When we repeated the prep, we used the acid elution described here, and all complex members were fully eluted without noticeable amounts of IgG. (The results for the postive Coomassie gel will be published elsewhere.)
7. The goal of the prep is to identify interacting proteins by mass spectrometry. To aid the spectrometrist, one has to be careful to avoid contaminating proteins (mostly keratins) during the IP and final gel.

References

1. Rieder, C. L. and Salmon, E. D. (1998) The vertebrate cell kinetochore and its roles during mitosis. *Trends Cell Biol.* **8,** 310–318.
2. Cheeseman, I. M., Anderson, S., Jwa, M., et al. (2002) Phospho-regulation of kinetochore-microtubule attachments by the Aurora kinase Ipl1p. *Cell* **111,** 163–172.
3. McCleland, M. L., Gardner, R. D., Kallio, M. J., et al. (2003) The highly conserved Ndc80 complex is required for kinetochore assembly, chromosome congression, and spindle checkpoint activity. *Genes Dev.* **17,** 101–114.
4. Zheng, L., Chen, Y., and Lee, W. H. (1999) Hec1p, an evolutionarily conserved coiled-coil protein, modulates chromosome segregation through interaction with SMC proteins. *Mol. Cell. Biol.* **19,** 5417–5428.
5. Wigge, P. A., Jensen, O. N., Holmes, S., Soues, S., Mann, M., and Kilmartin, J. V. (1998) Analysis of the *Saccharomyces* spindle pole by matrix-assisted laser desorption/ionization (MALDI) mass spectrometry. *J. Cell Biol.* **141,** 967–977.
6. Janke, C., Ortiz, J., Lechner, J., et al. (2001) The budding yeast proteins Spc24p and Spc25p interact with Ndc80p and Nuf2p at the kinetochore and are important for kinetochore clustering and checkpoint control. *EMBO J.* **20,** 777–791.
7. Kilmartin, J. V. and Janke, C. (2001) The budding yeast proteins Spc24p and Spc25p interact with Ndc80p and Nuf2p at the kinetochore and are important for kinetochore clustering and checkpoint control. *J. Cell Biol.* **152,** 349–360.
8. Wigge, P. A. and Kilmartin, J. V. (2001) The Ndc80p complex from Saccharomyces cerevisiae contains conserved centromere components and has a function in chromosome segregation. *J. Cell Biol.* **152,** 349–360.
9. DeLuca, J. G., Moree, B., Hickey, J. M., Kilmartin, J. V., and Salmon, E. D. (2002) hNuf2 inhibition blocks stable kinetochore-microtubule attachment and induces mitotic cell death in HeLa cells. *J. Cell Biol.* **159,** 549–555.
10. Martin-Lluesma, S., Stucke, V. M., and Nigg, E. A. (2002) Role of Hec1 in spindle checkpoint signaling and kinetochore recruitment of Mad1/Mad2. *Science* **297,** 2267–2270.
11. Harlow, E. and Lane, D. (1988) Antibodies: A laboratory manual. Cold Spring Harbor Laboratory Press, Cold Spring Harbor, NY.
12. Lohka, M. J. and Masui, Y. (1983) Formation in vitro of sperm pronuclei and mitotic chromosomes induced by amphibian ooplasmic components. *Science* **220,** 719–721.
13. Murray, A. W. and Kirschner, M. W. (1989) Cyclin synthesis drives the early embryonic cell cycle. *Nature* **339,** 275–280.
14. Murray, A. W., Solomon, M. J., and Kirschner, M. W. (1989) The role of cyclin synthesis and degradation in the control of maturation promoting factor activity. *Nature* **339,** 280–286.
15. Kuang, J., Penkala, J. E., Ashorn, C. L., Wright, D. A., Saunders, G. F., and Rao, P. N. (1991) Multiple forms of maturation-promoting factor in unfertilized *Xenopus* eggs. *Proc. Natl. Acad. Sci. USA* **88,** 11530–11534.

Index

About the Editors

Dr. Tim Humphrey

Tim Humphrey graduated from Heriot-Watt University, Edinburgh, with an honours degree in Brewing and Microbiology. He worked at the German Cancer Research Centre in Heidelberg before studying for his DPhil at the University of Oxford. As a postdoctoral research fellow at Harvard University he studied cell cycle responses to stress in fission yeast. In 1997 he became the Cell Cycle Group Leader and Director of Studies at the MRC Radiation and Genome Stability Unit, Harwell, Oxfordshire, UK. Tim was awarded tenure in 2003. His laboratory research is focused on the cellular responses to genotoxic and environmental stress using fission yeast as a model organism.

Professor Gavin Brooks

Gavin Brooks graduated with a first class honours degree in Pharmacy and obtained a PhD degree in the areas of organic chemistry and pharmacology at The School of Pharmacy, University of London. Following his PhD, he joined the Imperial Cancer Research Fund Laboratories in London as a postdoctoral fellow before becoming a Group Leader within the Department of Cardiovascular Research, The Rayne Institute, St. Thomas' Hospital, London. Gavin then worked for the cell cycle company, Prolifix Ltd., before returning to academia, where he now is Professor of Cardiovascular Research at the University of Reading. He was elected Head of the new Reading School of Pharmacy in January 2004. His current research interests focus on understanding the cell cycle mechanisms that control physiological and pathophysiological cardiovascular cell growth.

About the Editors

FSC
www.fsc.org
MIX
Papier | Fördert
gute Waldnutzung
FSC® C083411

Kolibri 360 GmbH
Ferdinand-Jühlke-Straße 7
99095 Erfurt, Deutschland
produktsicherheit@kolibri360.de